Harmonic Measure

During the last two decades several remarkable new results were discovered about
harmonic measure in the complex plane. This book provides a survey of these results
and an introduction to the branch of analysis that contains them. Many of these results,
due to Bishop, Carleson, Jones, Makarov, Wolff, and others, appear here in book form
for the first time.

The book is accessible to students who have completed standard graduate courses in
real and complex analysis. The first four chapters provide the needed background
material on univalent functions, potential theory, and extremal length, and each chapter
has many exercises to further inform and teach the reader.

JOHN B. GARNETT is Professor of Mathematics at the University of California,
Los Angeles.

DONALD E. MARSHALL is Professor of Mathematics at the University of
Washington.

Harmonic Measure

During the last two decades several remarkable new results were discovered about
harmonic measure in the complex plane. This book provides a survey of these results
and an introduction to the branch of analysis that contains them. Many of these results,
due to Bishop, Carleson, Jones, Makarov, Wolff, and others, appear here in book form
for the first time.
The book is accessible to students who have completed standard graduate courses in
real and complex analysis. The first four chapters provide the needed background
material on univalent functions, potential theory, and extremal length, and each chapter
has many exercises to further inform and teach the reader.

JOHN B. GARNETT is Professor of Mathematics at the University of California,
Los Angeles.

DONALD E. MARSHALL is Professor of Mathematics at the University of
Washington.

NEW MATH MONOGRAPHS

For information about Cambridge University Press mathematics publications
visit http://publishing.cambridge.org/stm/mathematics

Harmonic Measure

JOHN B. GARNETT

University of California, Los Angeles

DONALD E. MARSHALL

University of Washington

CAMBRIDGE
UNIVERSITY PRESS

Shaftesbury Road, Cambridge CB2 8EA, United Kingdom

One Liberty Plaza, 20th Floor, New York, NY 10006, USA

477 Williamstown Road, Port Melbourne, VIC 3207, Australia

314–321, 3rd Floor, Plot 3, Splendor Forum, Jasola District Centre, New Delhi – 110025, India

103 Penang Road, #05–06/07, Visioncrest Commercial, Singapore 238467

Cambridge University Press is part of Cambridge University Press & Assessment, a department of the University of Cambridge.

We share the University's mission to contribute to society through the pursuit of education, learning and research at the highest international levels of excellence.

www.cambridge.org
Information on this title: www.cambridge.org/9780521720601

© Cambridge University Press & Assessment 2005

First published 2005
Paperback edition published 2008

A catalogue record for this publication is available from the British Library

Library of Congress Cataloging-in-Publication data
Garnett, John B.
Harmonic measure / John B. Garnett, Donald E. Marshall.
    p.   cm. – (New math monographs)
Includes bibliographical references and indexes.
1. Functions of complex variables.   2. Potential theory (Mathematics).
I. Marshall, Donald E. (Donald Eddy), 1947–   II. Title.   III. New mathematical monographs.
QA331.7.G37   2005
515/.42 22   2004045893

ISBN   978-0-521-47018-6   Hardback
ISBN   978-0-521-72060-1   Paperback

To Dolores and Marianne

Contents

Preface

Several surprising new results about harmonic measure on plane domains have been proved during the last two decades. The most famous of these results are Makarov's theorems that harmonic measure on any simply connected domain is singular to Hausdorff measure Λ_α for all $\alpha > 1$ but absolutely continuous to Λ_α for all $\alpha < 1$. Also surprising was the extension by Jones and Wolff of Makarov's $\alpha > 1$ theorem to all plane domains. Further important new results include the work of Carleson, Jones and Wolff, and others on harmonic measure for complements of Cantor sets; the work by Carleson and Makarov, Bertilsson, Pommerenke, and others on Brennan's tantalizing conjecture that for univalent functions $\iint |\varphi'|^{2-p} dx dy < \infty$ if $\frac{4}{3} < p < 4$; several new geometric conditions that guarantee the existence of angular derivatives; and the Jones square sum characterization of subsets of rectifiable curves and its applications by Bishop and Jones to a variety of problems in function theory.

We wrote this book to explain these exciting new results and to provide beginning students with an introduction to this part of mathematics. We have tried to make the subject accessible to students who have completed graduate courses in real analysis from Folland [1984] or Wheeden and Zygmund [1977], for example, and in complex analysis from Ahlfors [1979] or Gamelin [2001], for example.

The first four chapters, along with the appendices on Hardy spaces, Hausdorff measures and martingales, provide a foundation that every student of function theory will need. In Chapter I we solve the Dirichlet problem on the half-plane and the disc and then on any simply connected Jordan domain by using the Carathéodory theorem on boundary continuity. Chapter I also includes brief introductions to hyperbolic geometry and univalent function theory. In Chapter II we solve the Dirichlet problem on domains bounded by finitely many Jordan curves and study the connection between the smoothness of a domain's boundary and the smoothness of its Poisson kernel. Here the main tools are two classical theorems about conjugate functions. Chapter II and the discussion in Chapter III of Wiener's solution of the Dirichlet problem on arbitrary domains follow the 1985 UCLA lecture course by Carleson. The introduction to extremal length in Chapter IV is based on the Institut Mittag–Leffler lectures of Beurling [1989]. Chapter V contains some applications of extremal length, such as Teichmüller's Modulsatz and some newer theorems about angular derivatives, that are not found in other books. Chapter VI is a blend of the classical theorems of F. and M. Riesz, Privalov, and Plessner and the more recent theorems of McMillan, Makarov, and Pommerenke on the comparison of harmonic measure and one dimensional Hausdorff measure for simply connected domains. Chapter VII surveys the beautiful circle of ideas around Bloch functions, univalent functions, quasicircles, and A^p weights. Chapter VIII is an exposition of Makarov's deeper results on the relations between harmonic measure in simply connected domains and Hausdorff measures and the work of Carleson and Makarov concerning Brennan's conjecture. Chapter IX discusses harmonic measure on infinitely connected plane domains. Chapter X begins by introducing the Lusin area function, the Schwarzian derivative, and the Jones square sums, and then applies these ideas to several problems about univalent functions and harmonic measures. The thirteen appendices at the end of the text provide further related material.

For space reasons we have not treated some important related topics. These include the connections between Chapters VIII and IX and thermodynamical formalism and several other connections between complex dynamics and harmonic measure. We have emphasized Wiener's solution of the Dirichlet problem instead of the Perron method. The beautiful Perron method can be found in Ahlfors [1973] and Tsuji [1959]. We also taken a few detours around the theory of prime ends. There are excellent discussions of prime ends in Ahlfors [1973], Pommerenke [1975], and Tsuji [1959]. Finally, the theory of harmonic measure in higher dimensions has a different character, and we have omitted it entirely.

At the end of each chapter there is a brief section of biographical notes and a section called "Exercises and Further Results". An exercise consisting of a stated result without a reference is meant to be homework for the reader. "Further results" are outlines, with detailed references, of theorems not in the text.

Results are numbered lexicographically within each chapter, so that Theorem 2.4 is the fourth item in Section 2 of the same chapter, while Theorem III.2.4 is from Section 2 of Chapter III. The same convention is used for formulas, so that (3.2) is in the same chapter, while (IV.6.4) refers to (6.4) from Chapter IV.

Many of the results that inspired us to write this book are also covered in

Pommerenke's excellent book [1991]. However, our emphasis differs from the one in Pommerenke [1991] and we hope the two books will complement each other.

Some unpublished lecture notes from a 1986 Nachdiplom Lecture course at Eidgenössische Technische Hochschule Zurich by the first listed author and the out-of-print monograph Garnett [1986] were preliminary versions of the present book.

The web page
    http://www.math.washington.edu/˜marshall/HMcorrections.html
will list corrections to the book. Though we have tried to avoid errors, the observant reader will no doubt find some. We would appreciate receiving email at marshall@math.washington.edu about any errors you come across.

Many colleagues, friends, and students have helped with their comments and suggestions. Among these, we particularly thank A. Baernstein, M. Benedicks, D. Bertilsson, C. J. Bishop, K. Burdzy, L. Carleson, S. Choi, R. Chow, M. Essèn, R. Gundy, P. Haissinski, J. Handy, P. Jones, P. Koosis, N. Makarov, P. Mateos, M. O'Neill, K. Øyma, R. Pérez-Marco, P. Poggi-Corradini, S. Rohde, I. Uriarte-Tuero, J. Verdera, S. Yang and S. Yoshinobu.

We gratefully acknowledge support during the writing of this book by the Royalty Research Fund of the University of Washington, the University of Washington–University of Bergen Faculty Exchange Program, the Institut des Hautes Etudes Scientifiques, the Centre de Recerca Matemàtica, Barcelona, and the National Science Foundation.

Los Angeles and Seattle   John B. Garnett
Seattle and Bergen   Donald E. Marshall

I

Jordan Domains

To begin we construct harmonic measure and solve the Dirichlet problem in the upper half-plane and the unit disc. We next prove the Fatou theorem on nontangential limits. Then we construct harmonic measure on domains bounded by Jordan curves, via the Riemann mapping theorem and the Carathéodory theorem on boundary correspondence. We review two topics from classical complex analysis, the hyperbolic metric and the elementary distortion theory for univalent functions. We conclude the chapter with the theorem of Hayman and Wu on lengths of level sets. Its proof is an elementary application of harmonic measure and the hyperbolic metric.

1. The Half-Plane and the Disc

Write $\mathbb{H} = \{z : \mathrm{Im}\, z > 0\}$ for the upper half-plane and $\mathbb{R}$ for the real line. Suppose $a < b$ are real. Then the function

$$\theta = \theta(z) = \arg\left(\frac{z-b}{z-a}\right) = \mathrm{Im}\log\left(\frac{z-b}{z-a}\right)$$

is harmonic on $\mathbb{H}$, and $\theta = \pi$ on (a, b) and $\theta = 0$ on $\mathbb{R} \setminus [a, b]$.

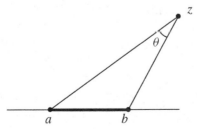

Figure I.1 The harmonic function $\theta(z)$.

1

Viewed geometrically, $\theta(z) = \operatorname{Re}\varphi(z)$ where $\varphi(z)$ is any conformal mapping from $\mathbb{H}$ to the strip $\{0 < \operatorname{Re}z < \pi\}$ which maps (a, b) onto $\{\operatorname{Re}z = \pi\}$ and $\mathbb{R} \setminus [a, b]$ into $\{\operatorname{Re}z = 0\}$. Let $E \subset \mathbb{R}$ be a finite union of open intervals and write $E = \bigcup_{j=1}^{n}(a_j, b_j)$ with $b_{j-1} < a_j < b_j$. Set

$$\theta_j = \theta_j(z) = \arg\left(\frac{z - b_j}{z - a_j}\right)$$

and define the **harmonic measure** of E at $z \in \mathbb{H}$ to be

$$\omega(z, E, \mathbb{H}) = \sum_{j=1}^{n} \frac{\theta_j}{\pi}. \tag{1.1}$$

Then

(i) $0 < \omega(z, E, \mathbb{H}) < 1$ for $z \in \mathbb{H}$,
(ii) $\omega(z, E, \mathbb{H}) \to 1$ as $z \to E$, and
(iii) $\omega(z, E, \mathbb{H}) \to 0$ as $z \to \mathbb{R} \setminus \overline{E}$.

The function $\omega(z, E, \mathbb{H})$ is the unique harmonic function on $\mathbb{H}$ that satisfies (i), (ii), and (iii). The uniqueness of $\omega(z, E, \mathbb{H})$ is a consequence of the following lemma, known as **Lindelöf's maximum principle**.

Lemma 1.1 (Lindelöf). *Suppose the function $u(z)$ is harmonic and bounded above on a region Ω such that $\overline{\Omega} \neq \mathbb{C}$. Let F be a finite subset of $\partial\Omega$ and suppose*

$$\limsup_{z \to \zeta} u(z) \leq 0 \tag{1.2}$$

for all $\zeta \in \partial\Omega \setminus F$. Then $u(z) \leq 0$ on Ω.

Proof. Fix $z_0 \notin \overline{\Omega}$. Then the map $1/(z - z_0)$ transforms Ω into a bounded region, and thus we may assume Ω is bounded. If (1.2) holds for all $\zeta \in \partial\Omega$, then the lemma is the ordinary maximum principle. Write $F = \{\zeta_1, \ldots, \zeta_N\}$, let $\varepsilon > 0$, and set

$$u_\varepsilon(z) = u(z) - \varepsilon \sum_{j=1}^{N} \log\left(\frac{\operatorname{diam}(\Omega)}{|z - \zeta_j|}\right).$$

Then u_ε is harmonic on Ω and $\limsup_{z \to \zeta} u_\varepsilon(z) \leq 0$ for all $\zeta \in \partial\Omega$. Therefore $u_\varepsilon \leq 0$ for all ε, and

$$u(z) \leq \lim_{\varepsilon \to 0} \varepsilon \sum_{j=1}^{N} \log\left(\frac{\operatorname{diam}(\Omega)}{|z - \zeta_j|}\right) = 0. \qquad \blacksquare$$

Lindelöf [1915] proved Lemma 1.1 under the weaker hypothesis that $\partial\Omega$ is infinite. See also Ahlfors [1973]. Exercise 3 and Exercise II.3 tell more about Lindelöf's maximum principle.

Given a domain Ω and a function $f \in C(\partial\Omega)$, the **Dirichlet problem** for f on Ω is to find a function $u \in C(\overline{\Omega})$ such that $\Delta u = 0$ on Ω and $u|_{\partial\Omega} = f$. Theorem 1.2 treats the Dirichlet problem on the upper half-plane $\mathbb{H}$.

Theorem 1.2. *Suppose $f \in C(\mathbb{R} \cup \{\infty\})$. Then there exists a unique function $u = u_f \in C(\overline{\mathbb{H} \cup \{\infty\}})$ such that u is harmonic on $\mathbb{H}$ and $u|_{\partial\mathbb{H}} = f$.*

Proof. We can assume f is real valued and $f(\infty) = 0$. For $\varepsilon > 0$, take disjoint open intervals $I_j = (t_j, t_{j+1})$ and real constants c_j, $j = 1, \ldots, n$, so that the step function

$$f_\varepsilon(t) = \sum_{j=1}^{n} c_j \chi_{I_j}$$

satisfies

$$\left\| f_\varepsilon - f \right\|_{L^\infty(\mathbb{R})} < \varepsilon. \tag{1.3}$$

Set

$$u_\varepsilon(z) = \sum_{j=1}^{n} c_j \omega(z, I_j, \mathbb{H}).$$

If $t \in \mathbb{R} \setminus \bigcup \partial I_j$, then

$$\lim_{\mathbb{H} \ni z \to t} u_\varepsilon(z) = f_\varepsilon(t)$$

by (ii) and (iii). Therefore by (1.3) and Lemma 1.1,

$$\sup_{\mathbb{H}} \left| u_{\varepsilon_1}(z) - u_{\varepsilon_2}(z) \right| < \varepsilon_1 + \varepsilon_2.$$

Consequently the limit

$$u(z) \equiv \lim_{\varepsilon \to 0} u_\varepsilon(z)$$

exists, and the limit $u(z)$ is harmonic on $\mathbb{H}$ and satisfies

$$\sup_{\mathbb{H}} |u(z) - u_\varepsilon(z)| \le 2\varepsilon.$$

We claim that

$$\limsup_{z \to t} |u_\varepsilon(z) - f(t)| \le \varepsilon \tag{1.4}$$

for all $t \in \mathbb{R}$. It is clear that (1.4) holds when $t \notin \bigcup \partial I_j$. To verify (1.4) at the endpoint $t_{j+1} \in \partial I_j \cap \partial I_{j+1}$, notice that by (ii), (iii), and Lemma 1.1,

$$\sup_{\mathbb{H}} \left| c_j \omega(z, I_j, \mathbb{H}) + c_{j+1}\omega(z, I_{j+1}, \mathbb{H}) - \left(\frac{c_j + c_{j+1}}{2}\right) \omega(z, I_j \cup I_{j+1}, \mathbb{H}) \right|$$

$$\leq \left| \frac{c_j - c_{j+1}}{2} \right|,$$

while

$$\lim_{z \to t_{j+1}} \left(\frac{c_j + c_{j+1}}{2}\right) \omega(z, I_j \cup I_{j+1}, \mathbb{H}) = \frac{c_j + c_{j+1}}{2}.$$

Hence all limit values of $u_\varepsilon(z)$ at t_{j+1} lie in the closed interval with endpoints c_j and c_{j+1}, and then (1.3) yields (1.4) for the endpoint t_{j+1}.

Now let $t \in \mathbb{R}$. By (1.4)

$$\limsup_{z \to t} |u(z) - f(t)| \leq \sup_{z \in \mathbb{H}} |u(z) - u_\varepsilon(z)| + \limsup_{z \to t} |u_\varepsilon(z) - f(t)| \leq 3\varepsilon.$$

The same estimate holds if $t = \infty$. Therefore u extends to be continuous on $\overline{\mathbb{H}}$ and $u|_{\partial \mathbb{H}} = f$. The uniqueness of u follows immediately from the maximum principle. ∎

For $a < b$, elementary calculus gives

$$\omega(x + iy, (a, b), \mathbb{H}) = \frac{1}{\pi}\left(\tan^{-1}\left(\frac{x - a}{y}\right) - \tan^{-1}\left(\frac{x - b}{y}\right)\right)$$

$$= \int_a^b \frac{y}{(t - x)^2 + y^2} \frac{dt}{\pi}.$$

If $E \subset \mathbb{R}$ is measurable, we define the **harmonic measure** of E at $z \in \mathbb{H}$ to be

$$\omega(z, E, \mathbb{H}) = \int_E \frac{y}{(t - x)^2 + y^2} \frac{dt}{\pi}. \tag{1.5}$$

When E is a finite union of open intervals this definition (1.5) is the same as definition (1.1). For $z = x + iy \in \mathbb{H}$, the density

$$P_z(t) = \frac{1}{\pi} \frac{y}{(x - t)^2 + y^2}$$

is called the **Poisson kernel** for $\mathbb{H}$. If $f \in C(\mathbb{R} \cup \{\infty\})$, the proof of Theorem 1.2 shows that

$$u_f(z) = \int_{\mathbb{R}} f(t) P_z(t) dt,$$

and for this reason u_f is also called the **Poisson integral** of f.

Note that the harmonic measure $\omega(z, E, \Omega)$ is a harmonic function in its first variable z and a probability measure in its second variable E. If $z_1, z_2 \in \mathbb{H}$ then

$$0 < C^{-1} \leq \frac{\omega(z_1, E, \mathbb{H})}{\omega(z_2, E, \mathbb{H})} \leq C < \infty,$$

where C depends on z_1 and z_2 but not on E. This inequality, known as **Harnack's inequality**, is easily proved by comparing the kernels in (1.5).

Now let $\mathbb{D}$ be the unit disc $\{z : |z| < 1\}$ and let E be a finite union of open arcs on $\partial\mathbb{D}$. Then we define the **harmonic measure** of E at z in $\mathbb{D}$ to be

$$\omega(z, E, \mathbb{D}) \equiv \omega(\varphi(z), \varphi(E), \mathbb{H}), \qquad (1.6)$$

where φ is any conformal map of $\mathbb{D}$ onto $\mathbb{H}$. This harmonic function satisfies conditions analogous to (i), (ii), and (iii), so that by Lemma 1.1 the definition (1.6) does not depend on the choice of φ. It follows by the change of variables $\varphi(z) = i(1 + z)/(1 - z)$ that

$$\omega(z, E, \mathbb{D}) = \int_E \frac{1 - |z|^2}{|e^{i\theta} - z|^2} \frac{d\theta}{2\pi}.$$

An equivalent way to find this function is by a construction similar to (1.1). This construction is outlined in Exercise 1.

Theorem 1.3. *Let* $f(e^{i\theta})$ *be an integrable function on* $\partial\mathbb{D}$ *and set*

$$u(z) = u_f(z) = \int_0^{2\pi} f(e^{i\theta}) \frac{1 - |z|^2}{|e^{i\theta} - z|^2} \frac{d\theta}{2\pi}. \qquad (1.7)$$

Then $u(z)$ *is harmonic on* $\mathbb{D}$. *If* f *is continuous at* $e^{i\theta_0} \in \partial\mathbb{D}$, *then*

$$\lim_{\mathbb{D} \ni z \to e^{i\theta_0}} u(z) = f(e^{i\theta_0}). \qquad (1.8)$$

Clearly (1.8) also holds if the integrable function f is changed on a measure zero subset of $\partial\mathbb{D} \setminus \{e^{i\theta_0}\}$. The function $u = u_f$ is called the **Poisson integral** of f and the kernel

$$P_z(\theta) = \frac{1}{2\pi} \frac{1 - |z|^2}{|e^{i\theta} - z|^2}$$

is the **Poisson kernel** for the disc. If $f \in C(\partial\mathbb{D})$ then

$$U(z) = \begin{cases} u_f(z), & z \in \mathbb{D} \\ f(z), & z \in \partial\mathbb{D} \end{cases}$$

is the solution of the **Dirichlet problem** for f on $\mathbb{D}$.

In the special case when $f(e^{i\theta})$ is continuous, Theorem 1.3 follows from Theorem 1.2 and a change of variables. Conversely, Theorem 1.3 shows that

Theorem 1.2 can be extended to $f \in L^1(dt/(1+t^2))$, again by changing variables.

Proof of Theorem 1.3. We may suppose f is real valued. From the identity

$$\operatorname{Re}\left(\frac{e^{i\theta} + z}{e^{i\theta} - z}\right) = 2\pi P_z(\theta),$$

we see that u is the real part of the analytic function

$$\int_0^{2\pi} f(e^{i\theta}) \frac{e^{i\theta} + z}{e^{i\theta} - z} \frac{d\theta}{2\pi},$$

and therefore that u is a harmonic function. One can also see u is harmonic by differentiating the integral (1.7).

Suppose f is continuous at $e^{i\theta_0}$ and let $\varepsilon > 0$. Then

$$|f(e^{i\theta}) - f(e^{i\theta_0})| < \varepsilon$$

on an interval $I = (\theta_1, \theta_2)$ containing θ_0. Setting

$$u_\varepsilon(z) = \int_{[0,2\pi]\setminus I} \frac{1 - |z|^2}{|e^{i\theta} - z|^2} f(e^{i\theta}) \frac{d\theta}{2\pi} + f(e^{i\theta_0})\omega(z, I, \mathbb{D}),$$

we have

$$|u(z) - u_\varepsilon(z)| = \left| \int_I \frac{1 - |z|^2}{|e^{i\theta} - z|^2} (f(e^{i\theta}) - f(e^{i\theta_0})) \frac{d\theta}{2\pi} \right| \leq \varepsilon\omega(z, I, \mathbb{D}) \leq \varepsilon.$$

However, $\lim_{z \to e^{i\theta_0}} u_\varepsilon(z) = f(e^{i\theta_0})$ by the definition of u_ε. Therefore

$$\limsup_{z \to e^{i\theta_0}} |u(z) - f(e^{i\theta_0})| < \varepsilon,$$

and (1.8) holds when f is continuous at $e^{i\theta_0}$. ∎

2. Fatou's Theorem and Maximal Functions

When $f \in L^1(\partial\mathbb{D})$ the limit (1.8) can fail to exist at every $\zeta \in \partial\mathbb{D}$; see Exercise 7. However, there is a substitute result known as **Fatou's theorem**, in which the approach $z \to \zeta$ is restricted to cones. For $\zeta \in \partial\mathbb{D}$ and $\alpha > 1$, we define the **cone**

$$\Gamma_\alpha(\zeta) = \{z : |z - \zeta| < \alpha(1 - |z|)\}.$$

The cone $\Gamma_\alpha(\zeta)$ is asymptotic to a sector with vertex ζ and angle $2\sec^{-1}(\alpha)$ that is symmetric about the radius $[0, \zeta]$. The cones $\Gamma_\alpha(\zeta)$ expand as α increases.

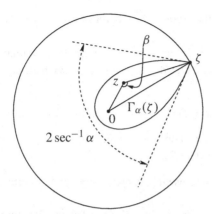

Figure I.2 The cone $\Gamma_\alpha(\zeta)$.

A function $u(z)$ on $\mathbb{D}$ has **nontangential limit** Λ at $\zeta \in \partial \mathbb{D}$ if

$$\lim_{\Gamma_\alpha(\zeta) \ni z \to \zeta} u(z) = A \tag{2.1}$$

for *every* $\alpha > 1$. A good example is the function $u(z) = e^{\frac{z+1}{z-1}}$. This function $u(z)$ is continuous on $\partial \mathbb{D} \setminus \{1\}$, and $|u(\zeta)| = 1$ on $\partial \mathbb{D} \setminus \{1\}$, but $u(z)$ has nontangential limit 0 at $\zeta = 1$. With fixed $\alpha > 1$, the **nontangential maximal function** of u at ζ is

$$u_\alpha^*(\zeta) = \sup_{\Gamma_\alpha(\zeta)} |u(z)|.$$

If u has a finite nontangential limit at ζ, then $u_\alpha^*(\zeta) < \infty$ for every $\alpha > 1$.

We write $|E|$ for the Lebesgue measure of $E \subset \partial \mathbb{D}$.

Theorem 2.1 (Fatou's theorem). *Let $f(e^{i\theta}) \in L^1(\partial \mathbb{D})$ and let $u(z)$ be the Poisson integral of f. Then at almost every $\zeta = e^{i\theta} \in \partial \mathbb{D}$,*

$$\lim_{\Gamma_\alpha(\zeta) \ni z \to \zeta} u(z) = f(\zeta) \tag{2.2}$$

for all $\alpha > 1$. Moreover, for each $\alpha > 1$

$$\left| \{ \zeta \in \partial \mathbb{D} : u_\alpha^*(\zeta) > \lambda \} \right| \le \frac{3 + 6\alpha}{\lambda} \|f\|_1. \tag{2.3}$$

When $u(z)$ is the Poisson integral of $f \in L^1(\partial \mathbb{D})$ the function $u = u_f$ is also called the **solution to the Dirichlet problem for** f, even though u converges to f on $\partial \mathbb{D}$ only nontangentially and only almost everywhere.

Inequality (2.3) says the operator $L^1(\partial \mathbb{D}) \ni f \to u_\alpha^*$ is **weak-type 1-1**. It follows from (2.2) that $u_\alpha^*(\zeta) < \infty$ almost everywhere, but (2.3) is a sharper, quantitative result. In the proof of the theorem we derive (2.2) from the estimate (2.3).

The proof of Fatou's theorem is a standard approximate identity argument from real analysis that derives almost everywhere convergence for all $f \in L^1(\partial \mathbb{D})$ from

(a) an estimate such as (2.3) for the maximal function, and

(b) the almost everywhere convergence (2.2) for all functions in a dense subset of $L^1(\partial \mathbb{D})$, such as $C(\partial \mathbb{D})$.

See Stein [1970]. We will use this approximate identity argument again later.

Proof. As promised, we first assume (2.3) and show (2.3) implies (2.2). Fix α temporarily. We may assume f is real valued. Set

$$W_f(\zeta) = \limsup_{\Gamma_\alpha \ni z \to \zeta} |u_f(z) - f(\zeta)|.$$

Then $W_f(\zeta) \leq u_\alpha^*(\zeta) + |f(\zeta)|$. Chebyshev's inequality gives

$$\left| \{ \zeta : |f(\zeta)| > \lambda \} \right| \leq \frac{\|f\|_1}{\lambda},$$

so that by (2.3),

$$\begin{aligned} \left| \{ \zeta : W_f(\zeta) > \lambda \} \right| &\leq \left| \{ \zeta : u_\alpha^*(\zeta) > \lambda/2 \} \right| + \left| \{ \zeta : |f(\zeta)| > \lambda/2 \} \right| \\ &\leq \frac{8 + 12\alpha}{\lambda} \|f\|_1. \end{aligned} \tag{2.4}$$

Fix $\varepsilon > 0$ and let $g \in C(\partial \mathbb{D})$ be such that $\|f - g\|_1 \leq \varepsilon^2$. Now $W_g(\zeta) = 0$ by Theorem 1.3, and hence

$$W_f(\zeta) = W_{f-g}(\zeta).$$

Applying (2.4) to $f - g$ then gives

$$\left| \{ \zeta : W_f(\zeta) > \varepsilon \} \right| \leq \frac{(8 + 12\alpha)\varepsilon^2}{\varepsilon} = (8 + 12\alpha)\varepsilon.$$

Therefore, for any fixed α, (2.2) holds almost everywhere. Because the cones Γ_α increase with α, it follows that (2.2) holds for every $\alpha > 1$, except for ζ in a set of measure zero.

To prove (2.3) we will dominate the nontangential maximal function with a second, simpler maximal function. Let $f \in L^1(\partial \mathbb{D})$ and write

$$Mf(\zeta) = \sup_{I \ni \zeta} \frac{1}{|I|} \int_I |f| \, d\theta$$

for the maximal average of $|f|$ over subarcs $I \subset \partial\mathbb{D}$ that contain ζ. The function Mf is called the **Hardy–Littlewood maximal function** of f. The function Mf is simpler than u_α^* because it features characteristic functions of intervals instead of Poisson kernels.

Lemma 2.2. *Let $u(z)$ be the Poisson integral of $f \in L^1(\partial\mathbb{D})$ and let $\alpha > 1$. Then*

$$u_\alpha^*(\zeta) \le (1 + 2\alpha)Mf(\zeta). \tag{2.5}$$

Proof. Assume $\zeta = 1$. Fix z so that $\theta_0 = \arg z$ has $|\theta_0| \le \pi$. Define

$$P_z^*(\theta) = \sup\{P_z(\varphi) : |\theta| \le |\varphi| \le \pi\}$$

$$= \begin{cases} \dfrac{1}{2\pi} \dfrac{1+|z|}{1-|z|}, & |\theta| \le |\theta_0| \\[2mm] \max(P_z(\theta), P_z(-\theta)), & |\theta_0| < |\theta| \le \pi. \end{cases}$$

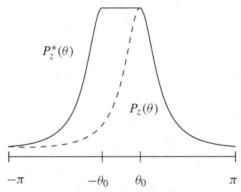

Figure I.3 The function P_z^*.

The function P_z^* satisfies

(i) $P_z^*(\theta)$ is an even function of $\theta \in [-\pi, \pi]$,
(ii) $P_z^*(\theta)$ is decreasing on $[0, \pi]$, and
(iii) $P_z^*(\theta) \ge P_z(\theta)$.

The even function P_z^* is the smallest decreasing majorant of P_z on $[0, \pi]$. We may assume $f(e^{i\theta}) \ge 0$, so that

$$\int f(e^{i\theta})P_z(\theta)d\theta \le \int f(e^{i\theta})P_z^*(\theta)d\theta.$$

Then properties (i) and (ii) imply

$$\int f(e^{i\theta}) P_z^*(\theta) d\theta \le ||P_z^*||_1 M f(1) \qquad (2.6)$$

because P_z^* is the increasing limit of a sequence of functions of the form

$$\sum c_j \left(\frac{1}{2\theta_j} \chi_{(-\theta_j,\theta_j)}(\theta) \right)$$

with $c_j \ge 0$ and $\sum c_j \le ||P_z^*||_1$.

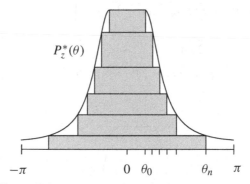

$P_z^*(\theta)$

$-\pi$ $0 \;\; \theta_0$ θ_n π

Figure I.4 Approximating P_z^* by a step function.

Now we claim that when $z \in \Gamma_\alpha(1)$,

$$||P_z^*||_1 \le (1 + 2\alpha). \qquad (2.7)$$

Note that (iii), (2.6), and (2.7) imply (2.5). To prove (2.7) we first assume $-\pi/2 \le \theta_0 = \arg z \le \pi/2$. Then by the law of sines,

$$\frac{|\theta_0|}{1 - |z|} \le \alpha \frac{|\theta_0|}{|1 - z|} \le \frac{\pi \alpha}{2} \frac{|\sin \theta_0|}{|1 - z|} = \frac{\pi \alpha}{2} \frac{|\sin \beta|}{1} \le \frac{\pi \alpha}{2},$$

where $\beta = \arg(z - 1)/z$ is explained by Figure I.2. If $\pi/2 \le |\theta_0| \le \pi$ and $z \in \Gamma_\alpha(1)$, then $|1 - z| \ge 1$ and

$$\frac{|\theta_0|}{1 - |z|} \le \alpha \frac{|\theta_0|}{|1 - z|} \le \pi \alpha.$$

Hence (see Figure I.3)

$$||P_z^*||_1 = 2 \int_{|\theta_0|}^{\pi} P_z(\theta) d\theta + \frac{2|\theta_0|}{2\pi} \frac{1 + |z|}{1 - |z|} \le (1 + 2\alpha).$$

That proves (2.7) and therefore Lemma 2.2. ∎

By Lemma 2.2, the inequality (2.3) will follow from the simpler inequality

$$\left|\{\zeta \in \partial\mathbb{D} : Mf(\zeta) > \lambda\}\right| \leq \frac{3\|f\|_1}{\lambda}, \qquad (2.8)$$

which says that the operator $L^1 \ni f \to Mf$ is also weak-type 1-1.

To prove the inequality (2.8), we use a covering lemma.

Lemma 2.3. *Let μ be a positive Borel measure on $\partial\mathbb{D}$ and let $\{I_j\}$ be a finite sequence of open intervals in $\partial\mathbb{D}$. Then $\{I_j\}$ contains a pairwise disjoint subfamily $\{J_k\}$ such that*

$$\sum \mu(J_k) \geq \frac{1}{3}\mu\left(\bigcup I_j\right). \qquad (2.9)$$

Proof. Because the family $\{I_j\}$ is finite, we may assume that no I_j is contained in the union of the others. Writing $I_j = \{e^{i\theta} : \theta \in (a_j, b_j)\}$, we may also assume that

$$0 \leq a_1 < a_2 < \ldots < a_n < 2\pi.$$

Then $b_{j+1} > b_j$, because otherwise $I_{j+1} \subset I_j$, and $b_{j-1} < a_{j+1}$, because otherwise $I_j \subset I_{j-1} \cup I_{j+1}$. If $n > 1$, then $b_n < b_1 + 2\pi$ and $b_{n-1} < a_1 + 2\pi$. Consequently, the family of even-numbered intervals I_j is pairwise disjoint. The family of odd-numbered intervals I_j is almost pairwise disjoint; only the first and last intervals can intersect. If

$$\sum_{j \text{ even}} \mu(I_j) \geq \frac{1}{3}\mu\left(\bigcup I_j\right),$$

we take the even-numbered intervals to be the subfamily $\{J_k\}$. Otherwise

$$\sum_{j \text{ odd}} \mu(I_j) \geq \frac{2}{3}\mu\left(\bigcup I_j\right).$$

In that case, if

$$\mu(I_1) \leq \frac{1}{2}\sum_{j \text{ odd}} \mu(I_j),$$

we take for $\{J_k\}$ the family of odd-numbered intervals, omitting the first interval I_1, while if

$$\mu(I_1) > \frac{1}{2}\sum_{j \text{ odd}} \mu(I_j),$$

we take $\{J_k\} = \{I_1\}$. Then in each case (2.9) holds for the subfamily $\{J_k\}$. ∎

Lemma 2.4. *The operator* $f \to M(f)$ *is weak-type 1-1: If* $f \in L^1$, *then*

$$\left|\{\zeta \in \partial\mathbb{D} : Mf(\zeta) > \lambda\}\right| \le \frac{3\|f\|_1}{\lambda}. \tag{2.8}$$

Proof. Let K be a compact subset of $E_\lambda = \{\zeta \in \partial\mathbb{D} : Mf(\zeta) > \lambda\}$. For each $\zeta \in E_\lambda$ there is an open interval I such that $\zeta \in I$ and

$$\frac{1}{|I|}\int_I |f|d\theta > \lambda,$$

so that

$$|I| < \frac{1}{\lambda}\int_I |f|d\theta.$$

Cover K by finitely many such intervals $\{I_j : 1 \le j \le n\}$, and let $\{J_k\}$ be the pairwise disjoint subfamily given by Lemma 2.3. Then

$$|K| \le \left|\bigcup I_j\right| \le 3\sum |J_k| \le \tfrac{3}{\lambda}\sum \int_{J_k}|f|d\theta \le \tfrac{3}{\lambda}\|f\|_1,$$

and letting $|K|$ tend to $|E_\lambda|$ establishes Lemma 2.4. ∎

By (2.5) and (2.8), inequality (2.3) holds with constant $3 + 6\alpha$, and the proof of Fatou's theorem is complete. ∎

Corollary 2.5. *If* u *is a bounded harmonic function on* $\mathbb{D}$, *then for every* $\alpha > 1$ *and for almost every* $\zeta = e^{i\theta} \in \partial\mathbb{D}$,

$$f(\zeta) = \lim_{\Gamma_\alpha(\zeta)\ni z\to\zeta} u(z)$$

exists, $u(z)$ *is the Poisson integral of* f, *and* $\|f\|_\infty = \sup_{z\in\mathbb{D}}|u(z)|$.

Proof. Let $r_n \to 1$ and let $f_n(e^{i\theta}) = u(r_n e^{i\theta})$. By the Banach–Alaoglu theorem, the sequence $\{f_n\}$ has a weak-star cluster point $f \in L^\infty(\partial\mathbb{D})$ satisfying $\|f\|_\infty \le \limsup \|f_n\|_\infty \le \sup_\mathbb{D}|u(z)|$. Since $u(r_n z)$ is the Poisson integral of f_n, and Poisson kernels are in L^1, u must be the Poisson integral of f. But then $|u(z)| \le \|f\|_\infty$. The corollary now follows from Fatou's theorem. ∎

In particular, the corollary implies that for any measurable $E \subset \partial\mathbb{D}$, there exists a unique bounded harmonic function $u(z)$ on $\mathbb{D}$ such that $u(z)$ has non-tangential limit χ_E almost everywhere. It is the function $u(z) = \omega(z, E, \mathbb{D})$.

3. Carathéodory's Theorem

Let Ω be a simply connected domain in the extended plane $\mathbb{C}^*$. We say Ω is a **Jordan domain** if $\Gamma = \partial\Omega$ is a Jordan curve in $\mathbb{C}^*$.

Theorem 3.1 (Carathéodory). *Let φ be a conformal mapping from the unit disc $\mathbb{D}$ onto a Jordan domain Ω. Then φ has a continuous extension to $\overline{\mathbb{D}}$, and the extension is a one-to-one map from $\overline{\mathbb{D}}$ onto $\overline{\Omega}$.*

Because φ maps $\mathbb{D}$ onto Ω, the continuous extension (also denoted by φ) must map $\partial\mathbb{D}$ onto $\Gamma = \partial\Omega$, and because φ is one-to-one on $\partial\mathbb{D}$, $\varphi(e^{i\theta})$ parameterizes the Jordan curve Γ.

Before we prove Carathéodory's theorem, we use it to solve the Dirichlet problem on a Jordan domain Ω. Let f be Borel function on Γ such that $f \circ \varphi$ is integrable on $\partial\mathbb{D}$. If $w = \varphi^{-1}(z)$, then

$$u(z) = u_f(z) = \int_0^{2\pi} f \circ \varphi(e^{i\theta}) \frac{1 - |w|^2}{|e^{i\theta} - w|^2} \frac{d\theta}{2\pi} \tag{3.1}$$

is harmonic on Ω, and by Theorem 3.1 and Theorem 1.3,

$$\lim_{\Omega \ni z \to \zeta} u(z) = f(\zeta) \tag{3.2}$$

whenever $\varphi^{-1}(\zeta) \in \partial\mathbb{D}$ is a point of continuity of $f \circ \varphi$. In particular, if f is continuous on Γ then (3.2) holds for all $\zeta \in \Gamma$ and $u(z) = u_f(z)$ solves the **Dirichlet problem** for f on Ω.

If f is a bounded Borel function on Γ, then $f \circ \varphi$ is Borel and the integral (3.1) is defined. For any Borel set $E \subset \Gamma$ we use (3.1) with $f = \chi_E$ to define the **harmonic measure** of E relative to Ω by:

$$\omega(z, E, \Omega) = \omega(w, \varphi^{-1}(E), \mathbb{D}) = \int_{\varphi^{-1}(E)} \frac{1 - |w|^2}{|e^{i\theta} - w|^2} \frac{d\theta}{2\pi}. \tag{3.3}$$

Then $E \to \omega(z, E)$ is a Borel measure on $\partial\Omega$, and (3.1) can be rewritten as

$$u(z) = \int_{\partial\Omega} f(\zeta) d\omega(z, \zeta). \tag{3.4}$$

Equations (3.3) and (3.4) do not depend on the choice of φ, because for every conformal self map T of $\mathbb{D}$,

$$\omega(T(w), T(\varphi^{-1}(E)), \mathbb{D}) = \omega(w, \varphi^{-1}(E), \mathbb{D}).$$

When f is a bounded Borel function on $\partial\Omega$, (3.4) and Fatou's theorem give

$$\sup_{z \in \Omega} |u(z)| = \|f\|_{L^\infty(\omega)}.$$

Moreover, Corollary 2.5 shows that every bounded harmonic function on Ω can be expressed in the form (3.4).

The principal goal of this book is to find geometric properties of the harmonic measure $\omega(z, E)$ more explicit than the definition (3.3). But (3.3) already points out the key issue: for a Jordan domain questions about harmonic measure are equivalent to questions about the boundary behavior of conformal mappings.

Proof of Theorem 3.1. We may assume Ω is bounded. Fix $\zeta \in \partial \mathbb{D}$. First we show φ has a continuous extension at ζ. Let $0 < \delta < 1$, write

$$B(\zeta, \delta) = \{z : |z - \zeta| < \delta\}$$

and set $\gamma_\delta = \mathbb{D} \cap \partial B(\zeta, \delta)$. Then $\varphi(\gamma_\delta)$ is a Jordan arc having length

$$L(\delta) = \int_{\gamma_\delta} |\varphi'(z)| ds.$$

By the Cauchy–Schwarz inequality

$$L^2(\delta) \leq \pi \delta \int_{\gamma_\delta} |\varphi'(z)|^2 ds,$$

so that for $\rho < 1$,

$$\int_0^\rho \frac{L^2(\delta)}{\delta} d\delta \leq \pi \int \int_{\mathbb{D} \cap B(\zeta, \rho)} |\varphi'(z)|^2 dx\, dy \qquad (3.5)$$

$$= \pi \operatorname{Area}(\varphi(\mathbb{D} \cap B(\zeta, \rho))) < \infty.$$

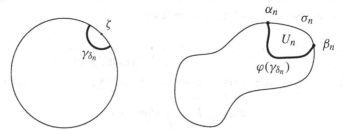

Figure I.5 The crosscuts γ_{δ_n} and $\varphi(\gamma_{\delta_n})$.

Therefore there is a sequence $\delta_n \downarrow 0$ such that $L(\delta_n) \to 0$. When $L(\delta_n) < \infty$, the curve $\varphi(\gamma_{\delta_n})$ has endpoints $\alpha_n, \beta_n \in \overline{\Omega}$, and both of these endpoints must lie on $\Gamma = \partial \Omega$. Indeed, if $\alpha_n \in \Omega$, then some point near α_n has two distinct preimages in $\mathbb{D}$ because φ maps $\mathbb{D}$ onto Ω, and that is impossible because φ is one-to-one. Furthermore,

$$|\alpha_n - \beta_n| \leq L(\delta_n) \to 0. \tag{3.6}$$

Let σ_n be that closed subarc of Γ having endpoints α_n and β_n and having smaller diameter. Then (3.6) implies $\mathrm{diam}(\sigma_n) \to 0$, because Γ is homeomorphic to the circle. By the Jordan curve theorem the curve $\sigma_n \cup \varphi(\gamma_{\delta_n})$ divides the plane into two regions, and one of these regions, say U_n, is bounded. Then $U_n \subset \Omega$, because $\mathbb{C}^* \setminus \overline{\Omega}$ is arcwise connected. Since

$$\mathrm{diam}(\partial U_n) = \mathrm{diam}\big(\sigma_n \cup \varphi(\gamma_{\delta_n})\big) \to 0,$$

we conclude that

$$\mathrm{diam}(U_n) \to 0. \tag{3.7}$$

Set $D_n = \mathbb{D} \cap \{z : |z - \zeta| < \delta_n\}$. We claim that for large n, $\varphi(D_n) = U_n$. If not, then by connectedness $\psi(\mathbb{D} \setminus \overline{D}_n) = U_n$ and

$$\mathrm{diam}(U_n) \geq \mathrm{diam}\big(\varphi(B(0, 1/2))\big) > 0,$$

which contradicts (3.7). Therefore $\mathrm{diam}(\varphi(D_n)) \to 0$ and $\bigcap \overline{\varphi(D_n)}$ consists of a single point, because $\varphi(D_{n+1}) \subset \varphi(D_n)$. That means φ has a continuous extension to $\{\zeta\} \cup \mathbb{D}$. It is an exercise to show that the union over ζ of these extensions defines a continuous map on $\overline{\mathbb{D}}$.

Let φ also denote the extension $\varphi : \overline{\mathbb{D}} \to \overline{\Omega}$. Since $\varphi(\mathbb{D}) = \Omega$, φ maps $\overline{\mathbb{D}}$ onto $\overline{\Omega}$. To show φ is one-to-one, suppose $\varphi(\zeta_1) = \varphi(\zeta_2)$ but $\zeta_1 \neq \zeta_2$. The argument used to show $\alpha_n \in \Gamma$ also shows that $\varphi(\partial \mathbb{D}) = \Gamma$, and so we can assume $\zeta_j \in \partial \mathbb{D}$, $j = 1, 2$. The Jordan curve

$$\{\varphi(r\zeta_1) : 0 \leq r \leq 1\} \cup \{\varphi(r\zeta_2) : 0 \leq r \leq 1\}$$

bounds a domain $W \subset \Omega$, and then $\varphi^{-1}(W)$ is one of the two components of

$$\mathbb{D} \setminus \Big(\{r\zeta_1 : 0 \leq r \leq 1\} \cup \{r\zeta_2 : 0 \leq r \leq 1\}\Big).$$

But since $\varphi(\partial \mathbb{D}) \subset \Gamma$,

$$\varphi(\partial \mathbb{D} \cap \partial \varphi^{-1}(W)) \subset \partial W \cap \partial \Omega = \{\varphi(\zeta_1)\}$$

and φ is constant on an arc of $\partial \mathbb{D}$. It follows that φ is constant, either by Schwarz reflection principle or by the Jensen formula, and this contradiction shows $\varphi(\zeta_1) \neq \varphi(\zeta_2)$. ∎

One can also prove φ is one-to-one by repeating for φ^{-1} the proof that φ is continuous. See Exercise 13(b). Exercise 12 gives the necessary and sufficient condition that φ have a continuous extension to $\overline{\mathbb{D}}$.

The Cauchy–Schwarz trick used to prove (3.5) is known as a **length–area** argument. The length–area method is the cornerstone of the theory of extremal length. See Chapter IV.

4. Distortion and the Hyperbolic Metric

Let $\mathbb{D}$ be the open unit disc. The **hyperbolic distance** from $z_1 \in \mathbb{D}$ to $z_2 \in \mathbb{D}$ is

$$\rho(z_1, z_2) = \rho_{\mathbb{D}}(z_1, z_2) = \inf \int_{z_1}^{z_2} \frac{|dz|}{1 - |z|^2}, \tag{4.1}$$

where the infimum is taken over all arcs in $\mathbb{D}$ connecting z_1 to z_2. Let $\mathcal{M}$ denote the set of conformal self maps of $\mathbb{D}$:

$$T(z) = \lambda \frac{z - a}{1 - \bar{a}z}, \quad a \in \mathbb{D}, \quad |\lambda| = 1.$$

When $T \in \mathcal{M}$, we have

$$\frac{|T'(z)|}{1 - |T(z)|^2} = \frac{1}{1 - |z|^2},$$

and thus the hyperbolic distance is conformally invariant,

$$\rho(T(z_1), T(z_2)) = \rho(z_1, z_2), \quad T \in \mathcal{M}. \tag{4.2}$$

This conformal invariance is the main reason we are interested in the hyperbolic distance.

The **hyperbolic metric** is the infinitesimal form $|dz|/(1 - |z|^2)$ of the hyperbolic distance. Taking

$$T(z) = \frac{z - z_1}{1 - \bar{z_1}z}$$

gives

$$\rho(z_1, z_2) = \rho(0, T(z_2)) = \int_0^{T(z_2)} \frac{|dz|}{1 - |z|^2}.$$

Therefore the hyperbolically shortest arc from 0 to $T(z_2)$ is the radius $[0, T(z_2)]$, and its hyperbolic length is

$$\rho(0, T(z_2)) = \frac{1}{2} \log\left(\frac{1 + |T(z_2)|}{1 - |T(z_2)|}\right).$$

In general

$$\rho(z_1, z_2) = \frac{1}{2} \log \left(\frac{1 + \left| \frac{z_2 - z_1}{1 - \overline{z_1} z_2} \right|}{1 - \left| \frac{z_2 - z_1}{1 - \overline{z_1} z_2} \right|} \right), \tag{4.3}$$

and the hyperbolically shortest curve, or the **geodesic**, from z_1 to z_2 is a segment of a diameter of $\mathbb{D}$ or an arc of a circle in $\mathbb{D}$ orthogonal to $\partial \mathbb{D}$.

By (4.3) we have

$$\left| \frac{z_2 - z_1}{1 - \overline{z_1} z_2} \right| = \frac{e^{2\rho(z_1, z_2)} - 1}{e^{2\rho(z_1, z_2)} + 1} = \tanh \rho(z_1, z_2).$$

Write

$$t = t(d) = \tanh(d) = \frac{e^{2d} - 1}{e^{2d} + 1}.$$

Then the hyperbolic ball $B = \{z : \rho(z, a) < d\}$ is the euclidean disc

$$\{z : \left| \frac{z - a}{1 - \overline{a} z} \right| < t\},$$

and a calculation shows that B has euclidean radius

$$r(a, d) = \frac{t(1 - |a|^2)}{1 - t^2 |a|^2} \tag{4.4}$$

and euclidean distance to $\partial \mathbb{D}$

$$\text{dist}(B, \partial \mathbb{D}) = \left(\frac{1 - t}{1 + |a|t} \right)(1 - |a|). \tag{4.5}$$

Therefore, if d is fixed, the euclidean distance $\text{dist}(B, \partial \mathbb{D})$ and the euclidean diameter of B are both comparable to $\text{dist}(a, \partial \mathbb{D})$. However, for $a \neq 0$ the euclidean center of B is not a. Figure I.6 shows two hyperbolic balls with the same hyperbolic radius and two geodesics with the same hyperbolic length.

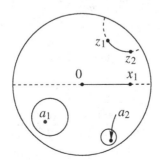

Figure I.6 Hyperbolic balls and geodesics.

Now assume $\psi(z)$ is a **univalent function** in $\mathbb{D}$, that is, assume ψ is analytic and one-to-one on $\mathbb{D}$. After dilating, translating, and rotating the domain $\psi(\mathbb{D})$, ψ is **normalized** by $\psi(0) = 0$ and $\psi'(0) = 1$, so that

$$\psi(z) = z + a_2 z^2 + \cdots. \tag{4.6}$$

Important examples are the **Koebe functions**

$$\psi(z) = \psi_\lambda(z) = \frac{z}{(1 - \lambda z)^2}, \quad |\lambda| = 1. \tag{4.7}$$

Note that

$$\psi_\lambda(z) = \sum_{n=1}^{\infty} n\lambda^{n-1} z^n$$

maps $\mathbb{D}$ to the complement of the radial slit $[-\bar\lambda/4, \infty]$.

Theorem 4.1 (Koebe one-quarter theorem). *Assume $\psi(z)$ is a univalent function on $\mathbb{D}$. If $\psi(z)$ has the form (4.6) then*

$$|a_2| \leq 2 \tag{4.8}$$

and

$$\mathrm{dist}(0, \partial\psi(\mathbb{D})) \geq \frac{1}{4}. \tag{4.9}$$

Equality holds in (4.8) or (4.9) if and only if ψ is a Koebe function (4.7).

Proof. First note that (4.8) implies (4.9). Indeed, suppose $w \notin \psi(\mathbb{D})$. Then (4.8) holds for the univalent function

$$g(z) = \frac{w\psi(z)}{w - \psi(z)} = z + (a_2 + \frac{1}{w})z^2 + \cdots,$$

so that

$$\left| a_2 + \frac{1}{w} \right| \leq 2, \tag{4.10}$$

and together (4.10) and (4.8) give

$$|w| \geq \frac{1}{4}.$$

To prove (4.8) we form the odd function

$$f(z) = z\sqrt{\frac{\psi(z^2)}{z^2}} = z + \frac{a_2}{2} z^3 + \cdots.$$

Then f is univalent because ψ is univalent, and the $\mathbb{C}^* = \mathbb{C} \cup \{\infty\}$ valued function

$$F(z) = \frac{1}{f(z)} = \frac{1}{z} - \frac{a_2}{2}z + \cdots = \frac{1}{z} + \sum_{n=1}^{\infty} b_n z^n \qquad (4.11)$$

is also univalent in $\mathbb{D}$. To complete the proof we use the following lemma, called the **area theorem**, which we will prove after we establish (4.8).

Lemma 4.2 (area theorem). *If the univalent function $F(z)$ satisfies (4.11), then*

$$\sum_{n=1}^{\infty} n|b_n|^2 \leq 1. \qquad (4.12)$$

To establish (4.8) we apply (4.12) to $F = 1/f$. Since $b_1 = -a_2/2$ and $|b_1| \leq 1$, we have $|a_2| \leq 2$. ∎

The reader can verify that equality in either of (4.8) or (4.9) implies ψ is a Koebe function.

Proof of Lemma 4.2. The lemma is called the area theorem because of its proof. For $r < 1$, the Jordan curve $\Gamma_r = \{F(re^{i\theta}) : 0 \leq \theta \leq 2\pi\}$ encloses an area $A(r)$, and by Green's theorem,

$$A(r) = \frac{-i}{2} \int_{\Gamma_r} w d\overline{w} = \frac{-i}{2} \int_0^{2\pi} F(re^{i\theta}) \frac{\partial \overline{F}}{\partial \theta}(re^{i\theta}) d\theta.$$

Therefore by (4.11) and Fourier series,

$$A(r) = \pi \left(\frac{1}{r^2} - \sum_{n=1}^{\infty} n|b_n|^2 r^{2n} \right)$$

and

$$1 - \sum_{n=1}^{\infty} n|b_n|^2 = \lim_{r \to 1} \frac{A(r)}{\pi} \geq 0,$$

which yields (4.12). ∎

Theorem 4.3 (Koebe's estimate). *Let $\varphi(z)$ be a conformal mapping from the unit disc $\mathbb{D}$ onto a simply connected domain Ω. Then for all $z \in \mathbb{D}$*

$$\frac{1}{4}|\varphi'(z)|(1 - |z|^2) \leq \text{dist}(\varphi(z), \partial\Omega) \leq |\varphi'(z)|(1 - |z|^2). \qquad (4.13)$$

Proof. Fix $z_0 \in \mathbb{D}$. Then the univalent function

$$\psi(z) = \frac{\varphi\left(\frac{z+z_0}{1+\bar{z}_0 z}\right) - \varphi(z_0)}{\varphi'(z_0)\left(1 - |z_0|^2\right)}$$

satisfies $\psi(0) = 0$ and $\psi'(0) = 1$. Hence if $w \notin \varphi(\mathbb{D})$, then

$$\left| \frac{w - \varphi(z_0)}{\varphi'(z_0)(1 - |z_0|^2)} \right| \geq \frac{1}{4}$$

by (4.9), and this gives the left-hand inequality in (4.13).

To prove the right-hand inequality, fix $z \in \mathbb{D}$, take

$$f(w) = \varphi^{-1}\Big(\varphi(z) + \mathrm{dist}\big(\varphi(z), \partial\Omega\big)w\Big),$$

and apply the Schwarz lemma at $w = 0$ to the function

$$g(w) = \frac{f(w) - z}{1 - \bar{z} f(w)}. \qquad \blacksquare$$

We will often use the invariant form of (4.13):

Corollary 4.4. *Let ψ be a conformal mapping from a simply connected domain Ω_1 onto a simply connected domain Ω_2, and let $\psi(z_0) = w_0$. Then*

$$\frac{|\psi'(z_0)|}{4} \leq \frac{\mathrm{dist}(w_0, \partial\Omega_2)}{\mathrm{dist}(z_0, \partial\Omega_1)} \leq 4|\psi'(z_0)| \qquad (4.14)$$

Proof. Applying (4.13) to $\varphi(z) = \psi(z_0 + \mathrm{dist}(z_0, \partial\Omega_1)z)$ gives the left-hand inequality and the same argument with ψ^{-1} gives the right-hand inequality. $\blacksquare$

In a simply connected domain $\Omega \neq \mathbb{C}$, the **hyperbolic distance** is defined by moving back to $\mathbb{D}$ via a conformal map $\varphi : \mathbb{D} \to \Omega$. We write

$$\rho_\Omega(w_1, w_2) = \rho_{\mathbb{D}}(z_1, z_2)$$

when $w_j = \varphi(z_j)$. By (4.2), $\rho_\Omega(w_1, w_2)$ does not depend on the choice of the conformal map φ. The **quasihyperbolic distance** from $w_1 \in \Omega$ to $w_2 \in \Omega$ is

$$Q_\Omega(w_1, w_2) = \inf \int_{w_1}^{w_2} \frac{|dw|}{\mathrm{dist}(w, \partial\Omega)},$$

in which the infimum is taken over all arcs in Ω joining w_1 to w_2. Since (4.13) can be written as

$$\frac{|dz|}{1 - |z|^2} \leq \frac{|dw|}{\mathrm{dist}(w, \partial\Omega)} \leq \frac{4|dz|}{1 - |z|^2},$$

where $w = \varphi(z)$, we have

$$\rho_\Omega(w_1, w_2) \le Q_\Omega(w_1, w_2) \le 4\rho_\Omega(w_1, w_2). \qquad (4.15)$$

Consequently the geometric statement following (4.4) and (4.5) about hyperbolic distances near $\partial\mathbb{D}$ remains approximately true in every simply connected domain with nontrivial boundary.

Let Ω be any proper open subset of $\mathbb{C}$. Then there exist closed squares $\{S_j\}$, having pairwise disjoint interiors and sides parallel to the axes, such that

(i) S_j has sidelength $\ell(S_j) = 2^{-n_j}$
(ii) $\Omega = \bigcup S_j$, and
(iii) $\mathrm{diam}(S_j) \le \mathrm{dist}(S_j, \partial\Omega) < 4\,\mathrm{diam}(S_j)$.

The squares $\{S_j\}$ are called **Whitney squares**. Here is one way to construct Whitney squares in the case $\mathrm{diam}(\Omega) < \infty$: Let $2^{-N+1} \le \mathrm{diam}(\Omega) < 2^{-N+2}$ and partition the plane into squares having sides parallel to the axes and sidelength 2^{-N}. We call these squares "2^{-N} squares". Include in the family $\{S_j\}$ any 2^{-N} square $S \subset \Omega$ which satisfies (iii), and divide each of the remaining 2^{-N} squares into four squares of side 2^{-N-1}. Next include in $\{S_j\}$ any of these new 2^{-N-1} squares contained in Ω and satisfying (iii), and continue. (See Figure I.7). Finding Whitney squares in the case $\mathrm{diam}(\Omega) = \infty$ is an exercise for the reader. Whitney squares can be viewed as replacements for hyperbolic balls since there are universal constants $c_1 < c_2$ such that each T_j contains a hyperbolic ball of radius c_1 and is contained in a hyperbolic ball of radius c_2.

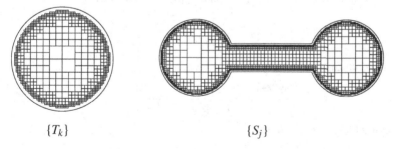

$\{T_k\}$ $\qquad\qquad\qquad\qquad\qquad$ $\{S_j\}$

Figure I.7 Whitney squares in $\mathbb{D}$ and Ω.

Whitney squares are almost conformally invariant in the following sense: Assume Ω is simply connected, let $\varphi : \mathbb{D} \to \Omega$ be a conformal mapping, let $\{S_j\}$ be the Whitney squares for Ω and let $\{T_k\}$ be the Whitney squares for $\mathbb{D}$. Then by (4.15) there is a constant M, not depending on the map φ, such that for each k, $\varphi(T_k)$ is contained in at most M squares S_j, and $\varphi^{-1}(S_k)$ is contained in at most M squares T_j. In particular, for each $d > 0$, there is $M(d)$ such that every hyperbolic ball $\{z : \rho_\Omega(z, a) < d\}$ in Ω is covered by $M(d)$ Whitney squares.

Theorem 4.5. *Let* $\psi(z)$ *be a univalent function satisfying* $\psi(0) = 0$ *and* $\psi'(0) = 1$. *Then*

$$\frac{|z|}{(1+|z|)^2} \le |\psi(z)| \le \frac{|z|}{(1-|z|)^2}, \tag{4.16}$$

and

$$\frac{1-|z|}{(1+|z|)^3} \le |\psi'(z)| \le \frac{1+|z|}{(1-|z|)^3}. \tag{4.17}$$

Moreover,

$$\frac{1-|z|}{|z|(1+|z|)} \le \frac{|\psi'(z)|}{|\psi(z)|} \le \frac{1+|z|}{|z|(1-|z|)}. \tag{4.18}$$

Inequality (4.16) is known as the **growth theorem**, while (4.17) is called the **distortion theorem**.

Proof. The critical inequalities are (4.17) and we prove them first. Fix $z_0 \in \mathbb{D}$ and take

$$f(z) = \frac{\psi\left(\frac{z+z_0}{1+\bar{z}_0 z}\right) - \psi(z_0)}{\psi'(z_0)\left(1 - |z_0|^2\right)}. \tag{4.19}$$

Then f is univalent on $\mathbb{D}$, $f(0) = 0$ and $f'(0) = 1$. By (4.8),

$$|f''(0)| = \left|\psi''(z_0)\frac{1-|z_0|^2}{\psi'(z_0)} - 2\bar{z}_0\right| \le 4,$$

so that when $z_0 = re^{i\theta}$,

$$\left|\frac{e^{i\theta}\psi''(z_0)}{\psi'(z_0)} - \frac{2|z_0|}{1-|z_0|^2}\right| \le \frac{4}{1-|z_0|^2}. \tag{4.20}$$

Applying the general formula

$$\text{Re}\frac{zg'(z)}{|z|} = \frac{\partial \text{Re} g}{\partial r}$$

to $g = \log \psi'$, we then obtain

$$\frac{2r-4}{1-r^2} \le \frac{\partial}{\partial r}\log|\psi'(z_0)| \le \frac{2r+4}{1-r^2}. \tag{4.21}$$

Integrating (4.21) along the radius $[0, z]$ gives both inequalities in (4.17).

To prove the upper bound in (4.16), integrate the upper bound in (4.17) along $[0, z]$. To prove the lower bound, we can assume that $|\psi(z)| < \frac{1}{4}$, because $\frac{|z|}{(1+|z|)^2} < \frac{1}{4}$. Then by (4.9) there exists an arc $\gamma \in \mathbb{D}$ with $\psi(\gamma) = [0, \psi(z)]$, and integrating $|\psi'(z)||dz|$ along γ then gives the lower bound in (4.16).

To prove (4.18), apply (4.16) at $-z_0$ to the function f defined by (4.19). $\blacksquare$

Again, equality any place in (4.16), (4.17), (4.18), or (4.21) implies that ψ is a Koebe function.

5. The Hayman–Wu Theorem

We give a very elementary proof, based on an idea of the late K. Øyma [1992], of the theorem of Hayman and Wu. The Hayman–Wu theorem will be a recurrent topic throughout this book.

Theorem 5.1 (Hayman–Wu). *Let φ be a conformal mapping from $\mathbb{D}$ to a simply connected domain Ω and let L be any line. Then*

$$\text{length}\left(\varphi^{-1}(L \cap \Omega)\right) \leq 4\pi. \tag{5.1}$$

Hayman and Wu [1981] gave the first proof of (5.1) with 4π replaced by some large unknown constant. Øyma [1992] obtained the constant 4π, Rohde [2002] proved that the best constant in (5.1) is strictly smaller than 4π, and Øyma [1993] proved that the best constant is at least π^2. The sharp constant in (5.1) is not known. See Exercises 24 and VI.3. We present Øyma's elementary proof, as modified by Rohde.

For the proof it will be convenient to replace the hyperbolic metric $\rho(z_1, z_2)$ by the **pseudohyperbolic metric**, defined in $\mathbb{D}$ by

$$\delta_{\mathbb{D}}(z_1, z_2) = \left| \frac{z_1 - z_2}{1 - \bar{z}_1 z_2} \right| = \tanh \rho(z_1, z_2),$$

and in Ω by

$$\delta_{\Omega}(w_1, w_2) = \delta_{\mathbb{D}}(\varphi^{-1}(w_1), \varphi^{-1}(w_2)).$$

Proof. We can assume that φ is analytic and one to one in a neighborhood of $\overline{\mathbb{D}}$ and that $L = \mathbb{R}$. Let L_k denote the components of $\Omega \cap L$ and let Ω_k be that component of

$$\Omega \cap \{\bar{z} : z \in \Omega\}$$

such that $L_k \subset \Omega_k$. Then Ω_k is a Jordan domain symmetric about $\mathbb{R}$. When $k \neq j$, $\partial\Omega_k \cap \partial\Omega_j \subset \mathbb{R}$ and $\partial\Omega_k \cap \partial\Omega_j$ contains at most one point, because Ω

is simply connected.

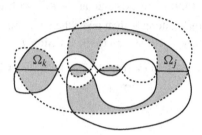

Figure I.8 The domains Ω_k are shaded.

By symmetry there is a conformal mapping $\psi_k : \Omega_k \to -i\mathbb{H}$ such that $\psi_k(L_k) = \mathbb{R}^+$ and ψ_k extends continuously to $\overline{\Omega_k}$. For $\zeta \in \partial\varphi^{-1}(\Omega_k) \cap \partial\mathbb{D}$, set

$$\alpha = \varphi(\zeta), \quad x = |\psi_k(\alpha)|, \quad \beta = \psi_k^{-1}(x), \quad \text{and} \quad z = \varphi^{-1}(\beta).$$

Then the composition

$$\Phi \equiv \varphi^{-1} \circ \psi_k^{-1}(|\psi_k \circ \varphi|)$$

is a smooth map of $\varphi^{-1}(\bigcup \partial\Omega_k \cap \partial\Omega \setminus P) \subset \partial\mathbb{D}$ onto $\varphi^{-1}(\bigcup L_k) \setminus P'$ where P and P' are finite sets.

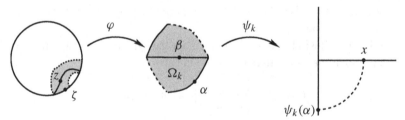

Figure I.9 The map $z = \Phi(\zeta)$

To prove Theorem 5.1, it suffices to show that

$$|\nabla\Phi| \leq 2. \tag{5.2}$$

To prove (5.2), suppose that $I = (\zeta, \zeta')$ is an open interval contained in $\varphi^{-1}(\partial\Omega_k) \cap \partial\mathbb{D}$. Set

$$\alpha' = \varphi(\zeta'), \quad x' = |\psi_k(\alpha')|, \quad \beta' = \psi_k^{-1}(x'), \quad \text{and} \quad z' = \varphi^{-1}(\beta').$$

Then by Pick's theorem (see Exercise 14(a))

$$\delta_{\mathbb{D}}(\Phi(\zeta), \Phi(\zeta')) = \delta_{\Omega}(\beta, \beta') \leq \delta_{\Omega_k}(\beta, \beta') = \delta_{-i\mathbb{H}}(x, x') = \left|\frac{x - x'}{x + x'}\right|,$$

and by Lindelöf's maximum principle

$$\omega(x, \psi_k(\varphi(I)), -i\mathbb{H}) = \omega(z, I, \varphi^{-1}(\Omega_k)) \leq \omega(\Phi(\zeta), I, \mathbb{D}).$$

Letting $\zeta' \to \zeta$ we obtain the inequality

$$\frac{|\nabla\Phi|}{1 - |\Phi(\zeta)|^2} \leq \pi \cdot \frac{1}{2\pi} \frac{1 - |\Phi(\zeta)|^2}{|\zeta - \Phi(\zeta)|^2}, \tag{5.3}$$

which easily implies (5.2). This completes the proof of the Hayman–Wu theorem. ∎

Because it does not depend on the mapping φ, the conclusion of Theorem 5.1 is actually stronger than (5.1). See Exercise 20.

In Chapter VII we will see that for some $1 < p < 2$, independent of φ

$$\int_{L \cap \Omega} \left|(\varphi^{-1})'(w)\right|^p |dw| < C_p, \tag{5.4}$$

but the largest permissible p is unknown. A slit disc shows (5.4) fails at $p = 2$, and a counterexample for some $p < 2$, due to Baernstein, is given in Chapter VIII. In Chapter X we will determine the class of curves L for which (5.1) holds.

Notes

This chapter is a survey of elementary material which can be found in many other sources. We learned Lemma 2.3 from Garsia [1970], who attributed it to W. H. Young. The proof of Lemma 2.3 also shows that every open cover has a subcover $\{I_k'\}$ such that $\sum \chi_{I_k'} \leq 2$. In Wheeden and Zygmund [1977] the higher dimensional formulation of this statement is called the "simple Vitali lemma". The classical Vitali covering lemma is an easy consequence of this or Lemma 2.3. See Exercise 9 or Mattila's book [1995], which has much about covering lemmas. Stein's classic book [1970] also gives an excellent discussion of the relation between covering lemmas, maximal functions, and almost everywhere convergence and a clear introduction to Whitney squares.

Section 4 follows Duren [1983], Hayman [1958], and Goluzin [1952]. We call (4.13) Koebe's estimate and we will refer to it as such many times, but in

[1907] Koebe only proved the left half of (4.13) with a constant $\rho \leq \frac{1}{4}$. Later, in [1916] Bieberbach showed $\rho = \frac{1}{4}$. See Duren [1983].

Exercises and Further Results

1. (a) If E is the arc $(e^{i\theta_1}, e^{i\theta_2})$ of $\partial \mathbb{D}$ in Figure I.10, then

$$2\pi \omega(z, E, \mathbb{D}) = 2 \arg \left(\frac{e^{i\theta_2} - z}{e^{i\theta_1} - z} \right) - (\theta_2 - \theta_1) = 2\varphi - (\theta_2 - \theta_1), \quad \text{(E.1)}$$

where φ is the angle subtended at z by the arc $(e^{i\theta_1}, e^{i\theta_2})$.
Hint: Verify (i), (ii), and (iii) from Section 1 via elementary geometry and use Lemma 1.1. An alternative approach from the Poisson integral formula is to write

$$\frac{e^{i\theta} + z}{e^{i\theta} - z} = \frac{2e^{i\theta}}{e^{i\theta} - z} - 1 = \frac{d}{d\theta}\left[\frac{2}{i} \log(e^{i\theta} - z) - \theta \right]$$

and integrate the real part of this expression over the interval (θ_1, θ_2).
(b) We also have

$$\omega(z, E) = \frac{1}{2\pi}(\varphi_2 - \varphi_1),$$

where $e^{i\varphi_j}$ is the other endpoint of the chord extending from $e^{i\theta_j}$ through z. This is easy from (E.1) and elementary geometry.

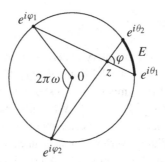

Figure I.10 Harmonic measure of an arc.

(c) Using $\varphi(z) = i(1 - z)/(1 + z)$, and a change of variables, derive the Poisson integral formula for $\mathbb{D}$ from the Poisson integral formula for $\mathbb{H}$.
Hints:

$$\frac{1 - |z|^2}{|e^{i\theta} - z|^2} = \operatorname{Re}\left(\frac{e^{i\theta} + z}{e^{i\theta} - z} \right) \quad \text{and} \quad \frac{y}{(x - t)^2 + y^2} = \operatorname{Im} \frac{1}{t - z}.$$

2. If $e^{i\theta} \in \partial \mathbb{D}$ is fixed, the level curves of $z \to P_z(e^{i\theta})$ are circles in $\mathbb{D}$ tangent to $\partial \mathbb{D}$ at $e^{i\theta}$. If E is an arc of $\partial \mathbb{D}$, the level set $\{\omega(z, E) = t\}$ is the arc of a circle cutting $\partial \mathbb{D}$ at the endpoints of E in the angle πt.

3. (a) The Lindelöf maximum principle, Lemma 1.1, is false if $u(z)$ is not assumed to be bounded above.

(b) But Lindelöf's maximum principle is also true for subharmonic functions: If $u(z)$ is a subharmonic function on a domain Ω such that $\overline{\Omega} \neq \mathbb{C}$, if $u(z)$ is bounded above, and if

$$\limsup_{z \to \zeta} u(z) \leq 0,$$

for all $\zeta \in \partial \Omega \setminus F$, where F is a finite subset of $\partial \Omega$, then $u(z) \leq 0$ on Ω. See Ahlfors [1973].

(c) Show that the conclusion in 3(b) still holds if the hypothesis $\overline{\Omega} \neq \mathbb{C}$ is replaced by the weaker hypothesis $\partial \Omega \setminus F \neq \emptyset$.

4. Let u be a real valued continuous function on a region Ω. Prove that the following three conditions are equivalent:

(a) For each sufficiently small disc $D \subset \Omega$ and each v harmonic on D, $u - v$ and $v - u$ satisfy the maximum principle on D.

(b) u satisfies the mean value property for each sufficiently small disc D contained in Ω.

(c) u is harmonic on Ω.

Hint: Use the Poisson integral on D.

5. (a) **Harnack's inequality.** Let p and q be points of an arbitrary plane domain Ω. Prove there is a constant $C_{p,q}$ such that whenever $u(z)$ is positive and harmonic on Ω,

$$\frac{1}{C_{p,q}} \leq \frac{u(p)}{u(q)} \leq C_{p,q}.$$

(b) If p and q remain in a compact subset $K \subset \Omega$, then there is a constant $C_K < \infty$ such that $C_{p,q} \leq C_K$.

(c) **Harnack's principle.** Let $\{\Omega_n\}$ be a sequence of domains and let u_n be a harmonic function on Ω_n. Assume Ω is a domain such that every $z \in \Omega$ has a neighborhood V_z such that for n sufficiently large, $V_z \subset \Omega_n$, and $u_n \leq u_{n+1}$ on V_z. Then either u_n tends to ∞ uniformly on every compact subset of Ω, or there is a harmonic function u on Ω and u_n converges to u uniformly on every compact subset of Ω.

(d) Let Ω_n and Ω be as in (c), and let u_n be a harmonic function on Ω_n. Now assume there is $M_z < \infty$ such that for n sufficiently large, $\sup_{V_z} |u_n| \leq M_z$. Prove $\{u_n\}$ has a subsequence converging uniformly on all compact subsets of Ω to a function harmonic on Ω.

(e) If u is harmonic on $\mathbb{D}$ and $|u| \leq M$ then

$$|\nabla u| \leq \frac{2M}{1 - |z|^2}.$$

Hint: $P_z(t) = \frac{1}{2\pi} \mathrm{Re}\left(\frac{e^{it} + z}{e^{it} - z}\right)$.

6. **The boundary Harnack inequality.** Let u and v be positive harmonic functions on $\mathbb{D}$ with $u(0) = v(0)$, let $I \subset \partial\mathbb{D}$ be an open arc and assume

$$\lim_{z \to \zeta} u(z) = \lim_{z \to \zeta} v(z) = 0$$

for all $\zeta \in I$. Prove that for every compact $K \subset \mathbb{D} \cup I$ there is a constant $C(K)$ independent of u and v such that on $K \cap \mathbb{D}$

$$\frac{1}{C(K)} \leq \frac{u(z)}{v(z)} \leq C(K).$$

Hint: Using a conformal map of a subregion of $\mathbb{D}$ onto $\mathbb{D}$, you may assume u and v are continuous on $\overline{\mathbb{D}}$. Estimate the Poisson kernel on $\partial\mathbb{D} \setminus I$. See Wu [1978] for a more general result.

7. Show there is a Borel set $A \subset \partial\mathbb{D}$ such that any function almost everywhere equal to χ_A is not continuous at any point $\zeta \in \partial\mathbb{D}$. Then show that for $f = \chi_A$, (1.8) fails for all $\zeta \in \partial\mathbb{D}$.

8. Suppose $G(z)$ is continuous on the closed disc and analytic on the open disc. Use the Poisson kernel to show that if $dG(e^{i\theta})/d\theta$ exists at θ_0, then $G'(z)$ has nontangential limit at $e^{i\theta_0}$.

9. A **Vitali cover** of a measurable set $E \subset \partial\mathbb{D}$ is a family $\mathcal{I}$ of arcs I such that for every $\zeta \in E$ and every $\delta > 0$ there exists $I \in \mathcal{I}$ such that $\zeta \in I$ and $|I| < \delta$. Prove the **Vitali covering lemma**: If $\mathcal{I}$ is a Vitali cover of E and if $\varepsilon > 0$ there exists a pairwise disjoint sequence $\{I_j\} \subset \mathcal{I}$ such that $|E \setminus \bigcup I_j| < \varepsilon$.

10. Assume $u(z)$ is the Poisson integral of a finite positive measure μ on $\partial\mathbb{D}$, and let $d\mu = f(\theta)d\theta + d\mu_s$ be the Lebesgue decomposition of μ. Prove that almost everywhere $d\theta$, u has nontangential limit $f(\theta)$, and that almost everywhere $d\mu_s$, u has nontangential limit ∞. Hint: Use Lemma 2.3.

11. Let $u(z)$ be real and harmonic in $\mathbb{D}$.

(a) If $1 < \alpha < \beta$, then

$$u_\alpha^* \leq u_\beta^* \leq C_{\alpha,\beta} M(u_\alpha^*).$$

Moreover, if $0 < p < \infty$, then

$$\|u_\beta^*\|_{L^p(\partial\mathbb{D})} \leq C'_{\alpha,\beta} \|u_\alpha^*\|_{L^p(\partial\mathbb{D})}.$$

(b) If $1 \le p < \infty$ and if u is the Poisson integral of $f \in L^p$, then

$$\lim_{r \to 1} \int |u(re^{i\theta}) - f(e^{i\theta})| d\theta = 0.$$

(c) If $p > 1$ show u is the Poisson integral of $f \in L^p(\partial \mathbb{D})$ if and only if

$$\sup_{r<1} \int |u(re^{i\theta})|^p \frac{d\theta}{2\pi} < \infty.$$

(d) Show u is the Poisson integral of a probability measure μ on $\partial \mathbb{D}$

$$u(z) = \int P_z(\theta) d\mu(\theta),$$

if and only if $u > 0$ and $u(0) = 1$.

(e) Show u is the Poisson integral of a finite (signed) Borel measure on $\partial \mathbb{D}$ if and only if

$$\sup_{r<1} \int |u(re^{i\theta})| \frac{d\theta}{2\pi} < \infty. \tag{E.2}$$

If (E.2) holds, then u has an almost everywhere nontangential limit $f(\theta)$, but $u(z)$ need not be the Poisson integral of f. See Exercise 10.

(f) If u is the Poisson integral of a finite signed measure v on $\partial \mathbb{D}$, then the measures $u(re^{i\theta}) \frac{d\theta}{2\pi}$ converge weak-star to v as $r \to 1$. It follows that u determines the measure v uniquely.

(g) Show u is the Poisson integral of $f \in L^1(\partial \mathbb{D})$ if and only if the family $\{u(re^{i\theta}) : 0 < r < 1\}$ is uniformly integrable in θ.

(h) Let $1 \le p \le \infty$. If $u_\alpha^* \in L^p(\partial \mathbb{D})$, then u is the Poisson integral of some $f \in L^p(\partial \mathbb{D})$.

See Appendix A for the corresponding results when $p < 1$.

12. A compact set K is **locally connected** if whenever U is a relatively open subset of K and $z \in U \subset K$ there is a relatively open subset V of K such that V is connected and $z \in V \subset U$. Let Ω be a simply connected domain such that $\partial \Omega$ contains at least two points. Prove $\partial \Omega$ is locally connected if and only if the Riemann map $\varphi : \mathbb{D} \to \Omega$ extends continuously to $\overline{\mathbb{D}}$. Hint: For one direction, use the uniform continuity of φ. For the other direction follow the proof of Carathéodory's theorem.

13. (a) Let Ω be a simply connected domain and let $\sigma \subset \Omega$ be a **crosscut**, that is, a Jordan arc in Ω having distinct endpoints in $\partial \Omega$. Prove that $\Omega \setminus \sigma$ has two components Ω_1 and Ω_2, each simply connected, and $\beta_j = \partial \Omega_j \setminus \sigma$ is connected.

(b) Let Ω be simply connected, let $\psi : \Omega \to \mathbb{D}$ be conformal and fix $z \in \Omega$ and $\zeta \in \partial \Omega$. Assume ζ and z can be separated by a sequence of crosscuts

$\gamma_n \subset \Omega$ such that length$(\gamma_n) \to 0$. Let U_n be the component of $\Omega \setminus \gamma_n$ such that $z \notin U_n$. Prove that $\psi(U_n) \to \alpha \in \partial\mathbb{D}$.

14. Let $f(z)$ be an analytic function on $\mathbb{D}$ and assume $|f(z)| \leq 1$.

    (a) Prove Pick's theorem:

$$\left| \frac{f(z) - f(w)}{1 - \overline{f(w)}f(z)} \right| \leq \left| \frac{z - w}{1 - \overline{w}z} \right|.$$

Equivalently, prove

$$\rho(f(z), f(w)) \leq \rho(z, w).$$

In other words, analytic maps from $\mathbb{D}$ to $\mathbb{D}$ are Lipschitz with respect to the hyperbolic metric.

Hint: Repeat the proof of the Schwarz lemma, using $(S \circ f)/T$ for suitable S and T in $\mathcal{M}$.

    (b) Deduce that

$$\frac{|f'(z)|}{1 - |f(z)|^2} \leq \frac{1}{1 - |z|^2}.$$

    (c) If equality holds in (a) at any two points z and w, or if equality holds in (b) at any point, then $f \in \mathcal{M}$.

15. (a) Prove (4.4) and (4.5).

    (b) Find the euclidean center of the hyperbolic ball $\{z : \rho(z, a) < d\}$.

16. Let $\varphi(z)$ be univalent on $\mathbb{D}$ and assume $\varphi(z) \neq 0$ for every z.

    (a) Prove

$$|\varphi'(0)| \leq 4|\varphi(0)|,$$

using (4.9).

    (b) Prove

$$\left(\frac{1 - |z|}{1 + |z|}\right)^2 \leq \frac{|\varphi(z)|}{|\varphi(0)|} \leq \left(\frac{1 + |z|}{1 - |z|}\right)^2.$$

Hint: Fix z and apply (a) to

$$\psi(w) = \varphi\left(\frac{w + z}{1 + \overline{z}w}\right)$$

to get

$$\left|\frac{\varphi'(z)}{\varphi(z)}\right| \leq \frac{4}{1 - |z|^2},$$

and integrate along $[0, z]$.

17. (a) Show that if equality holds anywhere in (4.8), (4.9), (4.16), (4.17), (4.18), or (4.21) then ψ is a Koebe function, defined by (4.7).

(b) On the other hand, equality in (4.12) holds if and only if $\mathbb{C} \setminus F(\mathbb{D})$ has area zero.

18. (a) Find an upper bound for the constant M mentioned in the remarks preceding Figure I.7.

(b) Give a construction of Whitney squares in the case $\text{diam}(\Omega) = \infty$.

19. A region of the form

$$Q = \{z = re^{i\theta} : \theta_0 < \theta < \theta_0 + \ell(Q), \ 1 - \ell(Q) \le r < 1\}$$

is called a **box** or a **Carleson box**, and the arc

$$\{e^{i\theta} : \theta_0 < \theta < \theta_0 + \ell(Q)\}$$

is the **base** of Q. See Figure I.11. A finite measure on $\mathbb{D}$ is a **Carleson measure** if there is a constant C such that

$$\mu(Q) \le C\ell(Q)$$

for every Carleson box.

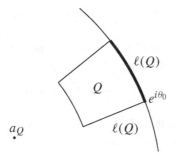

Figure I.11 A Carleson box.

(a) Let $\alpha > 1$. Prove μ is a Carleson measure if and only if there is C_α such that

$$\mu(\{z \in \mathbb{D} : u(z) > \lambda\}) \le C_\alpha \left|\{\zeta \in \partial\mathbb{D} : u_\alpha^* > \lambda\}\right|$$

whenever u is the Poisson integral of $f \in L^1(\partial\mathbb{D})$. Hint: Write the set $\{u_\alpha^* > \lambda\} = \bigcup I_j$ where the I_j are pairwise disjoint open arcs on $\partial\mathbb{D}$, and define $T_j = \mathbb{D} \setminus \bigcup_{\zeta \notin I_j} \Gamma_\alpha(\zeta)$. The set T_j looks like a tent with base I_j.

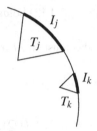

Figure I.12 Tents.

Then $\mu(T_j) \le C_\alpha(I_j)$ and $|u| \le \lambda$ on $\mathbb{D} \setminus \bigcup T_j$. Conversely, let Q be the box with base I, and let $u(z) = 4\lambda\omega(z, I, \mathbb{D})$. Then $u(z) > \lambda$ on Q while by (2.3),

$$\left|\{u_\alpha^* > \lambda\}\right| \le C_\alpha |I|.$$

(b) If μ is a finite measure, then μ is a Carleson measure if and only if

$$\sup_{a \in \mathbb{D}} \int \frac{1 - |a|^2}{|1 - \overline{a}z|^2} d\mu(z) < \infty.$$

Hint: Let a be the center of Q. See page 239 of Garnett [1981].

(c) Let Γ be a countable union of rectifiable curves in $\mathbb{D}$. Then

$$\sup_{T \in \mathcal{M}} \text{ length}(T(\Gamma)) < \infty$$

if and only if arc length on Γ is a Carleson measure.

(d) A region in the upper half-plane $\mathbb{H}$ of the form

$$Q = \{(x, y) : a \le x \le a + \ell(Q), \ 0 < y \le \ell(Q)\}$$

is also called a box or Carleson box and a finite measure μ on $\mathbb{H}$ is called a Carleson measure if there is a constant C such that

$$\mu(Q) \le C\ell(Q)$$

for every Carleson box in $\mathbb{H}$. A box Q is called **dyadic** if $\ell(Q) = 2^{-k}$ and $a = j2^{-k}$ for some integers j and k. If $\mu(Q) \le C\ell(Q)$ for all dyadic boxes, prove that μ is a Carleson measure on $\mathbb{H}$.

(e) If μ is a Carleson measure on $\mathbb{D}$ and τ is a conformal map of $\mathbb{H}$ onto $\mathbb{D}$, then prove $\mu \circ \tau$ is a Carleson measure on $\mathbb{H}$. Conversely, if μ is a Carleson measure on $\mathbb{H}$ supported in the unit box $[0, 1] \times [0, 1]$, then $\mu \circ \tau^{-1}$ is a Carleson measure on $\mathbb{D}$.

Part (a) is due to E. M. Stein (unpublished).

20. Let φ be a conformal mapping from $\mathbb{D}$ to a simply connected domain Ω, let L be any line and set $\Gamma = \varphi^{-1}(\Omega \cap L)$. Then
(a) Arc length on Γ is a Carleson measure; in other words,

$$\text{length}(\Gamma \cap Q) \leq C\ell(Q)$$

for any box $Q = \{re^{i\theta} : \theta_0 < \theta < \theta_0 + \ell(Q), 1 - \ell(Q) \leq r < 1\}$.
(b) Moreover, for any disc $B(z, r)$,

$$\text{length}(\Gamma \cap B(z, r)) \leq C'r. \tag{E.3}$$

Hint: If $r < 1 - |z|$ or $r \geq 0.4$ use a linear map and the Hayman-Wu theorem. If $1 - |z| < r < 0.4$ then there is a Carleson box $Q \supset \mathbb{D} \cap B(z, r)$ such that $\ell(Q) \leq Cr$, and part (a) gives (E.3).

21. Let $\varphi(z)$ be univalent on $\mathbb{D}$ and assume $\varphi(z) \neq 0$ for every z.
(a) There is $C_1 < \infty$, independent of φ, such that

$$\iint_{\mathbb{D}} \left| \frac{\varphi'(z)}{\varphi(z)} \right| dxdy \leq C_1.$$

Hint: Write $w = u + iv = \varphi(z)$, $w \in \Omega$. Then

$$\left(\iint_{\mathbb{D}} \left| \frac{\varphi'(z)}{\varphi(z)} \right| dxdy \right)^2 \leq \iint_{\mathbb{D}} \left(\frac{1}{1-|z|} \right)^{\frac{2}{3}} dxdy \iint_{\mathbb{D}} \left| \frac{\varphi'(z)}{\varphi(z)} \right|^2 (1 - |z|)^{\frac{2}{3}} dxdy$$

$$\leq C_2 \iint_{\Omega} \left(1 - |\varphi^{-1}(w)| \right)^{\frac{2}{3}} \frac{dudv}{|w|^2}.$$

Now integrate over $\Omega \cap \{|w| > 1\}$ and $\Omega \cap \{|w| \leq 1\}$ separately, and use Exercise 16.
(b) Let $f = \log \varphi$. There is $C_2 < \infty$, independent of φ such that

$$\iint_{Q} |f'| dxdy \leq C_2 \ell(Q) \tag{E.4}$$

for any Carleson box

$$Q - \{z = re^{i\theta} : \theta_0 \leq \theta \leq \theta_0 + \ell(Q), \ 1 - \ell(Q) \leq r < 1\}.$$

In other words, $|f'(z)| dxdy$ is a Carleson measure. For the proof we may assume $\ell(Q)$ is small. Let $a = a_Q = (1 - 2\ell(Q))e^{i(\theta_0 + \frac{\ell(Q)}{2})}$ and use (a) with $w = (z - a)/(1 - \bar{a}z)$. See Figure I.11.

(c) It follows from (b) that almost everywhere and in L^1 norm,

$$\lim_{r \to 1} f(re^{i\theta}) = f(e^{i\theta})$$

exists and $f(z)$ is the Poisson integral of $f(e^{i\theta})$. In fact, $f(e^{i\theta}) \in$ BMO (for **bounded mean oscillation**), which by definition means there exists a constant $C_3 < \infty$, independent of φ, such that for every arc $I \subset \partial \mathbb{D}$,

$$\inf_{\alpha \in \mathbb{C}} \frac{1}{|I|} \int_I |f - \alpha| d\theta \leq C_3. \tag{E.5}$$

Indeed, if Q has base I, then one can show

$$\frac{1}{|I|} \int_I |f - f(a_Q)| d\theta \leq C \iint_Q |f'| dx dy.$$

See Jones [1980]. Baernstein [1976] gave the first proof of (E.5). Cima and Schober [1976] and Pommerenke [1976] have related results. When $e^{f(z)}$ is not univalent, (E.4) is not equivalent to (E.5). However, in general (E.5) is equivalent to

$$\iint_Q |f'|^2 (1 - |z|) dx dy \leq C\ell(Q).$$

See Fefferman and Stein [1972], Baernstein [1980], Garnett [1981], or Appendix F.

22. Again let φ be univalent and assume $\varphi(z) \neq 0$ for all z. Prove that for $r \geq 1/2$

$$I(r) = \int \left| \frac{\varphi'(re^{i\theta})}{\varphi(re^{i\theta})} \right|^2 d\theta \leq \frac{C_4}{1 - r} \log\left(\frac{2}{1 - r}\right), \tag{E.6}$$

where C_4 does not depend on φ. Hayman [1980] attributes the following argument to P. L. Duren: By Exercise 16,

$$\iint_{\{|z| < r\}} \left| \frac{\varphi'(re^{i\theta})}{\varphi(re^{i\theta})} \right|^2 dx dy \leq 8\pi \log\left(\frac{1}{1 - r}\right) + 8\pi \log 2.$$

Take $r < R < 1$. Since $I(r)$ increases with r,

$$I(r) \leq \frac{2}{R^2 - r^2} \iint_{\{r < |z| < R\}} \left| \frac{\varphi'(re^{i\theta})}{\varphi(re^{i\theta})} \right|^2 dx dy$$

$$\leq \frac{2}{R^2 - r^2} \left(8\pi \log\left(\frac{1}{1 - R}\right) + 8\pi \log 2 \right).$$

Now take

$$1 - R^2 = \frac{1}{2}(1 - r^2).$$

Hayman [1980] shows that the logarithm in (E.6) is necessary.

23. Let $0 < p < \infty$. An analytic function $f(z)$ on the disc $\mathbb{D}$ is in the **Hardy space H^p** if

$$\sup_{0<r<1} \int_0^{2\pi} |f(re^{i\theta})|^p d\theta = \|f\|_{H_p}^p < \infty.$$

Let $\psi(z)$ be a univalent function on $\mathbb{D}$, normalized by

$$\psi(z) = z + a_2 z^2 + \dots.$$

(a) For $0 < p < \infty$, verify the identity

$$\Delta(|\psi|^p) = p^2 |\psi|^{p-2} |\psi'|^2.$$

(b) If $0 < p < \frac{1}{2}$ then $\psi \in H^p$ and

$$\|\psi\|_{H^p}^p \le \frac{C}{1 - 2p},$$

where C is independent of p. This result is due to Prawitz [1927]. Green's theorem and (a) give

$$\frac{d}{dr}\left(\frac{1}{2\pi} \int_0^{2\pi} |\psi(re^{i\theta})|^p d\theta\right) = \frac{p^2}{2\pi r} \iint\limits_{\{|z|<r\}} |\psi(z)|^{p-2} |\psi'|^2 dx dy$$

$$= \frac{p^2}{2\pi r} \iint\limits_{\psi(\{|z|<r\})} |w|^{p-2} dA(w).$$

Write $M(r) = \sup_{|z|\le r} |\psi(z)|$. Then

$$\frac{d}{dr}\left(\frac{1}{2\pi} \int_0^{2\pi} |\psi|^p d\theta\right) \le \frac{p^2}{2\pi r} \iint\limits_{|w|<M(r)} |w|^{p-2} dA(w) = \frac{p}{r}(M(r))^p,$$

while by the distortion theorem,

$$M(r) < \frac{r}{(1-r)^2}.$$

Therefore

$$\frac{1}{2\pi} \int_0^{2\pi} |\psi|^p d\theta \le p \int_0^r \frac{(M(t))^p}{t} dt$$

$$\le p \int_0^r t^{p-1}(1-t)^{-2p} dt = O\left(\frac{1}{1-2p}\right).$$

(c) If $p > \frac{1}{2}$, a similar argument gives

$$\frac{1}{2\pi} \int_0^{2\pi} |\psi(re^{i\theta})|^p d\theta \leq C_p \frac{1}{(1-r)^{2p-1}}.$$

(d) If $p > \frac{1}{2}$, (c) and (4.18) imply

$$\frac{1}{2\pi} \int_0^{2\pi} |\psi'(re^{i\theta})|^p d\theta \leq C_p' \frac{1}{(1-r)^{3p-1}}.$$

For $p = 1$, (c) implies (d) if ψ is analytic but not necessarily univalent.

(e) The Koebe function $\psi(z) = z/(1-z)^2$ is not in $H^{\frac{1}{2}}$.

In a deep paper [1974] Baernstein proved that the extremal functions for parts (b) and (c) are the rotates of the Koebe function. See Duren [1983] for these results and more, and see Appendix A for the basic theory of Hardy spaces.

24. (a) Show that the best constant in the Hayman–Wu theorem is smaller than 4π (Rohde [2002]). Hint: By (5.3) if $|\nabla\Phi(\zeta)| > 2 - \varepsilon$ then $\Phi(\zeta)$ lies in a cone at ζ with opening $< C\varepsilon$ and $|\Phi| > 1 - C\varepsilon$. Moreover the curve $\alpha = \Phi(\partial\mathbb{D})$ makes an angle with the radius through $\Phi(\zeta)$ approximately $\tan^{-1}(1/2)$. Then use (4.20) to show that if α makes a larger angle with the radius at a point z then $|\nabla\Phi| < 2 - \varepsilon$ on an interval centered at $z/|z|$ of length $\delta(1 - |z|)$. Use this to show that there is a $\delta > 0$, independent of φ so that $|\nabla\Phi| < 2 - \varepsilon$ on a subset of the circle of length at least δ.

(b) Use (5.3) to show that if we also require $\varphi(0) \in L$ in the Hayman–Wu theorem, then the best constant is less than $4\pi - 2$.

25. (a) Let $\Omega_1 \subset \Omega_2$ be Jordan domains, and let $E = \partial\Omega_1 \cap \partial\Omega_2$. Then on Ω_1,

$$\omega(z, E, \Omega_1) \leq \omega(z, E, \Omega_2).$$

(b) Let $0 < \delta < 1$ and let Ω be a Jordan domain such that

$$\{|z| < 1 - \delta\} \subset \partial\Omega \subset \{|z| < 1 + \delta\}.$$

Let $E = \partial\Omega \cap \{\theta_1 \leq \arg z \leq \theta_2\}$, where $\theta_2 < \theta_1 + 2\pi$. Show

$$\left|\omega(0, E, \Omega) - \frac{(\theta_2 - \theta_1)}{2\pi}\right| \leq C\delta \log\left(\frac{1}{\delta}\right).$$

Hint: By (a) the worst case in (b) is for regions of the following form. Let $I = [e^{i\theta_1}, e^{i\theta_2}]$, let $\Omega = \mathbb{D} \setminus \{rz : z \in I, \ r \geq 1 - \delta\}$ where $\delta < |I|$, and let $E = \partial\Omega \cap \{|z| = 1 - \delta\}$. Show that if $|z| = 1 - \delta$, and dist$(z, E) \geq \delta$, then

$$\omega(z, E, \Omega) \leq 2\omega(z, I, \partial\mathbb{D}).$$

(c) Show the estimate in (b) is sharp, except for the value of C.

II

Finitely Connected Domains

In this chapter we solve the Dirichlet problem on a domain bounded by a finite number of Jordan curves. For a simply connected Jordan domain the problem was solved in Chapter I via the theorem of Carathéodory. For a multiply connected domain the problem will be reduced to the simply connected case using the Schwarz alternating method.

Solving the Dirichlet problem on a domain Ω is equivalent to constructing harmonic measure on $\partial \Omega$. In Section 2 we describe harmonic measure in terms of the normal derivative of Green's function in the case when $\partial \Omega$ consists of analytic curves. In Section 4 we study the relation between the smoothness of $\partial \Omega$ and the smoothness of the Poisson kernel (the Radon–Nikodym derivative of harmonic measure against arc length). This relation hinges on two classical estimates for conjugate functions which we prove in Section 3.

1. The Schwarz Alternating Method

Let Ω be a plane domain such that $\partial \Omega$ is a finite union of pairwise disjoint Jordan curves

$$\partial \Omega = \Gamma_1 \cup \Gamma_2 \cup \ldots \cup \Gamma_p.$$

We say Ω is a **finitely connected Jordan domain**. A bounded function f on $\partial \Omega$ is **piecewise continuous** if there is a finite set $E \subset \partial \Omega$ such that f is continuous on $\partial \Omega \setminus E$ and f has left and right limits at each point of E. In this section we solve the Dirichlet problem for piecewise continuous boundary data on a finitely connected Jordan domain.

Theorem 1.1. *Let Ω be a finitely connected Jordan domain, and let f be a bounded piecewise continuous function on $\partial \Omega$. Then there is a unique function*

$u(z) = u_f(z)$, *bounded and harmonic on* Ω, *such that*

$$\lim_{z \to \zeta} u(z) = f(\zeta) \qquad (1.1)$$

at every point of continuity ζ *of* f. *Moreover*,

$$\sup_{\Omega} |u| \le \sup_{\partial \Omega} |f|.$$

Proof. We may assume Ω is bounded. The uniqueness of u_f is immediate from Lindelöf's maximum principle, Lemma I.1.1. The existence of u_f in case $p = 1$ was treated in Section I.3, and so we assume $p > 1$. Take a Jordan arc σ with endpoints a, $b \notin \overline{\Omega}$ such that $\Omega_1 = \Omega \setminus \sigma$ is simply connected and such that $\sigma \cap \partial\Omega$ is a finite set. See Figure II.1. Then $\varphi(z) = \sqrt{\frac{z-a}{z-b}}$ has a single valued analytic branch defined on Ω_1, and we can solve the Dirichlet problem on Ω_1 by transplanting it to the Jordan region $\varphi(\Omega_1)$. Take a second Jordan arc σ' such that $\Omega_1' = \Omega \setminus \sigma'$ is simply connected, $\sigma' \cap \partial\Omega$ is a finite set, and $\sigma \cap \sigma' = \emptyset$. We can also solve the Dirichlet problem on Ω_1'.

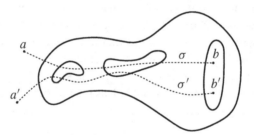

Figure II.1 The proof of Theorem 1.1.

Let $E \subset \partial\Omega$ be a finite set and set $F = E \cup (\sigma \cap \partial\Omega) \cup (\sigma' \cap \partial\Omega)$. Suppose without loss of generality that $f \in C(\partial\Omega \setminus E)$ is positive and bounded. To start, let u_1 be the solution to the Dirichlet problem on Ω_1 with boundary data

$$u_1(\zeta) = \begin{cases} f(\zeta), & \zeta \in \partial\Omega \\ \max_{\partial\Omega} f, & \zeta \in \sigma. \end{cases}$$

Then u_1 is harmonic on Ω_1 and continuous on $\overline{\Omega} \setminus F$ and u_1 matches its boundary data on $\partial\Omega_1 \setminus F$. Next let u_1' be the solution to the Dirichlet problem on Ω_1' with boundary data $u_1(\zeta)$, $\zeta \in \partial\Omega_1'$. Then u_1' is harmonic on Ω_1' and continuous on $\overline{\Omega} \setminus F$ and u_1' matches its boundary data on $\partial\Omega_1' \setminus F$. In particular, $u_1' = u_1$ on $\partial\Omega \setminus F$. By the Lindelöf maximum principle we have $u_1' \le \max_{\partial\Omega} f = u_1$ on $\sigma \cap \Omega$, and therefore $u_1' \le u_1$ on $\overline{\Omega} \setminus F$. Now let u_2 be the solution to the Dirichlet problem on Ω_1 with boundary data $u_1'(\zeta)$, $\zeta \in \partial\Omega_1$. On $\sigma \cap \Omega$ we have $u_2 = u_1' \leq u_1$, while on $\partial\Omega \setminus F$ we have $u_2 = f = u_1$. Therefore $u_2 \leq u_1$ on Ω_1, again by Lemma I.1.1. Consequently $u_2 \leq u_1' = u_1$ on $\sigma' \cap \Omega$, and by the maximum principle, $u_2 \leq u_1'$ on $\overline{\Omega} \setminus F$.

Continuing in this way we obtain a decreasing sequence

$$u_1 \geq u_1' \geq u_2 \geq u_2' \geq u_3 \ldots$$

of positive functions, alternately harmonic on Ω_1 and Ω_1'. By Harnack's principle, Exercise I.5, the limit

$$u(z) = \lim_{n\to\infty} u_n(z) = \lim_{n\to\infty} u_n'(z)$$

is a bounded harmonic function on Ω such that $u \leq u_1 \leq \max_{\partial\Omega} f$.

To complete the proof we show

$$\lim_{z\to\zeta} u(z) = f(\zeta) \tag{1.1}$$

whenever $\zeta \in \partial\Omega$ is a point of continuity of f. We may assume $\zeta \notin \sigma \cup E$. Take a neighborhood V of ζ such that $W = V \cap \Omega$ is a Jordan domain and such that $W \cap (E \cup \sigma) = \emptyset$. Let φ be a conformal map of $\mathbb{D}$ onto W. By Carathéodory's theorem $\varphi(\partial\mathbb{D}) = \partial W$, and for $w = \varphi(z) \in W$

$$u_n(w) = \int_{\varphi^{-1}(\partial\Omega)} \frac{1 - |z|^2}{|e^{i\theta} - z|^2} f \circ \varphi(e^{i\theta}) \frac{d\theta}{2\pi}$$

$$+ \int_{\partial\mathbb{D}\setminus\varphi^{-1}(\partial\Omega)} \frac{1 - |z|^2}{|e^{i\theta} - z|^2} u_n \circ \varphi(e^{i\theta}) \frac{d\theta}{2\pi}.$$

Because $\varphi^{-1}(\partial\Omega)$ is a neighborhood of $\varphi^{-1}(\zeta)$ in $\partial\mathbb{D}$, the first integral approaches $f(\zeta)$ as $w \to \zeta$. Because $|u_n| \leq \sup_{\partial\Omega} |f|$, the second integral tends to zero, uniformly in n, as $w \to \zeta$. Therefore (1.1) holds and the theorem is proved. ∎

If Ω is a finitely connected Jordan domain, Theorem 1.1 shows that the map $f \to u_f(z)$ is a bounded linear functional on $C(\partial\Omega)$. Define the **harmonic measure** of a relatively open subset $U \subset \partial\Omega$ by

$$\omega(z, U) \equiv \omega(z, U, \Omega) = \sup\{u_f(z) : f \in C(\partial\Omega), 0 \leq f \leq \chi_U\},$$

and for an arbitrary subset E of $\partial\Omega$, set

$$\omega(z, E) \equiv \omega(z, E, \Omega) = \inf\{\omega(z, U) : U \text{ open in } \partial\Omega, U \supset E\}.$$

This procedure, which mimics the usual proof of the Riesz representation theorem, shows $\omega(z, E)$ is a Borel measure on $\partial\Omega$ such that

$$u_f(z) = \int_{\partial\Omega} f(\zeta)d\omega(z, \zeta) \tag{1.2}$$

for continuous f. When Ω is simply connected, this definition of harmonic measure agrees with the earlier definition (I.3.3).

For every $z_1, z_2 \in \Omega$ there exists, by virtue of Harnack's inequality, a constant $c = c(z_1, z_2)$ such that

$$\frac{1}{c}\omega(z_1, E) \le \omega(z_2, E) \le c\omega(z_1, E), \tag{1.3}$$

and the constants $c(z_1, z_2)$ remain uniformly bounded if z_1 and z_2 remain in a compact subset of Ω. If $f \in L^1(\partial\Omega, d\omega)$, and in particular if f is bounded and Borel, there is a sequence $\{f_n\}$ in $C(\partial\Omega)$ such that for some fixed $z_0 \in \Omega$,

$$\int |f_n(\zeta) - f(\zeta)| d\omega(z_0, \zeta) \to 0.$$

Write

$$u_n(z) = \int f_n(\zeta)d\omega(z, \zeta)$$

and

$$u(z) = u_f(z) = \int f(\zeta)d\omega(z, \zeta). \tag{1.4}$$

Then by (1.3)

$$u_n(z) \to u(z)$$

for all $z \in \Omega$. By (1.3) we also see that the harmonic functions $u_n(z)$ are uniformly bounded on compact subsets of Ω. Thus by Harnack's principle the limit function $u(z)$ is harmonic on Ω. We call u the **solution to the Dirichlet problem for** f on Ω. If f is bounded, then we also have

$$\sup_{z\in\Omega} |u_f(z)| \le \|f\|_{L^\infty(\Omega, d\omega)}.$$

Moreover, if f is bounded and continuous at $\zeta \in \partial\Omega$, then (1.1) also holds at ζ. See Theorem 2.7 or Exercise 1(b) for a proof.

Note however that the Schwarz alternating method cannot be applied directly to a bounded Borel function because the hypothesis of Lindelöf's maximum principle holds only for piecewise continuous functions.

The next section will give a much more explicit description of the measure $\omega(z, E)$ when $\partial\Omega$ has some additional smoothness.

2. Green's Functions and Poisson Kernels

Again let Ω be a finitely connected Jordan domain, and assume Ω is bounded. For fixed $w \in \Omega$, let $h(z, w)$ be the solution to the Dirichlet problem for the boundary value $f(\zeta) = \log |\zeta - w| \in C(\partial \Omega)$, and define **Green's function with pole w** to be

$$g(z, w) = \log \frac{1}{|z - w|} + h(z, w). \qquad (2.1)$$

Then $g(z, w)$ is continuous in $z \in \overline{\Omega} \setminus \{w\}$, and

$$g(z, w) > 0 \qquad \text{on } \Omega, \qquad (2.2)$$

$$g(\zeta, w) = 0 \qquad \text{on } \partial \Omega, \qquad (2.3)$$

$$z \to g(z, w) \text{ is harmonic on } \Omega \setminus \{w\}, \qquad (2.4)$$

$$z \to g(z, w) - \log \frac{1}{|w - z|} \text{ is harmonic at } w. \qquad (2.5)$$

These properties are easily derived from Theorem 1.1 and the definition (2.1). By the maximum principle, properties (2.3), (2.4), and (2.5) determine $g(z, w)$ uniquely.

When Ω is unbounded we fix $a \notin \overline{\Omega}$. For $w \neq \infty$, we let $h(z, w)$ solve the Dirichlet problem on Ω for $f(\zeta) = \log \left| \frac{\zeta - w}{\zeta - a} \right|$, and define

$$g(z, w) = \log \left| \frac{z - a}{z - w} \right| + h(z, w).$$

For $w = \infty$, we instead use $f(\zeta) = \log \left| \frac{1}{\zeta - a} \right|$ to define $h(z, \infty)$ and set $g(z, \infty) = \log |z - a| + h(z, \infty)$. These definitions are independent of the choice of a, and with them (2.2)–(2.5) still hold and determine $g(z, w)$.

Suppose φ is a conformal mapping from one finitely connected Jordan domain Ω onto another finitely connected Jordan domain Ω'. Then $\varphi(z) \to \partial \Omega'$ whenever $z \to \partial \Omega$, because $\varphi : \Omega \to \Omega'$ is a homeomorphism. It follows that

$$g_{\Omega'}(\varphi(z), \varphi(w)) = g_{\Omega}(z, w) \qquad (2.6)$$

because Green's functions are characterized by (2.3), (2.4), and (2.5).

If Ω is the unit disc $\mathbb{D}$ then clearly

$$g(z, w) = \log \left| \frac{1 - z \overline{w}}{z - w} \right|. \qquad (2.7)$$

Consequently Green's function for any simply connected Jordan domain Ω can be expressed in terms of the conformal mapping $\psi : \Omega \to \mathbb{D}$.

Theorem 2.1. *Let Ω be a simply connected domain bounded by a Jordan curve,
let $w \in \Omega$ and let $\psi : \Omega \to D$ be a conformal map with $\psi(w) = 0$. Then*

$$g(z, w) = - \log |\psi(z)|.$$

Proof. Immediate from (2.6) and (2.7). ∎

By definition, an **analytic arc** is the image $\psi((-1, 1))$ of the open interval
$(-1, 1)$ under a one-to-one and analytic map ψ defined on a neighborhood of
$(-1, 1)$. An **analytic Jordan curve** is a Jordan curve that is a finite union of
(open) analytic arcs.

Lemma 2.2. *Let Ω be a finitely connected Jordan domain. Then there exists a
finitely connected Jordan domain Ω^* such that $\partial\Omega^*$ consists of finitely many
pairwise disjoint analytic Jordan curves and there exists a conformal map from
Ω onto Ω^* which extends to be a homeomorphism from $\overline{\Omega}$ to $\overline{\Omega^*}$.*

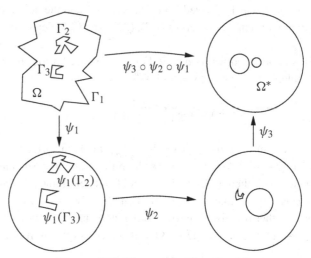

Figure II.2 The proof of Lemma 2.2.

Proof. Write $\partial\Omega = \Gamma_1 \cup \ldots \cup \Gamma_p$, where each Γ_j is a Jordan curve. Let Ω_1 be
the component of $\mathbb{C}^* \setminus \Gamma_1$ containing Ω, where $\mathbb{C}^*$ is the extended plane, and let
ψ_1 be a conformal map of Ω_1 onto $\mathbb{D}$. Let Ω_2 be the component of $\mathbb{C}^* \setminus \psi_1(\Gamma_2)$
containing $\psi_1(\Omega)$, and let ψ_2 be a conformal map of Ω_2 onto $\mathbb{D}$. Repeating this
process for each boundary curve, we obtain a conformal map ψ_p from Ω to
a region Ω^* such that $\partial\Omega^*$ consists of finitely many pairwise disjoint analytic

Jordan curves. Applying Carathéodory's theorem to each ψ_k, we see that ψ_p extends to be a homeomorphism from $\overline{\Omega}$ to $\overline{\Omega^*}$. ∎

In Exercise 5 the proof of Lemma 2.2 is used to define Green's function in non-Jordan, finitely connected domains.

Theorem 2.3. *Let Ω be a finitely connected Jordan domain and let z_1, $z_2 \in \Omega$. Then*

$$g(z_1, z_2) = g(z_2, z_1). \tag{2.8}$$

Proof. By Lemma 2.2 we may assume $\partial\Omega$ consists of analytic Jordan curves. When $\partial\Omega$ consists of analytic curves, an argument with Schwarz reflection, which we will use many times and prove in Lemma 2.4, shows there is a neighborhood V of $\partial\Omega$ to which $z \to g(z, w)$ has a harmonic extension. Hence $g(z, w)$ is C^∞ on some neighborhood V of $\partial\Omega$ and we can use Green's theorem in the form

$$\iint_{\mathcal{U}} (u\Delta v - v\Delta u)dxdy = \int_{\partial\mathcal{U}} \left(u\frac{\partial v}{\partial n} - v\frac{\partial u}{\partial n}\right)ds,$$

where n is the unit normal vector pointing out from the domain $\mathcal{U}$.

Fix distinct $z_1, z_2 \in \Omega$. We apply Green's theorem on the domain

$$\Omega_\varepsilon = \Omega \setminus \left(\{|z - z_1| \le \varepsilon\} \cup \{|z - z_2| \le \varepsilon\}\right),$$

when ε is small, with $u(z) = g(z, z_1)$ and $v(z) = g(z, z_2)$. Because $u = v = 0$ on $\partial\Omega$,

$$\int_{\partial\Omega} \left(u\frac{\partial v}{\partial n} - v\frac{\partial u}{\partial n}\right)ds = 0,$$

and because u and v are harmonic on Ω_ε, the area integral in Green's theorem vanishes. We conclude that

$$
\begin{aligned}
\varepsilon &\int_0^{2\pi} g(z_1 + \varepsilon e^{i\theta}, z_1)\frac{\partial g}{\partial r}(z_1 + \varepsilon e^{i\theta}, z_2)\frac{d\theta}{2\pi} \\
&- \varepsilon \int_0^{2\pi} g(z_1 + \varepsilon e^{i\theta}, z_2)\frac{\partial g}{\partial r}(z_1 + \varepsilon e^{i\theta}, z_1)\frac{d\theta}{2\pi} \\
&= \varepsilon \int_0^{2\pi} (g(z_2 + \varepsilon e^{i\theta}, z_2)\frac{\partial g}{\partial r}(z_2 + \varepsilon e^{i\theta}, z_1)\frac{d\theta}{2\pi} \\
&- \varepsilon \int_0^{2\pi} g(z_2 + \varepsilon e^{i\theta}, z_1)\frac{\partial g}{\partial r}(z_2 + \varepsilon e^{i\theta}, z_2)\frac{d\theta}{2\pi}.
\end{aligned}
\tag{2.9}
$$

For ε small, $g(z_j + \varepsilon e^{i\theta}, z_j) \leq 2\log(1/\varepsilon)$ and for $k \neq j$, $g(z, z_j)$ has bounded derivatives near z_k. That means the first and third integrals in (2.9) approach 0 as $\varepsilon \to 0$. By (2.1),

$$-\frac{\partial g}{\partial r}(z_j + \varepsilon e^{i\theta}, z_j) = \frac{\partial \log \varepsilon}{\partial \varepsilon} + O(1),$$

so that, as $\varepsilon \to 0$, the second integral tends to $g(z_1, z_2)$ and the fourth integral tends to $g(z_2, z_1)$. Therefore (2.8) holds. ∎

Lemma 2.4. *Suppose Ω is a finitely connected Jordan domain and suppose $\gamma \subset \partial\Omega$ is an analytic arc. Let $u(z)$ be a harmonic function in Ω. If*

$$\lim_{z \to \zeta} u(z) = 0$$

for all $\zeta \in \gamma$, then there is an open set $W \supset \gamma \cup \Omega$ such that u extends to be harmonic on W. If also $u(z) > 0$ in Ω, then

$$\frac{\partial u}{\partial n}(\zeta) < 0 \tag{2.10}$$

for all $\zeta \in \gamma$.

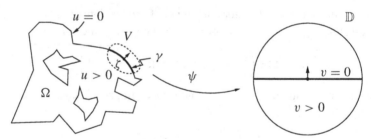

Figure II.3 Straightening an analytic arc in $\partial\Omega$.

Proof. Let $\zeta \in \Gamma$. Because γ is an analytic arc, there is a neighborhood V of ζ and a conformal map $\psi : V \to \mathbb{D}$ such that $\psi(\zeta) = 0$,

$$\psi(V \cap \Omega) = \mathbb{D}^- = \mathbb{D} \cap \{\operatorname{Im} w < 0\},$$

and

$$\psi(\gamma \cap V) = (-1, 1).$$

Set

$$v(w) = \begin{cases} u(\psi^{-1}(w)), & w \in \mathbb{D}^- = \mathbb{D} \cap \{\mathrm{Im}\, w < 0\}, \\ -u(\psi^{-1}(\overline{w})), & w \in \mathbb{D}^+ = \mathbb{D} \cap \{\mathrm{Im}\, w > 0\}, \\ 0, & w \in (-1, 1). \end{cases}$$

Then v is continuous in $\mathbb{D}$ and v has the mean value property over sufficiently small circles centered at any $w \in \mathbb{D}$. Hence v is harmonic in $\mathbb{D}$ and $U = v \circ \psi$ defines a harmonic extension of u to V.

If U_1 and U_2 are extensions of u to neighborhoods V_1 and V_2 such that $V_1 \cap V_2 \cap \gamma$ is connected, then $U_1 = U_2$ in the component of $V_1 \cap V_2$ that contains $V_1 \cap V_2 \cap \gamma$. It follows that u has a harmonic extension to some open set $W \supset \gamma \cup \Omega$.

If $u > 0$ in Ω, then clearly $\partial u / \partial n \leq 0$ on γ. The strict inequality (2.10) will hold at ζ if and only if $(\partial v / \partial y)(0) < 0$. On $\mathbb{D}$, there is an analytic function $h = \tilde{v} - iv$ with $\mathrm{Im}\, h = -v$ and $h(0) = 0$. The Taylor expansion of h at 0 is

$$h(w) = a_n w^n + \mathrm{O}(|w^{n+1}|),$$

with $a_n \neq 0$. But if $n \geq 2$, then $h(\mathbb{D}^-) \cap \mathbb{D}^+ \neq \emptyset$, which is a contradiction. Hence $a_1 = h'(0) = -(\partial v / \partial y)(0) \neq 0$ and (2.10) holds. ∎

When $\partial \Omega$ consists of *analytic* curves, Green's function provides a formula for harmonic measure that generalizes the Poisson integral formula for $\mathbb{D}$.

Theorem 2.5. *Assume $\partial \Omega$ consists of finitely many pairwise disjoint analytic Jordan curves, and let $z \in \Omega$. Then Green's function $g(\zeta, z)$ extends to be harmonic (and hence real analytic) on a neighborhood of $\partial \Omega$ and*

$$\frac{-\partial g(\zeta, z)}{\partial n_\zeta} > 0 \qquad (2.11)$$

on $\partial \Omega$, where n_ζ is the unit outer normal at $\zeta \in \partial \Omega$. If $u \in C(\overline{\Omega})$ is harmonic on Ω, then

$$u(z) = \int_{\partial \Omega} \frac{-\partial g(\zeta, z)}{\partial n_\zeta} u(\zeta) \frac{ds(\zeta)}{2\pi}. \qquad (2.12)$$

Because of (2.12),

$$P_z(\zeta) = -\frac{1}{2\pi} \frac{\partial g(\zeta, z)}{\partial n_\zeta}$$

is called the **Poisson kernel** for Ω.

Proof. Fix $z \in \Omega$. By Lemma 2.4, $g(\zeta, z)$ extends to be harmonic (and real analytic) on some neighborhood of $\partial\Omega$ and then (2.11) follows from (2.10).

To prove (2.12) we first assume that u is C^∞ on a neighborhood of $\partial\Omega$. We then apply Green's theorem on $\Omega_\varepsilon = \Omega \setminus \{w : |w - z| < \varepsilon\}$ with ε small and $v(w) = g(w, z)$. Because $\Delta_w\, g(w, z) = \Delta u = 0$ on Ω_ε and because $g = 0$ on $\partial\Omega$, Green's theorem collapses into

$$\int_{\partial\Omega} -\frac{\partial g(\zeta, z)}{\partial n_\zeta} u(\zeta) \frac{ds(\zeta)}{2\pi} = \varepsilon \int_0^{2\pi} g(z + \varepsilon e^{i\theta}, z) \frac{\partial u(z + \varepsilon e^{i\theta})}{\partial r} \frac{d\theta}{2\pi}$$
$$- \varepsilon \int_0^{2\pi} u(z + \varepsilon e^{i\theta}) \frac{\partial g(z + \varepsilon e^{i\theta}, z)}{\partial r} \frac{d\theta}{2\pi}.$$

Since $g(z + \varepsilon e^{i\theta}, z) \le 2 \log 1/\varepsilon$ for small ε, and since u is C^∞, we have

$$\lim_{\varepsilon \to 0} \varepsilon \int_0^{2\pi} g(z + \varepsilon e^{i\theta}, z) \frac{\partial u(z + \varepsilon e^{i\theta})}{\partial r} \frac{d\theta}{2\pi} = 0.$$

As we have seen, (2.1) yields

$$\frac{\partial g(z + \varepsilon e^{i\theta}, z)}{\partial r} = -\frac{\partial \log \varepsilon}{\partial \varepsilon} + O(1),$$

while $u(z + \varepsilon e^{i\theta}) = u(z) + O(\varepsilon)$. Hence

$$-\varepsilon \int_0^{2\pi} u(z + \varepsilon e^{i\theta}) \frac{\partial g}{\partial r}(z + \varepsilon e^{i\theta}, z) \frac{d\theta}{2\pi} = u(z) + O(\varepsilon),$$

and that gives (2.12) when u is C^∞ on a neighborhood of $\partial\Omega$.

To prove (2.12) in general, set $\Omega^\delta = \{w \in \Omega : g(w, z) > \delta\}$ where δ is small. By uniqueness, Ω^δ has Green's function $g_\delta(w, z) = g(w, z) - \delta$. Therefore

$$\frac{\partial g_\delta}{\partial n_\zeta}(\zeta, z) = \frac{\partial g}{\partial n_\zeta}(\zeta, z)$$

on $\partial\Omega^\delta$. In a neighborhood of a point $\zeta_0 \in \partial\Omega$, the function $\varphi = g + i\widetilde{g}$ is a conformal map and

$$\frac{1}{2\pi} \int_{N \cap \partial\Omega^\delta} \frac{-\partial g(\zeta, z)}{\partial n_\zeta} u(\zeta) ds = \int_{\{Re z = \delta\} \cap \varphi(N)} u \circ \varphi^{-1} ds,$$

which converges to

$$\int_{\{Re z = 0\} \cap \varphi(N)} u \circ \varphi^{-1} ds = \frac{1}{2\pi} \int_{\partial\Omega} \frac{-\partial g(\zeta, z)}{\partial n_\zeta} u(\zeta) ds.$$

Therefore

$$\frac{1}{2\pi} \int_{\partial\Omega^\delta} \frac{-\partial g(\zeta, z)}{\partial n_\zeta} u(\zeta) ds \to \frac{1}{2\pi} \int_{\partial\Omega} \frac{-\partial g(\zeta, z)}{\partial n_\zeta} u(\zeta) ds \qquad (2.13)$$

as $\delta \to 0$. But (2.12) holds for u on Ω^δ because u is C^∞ on a neighborhood of $\partial\Omega_\delta$, and hence (2.13) yields (2.12) for Ω. $\blacksquare$

Corollary 2.6. *If $\partial\Omega$ consists of finitely many pairwise disjoint analytic Jordan curves and if $z \in \Omega$, then*

$$d\omega(z, \zeta) = -\frac{\partial g(z, \zeta)}{\partial n_\zeta} \frac{ds(\zeta)}{2\pi}. \qquad (2.14)$$

In other words, harmonic measure for $z \in \Omega$ is absolutely continuous to arc length on $\partial\Omega$, the density

$$\frac{d\omega}{ds} = -\frac{1}{2\pi} \frac{\partial g(z, \zeta)}{\partial n_\zeta} = P_z(\zeta)$$

is real analytic on $\partial\Omega$, and

$$c_1 < \frac{d\omega}{ds} < c_2 \qquad (2.15)$$

for positive constants c_1 and c_2.

Proof. The identity (2.14) follows from (1.2) and (2.12) and the fact that Borel measures are determined by their actions on continuous functions. The inequalities (2.15) are immediate from (2.11). $\blacksquare$

One objective of this book is to compare harmonic measure for general domains to more geometrical measures such as arc length, and Corollary 2.6 is the first result of this kind.

Theorem 2.7. *Assume $\partial\Omega$ consists of finitely many pairwise disjoint analytic Jordan curves and for $\zeta \in \partial\Omega$ and $\alpha > 1$ define*

$$\Gamma_\alpha(\zeta) = \{z \in \Omega : |z - \zeta| < \alpha \operatorname{dist}(z, \partial\Omega)\}.$$

If $u(z)$ is a bounded harmonic function on Ω, then for ds almost all $\zeta \in \partial\Omega$ the limit

$$\lim_{\Gamma_\alpha(\zeta)\ni z\to\zeta} u(z) = f(\zeta) \qquad (2.16)$$

exists,

$$u(z) = \int_{\partial\Omega} P_z(\zeta) f(\zeta) ds(\zeta), \qquad (2.17)$$

and

$$\sup_\Omega |u(z)| = \|f\|_{L^\infty}. \qquad (2.18)$$

Conversely, if f is a bounded Borel function on $\partial\Omega$, then (2.17) defines a bounded harmonic function $u(z)$ on Ω such that (2.18) holds and (2.16) holds ds almost everywhere. Moreover, if f is continuous at $\zeta_0 \in \partial\Omega$, then

$$\lim_{z \to \zeta_0} u(z) = f(\zeta_0). \qquad (2.19)$$

By (2.18), (2.16) and (2.17) establish an isometry between the space of bounded harmonic functions on Ω and $L^\infty(\partial\Omega, ds)$ when $\partial\Omega$ consists of analytic curves.

Proof. A simple localization argument gives the existence of the nontangential limit f. If I is an open arc on $\partial\Omega$, there exists a neighborhood $V \supset I$ such that $V \cap \partial\Omega = I$ and $V \cap \Omega$ is simply connected, and there exists a conformal mapping ψ defined on V such that $\psi(V \cap \Omega) = \mathbb{D}$ and $\psi(I)$ is an arc on $\partial\mathbb{D}$. It follows that ψ maps conical approach regions at $\zeta \in V \cap \partial\Omega$ into cones at $\psi(\zeta)$:

$$\psi\big(V \cap \Gamma_\alpha(\zeta) \cap B_\delta(\zeta)\big) \subset \Gamma_{\beta(\alpha)}(\psi(\zeta)),$$

where $B_\delta(\zeta) = \{z : |z - \zeta| < \delta = \delta(\zeta)\}$. Then if $u(z)$ is a bounded harmonic function on Ω, we can apply Fatou's theorem to $u \circ (\psi^{-1})$ to obtain (2.16) ds almost everywhere on $V \cap \partial\Omega$.

The proof of (2.17) is exactly the same as the proof of (2.12) except that the dominated convergence theorem is applied in (2.13). By (2.16) we have $\|f\|_\infty \leq \sup_\Omega |u(z)|$, and because $P_z \geq 0$ and $\int_{\partial\Omega} P_z ds = 1$, we have $\sup_\Omega |u(z)| \leq \|f\|_\infty$. Hence (2.16) and (2.17) imply (2.18).

To prove the converse, let $f \in L^\infty(\partial\Omega, ds)$. Then the discussion following (1.4) shows (2.17) defines a bounded harmonic function $u(z)$ on Ω and $\sup_\Omega |u(z)| \leq \|f\|_\infty$. Therefore by (2.16), u has almost everywhere a nontangential limit, which we will temporarily call F, and u is the Poisson integral of F. What we must prove is that $F = f$ almost everywhere. Again let V be a neighborhood of an open arc $I = V \cap \partial\Omega$ such that $V \cap \Omega$ is simply connected. For $h \in L^\infty(\partial\Omega, ds)$, define

$$v_h(z) = \int_{\partial\Omega} P_z(\zeta)h(\zeta)ds(\zeta) - \int_I P_z(\zeta, V)h(\zeta)ds(\zeta),$$

where $P_z(\zeta, V)$ is the Poisson kernel for $z \in V \cap \Omega$. If $h \in C(\partial\Omega)$ then by (1.2), Theorem 1.1, (2.12), and (3.2) of Chapter I, $\lim_{z \to \zeta} v_h(z) = 0$ for all $\zeta \in I$, and hence by Lemma 2.4, v_h extends to be harmonic in a neighborhood W of I which does not depend on h. Thus by Exercise I.5(e) or (3.5) below, if J is a compact subset of I and if $\varepsilon > 0$ then there is a neighborhood N of J, depending only on $\|h\|_\infty$ and ε, so that $|v_h| < \varepsilon$ in N. Now take $h_n \in C(\partial\Omega)$

so that h_n converges to f in L^1, and $||h_n||_\infty \le ||f||_\infty$. For each $z \in V \cap N$, $v_{h_n}(z)$ converges to $v_f(z)$ and so $|v_f(z)| < \varepsilon$. Because $\varepsilon > 0$ was arbitrary, we conclude that $v_f(z) \to 0$ as $z \to \zeta \in J$. But by Theorem I.1.3,

$$F(\zeta) - f(\zeta) = \lim_{\Gamma_{\alpha(\zeta)} \ni z \to \zeta} v_f(z) = 0$$

almost everywhere in J. Consequently $F = f$ almost everywhere and (2.16) and (2.18) hold for all $f \in L^\infty(\partial\Omega, ds)$.

Finally, if f is continuous at $\zeta_0 \in I$ then $\int_I P_z(\zeta, V)f(\zeta)ds(\zeta)$ is continuous at ζ_0 by (3.2) of Chapter I and so by the continuity of v_f, (2.19) holds. ∎

The converse can be proved another way. Using the real analyticity of $g(w, z)$, one can refine the proof of Lemma I.2.2 and show that

$$\sup_{\Gamma_\alpha(\zeta)} |u(z)| \le C(\alpha, \Omega)M_s f(\zeta), \tag{2.20}$$

where u is the Poisson integral (2.17) and where the maximal function $M_s f(\zeta)$ is the supremum of the averages of f over arcs $\gamma \subset \partial\Omega$ with $\zeta \in \gamma$:

$$M_s f(\zeta) = \sup_{\gamma \ni \zeta} \frac{1}{\ell(\gamma)} \int_\gamma |f|ds.$$

A variation on the covering lemma shows that M_s is weak-type 1-1, and an approximation, as in the proof of Theorem I.2.1, then yields (2.16) for the Poisson integral of f. This is the argument that must be used in the Euclidean spaces $\mathbb{R}^d$, with $d \ge 3$, and in other situations.

With some care, the conformal mapping proof of (2.16) in the text can also be parlayed into a proof of the maximal estimate (2.20). See Exercise 9.

Let Ω be any finitely connected Jordan domain and let φ be the conformal map, given in Lemma 2.2, of Ω onto a domain Ω^*, where $\partial\Omega^*$ consists of analytic Jordan curves. Since φ is a homeomorphism of $\overline{\Omega}$ onto $\overline{\Omega^*}$, harmonic measure can be transplanted from Ω^* to Ω via φ, just as it was in Section I.3 for simply connected Jordan domains. This gives an alternate but equivalent definition of harmonic measure for Ω.

In Section 4 we will consider two questions. Let Ω be a finitely connected Jordan domain.

Question 1. If $\partial\Omega$ has some degree of differentiability and if $f \in C(\partial\Omega)$ also has some degree of differentiability along $\partial\Omega$, how smooth is the solution $u_f(z)$, as z approaches $\partial\Omega$?

Question 2. What smoothness conditions on $\partial\Omega$, weaker than real analyticity, will ensure that $\partial g(z, \zeta)/\partial n_\zeta$ exists on $\partial\Omega$ and that (2.14) and (2.15) still hold?

These two questions are equivalent. Their answers will depend on Kellogg's theorem about the boundary behavior of conformal mappings. The proof of Kellogg's theorem in turn depends on the estimates for conjugate functions in the next section.

3. Conjugate Functions

Let $f \in L^1(\partial \mathbb{D})$ be real. For convenience we write $f(\theta)$ for $f(e^{i\theta})$. If $u(z)$ is the Poisson integral of f on $\mathbb{D}$, then $u(z)$ is harmonic and real and there exists a unique harmonic function $\widetilde{u}(z)$ such that $\widetilde{u}(0) = 0$ and

$$F = u + i\widetilde{u}$$

is analytic on $\mathbb{D}$. The function $\widetilde{u}$ is called the **conjugate function** or **harmonic conjugate** of u. The nontangential limit

$$\widetilde{f}(\theta) = \lim_{\Gamma_\alpha(e^{i\theta}) \ni z \to e^{i\theta}} \widetilde{u}(z) \tag{3.1}$$

exists almost everywhere, and this has an easy proof from Fatou's theorem: We may assume $f \geq 0$, so that

$$G(z) = e^{-(u(z) + i\widetilde{u}(z))}$$

is bounded and analytic on $\mathbb{D}$. By Corollary I.2.5, G has a nontangential limit $G(e^{i\theta})$ almost everywhere. Since $|G(e^{i\theta})| = e^{-f(\theta)}$ and $f \in L^1$, $|G(e^{i\theta})| > 0$ almost everywhere. At such $e^{i\theta}$, G is continuous and non-zero on the cone $K = \overline{\Gamma_\alpha(e^{i\theta})}$. Consequently $\log G = -(u + i\widetilde{u})$ has a continuous extension to $K \cap \{z : |G(z) - G(e^{i\theta})| < \frac{1}{2}|G(e^{i\theta})|\}$ and the limit (3.1) exists at $e^{i\theta}$.

There is a close connection between conjugate functions and conformal mappings. If u is harmonic and if $|u| < \pi/2$, then

$$\varphi(z) = \int_0^z e^{i(u+i\widetilde{u})(\zeta)} d\zeta$$

is a conformal map from $\mathbb{D}$ to a simply connected domain and $u = \arg \varphi'$. Indeed, if $a \neq b \in \mathbb{D}$, then

$$\varphi(b) - \varphi(a) = (b - a) \int_0^1 \varphi'(a + t[b - a])dt \neq 0,$$

because $\text{Re}(\varphi') > 0$.

When f is bounded, or even continuous, it can happen that $\widetilde{f}$ is not bounded. For an example, let $u + i\widetilde{u}$ be the conformal map of $\mathbb{D}$ onto the region

$$\{0 < x < 1/(1 + |y|)\}.$$

Then u is continuous on $\overline{\mathbb{D}}$ by Carathéodory's theorem, but $\tilde{u}$ is not bounded. For a second example see Exercise 11. The next two theorems get around the obstruction that $\tilde{f}$ may be unbounded even when f is continuous.

Theorem 3.1 (Zygmund). *Let $f \in L^\infty(\partial \mathbb{D})$ be real with $\|f\|_\infty \le 1$.*
(a) *For $0 < \lambda < \pi/2$ there is a constant C_λ, depending only on λ, such that*

$$\int e^{\lambda|\tilde{f}(\theta)|} \frac{d\theta}{2\pi} \le C_\lambda.$$

(b) *If $f \in C(\partial \mathbb{D})$, then for all $\lambda < \infty$*

$$\sup_{0 < r < 1} \int e^{\lambda|\tilde{u}(re^{i\theta})|} \frac{d\theta}{2\pi} < \infty.$$

Proof. Let $u(z)$ be the Poisson integral of f in $\mathbb{D}$ and consider the analytic function $g(z) = \tilde{u}(z) - iu(z)$. For $r < 1$, $g(z)$ satisfies

$$e^{\lambda g(0)} = \int_0^{2\pi} e^{\lambda g(re^{i\theta})} \frac{d\theta}{2\pi},$$

and because $\tilde{u}(0) = 0$, that gives

$$\cos \lambda u(0) = \int_0^{2\pi} e^{\lambda \tilde{u}(re^{i\theta})} \cos \lambda u(re^{i\theta}) \frac{d\theta}{2\pi}.$$

But if $0 < \lambda < \pi/2$, then

$$0 < \cos \lambda < \cos \lambda u \le 1$$

because $|u| \le 1$, so that

$$\int_0^{2\pi} e^{\lambda \tilde{u}(re^{i\theta})} \frac{d\theta}{2\pi} \le \sec \lambda.$$

Then by (3.1) and Fatou's lemma

$$\int_0^{2\pi} e^{\lambda \tilde{f}(\theta)} \frac{d\theta}{2\pi} \le \sec \lambda.$$

By repeating this argument with $-f(\theta)$, we then obtain part (a) with constant $C_\lambda = 2 \sec \lambda$.

To prove (b), fix $\lambda < \infty$ and take a trigonometric polynomial

$$p(\theta) = \sum_{n=0}^{N} (a_n \cos n\theta + b_n \sin n\theta)$$

such that $\|f - p\|_\infty < \pi/(2\lambda)$. Then

$$\widetilde{p}(re^{i\theta}) = \sum_{n=0}^{N} r^n (a_n \sin n\theta - b_n \cos n\theta)$$

is bounded, while part (a) gives

$$B_\lambda = \sup_{0 < r < 1} \int e^{\lambda |\widetilde{(u-p)}(re^{i\theta})|} \frac{d\theta}{2\pi} < \infty.$$

Therefore, since $|\widetilde{u}| \le |\widetilde{p}| + |\widetilde{(u - p)}|$, we have

$$\sup_{0 < r < 1} \int e^{\lambda |\widetilde{u}(re^{i\theta})|} \frac{d\theta}{2\pi} \le B_\lambda e^{\lambda \|\widetilde{p}\|_\infty} < \infty,$$

and (b) is proved.   ∎

Let $0 < \alpha < 1$. The **Lipschitz class** C^α is

$$C^\alpha = \left\{ f \in L^\infty(\partial\mathbb{D}) : \sup_{t > 0} \frac{\|f(\theta + t) - f(\theta)\|_\infty}{t^\alpha} < \infty \right\}.$$

Every $f \in C^\alpha$ agrees almost everywhere with a function continuous on $\partial\mathbb{D}$. The class C^α is given the **Lipschitz norm**

$$\|f\|_{C^\alpha} = \|f\|_\infty + \sup_{t > 0} \frac{\|f(\theta + t) - f(\theta)\|_\infty}{t^\alpha}. \tag{3.2}$$

In a moment we shall prove Privalov's theorem that $\widetilde{f} \in C^\alpha$ whenever $f \in C^\alpha$.
   In the Poisson integral formula

$$u(z) = \mathrm{Re} \int_{-\pi}^{\pi} \frac{e^{it} + z}{e^{it} - z} f(t) \frac{dt}{2\pi},$$

the kernel $(e^{it} + z)/(e^{it} - z)$ is analytic in $z \in \mathbb{D}$ and real at $z = 0$. Then the uniqueness of $\widetilde{u}$ shows

$$u(z) + i\widetilde{u}(z) = F(z) = \int_{-\pi}^{\pi} \frac{e^{it} + z}{e^{it} - z} f(t) \frac{dt}{2\pi}. \tag{3.3}$$

The analytic function $F(z)$ in (3.3) is called the **Herglotz integral** of f. Write

$$\nabla u = \left(\frac{\partial u}{\partial x}, \frac{\partial u}{\partial y} \right)$$

when u is differentiable on an open plane set. If u is harmonic and bounded on $\mathbb{D}$ then u is the Poisson integral of some $f \in L^\infty(\partial\mathbb{D})$ and by (3.3) and the

Cauchy–Riemann equations,

$$|\nabla u(z)| = |F'(z)| = \left| \int_{-\pi}^{\pi} \frac{e^{it} f(t)}{(e^{it} - z)^2} \frac{dt}{\pi} \right|. \tag{3.4}$$

Therefore (as in Exercise I.5(e))

$$|\nabla u(z)| \leq \left(\int_{-\pi}^{\pi} \frac{1}{|e^{it} - z|^2} \frac{dt}{\pi} \right) \|f\|_{\infty} \leq 2(1 - |z|)^{-1} \|f\|_{\infty}. \tag{3.5}$$

We will often use the following consequence of (3.5): If $u(z)$ is harmonic and $|u(z)| \leq M$ on $B(z, R)$, then

$$\sup_{B(z, \frac{R}{2})} |\nabla u| \leq \frac{4M}{R}. \tag{3.6}$$

To prove (3.6) simply apply (3.5) to $U(w) = u(z + Rw)$.

The next theorem shows that $f \in C^{\alpha}$ if and only if the estimate (3.5) can be upgraded to

$$|\nabla u| = O\big((1 - |z|)^{\alpha - 1}\big).$$

Theorem 3.2. *Let $0 < \alpha < 1$, let $f \in L^{\infty}(\partial \mathbb{D})$ be real, and let $u(z)$ be the Poisson integral of f. Then the following conditions are equivalent:*
(a) $f \in C^{\alpha}$.
(b) $\tilde{f} \in C^{\alpha}$.
(c) $|\nabla u(z)| = O\big((1 - |z|)^{\alpha - 1}\big)$.
(d) $u \in C^{\alpha}(\overline{\mathbb{D}})$; *that is, for all $z_1, z_2 \in \mathbb{D}$,*

$$|u(z_1) - u(z_2)| = O\big(|z_1 - z_2|^{\alpha}\big).$$

Moreover there is a constant C_1, independent of α, such that

$$\|\tilde{f}\|_{C^{\alpha}} \leq \frac{C_1}{\alpha(1 - \alpha)} \|f\|_{C^{\alpha}}, \tag{3.7}$$

$$|\nabla u(z)| \leq \frac{C_1}{(1 - \alpha)} (1 - |z|)^{\alpha - 1} \|f\|_{C^{\alpha}}, \tag{3.8}$$

and

$$\sup_{z_1 \neq z_2} \frac{|u(z_1) - u(z_2)|}{|z_1 - z_2|^{\alpha}} \leq \frac{C_1}{\alpha} \sup_{|z| < 1} \big\{ (1 - |z|)^{1 - \alpha} |\nabla u(z)| \big\}. \tag{3.9}$$

The equivalence (a) $\Longleftrightarrow$ (b) was first proved by Privalov [1916] who worked directly with the imaginary part of the integral (3.3); (a) $\Longleftrightarrow$ (c) was proved by Hardy and Littlewood in [1931].

Proof. Clearly (d) $\Longrightarrow$ (a) because if (d) holds, u is uniformly continuous on $\mathbb{D}$. We first show (a) $\Longrightarrow$ (c) and establish (3.8) and then show (c) $\Longrightarrow$ (d) and establish (3.9). Then (a) $\Longleftrightarrow$ (b) and inequality (3.7) will follow because $|\nabla \tilde{u}| = |\nabla u|$ by the Cauchy–Riemann equations.

Assume (a) holds, that is, assume u is the Poisson integral of $f \in C^\alpha$. We prove (3.8). By (3.5) we may assume $|z| \geq 1/2$. Let $e^{it_0} = z/|z|$. Because $\int e^{it}/(e^{it} - z)^2 dt = 0$, (3.4) yields

$$|\nabla u(z)| = \left| \int_{-\pi}^{\pi} \frac{e^{it}(f(t) - f(t_0))}{(e^{it} - z)^2} \frac{dt}{\pi} \right|,$$

so that

$$|\nabla u(z)| \leq \int_{\pi}^{\pi} \frac{|f(t) - f(t_0)|}{|e^{it} - z|^2} \frac{dt}{\pi}$$

$$= \int_{|t-t_0|<1-|z|} + \int_{1-|z|\leq|t-t_0|<\pi}.$$

The inequality $1 - |z| \leq |e^{it} - z|$, valid for all t, gives

$$\int_{|t-t_0|<1-|z|} \frac{|f(t) - f(t_0)|}{|e^{it} - z|^2} \frac{dt}{\pi} \leq \frac{2\|f\|_\alpha}{(1-|z|)^2} \int_0^{1-|z|} t^\alpha \frac{dt}{\pi}$$

$$= \frac{2\|f\|_\alpha}{\pi(1+\alpha)} (1 - |z|)^{\alpha-1}.$$

When $1 - |z| \leq |t - t_0| < \pi$, the inequality $|t - t_0|^2 \leq c|e^{it} - z|^2$, valid with c independent of z, gives

$$\int_{1-|z|<|t-t_0|<\pi} \frac{|f(t) - f(t_0)|}{|e^{it} - z|^2} \frac{dt}{\pi} \leq 2c\|f\|_\alpha \int_{1-|z|}^{\pi} t^{\alpha-2} \frac{dt}{\pi}$$

$$\leq \frac{2c\|f\|_\alpha}{\pi(1-\alpha)} (1 - |z|)^{\alpha-1}.$$

Therefore we obtain (3.8) with

$$C_1 = \sup_{0<\alpha<1} \frac{2}{\pi} \left(c + \frac{(1-\alpha)}{(1+\alpha)} \right) = \frac{2(c+1)}{\pi}.$$

If (c) holds, we may assume $(1 - |z|)^{1-\alpha} |\nabla u(z)| \leq 1$. Let $z_j = r_j e^{i\theta_j} \in \mathbb{D}$, as in Figure II.4. We may assume $|z_j| \geq 1/2$ and $\delta = |z_1 - z_2| \leq 1/2$. Set $w_j = (1 - \delta)z_j$. Then

$$|u(z_1) - u(z_2)| \leq |u(z_1) - u(w_1)| + |u(z_2) - u(w_2)| + |u(w_1) - u(w_2)|.$$

However,

$$|u(z_j) - u(w_j)| = \left| \int_{(1-\delta)r_j}^{r_j} \frac{\partial u}{\partial t}(te^{i\theta_j})dt \right| \leq \int_{1-\delta}^{1} (1-t)^{\alpha-1}dt \leq \frac{\delta^\alpha}{\alpha},$$

while

$$|u(w_1) - u(w_2)| \leq |w_1 - w_2| \text{Max}_j \left(1 - |w_j|\right)^{\alpha-1} \leq \delta^\alpha.$$

Therefore (d) and (3.9) hold.

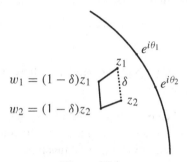

$$w_1 = (1-\delta)z_1$$
$$w_2 = (1-\delta)z_2$$

Figure II.4

Now suppose (a) holds. Then (c) holds and $|\nabla \widetilde{u}| = O((1 - |z|)^{\alpha-1})$. Therefore (d) and (3.9) hold for $\widetilde{u}$, and $\widetilde{u}$ extends continuously to $\partial \mathbb{D}$ where $\widetilde{u}$ has boundary value $\widetilde{f}$. It follows that $\widetilde{f} \in C^\alpha$ because (d) implies (a). Finally since $\widetilde{f} = -f + u(0)$, it follows that (a) $\Longleftrightarrow$ (b) and that (3.7) holds. ∎

Let k be a non-negative integer, let $0 \leq \alpha < 1$ and let $f \in C(\partial \mathbb{D})$. We say $f \in C^{k+\alpha}$ if f is k times continuously differentiable on $\partial \mathbb{D}$ and $(d/d\theta)^k f \in C^\alpha$ if $\alpha > 0$. If $F(z)$ is analytic on $\mathbb{D}$ we say $F \in C^{k+\alpha}(\overline{\mathbb{D}})$ if F and its first k derivatives $F', F'', \cdots, F^{(k)}$ extend continuously to $\overline{\mathbb{D}}$ and if there is C such that

$$|F^{(k)}(z_1) - F^{(k)}(z_2)| \leq C|z_1 - z_2|^\alpha$$

for all $z_1, z_2 \in \overline{\mathbb{D}}$.

Corollary 3.3. *Assume k is a non-negative integer and assume $0 < \alpha < 1$. Let $f \in C(\partial \mathbb{D})$ be real and let $F(z) = u(z) + i\widetilde{u}(z)$ be the Herglotz integral of f. Then $F \in C^{k+\alpha}(\overline{\mathbb{D}})$ if and only if $f \in C^{k+\alpha}(\partial \mathbb{D})$.*

Proof. Because $f = \text{Re} F$, it is clear that $f \in C^{k+\alpha}(\partial \mathbb{D})$ if $F \in C^{k+\alpha}(\overline{\mathbb{D}})$. Assume $f \in C^{k+\alpha}$. If $f(\theta)$ has Fourier series

$$f(\theta) \sim \sum_{-\infty}^{\infty} a_n e^{in\theta},$$

then F has Taylor series

$$F(z) = a_0 + 2 \sum_{n=1}^{\infty} a_n z^n.$$

Therefore $df/d\theta$, which has Fourier series

$$\frac{df}{d\theta} \sim \sum_{-\infty}^{\infty} in a_n e^{in\theta},$$

has Herglotz integral $izF'(z)$ and the corollary follows from the special case $k = 0$, which is Theorem 3.2.   ∎

Theorem 3.2 fails when $\alpha = 1$. The harmonic conjugate of a continuously differentiable function on $\partial \mathbb{D}$ need not have a continuous derivative, and the conjugate of a **Lipschitz function**, that is, a function satisfying

$$\left| f(\theta + t) - f(\theta) \right| \leq M|t|,$$

need not be a Lipschitz function. See Exercise 13. For the same reason, Corollary 3.3 is also false when $\alpha = 0$ and k is a non-negative integer. However, conjugation does preserve the **Zygmund class** Z^* of continuous f on $\partial \mathbb{D}$ such that

$$\sup_{t>0} \frac{\| f(\theta + t) + f(\theta - t) - 2f(\theta) \|_\infty}{t} < \infty.$$

The Zygmund class has norm

$$\| f \|_{Z^*} = \| f \|_\infty + \sup_{t>0} \frac{\| f(\theta + t) + f(\theta - t) - 2f(\theta) \|_\infty}{t}.$$

When $f \in Z^*$, we say f is a **Zygmund function**.

Define

$$\left| \nabla_2 u(z) \right| = \left| \nabla u_x \right| = \left(|u_{xx}(z)|^2 + |u_{yx}(z)|^2 \right)^{\frac{1}{2}}$$

$$= \left| F''(z) \right| = \left| \frac{2}{\pi} \int_{-\pi}^{\pi} \frac{e^{it} f(t)}{(e^{it} - z)^3} dt \right|,$$

where $F = u + i\widetilde{u}$ is the Herglotz integral of $f \in L^1(\partial\mathbb{D})$. If $f \in C^\alpha$, then (3.8) and (3.6), applied to u_x in the disc $B(z, \frac{1-|z|}{2})$, give us

$$\left|\nabla_2 u(z)\right| = O\big((1 - |z|)^{\alpha-2}\big). \tag{3.10}$$

Conversely, if (3.10) holds, then integrating F'' along radii shows that $\left|\nabla u(z)\right| = O\big((1 - |z|)^{\alpha-1}\big)$, and $f \in C^\alpha$. Thus (3.10) provides yet another characterization of C^α functions, in terms of second derivatives. Zygmund functions have a similar characterization.

Theorem 3.4. *Let $f \in L^\infty(\partial\mathbb{D})$ be real and let $u(z)$ be the Poisson integral of f. Then the following are equivalent:*
(a) $f \in Z^*$.
(b) $\widetilde{f} \in Z^*$.
(c) $\left|\nabla_2 u(z)\right| = O\big((1 - |z|)^{-1}\big)$.
There is a constant C such that

$$\|\widetilde{f}\|_{Z^*} \le C\|f\|_{Z^*} \tag{3.11}$$

and

$$\frac{1}{C}\|f\|_{Z^*} \le \|f\|_\infty + \sup_{z\in\mathbb{D}}\big\{(1 - |z|)\left|\nabla_2 u(z)\right|\big\} \le C\|f\|_{Z^*}. \tag{3.12}$$

Notice that by (3.12) and (3.10), $Z^* \subset C^\alpha$ for all $\alpha < 1$. In particular, if $f \in Z^*$, then f and $\widetilde{f}$ are continuous. On the other hand, if f is Lipschitz, then clearly $f \in Z^*$. Thus we have the increasing scale of spaces

$$C^1 \subset \text{Lipschitz} \subset Z^* \subset C^\alpha \subset C,$$

$\alpha < 1$, where C^1 denotes the space of continuously differentiable functions and C is the space of continuous functions.

Proof. The logic is the same as in the proof of Theorem 3.2. First assume (a) holds. Fix $z \in \mathbb{D}$. We may assume $z = |z| = \text{Re}z$ and $|z| > \frac{1}{2}$. Because

$$\int \frac{e^{i\theta}}{(e^{i\theta} - z)^3} d\theta = 0$$

and $f(-\theta)$ has Herglotz integral $\overline{F(\overline{z})}$, we have

$$\frac{\partial^2 u}{\partial x^2}(z) = \text{Re}F''(z) = \frac{1}{\pi}\int \frac{e^{it}\big(f(t) + f(-t) - 2f(0)\big)}{(e^{it} - |z|)^3} dt$$

and

$$\left| \frac{\partial^2 u}{\partial x^2}(z) \right| = \left| \operatorname{Re} F''(z) \right|$$

$$\leq \frac{1}{\pi} \int_{|t| \leq 1-|z|} \frac{\|f\|_{Z^*} |t|}{(1-|z|)^3} dt + \frac{c}{\pi} \int_{1-|z|<|t|\leq\pi} \frac{\|f\|_{Z^*}}{|t|^2} dt$$

$$\leq C(1-|z|)^{-1} \|f\|_{Z^*}.$$

Unfortunately this trick does not us help with $\left| \operatorname{Im} F''(z) \right| = \left| \frac{\partial^2 u}{\partial x \partial y} \right|$. Instead, we apply (3.6) to $U = \partial^2 u / \partial x^2$ on $B(z, \frac{(1-|z|)}{2})$, getting

$$\left| \frac{\partial^3 u}{\partial y \partial^2 x} \right| \leq \frac{C \|f\|_{Z^*}}{(1-|z|)^2}.$$

An integration then yields

$$\left| \frac{\partial^2 u}{\partial y \partial x}(z) \right| \leq \left| F''(0) \right| + \int_0^{|z|} \frac{C \|f\|_{Z^*}}{(1-s)^2} ds \leq C' \frac{\|f\|_{Z^*}}{1-|z|}.$$

That proves (c), and the right inequality in (3.12).

Now assume (c) holds. Then by (3.10), $f \in C^\alpha$ and f and $\widetilde{f}$ are continuous. Fix θ and t with $0 < t \leq \pi$ and set $r = 1 - (t/\pi)$. Then

$$
\begin{aligned}
f(\theta+t) + f(\theta-t) - 2f(\theta) &= f(\theta+t) - u(re^{i(\theta+t)}) \\
&+ f(\theta-t) - u(re^{i(\theta-t)}) \\
&+ 2u(re^{i\theta}) - 2f(\theta) \qquad (3.13) \\
&+ u(re^{i(\theta+t)}) - u(re^{i\theta}) \\
&+ u(re^{i(\theta-t)}) - u(re^{i\theta}).
\end{aligned}
$$

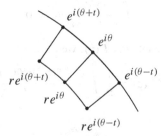

Figure II.5

Because

$$\left|u(re^{i(\theta+t)}) + u(re^{i(\theta-t)}) - 2u(re^{i\theta})\right| \le |t|^2 \sup_{|w|=r} \left|\frac{\partial^2 u}{\partial\theta^2}(w)\right|,$$

the last two terms on the right side of (3.13) are $O(|t|)$ by (c). Then because $\lim_{s\to 1}(1-s)\frac{\partial u}{\partial s}(se^{i\alpha}) = 0$ by (c), an integration by parts shows

$$f(\alpha) - u(re^{i\alpha}) = \int_r^1 (1-s)\frac{\partial^2 u}{\partial s^2}(se^{i\theta})ds + (1-r)\frac{\partial u}{\partial r}(re^{i\alpha}).$$

Therefore when (c) holds, the sum of the first three lines of the right side of (3.13) is

$$\frac{t}{\pi}\left(u_r(re^{i(\theta+t)}) + u_r(re^{i(\theta-t)}) - 2u_r(re^{i\theta})\right) + O(t)$$

$$\le \frac{t}{\pi}\left|u_r(re^{i(\theta+t)}) - u_r(re^{i\theta})\right| + \frac{t}{\pi}\left|u_r(re^{i(\theta-t)}) - u_r(re^{i\theta})\right| + O(t)$$

$$\le \frac{2t^2}{\pi} \sup_{|w|=r} \left|\frac{\partial^2 u}{\partial r\partial\theta}(w)\right| + O(t) \le C''t.$$

That proves (a), and the left-hand inequality in (3.12) can be seen by following the constants in the previous argument.

Since $\left|\nabla_2\tilde{u}\right| = \left|\nabla_2 u\right|$, it follows that (a) $\Longleftrightarrow$ (b) and that (3.11) holds. ∎

4. Boundary Smoothness

Let Ω be a Jordan domain with boundary Γ and let φ be a conformal mapping from $\mathbb{D}$ onto Ω, so that φ extends to a homeomorphism from $\partial\mathbb{D}$ to the Jordan curve $\Gamma = \partial\Omega$. In this section we examine the connection between the smoothness of Γ and the differentiability of φ on $\partial\mathbb{D}$. When Γ has some degree of smoothness, we also study the relation between the differentiability of $f \in C(\Gamma)$ and the differentiability of its solution u_f to the Dirichlet problem at points of Γ.

The results do not depend on the choice of the mapping $\varphi : \mathbb{D} \to \Omega$ because any other such map has the form $\varphi \circ T$, with $T \in \mathcal{M}$. We first show that the smoothness of φ in a neighborhood of $\varphi^{-1}(\zeta)$ depends only on the smoothness of Γ in a neighborhood of ζ.

Theorem 4.1. *Let Ω_1 and Ω_2 be Jordan domains such that $\Omega_1 \subset \Omega_2$ and let $\gamma \subset \partial\Omega_1 \cap \partial\Omega_2$ be an open subarc. Let φ_j be a conformal map of $\mathbb{D}$ onto Ω_j. Then $\psi = \varphi_2^{-1} \circ \varphi_1$ has an analytic continuation across $\varphi_1^{-1}(\gamma)$, and $\psi' \neq 0$ on $\varphi_1^{-1}(\gamma)$.*

Proof. The analytic function $\psi = \varphi_2^{-1} \circ \varphi_1$ from $\mathbb{D}$ into $\mathbb{D}$ has a continuous and unimodular extension to the arc $\varphi_1^{-1}(\gamma)$. By Schwarz reflection ψ has an analytic and one-to-one extension to a neighborhood of $\varphi_1^{-1}(\gamma)$ in $\mathbb{C}$ and hence $\varphi' \neq 0$ on $\varphi_1^{-1}(\gamma)$. ∎

Let Γ be an arc parameterized as $\{\zeta(t) : a < t < b\}$. We say Γ has a **tangent** at $\zeta_0 = \zeta(t_0)$ if

$$\lim_{t \downarrow t_0} \frac{\zeta(t) - \zeta_0}{|\zeta(t) - \zeta_0|} = e^{i\tau} \tag{4.1}$$

and

$$\lim_{t \uparrow t_0} \frac{\zeta(t) - \zeta_0}{|\zeta(t) - \zeta_0|} = -e^{i\tau}. \tag{4.2}$$

If (4.1) and (4.2) hold then Γ has unit tangent vector $e^{i\tau}$ at ζ_0. Except for reversals of orientation, the existence of a tangent at ζ_0 and its value $e^{i\tau}$ do not depend on the choice of the parameterization $t \to \zeta(t)$. We say Γ has a **continuous tangent** if Γ has a tangent at each $\zeta \in \Gamma$ and if $e^{i\tau(\zeta)}$ is continuous on Γ.

Theorem 4.2. *The curve Γ has a tangent at $\zeta = \varphi(e^{i\theta})$ if and only if the limit*

$$\lim_{\mathbb{D} \ni z \to e^{i\theta}} \arg\left(\frac{\varphi(z) - \zeta}{z - e^{i\theta}}\right) \tag{4.3}$$

exists and is finite. In that case,

$$\lim_{\mathbb{D} \ni z \to e^{i\theta}} \arg\left(\frac{\varphi(z) - \zeta}{z - e^{i\theta}}\right) = \tau(\zeta) - \theta - \frac{\pi}{2} \quad (\text{modulo } 2\pi). \tag{4.4}$$

The curve Γ has a continuous tangent if and only if $\arg\varphi'(z)$ has a continuous extension to $\overline{\mathbb{D}}$. If Γ has a continuous tangent, then $\varphi \in C^\alpha(\partial\mathbb{D})$ for all $\alpha < 1$ and Γ is rectifiable.

There exist Jordan domains Ω and conformal maps $\varphi : \mathbb{D} \to \Omega$ such that $\partial\Omega$ has a continuous tangent but $\varphi \notin C^1(\partial\mathbb{D})$. An example can be built from the connection in Section 3 between conjugate functions and conformal mappings: If $u(e^{i\theta})$ is continuous and $|u| < \frac{\pi}{2}$, then $u(e^{i\theta}) = \arg\varphi'(e^{i\theta})$, where φ is a conformal mapping onto a Jordan domain, but $\tilde{u}(e^{i\theta}) = -\log|\varphi'(e^{i\theta})|$ may not be bounded above or below.

Proof. Set

$$v(z) = \arg\left(\frac{\varphi(z) - \zeta}{z - e^{i\theta}}\right).$$

Then $v(z)$ is harmonic on $\mathbb{D}$ and v is continuous on $\overline{\mathbb{D}} \setminus \{e^{i\theta}\}$. If Γ has a tangent at $\zeta = \varphi(e^{i\theta})$ or if the limit (4.3) exists at $e^{i\theta}$, then $v(z)$ is bounded on $\mathbb{D}$. Thus in either case v is the Poisson integral of $v(e^{i\theta})$. Therefore v has a continuous extension to $e^{i\theta}$ if and only if $v|_{\partial\mathbb{D}}$ has a continuous extension to $e^{i\theta}$, and if and only if Γ has a tangent at ζ. Clearly, (4.4) holds when v is continuous at $e^{i\theta}$.

Now suppose Γ has a continuous tangent. Then $\tau \circ \varphi$ is continuous on $\partial\mathbb{D}$. If $h \neq 0$, then

$$A_h(z) = \arg\left(\frac{\varphi(ze^{ih}) - \varphi(z)}{z(e^{ih} - 1)}\right)$$

is continuous on $\overline{\mathbb{D}}$ and harmonic on $\mathbb{D}$. For $|z| < 1$,

$$\lim_{h \to 0} A_h(z) - \arg\varphi'(z).$$

For $|z| = 1$ there exists $k, 0 < |k| < |h|$, such that

$$\arg\left(\frac{\varphi(ze^{ih}) - \varphi(z)}{h}\right) = \arg\left(\varphi(ze^{ih}) - \varphi(z)\right) = \tau(\varphi(ze^{ik})),$$

by the proof of the mean value theorem from calculus. Consequently

$$\lim_{h \to 0} A_h(z) = \tau(\varphi(z)) - \arg z - \frac{\pi}{2},$$

with the convergence uniform on $\partial\mathbb{D}$, and hence $\arg(\varphi')$ is the Poisson integral of the continuous function

$$\tau(\varphi(e^{i\theta})) - \theta - \frac{\pi}{2}.$$

Now suppose $\arg\varphi'(z)$ has a continuous extension to $\overline{\mathbb{D}}$. For $r < 1$, the curve $\Gamma_r = \{\varphi(re^{i\theta}) : 0 \leq \theta \leq 2\pi\}$ has tangent $e^{i\tau(\varphi(re^{i\theta}))}$ satisfying

$$e^{i\tau(\varphi(re^{i\theta}))} = ie^{i\theta}e^{i\arg\varphi'(re^{i\theta})},$$

and $e^{i\tau(\varphi(z))}$ has a continuous extension to $\overline{\mathbb{D}} \setminus \{0\}$. Then for $h > 0$

$$\arg\left(\varphi(e^{i(\theta+h)}) - \varphi(e^{i\theta})\right) = \lim_{r \to 1} \arg\left(\varphi(re^{i(\theta+h)}) - \varphi(re^{i\theta})\right)$$

$$= \lim_{r \to 1} \tau(\varphi(re^{i(\theta+k_r)})),$$

where $0 < k_r < h$, again by the proof of the mean value theorem. Therefore

$$\lim_{h \downarrow 0} \arg\left(\varphi(e^{i(\theta+h)}) - \varphi(e^{i\theta})\right) = \lim_{z \to e^{i\theta}} \tau(\varphi(z)),$$

and a similar argument for $h < 0$ yields

$$\lim_{h \uparrow 0} \arg\left(\varphi(e^{i(\theta+h)}) - \varphi(e^{i\theta})\right) = \lim_{z \to e^{i\theta}} \tau(\varphi(z)) + \pi.$$

Thus Γ has a continuous tangent.

If Γ has a continuous tangent, then by part (b) of Theorem 3.1,

$$\sup_{r<1} \int |\varphi'(re^{i\theta})|^\lambda d\theta = \sup_{r<1} \int e^{-\lambda \widetilde{(\arg \varphi')}(re^{i\theta})} d\theta = B_\lambda < \infty$$

for all $\lambda < \infty$, where B_λ depends on λ and φ. Take $\lambda = 1/(1-\alpha)$, where $0 < \alpha < 1$. Let $a < b < a + \pi$. Then for any $r < 1$, Hölder's inequality gives

$$\int_a^b |\varphi'(re^{i\theta})| d\theta \le |b-a|^\alpha \left(\int_a^b |\varphi'(re^{i\theta})|^\lambda d\theta \right)^{1-\alpha} \le |b-a|^\alpha B_\lambda^{1-\alpha}.$$

Therefore if $a = \theta_0 < \theta_1 < \ldots < \theta_n = b$, then

$$\sum_{j=1}^n |\varphi(re^{i\theta_j}) - \varphi(re^{i\theta_{j-1}})| \le \int_a^b |\varphi'(re^{i\theta})| d\theta \le |b-a|^\alpha B_\lambda^{1-\alpha}, \qquad (4.5)$$

and letting r tend to 1, we see that Γ is rectifiable and that $\varphi \in C^\alpha$.   ∎

Let k be a non-negative integer and let $0 \le \alpha < 1$. We say the curve Γ is of class $C^{k+\alpha}$ if Γ is rectifiable and if, in the arc length parameterization

$$\Gamma = \{\gamma(s) : 0 \le s \le \ell(\Gamma) = \text{ length } (\Gamma)\},$$

the function γ is k times continuously differentiable and $d^k\gamma/ds^k \in C^\alpha$ if $\alpha > 0$.

Theorem 4.3 (Kellogg). *Let $k \ge 1$ and $0 < \alpha < 1$. Then the following conditions are equivalent:*
(a) *Γ is of class $C^{k+\alpha}$.*
(b) *$\arg \varphi' \in C^{k-1+\alpha}(\partial \mathbb{D})$.*
(c) *$\varphi \in C^{k+\alpha}(\overline{\mathbb{D}})$ and $\varphi' \ne 0$ on $\overline{\mathbb{D}}$.*

Note that if $\alpha = 0$ and $k \ge 1$ then (a) does not imply (c) but (a) is still equivalent to (b). We need the elementary lemma.

Lemma 4.4. *Let k be a positive integer, and let $0 \le \alpha < 1$. Let $f \in C^1([0, 1])$ satisfy $f' > 0$ and let $g \equiv f^{-1}$. Then $g \in C^{k+\alpha}$ if and only if $f \in C^{k+\alpha}$.*

Proof. The case $\alpha = 0$ and $k = 1$ is clear since $g'(y) = 1/f'(g(y)) > 0$. If $\alpha = 0$ and $k \ge 2$, the proof is by induction: if $f \in C^k$ and $g \in C^{k-1}$, then $f' \circ g \in C^{k-1}$ and because $f' > 0$, $g' = 1/(f' \circ g) \in C^{k-1}$. Hence $g \in C^k$. Now suppose $\alpha > 0$ and $k = 1$. If $f' \in C^\alpha$ then

$$|g'(y_1) - g'(y_2)| = \left| \frac{1}{f'(g(y_1))} - \frac{1}{f'(g(y_2))} \right|$$

$$\leq \frac{|f'(g(y_2)) - f'(g(y_1))|}{\min |f'|^2}$$

$$\leq C|g(y_2) - g(y_1)|^\alpha \leq C'|y_2 - y_1|^\alpha,$$

and so $g' \in C^\alpha$. Finally, suppose $\alpha > 0$ and $k \geq 2$. If $f \in C^{k+\alpha}$, then $g \in C^k$ and $g^{(k)}$ can be written as a sum of products of the functions

$$g^{(1)}, \ldots, g^{(k-1)}, f^{(2)} \circ g, \ldots, f^{(k)} \circ g.$$

All these functions are C^1, except perhaps $f^{(k)} \circ g$. But $f^{(k)} \circ g \in C^\alpha$, and thus $g^{(k)} \in C^\alpha$. The converse follows from interchanging g and f. ∎

Proof of Theorem 4.3. If $\arg \varphi' \in C^{k-1+\alpha}$, $0 < \alpha < 1$, then by Corollary 3.3, $\log |\varphi'| \in C^{k-1+\alpha}$, and by taking exponentials $\varphi' \in C^{k-1+\alpha}(\overline{\mathbb{D}})$ and $\varphi' \neq 0$. Conversely, if $\varphi' \in C^{k-1+\alpha}(\overline{\mathbb{D}})$ and $\varphi' \neq 0$, then $e^{i \arg \varphi'} = \varphi'/|\varphi'| \in C^{k-1+\alpha}$, and $\arg \varphi' \in C^{k-1+\alpha}$. Hence (b) and (c) are equivalent.

Assume (c) holds. If

$$s(\theta) = \int_0^\theta |\varphi'(e^{it})| \, dt,$$

then $s' > 0$ and $s'(\theta) = |\varphi'(e^{i\theta})| \in C^{k-1+\alpha}$. Thus $\theta'(s) \in C^{k-1+\alpha}$ by Lemma 4.4, and by (4.4)

$$\arg \frac{d\gamma}{ds} = \arg \varphi'(e^{i\theta(s)}) + \frac{\pi}{2} + \theta(s). \tag{4.6}$$

Since $\arg \varphi' \in C^{k-1+\alpha}$ and $\theta' \in C^{k-1+\alpha}$, we conclude from Corollary 3.3 that $d\gamma/ds \in C^{k-1+\alpha}$ and that Γ is of class $C^{k+\alpha}$. Therefore (a) holds.

Now assume (a) holds, that is, assume Γ is of class $C^{k+\alpha}$. Then by Theorem 4.2, $\arg \varphi' \in C$ and by (4.6), $d\gamma(s(\theta))/ds \in C$. If $k = 1$, so that $d\gamma/ds \in C^\alpha$, then

$$\left| \frac{d\gamma}{ds}(s(\theta_1)) - \frac{d\gamma}{ds}(s(\theta_2)) \right| \leq C|s(\theta_1) - s(\theta_2)|^\alpha \tag{4.7}$$

so that by (4.5)

$$\left| \frac{d\gamma}{ds}(s(\theta_1)) - \frac{d\gamma}{ds}(s(\theta_2)) \right| \leq C|\theta_1 - \theta_2|^{(1-\varepsilon)\alpha}$$

for any $\varepsilon > 0$. Thus $\arg \varphi' \in C^{(1-\varepsilon)\alpha}$ and by Corollary 3.3, $\varphi' \in C^{(1-\varepsilon)\alpha}$. Thus $s'(\theta) = |\varphi'(e^{i\theta})| \in C^{(1-\varepsilon)\alpha}$ and $|s(\theta_1) - s(\theta_2)| \leq K|\theta_1 - \theta_2|$. But then by (4.6) and (4.7), $\arg \varphi' \in C^\alpha$. That proves (b) when $k = 1$. Moreover, by Theorem 3.2 we have $s'(\theta) = |\varphi'(e^{i\theta})| \in C^\alpha$, and $s \in C^{1+\alpha}$. If $k \geq 2$, we use induction. If $d\gamma/ds \in C^{k-1+\alpha}$ and if $s(\theta) \in C^{k-1+\alpha}$, then by (4.6), $\arg \varphi' \in C^{k-1+\alpha}$, so that $s' = |\varphi'| \in C^{k-1+\alpha}$, again by Theorem 3.2, and $s \in C^{k+\alpha}$. That gives (b) in general.   ∎

Let l be a non-negative integer and let $0 \leq \beta < 1$. If Γ is of class $C^{k+\alpha}$ and if $f \in C(\Gamma)$ we say $f \in C^{l+\beta}(\Gamma)$ if $f(\gamma(s)) \in C^{l+\beta}$, when viewed as a function of arc length on Γ.

Corollary 4.5. *Suppose Γ is of class $C^{k+\alpha}$, where $k + \alpha > 1$, and suppose $f \in C^{l+\beta}(\Gamma)$. Set $n + \sigma = \min(k + \alpha, \, l + \beta)$, where $0 < \sigma < 1$ and n is a non-negative integer. Let φ be a conformal map of $\mathbb{D}$ onto Ω, let G be the Herglotz integral of $f \circ \varphi$, and let $F = G \circ \varphi^{-1}$. Then $F \in C^{n+\sigma}(\overline{\Omega})$.*

Proof. Use Corollary 3.3, Theorem 4.3, and Lemma 4.4.   ∎

The same result holds for finitely connected Jordan domains whose boundary curves are of class $C^{k+\alpha}$, except that conjugate functions and Herglotz integrals cannot be defined in multiply connected domains.

Corollary 4.6. *Let $\partial\Omega$ be a finite union of pairwise disjoint $C^{k+\alpha}$ Jordan curves, where $k + \alpha > 1$, and let $f \in C(\partial\Omega)$ be a $C^{l+\beta}$ function of arc length on each component of $\partial\Omega$. Set $n + \sigma = \min(k + \alpha, \, l + \beta)$, where $0 < \sigma < 1$ and n is a non-negative integer. Then $u(z) = u_f(z)$ and its first n partial derivatives extend continuously to $\overline{\Omega}$ and*

$$|D^n u(z_1) - D^n u(z_2)| \leq K|z_1 - z_2|^\sigma,$$

for all $z_1, z_2 \in \overline{\Omega}$, where D^n denotes any n-th partial derivative.

Proof. By Theorem 4.1 it is enough to work in some neighborhood of $\zeta \in \partial\Omega$. Let J be the component of $\partial\Omega$ such that $\zeta \in J$. Let Ω_1 be that component of $\mathbb{C}^* \setminus J$ such that $\Omega \subset \Omega_1$, and let u_1 be the solution to the Dirichlet problem on Ω_1 with boundary value f. Near J, u_1 has the required smoothness by Corollary 4.5. If φ_1 is a conformal map of $\mathbb{D}$ on Ω_1, then $v = (u - u_1) \circ \varphi_1$ is harmonic on $A = \{r < |z| < 1\}$, for some $r < 1$, v is continuous on $\overline{A}$, and $v = 0$ on $\{|z| = 1\}$. By reflection v extends to be harmonic across ∂D. Hence $v \in C^\infty$ and

$$u = u_1 + v \circ \varphi_1^{-1}$$

has as much smoothness as u_1 and φ_1 both have.   ∎

Corollary 4.6 answers Question 1 from the end of Section 2. The next corollary answers Question 2.

Corollary 4.7. *If $\partial\Omega$ consists of finitely many pairwise disjoint Jordan curves of class $C^{1+\alpha}$, where $\alpha > 0$, then*

$$d\omega(z, \zeta) = -\frac{\partial g(z, \zeta)}{\partial n_\zeta} \frac{ds(\zeta)}{2\pi}. \qquad (4.8)$$

In other words, harmonic measure for $z \in \Omega$ is absolutely continuous to arc length on $\partial\Omega$, and the density

$$\frac{d\omega}{ds} = -\frac{1}{2\pi} \frac{\partial g(\zeta, z)}{\partial n_\zeta}$$

is of class $C^\alpha(\partial\Omega)$ and satisfies

$$c_1 < \frac{d\omega}{ds} < c_2 \qquad (4.9)$$

for positive constants c_1 and c_2.

Proof. Let φ be a conformal map from Ω^* onto Ω, where $\partial\Omega^*$ consists of analytic Jordan curves. If $\zeta = \varphi(\zeta^*)$ and $z = \varphi(z^*)$, then

$$\frac{\partial g_\Omega(\zeta, z)}{\partial n_\zeta} = \frac{\partial g_{\Omega^*}(\zeta^*, z^*)}{\partial n_{\zeta^*}} \frac{1}{|\varphi'(\zeta^*)|}$$

and

$$|\varphi'(\zeta^*)| = \frac{ds(\zeta)}{ds(\zeta^*)},$$

by Corollary 4.6 and the uniqueness of Green's function. Since harmonic measure is conformally invariant, (4.8) now follows from the case when $\partial\Omega$ is analytic, as in Corollary 2.6. By Corollary 4.6, $\partial g/\partial n_\zeta \in C^\alpha(\partial\Omega)$. And (4.9) holds because $|\varphi'| > 0$ on $\partial\Omega^*$, by Theorem 4.3.   ■

This answers Question 2 from Section 2. When $\partial\Omega$ is of class C^1, harmonic measure is absolutely continuous to arc length, but the density may not be continuous, bounded or bounded below. See Exercise 14.

Exercise 20 shows how Green's theorem can be applied when $\partial\Omega$ is of class $C^{1+\alpha}$, without first mapping to a domain with real analytic boundary.

Notes

This chapter is based on Lennart Carleson's 1985 lectures at UCLA. The conformal maps in Figure II.2 were computed using the Zipper algorithm from Marshall [1993]. Stein [1970] gives a good introduction to the scale of Lipschitz spaces beyond C^α and Z^*. For more complete discussions see also Adams [1975], Adams and Hedberg [1996], Bennett and Sharpley [1988], Stein [1993], and Triebel [1983] and [1992]. See Zygmund [1959] for the classical picture. Theorem 4.3 and its corollaries are due to Kellogg [1929]. Other boundary problems are discussed in Appendix B and a different approach, through the Dirichlet principle, is presented in Appendix C.

Exercises and Further Results

1. Suppose f is real valued, bounded, and Borel measurable on $\partial\Omega$, where Ω is a finitely connected Jordan domain and let $u = u_f$ be the solution to the Dirichlet problem on Ω for f given by (1.4).

(a) Prove

$$\liminf_{\partial\Omega\ni\zeta\to\zeta_0} f(\zeta) \le \liminf_{\Omega\ni z\to\zeta_0} u(z) \le \limsup_{\Omega\ni z\to\zeta_0} u(z) \le \limsup_{\partial\Omega\ni\zeta\to\zeta_0} f(\zeta).$$

(b) If f is continuous at $\zeta \in \partial\Omega$, prove $\lim_{z\to\zeta} u(z) = f(\zeta)$.

Hint: Use Theorem I.1.3, Carathéodory's theorem, and Theorem 1.1 applied to a piecewise constant function.

2. A numerical routine for solving the Dirichlet problem in a finitely connected Jordan domain bounded by $C^{1+\alpha}$ curves can be based on the following construction: With u_n and u_{n+1} as the proof of Theorem 1.1, find functions h and k, explicitly in terms of the data f and Green's functions g and g' of the simply connected domains Ω_1 and Ω_1', such that

$$u_{n+1} = h + \int_{\sigma'} k u_n ds \quad \text{on } \sigma'.$$

It follows that u satisfies the integral equation

$$u = h + \int_{\sigma'} k u ds \quad \text{on } \Omega.$$

3. (a) Let Ω be a finitely connected Jordan domain, let E be a finite union of open arcs in $\partial\Omega$ and let $f(z)$ be a bounded analytic function on Ω. If

$$|f(z)| \le M, \qquad z \in \Omega,$$

and

$$\limsup_{z \to \zeta} |f(z)| \le m, \qquad \zeta \in E,$$

then

$$|f(z)| \le m^{\omega} M^{1-\omega}, \qquad z \in \Omega,$$

where $\omega = \omega(z, E, \Omega)$. Hint: Apply Exercise I.3(b) with $\log |f(z)|$.

(b) On the annulus $\{r_1 < |z| < r_2\}$ show the circle $\{|z| = r_2\}$ has harmonic measure

$$\frac{\log |z| - \log r_1}{\log r_2 - \log r_1}.$$

(c) (three circles theorem) Suppose that $f(z)$ is analytic on the annulus $\{r_1 < |z| < r_2\}$, and suppose that

$$m_j = \sup_{|\zeta|=r_j} \left\{ \limsup_{z \to \zeta} |f(z)| \right\}$$

is finite for $j = 1, 2$. Then

$$\sup_{|z|=r} |f(z)| \le m_1^{\lambda(r)} m_2^{1-\lambda(r)},$$

where

$$\lambda(r) = \frac{\log r_2 - \log r}{\log r_2 - \log r_1}.$$

(d) (Lindelöf) Suppose $f(z)$ is bounded and analytic on the disc $\mathbb{D}$. Let γ be an arc in $\mathbb{D}$ terminating at a point $\zeta \in \partial\mathbb{D}$, and assume

$$\lim_{\gamma \ni z \to \zeta} f(z) = a.$$

Then $f(z)$ has nontangential limit a at ζ. Hint: Take $a = 0$. If Γ is a nontangential cone terminating at ζ, then $\liminf_{z \in \partial\Gamma} \omega(z, \gamma) > 0$. Then (a) can be used.

(e) (Lindelöf) Suppose $f(z)$ is bounded and analytic on $\mathbb{D}$. If the nontangential limit $f(e^{i\theta})$ is continuous in some deleted arc $(\alpha - \delta, \alpha) \cup (\alpha, \alpha + \delta)$ and if the two limits

$$\lim_{\theta \uparrow 0} f(e^{i(\alpha+\theta)})$$

and

$$\lim_{\theta \downarrow 0} f(e^{i(\alpha+\theta)})$$

both exist, then these two limits are equal.

(f) Let φ be a univalent function on $\mathbb{D}$. If σ is an arc in $\mathbb{D}$ with endpoint $\zeta \in \partial\mathbb{D}$ and if

$$\lim_{\sigma \ni z \to \zeta} \varphi(z) = w,$$

then φ has nontangential limit w at ζ.

4. Suppose Ω is a finitely connected Jordan domain, and suppose Ω_n are finitely connected Jordan domains bounded by disjoint $C^{1+\alpha}$ Jordan curves such that $\overline{\Omega_n} \subset \Omega_{n+1} \subset \Omega$ and $\bigcup \Omega_n = \Omega$. Fix $z \in \Omega$ and let $d\omega_n$ denote harmonic measure for z with respect to the region Ω_n. Prove that $d\omega_n$ converges weak-star to $d\omega$ where $d\omega$ is harmonic measure for z with respect to Ω. In other words, if $f \in C(\overline{\Omega})$, then prove

$$\lim_n \int_{\partial\Omega_n} f \, d\omega_n = \int_{\partial\Omega} f \, d\omega.$$

Hint: If u_f is the harmonic extension of $f|_{\partial\Omega}$ to Ω, $\lim \int (u_f - f) d\omega_n = 0$ and thus any weak-star cluster point $d\mu$ of $d\omega_n$ satisfies $\int_{\partial\Omega} f \, d\mu = u_f(z)$. So by the Riesz representation theorem and (1.2), $d\mu = d\omega$.

5. Let Ω be a finitely connected domain such that $\partial\Omega$ has no isolated points.

(a) Prove there is a conformal map φ of Ω onto a domain Ω^* bounded by finitely many pairwise disjoint analytic Jordan curves.

(b) Define $g_\Omega(z, w) = g_{\Omega^*}(\varphi(z), \varphi(w))$, and verify (2.2), (2.4), (2.5), and

$$\lim_{\Omega \ni z \to \zeta \in \partial\Omega} g_\Omega(z, w) = 0. \tag{2.3'}$$

(c) Then (2.3'), (2.4), and (2.5) uniquely determine g_Ω and this definition of g_Ω is independent of the choice of Ω^* and φ.

(d) Using the preceding results and the Riemann mapping theorem, derive the conclusion of Theorem 2.1 for any simply connected domain Ω not bounded by a Jordan curve.

6. Let Ω be a simply connected plane domain and assume that there exists a (Green's) function $g_\Omega(z, w)$ on $\Omega \times \Omega \setminus \{z = w\}$ satisfying (2.3'), (2.4), and (2.5). Let $u(z) = g(z, w) + \log|z - w|$. Prove the Riemann mapping theorem by showing that

$$\varphi(z) = \varphi(z, w) = (z - w)e^{-(u(z) + i\tilde{u}(z))}$$

defines a conformal mapping from Ω onto $\mathbb{D}$. Hint: To show φ is one-to-one, compare $-\log|T \circ \varphi(z, w)|$ to $g(z, w')$ for some $T \in \mathcal{M}$.

7. (a) If Ω is a doubly connected Jordan domain, if E is one of the boundary

components, and if $\omega = \omega(z, E, \Omega)$, show that for some choice of $a > 0$,

$$\varphi = e^{a(\omega + i\tilde{\omega})}$$

defines a single valued conformal map of Ω onto an annulus.

(b) Let Ω be a domain such that $\mathbb{C}^* \setminus \Omega$ has exactly two components. Prove there is a conformal mapping from Ω onto a ring $\{z : r_1 < |z| < r_2\}$ where $0 \le r_1 < r_2 \le \infty$.

(c) Let Ω be a finitely connected Jordan domain and suppose $z_0 \in \Omega$. Set

$$u_\varepsilon(z) = \log(1/\varepsilon)\omega(z, B_\varepsilon, \Omega \setminus B_\varepsilon),$$

where $B_\varepsilon = \{z : |z - z_0| \le \varepsilon\}$. Then

$$|u_\varepsilon - g_\Omega(z, z_0)| \le C\omega(z, B_\varepsilon, \Omega \setminus B_\varepsilon),$$

and hence $u_\varepsilon \to g_\Omega(z, z_0)$, uniformly on compact subsets of $\Omega \setminus \{z_0\}$.

(d) Let Ω be a finitely connected Jordan domain bounded by $C^{1+\alpha}$ curves. The period of the conjugate $\tilde{g}_\Omega(z, z_0)$ of Green's function around a component Γ_j of $\partial\Omega$ is equal to the harmonic measure of Γ_j at z_0.

8. (a) Let $u(z)$ be the Poisson integral of $f \in L^2$. Use Green's theorem to show

$$\frac{1}{2\pi} \int |f - f(0)|^2 d\theta = \frac{2}{\pi} \iint_{\mathbb{D}} |\nabla u(z)|^2 \log\frac{1}{|z|} dx dy$$

$$\ge \frac{2}{\pi} \iint_{\mathbb{D}} |\nabla u(z)|^2 (1 - |z|) dx dy$$

$$\ge \frac{4}{3\pi} \iint_{\mathbb{D}} |\nabla u(z)|^2 \log\frac{1}{|z|} dx dy.$$

Hints: Replace f by $u(re^{i\theta})$, use the identity $\Delta(u^2) = 2|\nabla u|^2$ and Green's theorem, and send $r \to 1$. The final inequality is a direct calculation using polar coordinates and Fourier series. The identity on the left can also be proved with Fourier series.

(b) If $f \in H^2$ (see Appendix A), then by (a)

$$\frac{1}{2\pi} \int |f - f(0)|^2 d\theta = \frac{2}{\pi} \iint_{\mathbb{D}} |f'(z)|^2 \log\frac{1}{|z|} dx dy$$

$$\ge \frac{2}{\pi} \iint_{\mathbb{D}} |f'(z)|^2 (1 - |z|) dx dy$$

$$\ge \frac{4}{3\pi} \iint |f'(z)|^2 \log\frac{1}{|z|} dx dy.$$

In particular, the two area integrals are comparable.

9. (a) In Theorem 2.7, verify (2.16) when $f \in L^1(ds)$.

(b) Prove (2.20) by following the conformal mapping proof of (2.16) given in the text.

10. (a) Prove that every analytic Jordan curve is the image of $\partial \mathbb{D}$ under an analytic map that is one-to-one on a neighborhood of $\partial \mathbb{D}$.

(b) If Ω is a domain in $\mathbb{C}^*$ then we can find subdomains Ω_n, each bounded by finitely many pairwise disjoint analytic Jordan curves so that $\overline{\Omega_n} \subset \Omega_{n+1}$ and $\Omega = \bigcup \Omega_n$. Hint: Use the proof of Lemma 2.2.

11. (a) If $f \in L^1(\partial \mathbb{D})$, and if u is the Poisson integral of f, then

$$\tilde{u}(re^{i\varphi}) = \int_{-\pi}^{\pi} \frac{2r\sin(\varphi - \theta)}{1 - 2r\cos(\varphi - \theta) + r^2} f(e^{i\theta}) \frac{d\theta}{2\pi}.$$

(b) Almost everywhere on $\partial \mathbb{D}$,

$$\tilde{f}(\varphi) = \lim_{\varepsilon \to 0} \int_{|\theta - \varphi| > \varepsilon} \cot\left(\frac{\varphi - \theta}{2}\right) f(e^{i\theta}) \frac{d\theta}{2\pi}.$$

(c) Define the **modulus of continuity** of the function f as

$$\omega_f(\delta) = \sup\{|f(\theta) - f(\varphi)| : |\theta - \varphi| < \delta\}.$$

The function f is called **Dini continuous** if

$$\int_0 \frac{\omega_f(t)}{t} dt < \infty.$$

Prove that $\tilde{f}$ is continuous if f is Dini continuous, in fact,

$$\omega_{\tilde{f}}(\delta) \leq \int_0^{\delta} \frac{\omega(t)}{t} dt + \delta \int_{\delta}^{\pi} \frac{\omega(t)}{t^2} dt.$$

(d) Let $h(\theta)$ be odd and increasing on $(-\pi, \pi]$. Then $\lim_{r \to 1} \tilde{h}(r)$ exists and is finite if and only if

$$\int_0 \frac{h(\theta)}{\theta} d\theta < \infty.$$

(e) Use (d) to show (c) is sharp.

12. Let Ω be a simply connected domain. Write $d(w) = \text{dist}(w, \partial \Omega)$ and let $\rho(w, w_0)$ denote the hyperbolic distance in Ω. Let $\varphi : \mathbb{D} \to \Omega$ be conformal. Then $\varphi \in C^{\alpha}(\partial \mathbb{D})$ if and only if

$$\limsup_{d(w) \to 0} \left(2\rho(w, w_0) + \frac{1}{\alpha} \log d(w)\right) < \infty$$

for any $w_0 \in \Omega$. Hint: Use Theorem 3.2 and Theorem I.4.3. See Becker and Pommerenke [1982a].

13. Corollary 3.3 is false when $k = 1$ and $\alpha = 0$. Worse yet, there exists $f \in C^1(\partial\mathbb{D})$ such that $\tilde{f}$ is not a Lipschitz function.

14. There is a simply connected domain Ω such that $\partial\Omega \in C^1$ but such that no conformal map $\varphi : \mathbb{D} \to \Omega$ is class C^1 on $\overline{\mathbb{D}}$. Worse yet, $|\varphi'|$ can have infinite nontangential limit at some point on $\partial\mathbb{D}$.

15. In Theorem 4.2, if Γ has a continuous tangent, then $\varphi' \in H^p$ for all finite p. In other words,

$$\sup_{0<r<1} \int |\varphi'(re^{i\theta})|^p d\theta < \infty.$$

See Appendix A.

16. Let Ω be a Jordan domain with C^1 boundary Γ parameterized by arc length

$$\Gamma = \{\gamma(s) : 0 < s < \ell(\gamma)\},$$

and let φ be a conformal map from $\mathbb{D}$ onto Ω.
(a) If γ is Dini continuous, then φ' extends continuously to $\overline{\mathbb{D}}$ and $\varphi' \neq 0$. See Warschawski [1932a] and [1961].
(b) If $\gamma \in C^1$ and if γ' is Dini continuous, then φ'' extends continuously to $\overline{\mathbb{D}}$. See Warschawski [1932a].

17. Let Ω be a finitely connected Jordan domain with $n + 1$ boundary components $\Gamma_0, \Gamma_1, \ldots, \Gamma_n$ and let $u(z) = \omega(z, \Gamma_j, \Omega)$. Prove that $u(z)$ has $n - 1$ critical points in Ω. In other words, prove the analytic function $f(z) = u_x(z) - iu_y(z)$ has $n - 1$ zeros in Ω, when counted with multiplicities. Hint: We may assume each Γ_k is an analytic curve. Then $f \neq 0$ on $\partial\Omega$ and since u is constant on Γ_k, the vector $\nabla u = \overline{f}$ is parallel to the normal vector on Γ_k. Hence $\arg f$ decreases by 2π as Γ_k is traversed counterclockwise, and the argument principle then shows that f has $n - 1$ zeros in Ω.

18. Prove that conditions (a) and (b) of Theorem 4.3 are equivalent even when $\alpha = 0$.

19. (a) Formulate a notion of nontangential convergence for a finitely connected Jordan domain with C^1 boundary. Use it to show that $\omega(z, E, \Omega)$ is the unique bounded harmonic function on Ω with nontangential limits 1 on E and 0 on $\partial\Omega \setminus E$, almost everywhere with respect to arc length on $\partial\Omega$.
(b) If $\partial\Omega$ is $C^{1+\varepsilon}$ for some $\varepsilon > 0$, then under the conformal mapping cones $\Gamma_\alpha(\zeta)$ correspond to cones $\Gamma_\alpha(\varphi(\zeta))$, and then (2.16) holds for $f \in L^1(ds)$.

20. Green's theorem can be used on a finitely connected domain bounded by a finite number of pairwise disjoint $C^{1+\alpha}$ curves with functions u and v in $C^1(\partial\Omega) \cap C^2(\Omega)$. Hint: Use a finite number of Riemann maps and Theorem 4.3.

21. More generally, Green's theorem also holds for Lipschitz domains, which will be defined in this exercise. A K-**Lipschitz graph** is a set of the form

$$\Gamma = \{(x, \beta(x)) : |x| \le M\},$$

where β is a real valued Lipschitz continuous function:

$$|\beta(x_2) - \beta(x_1)| \le K|x_2 - x_1|,$$

for some constant K. A Lipschitz graph is rectifiable, and since β is absolutely continuous, arc length on Γ has the form

$$ds = \sqrt{1 + (\beta'(x))^2}dx,$$

and almost everywhere,

$$n = \frac{(\beta'(x), -1)}{\sqrt{1 + (\beta'(x))^2}}$$

is a vector normal to Γ. For $B > \sup_{[-M,M]} |\beta(x)|$, set

$$\Omega = \Omega_\beta = \{(x, y) : |x| < M, \ \beta(x) < y < B\}.$$

(a) Prove that Green's theorem holds on Ω in this form: Suppose $u \in C^1(\Omega)$ and $v \in C^2(\Omega)$, suppose u, ∇u, v, ∇v, and Δv are bounded on Ω, and suppose u and ∇v extend continuously to $\overline{\Omega}$. Then

$$\int\int_\Omega (u\Delta v + \nabla u \cdot \nabla v)dxdy = \int_{\partial\Omega} u\frac{\partial v}{\partial n}ds. \qquad (E.1)$$

Hint: Approximate Ω by C^∞ subdomains as follows: Let $X \in C^\infty(-1, 1)$ satisfy $X \ge 0$ and

$$\int_{-1}^{1} X dx = 1.$$

If $X_\delta(x) = \frac{1}{\delta}X(\frac{x}{\delta})$ and $\beta^{(\delta)}(x) = X_\delta * \beta(x) \in C^\infty$, $\beta^{(\delta)}(x) \to \beta(x)$ uniformly, and $\nabla\beta^{(\delta)}(x) \to \nabla\beta(x)$ almost everywhere. Show there are $\delta_n \downarrow 0$ such that $\beta_n = \delta_n + \beta^{(\delta_n)}$ satisfies $\beta_n > \beta_{n+1} > \beta$. Then derive (E.1) from Green's theorem for Ω_{β_n}.

(b) A finitely connected Jordan domain Ω is a **Lipschitz domain** if each $p \in \partial\Omega$ has a neighborhood V so that $V \cap \partial\Omega$ is the image of a Lipschitz graph under a linear mapping. Prove Green's theorem (E.1) holds for the Lipschitz domain Ω under the same smoothness assumptions on u and v that we made in (a). That is, $u \in C^1(\Omega)$; and $v \in C^2(\Omega)$; u, ∇u, v, ∇v, and Δv are bounded on Ω; and u and ∇v extend continuously to $\overline{\Omega}$.

III

Potential Theory

The goal of this chapter is to solve the Dirichlet problem on an arbitrary plane domain Ω. There are three traditional ways to solve this problem:

(i) The Wiener method is to approximate Ω from inside by subdomains Ω_n of the type studied in Chapter II and to show that the harmonic measures $\omega(z, E, \Omega_n)$ converge weak-star to a limit measure on $\partial\Omega$. With Wiener's method one must prove that the limit measure $\omega(z, E, \Omega)$ does not depend on the approximating sequence Ω_n.

(ii) The Perron method associates to any bounded function f on $\partial\Omega$ a harmonic function $\mathcal{P}_f$ on Ω. The function $\mathcal{P}_f$ is the upper envelop of a family of subharmonic functions constrained by f on $\partial\Omega$. Perron's method is elegant and general. With Perron's method the difficulty is linearity; one must prove that $\mathcal{P}_{-f} = -\mathcal{P}_f$, at least for f continuous.

(iii) The Brownian motion approach, originally from Kakutani [1944a], identifies $\omega(z, E, \Omega)$ with the probability that a randomly moving particle, starting at z, first hits $\partial\Omega$ in the set E. This method has considerable intuitive appeal, but it leaves many theorems hard to reach.

We follow Wiener and use the energy integral to prove that the limit $\omega(z, E, \Omega)$ is unique. This leads to the notions of capacity, equilibrium distribution, and regular point and to the characterization of regular points by Wiener series.

For the Perron method see Ahlfors [1979] or Tsuji [1959]. Appendix F below includes Kakutani's theorem for the discrete version of Brownian motion.

We conclude the chapter with some potential theoretic estimates for harmonic measure.

1. Capacity and Green's Functions

Let E be a compact plane set such that $\Omega = \mathbb{C}^* \setminus E$ is a finitely connected Jordan domain. By Chapter II and a conformal mapping, we see that Ω has Green's function $g_\Omega(z, \infty)$ with pole at ∞, and if $a \notin \overline{\Omega}$

$$g_\Omega(z, \infty) = \log|z - a| + h(z, \infty),$$

where $h(z, \infty)$ is harmonic on Ω and continuous on $\partial\Omega$ and

$$h(\zeta, \infty) = -\log|\zeta - a| \qquad \zeta \in \partial\Omega.$$

(Recall that $u(z)$ is harmonic at ∞ if $u(1/z)$ is harmonic on a neighborhood of 0.) The quantity

$$\gamma = \gamma(E) = h(\infty, \infty),$$

is called **Robin's constant** for E, and we have

$$g_\Omega(z, \infty) = \log|z| + \gamma + o(1), \qquad \text{as } z \to \infty. \tag{1.1}$$

Define the **logarithmic capacity** of E to be

$$\text{Cap}(E) = e^{-\gamma(E)}.$$

Thus $\text{Cap}(E) > 0$ in the case at hand.

Suppose Ω_1 and Ω_2 are finitely connected Jordan domains such that $\infty \in \Omega_j$ and set $E_j = \mathbb{C}^* \setminus \Omega_j$. Assume there is a conformal map ψ of Ω_1 onto Ω_2, such that for $|z|$ large

$$\psi(z) = az + b_0 + \frac{b_1}{z} + \dots$$

with $a > 0$. Then

$$g_{\Omega_1}(z, \infty) = g_{\Omega_2}(\psi(z), \infty),$$

so that by (1.1)

$$\gamma(E_1) = \gamma(E_2) + \log a$$

and

$$\text{Cap}(E_2) = a\text{Cap}(E_1). \tag{1.2}$$

In particular, the capacity of a closed disc is the radius of the disc, because $g_{\mathbb{C}^* \setminus \overline{\mathbb{D}}}(z, \infty) = \log|z|$.

Now let E be any compact plane set and write Ω for the component of $\mathbb{C}^* \setminus E$ such that $\infty \in \Omega$. Fix a sequence $\{\Omega_n\}$ of finitely connected domains such that

$$\infty \in \Omega_n \subset \overline{\Omega_n} \subset \Omega_{n+1} \subset \Omega,$$

such that

$$\Omega = \bigcup \Omega_n,$$

and such that $\partial \Omega_n$ consists of $C^{1+\alpha}$ Jordan curves for some $\alpha > 0$. See Exercise II.10(b). Define $E_n = \mathbb{C} \setminus \Omega_n$. As we shall soon see, all our results will be independent of the sequence $\{\Omega_n\}$.

Because $\overline{\Omega_n} \subset \Omega_{n+1}$, it follows from from the maximum principle that $g_{\Omega_{n+1}}(z, \infty) > g_{\Omega_n}(z, \infty)$ on Ω_n, and hence that $\gamma(E_{n+1}) > \gamma(E_n)$ and $\text{Cap}(E_{n+1}) < \text{Cap}(E_n)$. Now define

$$\text{Cap}(E) = \lim_n \text{Cap}(E_n). \tag{1.3}$$

Because $g_\Omega(z, w)$ is an increasing function of Ω, an interlacing of the domains Ω_n shows that the definition (1.3) does not depend on the choice of the sequence $\{\Omega_n\}$. Note that if $\widehat{E} = \mathbb{C} \setminus \Omega$, then by definition,

$$\text{Cap}(\partial E) = \text{Cap}(E) = \text{Cap}(\widehat{E}) = \text{Cap}(\partial \widehat{E}). \tag{1.4}$$

By definition, the quantity **Robin's constant**

$$\gamma(E) \equiv \log\left(\frac{1}{\text{Cap}(E)}\right) = \lim_n \gamma(E_n)$$

is **Robin's constant** for the arbitrary compact set E. If $E \subset F$ then by (1.3)

$$\text{Cap}(E) \leq \text{Cap}(F) \quad \text{and} \quad \gamma(E) \geq \gamma(F). \tag{1.5}$$

If $\text{Cap}(E) > 0$, then $\lim_n \gamma(E_n) = \gamma(E) < \infty$, and by Harnack's principle

$$g_\Omega(z, \infty) = \lim_n g_{\Omega_n}(z, \infty)$$

defines a positive harmonic function on Ω having expansion

$$g_\Omega(z, \infty) = \log |z| + \gamma(E) + o(1) \tag{1.6}$$

at ∞. When $z = \infty$, the symmetry (2.8) of Chapter II shows that

$$\lim_n g_{\Omega_n}(\infty, w) = \lim_n g_{\Omega_n}(w, \infty)$$

exists for all $w \in \Omega \setminus \{\infty\}$. Thus by Harnack's principle the limit

$$g_\Omega(z, w) = \lim_n g_{\Omega_n}(z, w) \tag{1.7}$$

exists for all $z, w \in \Omega$ with $z \neq w$, and $g_\Omega(z, w)$ satisfies conditions (2.2), (2.4), and (2.5) of Chapter II. The function $g_\Omega(z, w)$ is **Green's function** for Ω **with pole at** w.

Example 1.1. Suppose E is compact and connected. Let Ω be the component of $\mathbb{C}^* \setminus E$ such that $\infty \in \Omega$ and let $\psi : \Omega \to \mathbb{C}^* \setminus \overline{\mathbb{D}}$ be the conformal mapping such that for $|z|$ large,

$$\psi(z) = az + b_0 + \frac{b_1}{z} + \dots,$$

with $a > 0$. For $r > 1$,

$$\Omega_r = \{z : |\psi(z)| > r\}$$

is bounded by an analytic Jordan curve and $g_{\Omega_r}(z, \infty) = \log |\psi(z)/r|$. Then by (1.7), $g_\Omega(z, \infty) = \log |\psi(z)|$ and by (1.2),

$$1 = \operatorname{Cap}(\overline{\mathbb{D}}) = a\operatorname{Cap}(E).$$

For example, the normalized conformal map $\varphi = \psi^{-1}$ of $\mathbb{C}^* \setminus \overline{\mathbb{D}}$ onto the complement of the interval $[-2, 2]$ is

$$\varphi(z) = z + \frac{1}{z}$$

and so

$$\operatorname{Cap}([-2, 2]) = 1.$$

Consequently

$$\operatorname{Cap}([\alpha, \beta]) = \frac{\beta - \alpha}{4}$$

for every interval $[\alpha, \beta] \subset \mathbb{R}$, by (1.2).

Example 1.2. *If $E \subset \partial\mathbb{D}$, then $\operatorname{Cap}(E) \geq \sin(|E|/4)$.*

Proof. If E is an arc on $\partial\mathbb{D}$, then after a conformal mapping, Example 1.1 gives $\operatorname{Cap}(E) = \sin(|E|/4)$. Because of (1.3) we may assume E is a finite union of arcs. Define

$$F(z) = \frac{1}{4} \int_E \frac{e^{i\theta} + z}{e^{i\theta} - z} d\theta,$$

and let $\omega(z) = \omega(z, E, \mathbb{D})$. Then $\overline{F(1/\bar{z})} = -F(z)$ and $F = \frac{\pi}{2}(\omega + i\tilde{\omega})$ on $\mathbb{D}$. Therefore

$$-\frac{\pi}{2} \leq \operatorname{Re}F \leq \frac{\pi}{2}$$

on $\Omega = \mathbb{C}^* \setminus E$ and $F(0) = -F(\infty) = |E|/4$. Hence $H(z) = e^{iF(z)}$ maps Ω

into the right half-plane and $H(\infty) = \overline{H(0)}$. Now

$$G(z) = \log\left|z\frac{H(z) + \overline{H(0)}}{H(z) - H(0)}\right|$$

is superharmonic on Ω and

$$\liminf_{z \ni \Omega \to \partial\Omega} G(z) \geq 0.$$

By the maximum principle, $G(z) > 0$ in Ω so that $G(z) \geq g_{\Omega_n}(z, \infty)$ for all n. For $|z|$ large

$$G(z) = \log|z| - \log\sin\left(\frac{|E|}{4}\right) + \ldots,$$

so that by (1.1)

$$-\log\sin\left(\frac{|E|}{4}\right) \geq \gamma(\partial\Omega_n) \to \gamma(E)$$

and $\text{Cap}(E) \geq \sin(|E|/4)$. ∎

2. The Logarithmic Potential

Let μ be a finite, compactly supported signed (Borel) measure. The **logarithmic potential** of μ is the function

$$U_\mu(z) = \int \log\frac{1}{|\zeta - z|}d\mu(\zeta).$$

By Fubini's theorem, the integral U_μ is absolutely convergent for area almost every z.

Lemma 2.1. *If $\mu > 0$, the potential U_μ is lower semicontinuous and superharmonic.*

Proof. The lemma holds because for ζ fixed, the function

$$z \to \log\frac{1}{|\zeta - z|}$$

is lower semicontinuous and superharmonic. ∎

The next theorem connects the notions of logarithmic potential, Green's function, capacity, and harmonic measure. Suppose $\Omega = \mathbb{C}^* \setminus E$ is bounded by a finite family of disjoint $C^{1+\alpha}$ Jordan curves, $\alpha > 0$, write μ_E for the harmonic measure of ∞ relative to Ω,

$$d\mu_E = d\omega(\infty, \cdot, \Omega) = -\frac{1}{2\pi} \frac{\partial g(\zeta, \infty)}{\partial n_\zeta} ds = P_\infty(\zeta) ds.$$

Theorem 2.2. *If $\Omega = \mathbb{C}^* \setminus E$ is connected and bounded by finitely many pairwise disjoint $C^{1+\alpha}$ Jordan curves, then the integral $U_{\mu_E}(z)$ is absolutely convergent at every $z \in \mathbb{C}$. The potential U_{μ_E} is continuous on $\mathbb{C}$ and*

$$g(z, \infty) = \gamma(E) - U_{\mu_E}(z), \qquad z \in \overline{\Omega}, \tag{2.1}$$

$$U_{\mu_E}(z) < \gamma(E), \qquad z \in \Omega, \tag{2.2}$$

and

$$U_{\mu_E}(z) = \gamma(E), \qquad z \in E = \mathbb{C}^* \setminus \Omega. \tag{2.3}$$

Later we shall see that μ_E is the unique probability measure μ on E such that U_μ is constant on E. For this reason μ_E is called the **equilibrium distribution** of E.

Proof. We can assume $0 \notin \overline{\Omega}$. Clearly, the integral $U_\mu(z)$ is absolutely convergent at all $z \notin \partial\Omega$. Since

$$g(z, \infty) = \log|z| - \int_{\partial\Omega} \log|\zeta| \, d\omega(z, \zeta),$$

we have

$$\gamma(E) = -\int_{\partial\Omega} \log|\zeta| \, d\mu_E(\zeta).$$

On the other hand, for fixed $z_0 \in \Omega$,

$$g(z, z_0) = \log\left|\frac{z}{z - z_0}\right| - \int_{\partial\Omega} \log\left|\frac{\zeta}{\zeta - z_0}\right| d\omega(z, \zeta), \tag{2.4}$$

because the right side of (2.4) satisfies the conditions (2.3), (2.4), and (2.5) from Chapter II and those conditions determine Green's function. Then sending z to ∞ yields

$$g(\infty, z_0) = \gamma(E) - \int \log \frac{1}{|\zeta - z_0|} d\mu_E(\zeta)$$
$$= \gamma(E) - U_{\mu_E}(z_0).$$

For $z \in \Omega$, (2.1) is then a consequence of the symmetry of Green's function, $g(\infty, z) = g(z, \infty)$. Then because $g(z, \infty) > 0$ on Ω, (2.1) implies (2.2).

For $z \notin \overline{\Omega}$, $v(\zeta) = \log(|\zeta|/|\zeta - z|)$ is harmonic on a neighborhood of $\overline{\Omega}$ and

$$
\begin{aligned}
0 = v(\infty) &= \int \log|\zeta| d\mu_E(\zeta) + \int \log \frac{1}{|\zeta - z|} d\mu_E(\zeta) \\
&= -\gamma(E) + U_{\mu_E}(z).
\end{aligned}
$$

Therefore (2.3) holds for $z \notin \overline{\Omega}$.

If $z \in \partial\Omega$, the lower semicontinuity of U_μ gives

$$
U_\mu(z) \leq \liminf_{\Omega \ni w \to z} U_\mu(w) \leq \gamma(E) < \infty.
$$

Since the integrand is bounded below, that means the integral $U_\mu(z)$ converges absolutely. Because $\partial\Omega$ consists of $C^{1+\alpha}$ curves, Area$(\partial\Omega) = 0$. Then by the superharmonicity of U_μ and the continuity of $g(z, \infty)$,

$$
U_\mu(z) \geq \limsup_{\delta \to 0} \left\{ \int_{\Omega \cap B(z,\delta)} U_\mu(\zeta) \frac{d\xi d\eta}{\pi \delta^2} + \int_{B(z,\delta)\backslash\overline{\Omega}} \gamma \frac{d\xi d\eta}{\pi \delta^2} \right\} = \gamma.
$$

Consequently (2.1) and (2.3) hold at all $z \in \partial\Omega$, and it follows that U_μ is continuous in $\mathbb{C}$. ∎

Let E be a compact set with Cap$(E) > 0$, and let $E_n = \mathbb{C}^* \setminus \Omega_n$ be as in Section 1. Then by (1.7) and Theorem 2.2 any weak-star cluster point μ_E of the sequence $\{\mu_{E_n}\}$ satisfies both (2.1) and (2.2) on Ω. In Section 4 we will use the energy integral to show that there is a unique weak-star limit μ_E independent of the sequence E_n and to establish a version of (2.3) for U_{μ_E} on E. A different proof of the uniqueness of the weak-star limit $\{\mu_{E_n}\}$ for a bounded domain Ω is given in Exercise 4.

3. The Energy Integral

Let v be a signed measure with compact support. If

$$
\iint \left| \log \frac{1}{|z - \zeta|} \right| d|v|(\zeta) \, d|v|(z) < \infty, \tag{3.1}
$$

we say v has **finite energy** and define the **energy integral** $I(v)$ by

$$
I(v) = \iint \log \frac{1}{|z - \zeta|} dv(\zeta) dv(z) = \int U_v(z) dv(z).
$$

The energy integral has a very important property: *It is positive definite on the space of signed measures with finite energy and zero integral.*

Theorem 3.1. *If* (3.1) *holds and if $\int dv = 0$, then $I(v) \geq 0$. Moreover, if $I(v) = 0$, then $v \equiv 0$.*

Proof. Write $L(z) = \log \frac{1}{|z|}$. By Green's theorem

$$f(z) = -\frac{1}{2\pi} \iint L(z-w)\Delta f(w)du\,dv \qquad (3.2)$$

whenever $f \in C^\infty$ has compact support.

First consider the special case of an absolutely continuous signed measure $dv = h(z)dx\,dy$ where $h \in C^\infty$ has compact support and satisfies

$$\iint h(z)dx\,dy = 0. \qquad (3.3)$$

Then the convolution $U_v = L * h$ is also C^∞ and by (3.3),

$$|U_v(z)| \leq \frac{C}{|z|} \qquad (3.4)$$

and

$$|\nabla U_v(z)| \leq \frac{C}{|z|^2} \qquad (3.5)$$

for $|z|$ large. For any $f \in C^\infty$ with compact support, Green's theorem (3.2) and Fubini's theorem give

$$\iint \Delta U_v f dx\,dy = \iint U_v \Delta f dx\,dy = -2\pi \iint hf dx\,dy,$$

because h and f have compact support. Therefore

$$\Delta U_v = -2\pi h.$$

Now by (3.4), (3.5), and Green's theorem again

$$I(v) = \iint U_v h dx\,dy = -\frac{1}{2\pi} \iint U_v \Delta U_v dx\,dy = \frac{1}{2\pi} \iint |\nabla U_v|^2 dx\,dy.$$

That shows $I(v) \geq 0$ in this special case. Moreover, if $I(v) = 0$ then $\nabla U_v = 0$ and $h = -\frac{1}{2\pi}\Delta U_v = 0$.

To derive the full Theorem 3.1 from the special case we employ a standard **mollification argument**. Let $K \in C_c^\infty(\mathbb{C})$ be a compactly supported infinitely differentiable function such that

(i) K is radial, $K(z) = K(|z|)$,
(ii) $K \geq 0$, and
(iii) $\int K dx\,dy = 1$.

Set

$$K_\varepsilon(z) = \varepsilon^{-2} K(z/\varepsilon)$$

and let ν_ε be the absolutely continuous measure having density

$$h_\varepsilon(z) = K_\varepsilon * \nu(z) = \int K_\varepsilon(z - w) d\nu(w).$$

Then

$$\lim_{\varepsilon \to 0} \int f d\nu_\varepsilon = \int f d\nu, \qquad (3.6)$$

for all continuous f. Furthermore, $h_\varepsilon \in C^\infty$ has compact support and satisfies (3.3). By Fubini's theorem and (i)

$$I(\nu_\varepsilon) = \iint K_\varepsilon * K_\varepsilon * L(z - \zeta) d\nu(z) d\nu(\zeta),$$

where

$$K_\varepsilon * K_\varepsilon(z) = \int K_\varepsilon(z - w) K_\varepsilon(w) du dv$$

and

$$K_\varepsilon * K_\varepsilon * L(w) = \int \left(K_\varepsilon * K_\varepsilon(z) \right) L(w - z) dx dy.$$

Now because K_ε is a radial function, $K_\varepsilon * K_\varepsilon$ is also a radial function, and therefore since $L(z)$ is superharmonic,

$$K_\varepsilon * K_\varepsilon * L(z) \le L(z). \qquad (3.7)$$

Also, the continuity of $L(z)$ viewed as a map to $(-\infty, \infty]$, gives

$$K_\varepsilon * K_\varepsilon * L(z) \to L(z), \qquad \text{as } \varepsilon \to 0.$$

Since $\int d\nu = 0$,

$$\iint \log \frac{1}{|\alpha z - \alpha \zeta|} d\nu(\zeta) d\nu(z) = \iint \log \frac{1}{|z - \zeta|} d\nu(\zeta) d\nu(z),$$

for any $\alpha > 0$, and we can assume ν has support in $\{z : |z| < 1/2\}$. Therefore by (3.1), (3.7) and dominated convergence,

$$\lim_{\varepsilon \to 0} I(\nu_\varepsilon) = I(\nu),$$

and we see that $I(\nu) \ge 0$.

Now assume $I(\nu) = 0$. Write $U_\varepsilon = U_{\nu_\varepsilon}$. Then

$$\iint |\nabla U_\varepsilon|^2 dx dy = I(\nu_\varepsilon) \to 0. \qquad (3.8)$$

We also have $U_\varepsilon(z) = O\left(\frac{1}{|z|}\right)$ uniformly in ε, because all h_ε, $\varepsilon < 1$, satisfy (3.3) and vanish outside a common compact set. Then by (3.8) and Lemma 3.2 below,

$$\lim_{\varepsilon \to 0} \iint |U_\varepsilon(z)|^2 \, dx dy = 0.$$

Let $f \in C^\infty$ have compact support. Then by (3.6) and Green's theorem once more,

$$\int f dv = \lim_{\varepsilon \to 0} \int f dv_\varepsilon = \lim_{\varepsilon \to 0} \frac{-1}{2\pi} \int \Delta f \, U_\varepsilon dx dy = 0.$$

Therefore $v = 0$.   ∎

Lemma 3.2. *Assume* $U_n(z) \in C^\infty(\mathbb{C})$ *satisfy*

$$|U_n(z)| \leq C/|z|, \tag{3.9}$$

and

$$\|\nabla U_n\|_2 \to 0.$$

Then

$$\iint\limits_K |U_n|^2 dx dy \to 0, \tag{3.10}$$

for every compact set K.

Proof. Assume $K \subset [-L, L] \times [-R, R]$. Then

$$\int_{[-L,L]} |U_n(x, y)|^2 dx \leq 2 \int_{-L}^{L} |U_n(x, y) - U_n(x, -L)|^2 dx + 2C^2/L$$

$$\leq 2 \int_{-L}^{L} \left(\int_{-R}^{y} \left| \frac{\partial U_n}{\partial y} \right| dy \right)^2 dx + 2C^2/L$$

$$\leq 4L \|\nabla U_n\|_2^2 + 2C^2/L,$$

and a second integration gives (3.10) when L is large compared to R.   ∎

4. The Equilibrium Distribution

As promised at the end of Section 2, we use the energy integral to show that the sequence $\{\mu_{E_n}\}$ has a unique limit μ_E and that $U_{\mu_E} = \gamma(E)$ on $E \setminus A$, where $\text{Cap}(A) = 0$. Write $P(E)$ for the set of all Borel probability measures on the compact set E. Let Ω be the component of $\mathbb{C}^* \setminus E$ such that $\infty \in \Omega$, and recall the sequences $\{\Omega_n\}$ and $E_n = \mathbb{C}^* \setminus \Omega_n$ defined in Section 1. Also define $\widehat{E} = \mathbb{C} \setminus \Omega$.

Theorem 4.1. *Assume* $\mathrm{Cap}(E) > 0$. *Then sequence* $\{\mu_n\} = \{\mu_{E_n}\}$ *converges weak-star to some* $\mu = \mu_E \in P(E)$, *and the limit does not depend on the choice of the sequence* $\{E_n\}$. *Moreover,*

$$\gamma(E) = I(\mu_E) = \inf\{I(\sigma) : \sigma \in P(E)\} \tag{4.1}$$

and $I(\sigma) > \gamma(E)$ *for all* $\sigma \in P(E)$ *with* $\sigma \neq \mu_E$. *In other words,* μ_E *is the unique probability measure on* E *satisfying* $\gamma(E) = I(\mu_E)$.

It follows that the measure μ_E is the unique $\mu \in P(E)$ satisfying $U_\mu = \gamma(E)$ almost everywhere $d\mu$. By definition μ_E is the **equilibrium distribution** of E, and U_{μ_E} is called the **equilibrium potential**.

Lemma 4.2. *Let E be a compact set and let $\{\mu_n\}$ be a sequence in $P(E)$. If μ_n converges weak-star to $\mu \in P(E)$ then*

$$U_\mu(z) \leq \liminf U_{\mu_n}(z), \quad z \in \mathbb{C}, \tag{4.2}$$

and

$$I(\mu) \leq \liminf I(\mu_n). \tag{4.3}$$

Proof of Lemma 4.2. Write $h(z, \zeta) = \log 1/|z - \zeta|$ and for $N > 0$ define $h_N(z, \zeta) = \min(h(z, \zeta), N)$. Then each h_N is continuous, $h = \lim h_N$, and $h_N \leq h$. Therefore

$$\int h_N(z, \zeta) d\mu(\zeta) = \lim_n \int h_N(z, \zeta) d\mu_n(\zeta)$$

$$\leq \liminf_n \int h(z, \zeta) d\mu_n(\zeta)$$

$$= \liminf_n U_{\mu_n}.$$

Because h_N increases to h, this gives (4.2). Because $\mu_n \times \mu_n$ converges to $\mu \times \mu$ weak-star, a similar argument gives (4.3). ∎

Proof of Theorem 4.1. We first assume that $\Omega = \mathbb{C}^* \setminus E$ is bounded by finitely many pairwise disjoint $C^{1+\alpha}$ Jordan curves and in that case we prove μ_E is the unique minimizer for (4.1). For such E, $U_{\mu_E} = \gamma(E)$ on E by (2.3), so that $I(\mu_E) = \int U_{\mu_E} d\mu_E = \gamma(E)$. Thus if $\sigma \in P(E)$, then

$$\gamma(E) = \int U_{\mu_E} d\sigma = \int U_\sigma d\mu_E$$

by Fubini's theorem. But if $\sigma \neq \mu_E$, then by Theorem 3.1,

$$0 < I(\sigma - \mu_E)$$
$$= I(\sigma) + I(\mu_E) - \int U_{\mu_E} d\sigma - \int U_\sigma d\mu_E$$
$$= I(\sigma) - \gamma(E).$$

Therefore (4.1) holds, and $I(\sigma) = \gamma(E)$ if and only if $\sigma = \mu_E$.

Now suppose E is any compact set. By the Banach–Alaoglu theorem there is a subsequence $\{\mu_{n_j}\}$ of $\{\mu_{E_n}\}$ that converges weak-star to some $\mu \in P(E_1)$. By (4.3),

$$I(\mu) \leq \liminf I(\mu_{n_j}) = \liminf \gamma(E_{n_j}) = \gamma(E).$$

Moreover, any such μ has support contained in

$$\partial \left(\bigcap_n E_n \right) = \partial \widehat{E} \subset E, \qquad (4.4)$$

because μ_{E_n} has support ∂E_n. If $\sigma \in P(E) \subset P(E_n)$ then, because (4.1) holds for the sets E_n,

$$I(\sigma) \geq \lim_n \gamma(E_n) = \gamma(E).$$

Therefore $I(\sigma) \geq I(\mu) = \gamma(E)$.

Suppose $\sigma \in P(E)$ satisfies $I(\sigma) = \gamma(E)$ and $\sigma \neq \mu$. By Fubini's theorem $\int U_\mu d\sigma = \int U_\sigma d\mu$, and by Theorem 3.1

$$0 < I(\sigma - \mu) = I(\sigma) + I(\mu) - 2 \int U_\mu d\sigma. \qquad (4.5)$$

Therefore

$$\int U_\mu d\sigma < \gamma(E). \qquad (4.6)$$

But if

$$\sigma_t = t\sigma + (1-t)\mu, \qquad 0 < t < 1,$$

then $\sigma_t \in P(E)$ so that

$$\gamma(E) \leq I(\sigma_t) = t^2 I(\sigma) + (1-t)^2 I(\mu) + 2t(1-t) \int U_\mu d\sigma$$

$$\qquad (4.7)$$

$$\leq \gamma(E) + 2 \left(\int U_\mu d\sigma - \gamma(E) \right) t + O(t^2).$$

For t small (4.6) and (4.7) are in contradiction and hence μ is the unique extremal for the minimizing problem (4.1). The uniqueness implies $\{\mu_n\}$ converges weak-star, because the sequence $\{\mu_n\}$ has a unique weak-star accumulation point. The uniqueness also shows that the limit μ cannot depend on the choice of $\{E_n\}$. ∎

For any compact set E,

$$\partial \widehat{E} \subset \partial E \subset E \subset \widehat{E},$$

and by (4.4), μ_E is supported on $\partial \widehat{E}$. By (1.4) and the uniqueness in Theorem 4.1,

$$\mu_E = \mu_{\partial E} = \mu_{\widehat{E}} = \mu_{\partial \widehat{E}}.$$

The **capacity** of any Borel set A is defined as

$$\mathrm{Cap}(A) = \sup\{\mathrm{Cap}(E) : E \text{ compact}, E \subset A\}. \tag{4.8}$$

Corollary 4.3. *If E is compact and $\mathrm{Cap}(E) = 0$, then $I(\mu) = +\infty$ for all $\mu \in P(E)$. Consequently, if A is Borel with $\mathrm{Cap}(A) = 0$ and if μ is a positive measure with $I(\mu) < \infty$, then $\mu(A) = 0$.*

Proof. Suppose E is compact and $\mathrm{Cap}(E) = 0$ and suppose $\mu \in P(E)$. Then $P(E) \subset P(E_n)$ and $I(\mu) \geq \gamma(E_n) \to \infty$ by Theorem 4.1. If A is a Borel set with $\mathrm{Cap}(A) = 0$ and if σ is a positive measure with $\sigma(A) > 0$, then take $E \subset A$ compact, such that $\sigma(E) > 0$. Then $\mu = \left(\frac{\chi_E}{\sigma(E)}\right)\sigma \in P(E)$ and $I(\mu) = \infty$. Thus $I(\sigma) = \infty$. ∎

In particular, (4.8) and Corollary 4.3 imply that

$$\mathrm{Cap}(A) = 0 \implies \mathrm{Area}(A) = 0.$$

Sets of capacity zero are also called **polar sets**.

Theorem 4.4. *Let E be a compact set such that $\mathrm{Cap}(E) > 0$ and let μ_E be the equilibrium distribution for E. Then*

$$U_{\mu_E}(z) \leq \gamma(E), \quad z \in \mathbb{C} \tag{4.9}$$

$$U_{\mu_E}(z) = \gamma(E) - g_\Omega(z, \infty) < \gamma(E), \quad z \in \Omega = \mathbb{C}^* \setminus \widehat{E}, \tag{4.10}$$

and

$$\mathrm{Cap}\big(\{z \in \widehat{E} : U_{\mu_E}(z) \neq \gamma(E)\}\big) = 0. \tag{4.11}$$

Proof. Since $U_{\mu_{E_n}} \leq \gamma(E_n)$ and $\gamma(E_n) \to \gamma(E)$, (4.9) follows from Lemma 4.2. By (1.7), (2.1), and the weak-star convergence of μ_{E_n}, we conclude that

$U_{\mu_E}(z) = \gamma(E) - g_\Omega(z, \infty)$ for $z \in \Omega$. If $U_{\mu_E}(z) = \gamma(E)$ at some $z \in \Omega$, then by the maximum principle, U_{μ_E} is constant on Ω, but

$$\lim_{z \to \infty} U_{\mu_E}(z) = -\infty.$$

Therefore (4.10) holds.

To prove (4.11), let K be any compact subset of $\widehat{E}$ such that $\mathrm{Cap}(K) > 0$ and take $\sigma \in P(K)$ with $I(\sigma) < \infty$. For $0 < t < 1$, the inequality (4.7) still obtains, so that the coefficient of t in (4.7) is non-negative:

$$\int U_{\mu_E} d\sigma - \gamma(E) \geq 0.$$

But then by (4.9)

$$\int U_{\mu_E} d\sigma = \gamma(E)$$

and $U_{\mu_E}(z) = \gamma(E)$ almost everywhere with respect to σ. Thus $\mathrm{Cap}(K) = 0$ whenever $K \subset \{z \in \widehat{E} : U_{\mu_E}(z) \neq \gamma(E)\}$.   ∎

Theorem 4.5. *Suppose ψ is a contraction: $|\psi(z) - \psi(w)| \leq |z - w|$ for all $z, w \in E$. Then*

$$\mathrm{Cap}(\psi(E)) \leq \mathrm{Cap}(E).$$

Consequently, if $E^* = \{|z| : z \in E\}$ is the circular projection of E onto $\mathbb{R}$, then

$$\mathrm{Cap}(E) \geq \mathrm{Cap}(E^*) \geq \frac{|E^*|}{4}, \tag{4.12}$$

where $|E^*|$ denotes the length of E^*.

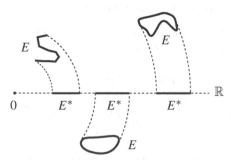

Figure III.1 Circular projection.

In Theorem 4.5, circular projections can also be replaced by vertical projections: $\widetilde{E} = \{x : x + iy \in E\}$. The proof is the same.

Proof. We may assume that $\mathrm{Cap}(\psi(E)) > 0$. Let $\mu \in P(\psi(E))$ be such that $I(\mu) < \infty$. By Lemma 4.6 below there is $\nu \in P(E)$ such that $\psi_*(\nu) = \mu$, where $\psi_*(\nu)$ is defined by

$$\psi_*(\nu)(A) = \nu(\psi^{-1}(A)) \tag{4.13}$$

for all Borel $A \subset \psi(E)$. Then

$$I(\mu) = \int_E \int_E \log \frac{1}{|\psi(z) - \psi(w)|} d\nu(z) d\nu(w) \geq I(\nu),$$

and by Theorem 4.1, $\gamma(E) \leq \gamma(\psi(E))$ and $\mathrm{Cap}(E) \geq \mathrm{Cap}(\psi(E))$.

In proving (4.12) we may assume that E is a finite union of smooth curves and E^* is a finite union of intervals in $\mathbb{R}$. Let F be the interval obtained by sliding the intervals of E^* together without altering their lengths. Then F is obtained from E by applying two contractions, and hence $\mathrm{Cap}(E) \geq \mathrm{Cap}(E^*) \geq \mathrm{Cap}(F)$. But then by Example 1.1, $\mathrm{Cap}(F) = |F|/4 = |E^*|/4$. ∎

Lemma 4.6. *Let E be a compact set and let ψ be a continuous map from E onto $\psi(E)$. For $\nu \in P(E)$ define $\psi_*(\nu)$ by (4.13) above. Then ψ_* maps $P(E)$ onto $P(\psi(E))$.*

Proof. By its definition, the map $\psi_* : P(E) \to P(\psi(E))$ is continuous, weak-star to weak-star. Given $\mu \in P(\psi(E))$, let μ_n be a sequence of atomic measures

$$\mu_n = \sum_{j=1}^{N_n} a_j^{(n)} \delta_{z_j^{(n)}},$$

where δ_z denotes the point-mass at z, such that $z_j^{(n)} \in \psi(E)$, such that $a_j^{(n)} > 0$ and $\sum a_j^{(n)} = 1$, and such that $\mu_n \longrightarrow \mu$ weak-star. Pick any $w_j^{(n)} \in \psi^{-1}(z_j^{(n)})$, and set

$$\nu_n = \sum_{j=1}^{N_n} a_j^{(n)} \delta_{w_j^{(n)}}.$$

Then $\psi_*(\nu_n) = \mu_n$, and if a subsequence of $\{\nu_n\}$ converges weak-star to ν, then $\nu \in P(E)$ and $\psi_*(\nu) = \mu$. ∎

In contrast to Theorem 4.5, there exist sets $E \subset \mathbb{R}$ with $|E| = 0$ but $\mathrm{Cap}(E) > 0$.

Example 4.7. The **Cantor set** is obtained from the unit interval $K_0 = [0, 1]$ as follows. The (closed) set K_n consists of 2^n intervals of length 3^{-n} obtained by removing the middle (open) one-third of each interval in K_{n-1}. The Cantor set is

$$K = \bigcap_{n=0}^{\infty} K_n,$$

and

$$\mathrm{Cap}(K) \geq \frac{1}{9}. \tag{4.14}$$

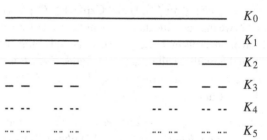

Figure III.2 The Cantor set.

To prove (4.14), note that by (1.3) $\mathrm{Cap}(K) = \lim_n \mathrm{Cap}(K_n)$. The set K_n consists of two subsets K_n^1 and K_n^2, with $\mathrm{dist}(K_n^1, K_n^2) = \frac{1}{3}$ such that $\mathrm{Cap}(K_n^1) = \mathrm{Cap}(K_n^2) = \mathrm{Cap}(K_{n-1})/3$. In other words,

$$\gamma(K_n^1) = \gamma(K_n^2) = \gamma(K_{n-1}) + \log 3.$$

Let $\mu = (\mu_1 + \mu_2)/2$ where μ_j is the equilibrium distribution for $K_n^j, j = 1, 2$. Then by Theorem 4.1,

$$I\left(\frac{\mu_1 + \mu_2}{2}\right) = \frac{\gamma(K_n^1)}{4} + \frac{\gamma(K_n^2)}{4} + \frac{1}{2}\iint \log\frac{1}{|z - \zeta|}d\mu_1(\zeta)d\mu_2(z),$$

so that by (4.1),

$$\gamma(K_n) \leq 2\left(\frac{\gamma(K_{n-1}) + \log 3}{4}\right) + \frac{\log 3}{2} = \frac{\gamma(K_{n-1}) + 2\log 3}{2}.$$

Thus $\gamma(K) = \lim \gamma(K_n) \leq 2\log 3$ and $\mathrm{Cap}(K) \geq 1/9$. ∎

Theorem 4.5 provides another proof of the Koebe one-quarter theorem I.4.1.

Corollary 4.8. *Let $\varphi(z)$ be a univalent function on the unit disc $\mathbb{D}$. If φ is normalized by $\varphi(0) = 0$ and $\varphi'(0) = 1$, then*

$$\varphi(\mathbb{D}) \supset \{w : |w| < 1/4\}. \tag{4.15}$$

Proof. Let $a = \inf\{|w| : w \notin \varphi(\mathbb{D})\}$. Then

$$\psi(z) = \frac{1}{\varphi(\frac{1}{z})} = z + b_0 + \frac{b_1}{z} + \cdots$$

is a conformal map from $\mathbb{C}^* \setminus \overline{D}$ to a domain Ω and $E = \mathbb{C}^* \setminus \Omega$ is a connected set with circular projection $E^* = [0, 1/a]$. By Theorem 4.5 and Example 1.1,

$$1 = \mathrm{Cap}(E) \geq \mathrm{Cap}(E^*) = \frac{1}{4a},$$

which gives (4.15).   ∎

5. Wiener's Solution to the Dirichlet Problem

Let Ω be a domain in $\mathbb{C}^*$ such that $\mathrm{Cap}(\mathbb{C} \setminus \Omega) > 0$, let $f \in C(\partial\Omega)$, and let $\{\Omega_n\}$ be the subdomains introduced in Section 1. Extend f to a continuous function on $\mathbb{C}^*$, still called f, and set

$$u_n(z) = \int_{\partial\Omega_n} f(\zeta)d\omega(z, \zeta, \Omega_n), \qquad z \in \Omega_n.$$

Then Theorem 4.1 yields the following fundamental result from Wiener [1924a].

Theorem 5.1 (Wiener). *If* $\mathrm{Cap}(\mathbb{C} \setminus \Omega) > 0$, *the sequence* $\{u_n(z)\}$ *converges uniformly on all compact subsets of* Ω *to a harmonic function* $u(z)$, *and* $u(z)$ *does not depend on the choice of the sequence* $\{\Omega_n\}$ *nor on the choice of the extension of* $f \in C(\partial\Omega)$.

We write

$$u(z) = u_f(z)$$

and we call $u_f(z)$ the **Wiener solution to the Dirichlet problem** with boundary values $f(\zeta)$.

Proof. Suppose $\infty \in \Omega$. As in Section 4, let $\mu_n = \mu_{E_n} = \omega(\infty, \cdot, \Omega_n)$ where $E_n = \mathbb{C} \setminus \Omega_n$. By Theorem 4.1, the sequence $\{\mu_n\}$ converges weak-star to μ_E, where $E = \mathbb{C} \setminus \Omega$. Hence

$$\lim_{n\to\infty} u_n(\infty) = \lim_{n\to\infty} \int_{\partial\Omega} f(\zeta)d\mu_{E_n}(\zeta) = \int_{\partial\Omega} f(\zeta)d\mu_E(\zeta), \qquad (5.1)$$

and the latter integral does not depend on the sequence $\{\Omega_n\}$. It also does not depend on the extension of f because the limit measure μ_E is supported on $\partial\Omega$.

For $z \in \Omega$ and $z \neq \infty$, let T be a Möbius transformation such that $Tz = \infty$ and note that

$$\mathrm{Cap}(\partial\Omega) > 0 \iff \mathrm{Cap}(T(\partial\Omega)) > 0. \tag{5.2}$$

Indeed, if $\mathrm{Cap}(\partial\Omega) > 0$, then by (1.7), $g_\Omega(w, T^{-1}(\infty))$ exists and

$$g_\Omega(w, T^{-1}(\infty)) = \lim_n g_{\Omega_n}(w, T^{-1}(\infty))$$
$$= \lim_n g_{T(\Omega_n)}(Tw, \infty) = g_{T(\Omega)}(Tw, \infty),$$

and that implies (5.2). The argument for $z = \infty$ shows $\lim_{n \to \infty} u_n(z) = u(z)$ exists for all $z \in \Omega$ and the limit $u(z)$ is independent of $\{\Omega_n\}$ and the extension of f. Because the functions $\{u_n(z)\}$ are uniformly bounded and harmonic, the convergence $u_n(z) \to u(z)$ is uniform on compact subsets of Ω and the limit $u(z)$ is harmonic in Ω.   ∎

Theorem 5.1 enables us to define harmonic measure for any domain Ω with $\mathrm{Cap}(\mathbb{C}^* \setminus \Omega) > 0$, because it says the weak-star limit

$$\omega(z, \cdot, \Omega) = \lim_{n \to \infty} \omega(z, \cdot, \Omega_n) \tag{5.3}$$

exists and is independent of the choice of $\{\Omega_n\}$. Then for any $f \in C(\partial\Omega)$,

$$u_f(z) = \int_{\partial\Omega} f(\zeta)\,d\omega(z, \zeta, \Omega),$$

where $u_f(z)$ is the Wiener solution to the Dirichlet problem for $f(\zeta)$. By definition, the limit measure $\omega(z, \cdot, \Omega)$ is the **harmonic measure** for z relative to Ω. By Exercise 4 of Chapter II, this definition of harmonic measure agrees with the definition in Chapter II for finitely connected Jordan domains. As in Chapter I, $\omega(z, A, \Omega)$ is a harmonic function of $z \in \Omega$ and

$$0 \le \omega(z, A, \Omega) \le 1.$$

By Harnack's inequality,

$$\omega(z_1, E, \Omega) \le C_{z_1,z_2}\omega(z_2, E, \Omega)$$

for every pair z_1, z_2 of points in Ω, where the constant $C_{z_1,z_2} > 0$ depends only on z_1, z_2, and Ω. Thus the measures $\omega(z, \cdot, \Omega), z \in \Omega$, are mutually absolutely continuous.

If $z = \infty$, then (5.1) shows that $\omega(\infty, \cdot, \Omega) = \mu_{\partial\Omega}$. If $z \in \Omega$ and if T is a Möbius transformation with $T(z) = \infty$, then for every Borel set A, $\omega(z, A, \Omega) = \mu_{T(\partial\Omega)}(T(A))$. By (5.2) and Corollary 4.3

$$\mathrm{Cap}(A) = 0 \implies \omega(z, A, \Omega) = 0 \tag{5.4}$$

for every Ω with $\text{Cap}(\mathbb{C} \setminus \Omega) > 0$ and every $z \in \Omega$.

If $\text{Cap}(\mathbb{C}^* \setminus \Omega) = 0$ then, as we shall see in Section 8, each bounded or positive harmonic function on Ω is constant. Therefore we cannot define harmonic measure or solve the Dirichlet problem when $\text{Cap}(\mathbb{C}^* \setminus \Omega) = 0$.

A major goal of this book is to understand the relation between harmonic measure and other measures. Every finite $F \subset \partial\Omega$ has harmonic measure zero. (One proof: $\text{Cap}(F) = 0$ by (4.1), so that $\omega(z, F, \Omega) = 0$ for all $z \in \Omega$ by (5.4).) We shall need the following quantitative form of this fact. Let $\{\Omega_n\}$ be the increasing sequence of domains introduced in Section 1.

Proposition 5.2. *If $E = \mathbb{C} \setminus \Omega$ satisfies $\text{Cap}(E) > 0$, then for any $\zeta \in \partial\Omega$, and any $\varepsilon > 0$,*

$$\lim_{\delta \to 0} \sup_{z \in \Omega_n \setminus B_\varepsilon(\zeta)} \omega(z, \partial\Omega_n \cap B_\delta(\zeta), \Omega_n) = 0 \tag{5.5}$$

uniformly in n.

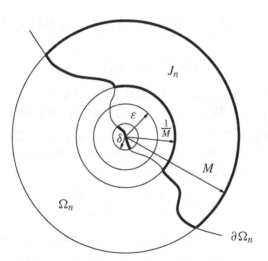

Figure III.3 Proof of Proposition 5.2.

Proof. Suppose $\infty \in \Omega$ and $\zeta = 0$. Let $J = E \cap \{z : 1/M \le |z| \le M\}$ with M so that $\text{Cap}(J) > 0$. For Ω_n as before, set $J_n = \{z : 1/M \le |z| \le M\} \setminus \Omega_n$. See Figure III.3. Let μ_n be the equilibrium distribution of J_n and let γ_n be Robin's constant for J_n. Then

$$U_n(z) \equiv \log \frac{1}{|z|} - U_{\mu_n}(z) = \int_{J_n} \log \left| \frac{z - \zeta}{z} \right| d\mu_n(\zeta)$$

is harmonic on $\mathbb{C}^* \setminus (J_n \cup \{0\})$. For $\delta < 1/M$ we have

$$U_n(z) \geq \log\left(\frac{1}{M\delta} - 1\right), \quad z \in \partial B_\delta(0),$$

and

$$U_n(z) \geq \log\frac{1}{M} - \gamma_n, \quad z \in \mathbb{C}.$$

For $\delta < \varepsilon$, these inequalities and the maximum principle give

$$\omega(z, \partial\Omega_n \cap B_\delta(0), \Omega_n) \leq \frac{U_n(z) + \log M + \gamma_n}{\log(\frac{1}{\delta} - M) + \gamma_n}$$

on $\{|z| > \delta\} \cap \Omega_n$. Hence if $|z| \geq \varepsilon \geq \delta$,

$$\omega(z, \partial\Omega_n \cap B_\delta(0), \Omega_n) \leq \frac{\log\left(1 + \frac{M}{\varepsilon}\right) + \log M + \gamma_n}{\log\left(\frac{1}{\delta} - M\right) + \gamma_n}.$$

But because M is fixed and $J \subset J_n \subset B_M(0)$, the quantities γ_n are bounded above and below, and therefore (5.5) holds.   ∎

Suppose $0 \notin \overline{\Omega}$. Then by Theorem 5.1 and the definition of $g(z, \infty)$ for finitely connected domains,

$$h(z, \infty) = g(z, \infty) - \log|z|$$

coincides with the solution to the Dirichlet problem in Ω with boundary value $-\log|z|$. The following proposition gives a similar result when $\overline{\Omega} = \mathbb{C}^*$.

Proposition 5.3. *Let $E \subset \{z : |z| < R\}$ be a compact set such that $\mathrm{Cap}(E) > 0$ and let $\Omega = \mathbb{C}^* \setminus E$. Then U_{μ_E} is the solution to the Dirichlet problem on $\Omega^R = \Omega \cap B(0, R)$ with boundary data*

$$f(\zeta) = \begin{cases} \gamma(E) & \text{for } \zeta \in \partial\Omega, \\ U_{\mu_E}(\zeta) & \text{for } |\zeta| = R, \end{cases}$$

and $g_\Omega(\cdot, \infty)$ is the solution to the Dirichlet problem on Ω^R with boundary data $\gamma(E) - f(\zeta)$.

Proof. Because $U_\mu(z) = \gamma(E) - g_\Omega(z, \infty)$, it is enough to show that U_{μ_E} is the solution of a Dirichlet problem. We may suppose the domains $\{\Omega_n\}$ satisfy $\{z : |z| \geq R\} \subset \Omega_1$. Extend f to $C(\mathbb{C})$ so that

$$f = \begin{cases} \gamma(E) & \text{on } E, \\ \gamma(E_n) & \text{on } \partial\Omega_n, \\ U_{\mu_E} & \text{on } |z| = R, \end{cases}$$

and let u_n be the Wiener solution of Dirichlet's problem on $\Omega_n^R = \Omega_n \cap B_R(0)$ for the boundary data $f|_{\partial \Omega_n^R}$. Then $u_n - U_{\mu_{E_n}}$ is continuous on $\overline{\Omega_n^R}$ and harmonic on Ω_n, and hence

$$u_n(z) - U_{\mu_{E_n}}(z) = \int_{|w|=R} \left(U_{\mu_E}(w) - U_{\mu_{E_n}}(w) \right) d\omega(z, w, \Omega_n^R)$$

on Ω_n^R. Because $U_{\mu_{E_n}} \to U_{\mu_E}$ uniformly on $\{|w| = R\}$, it follows that

$$\lim_n u_n(z) = \lim_n U_{\mu_{E_n}}(z) = U_{\mu_E}(z)$$

for $z \in \Omega^R$. Therefore U_{μ_E} is the Wiener solution to the Dirichlet problem in Ω^R for the boundary data $f|_{\partial \Omega^R}$. ∎

With the same proof, Proposition 5.3 is also valid if $B(0, R)$ is replaced by any bounded Jordan region Ω' such that $E \cap \partial \Omega' = \emptyset$.

6. Regular Points

Assume $\mathrm{Cap}(\mathbb{C}^* \setminus \Omega) > 0$, let $\zeta \in \partial \Omega$ and let $f \in C(\partial \Omega)$. In this section we determine when the Wiener solution u_f has boundary value $f(\zeta)$ at ζ. The point $\zeta \in \partial \Omega$ is a **regular point** for Ω if

$$\lim_{\Omega \ni z \to \zeta} u_f(z) = f(\zeta) \tag{6.1}$$

for all $f \in C(\partial \Omega)$. Equivalently, $\zeta \in \partial \Omega$ is a regular point if

$$\lim_{\Omega \ni z \to \zeta} \omega(z, \cdot, \Omega) = \delta_\zeta$$

in the weak-star topology, where δ_ζ is the point mass at ζ. A point $\zeta \in \partial \Omega$ not regular is called an **irregular point**. If $\partial \Omega$ consists of finitely many Jordan curves, then by Chapter II every boundary point is regular. However, if

$$\Omega = \{z : |z| < 1\} \setminus \{0\},$$

then every bounded harmonic function on Ω extends to be harmonic at 0. Therefore (6.1) fails at $\zeta = 0$ except in the rare cases where

$$f(0) = \int_0^{2\pi} f(e^{i\theta}) \frac{d\theta}{2\pi},$$

and therefore 0 is an irregular point of $\partial \Omega$.

Theorem 6.1. *The point $\zeta \in \partial \Omega$ is a regular point for Ω if and only if there is $\rho > 0$ such that for every $\varepsilon > 0$, there exists a function V_ε, harmonic in*

$\Omega \cap \{z : |z - \zeta| < \rho\}$, *satisfying*

$$V_\varepsilon(z) \geq 0, \qquad (6.2)$$

$$\limsup_{z \to \zeta} V_\varepsilon(z) \leq \varepsilon, \qquad (6.3)$$

and

$$V_\varepsilon(z) > 1/2 \quad on \quad \{z : \varepsilon < |z - \zeta| \leq \rho\} \cap \Omega. \qquad (6.4)$$

By definition a **barrier** is a harmonic function V on Ω such that

$$\lim_{z \to \zeta} V(z) = 0,$$

and

$$\liminf_{z \to \alpha} V(z) > 0$$

for $\alpha \in \partial\Omega \setminus \{\zeta\}$. If V is a barrier, then there are constants $C_\varepsilon > 0$ such that the functions $V_\varepsilon = C_\varepsilon V$ satisfy (6.2), (6.3), and (6.4).

Proof. Let ζ be a regular point. Fix $\varepsilon > 0$. For $0 < \delta < \varepsilon$ choose $f \in C(\partial\Omega)$ such that

$$0 \leq f \leq 1,$$

$$\{z \in \partial\Omega : f(z) = 0\} = \{\zeta\},$$

and

$$f(z) = 1, \quad when \quad |z - \zeta| > \delta.$$

Then clearly $V_\varepsilon(z) = u_f(z)$ satisfies (6.2) and (6.3), and when δ is small enough (6.4) also holds by Proposition 5.2.

Now suppose the functions $V_\varepsilon(z)$ exist. Let $f \in C(\mathbb{C}^*)$ satisfy

$$0 \leq f \leq 1/2 \quad and \quad f^{-1}(0) = \zeta \qquad (6.5)$$

and let Ω_n be as above. If $\delta > 0$, then for ε sufficiently small,

$$f(z) \leq V_\varepsilon(z) + \delta$$

on $\{z : |z - \zeta| < \rho\} \cap \partial\Omega_n$. Let u_n solve the Dirichlet problem on the subdomain Ω_n for the boundary value f. Then

$$u_n(z) \leq \sup f \leq 1/2 \leq V_\varepsilon(z)$$

on $\Omega_n \cap \{z : |z - \zeta| = \rho\}$, and by the maximum principle

$$u_n(z) \leq V_\varepsilon(z) + \delta$$

on $\Omega_n \cap \{|z - \zeta| < \rho\}$. Thus

$$0 \leq u_f(z) \leq V_\varepsilon(z) + \delta$$

on $\Omega \cap \{|z - \zeta| < \rho\}$. Because δ is arbitrary, (6.3) now yields

$$\lim_{z \to \zeta} u_f(z) = 0.$$

Because every real $g \in C(\mathbb{C}^*)$ is a constant plus a linear combination of two functions satisfying (6.5), it follows that ζ is a regular point. ∎

Theorem 6.1 implies that the regularity of ζ is a local question; it depends only on the behavior of Ω in some neighborhood of ζ.

Corollary 6.2. *Let $\zeta \in \partial\Omega$. If the connected component of $\partial\Omega$ containing ζ consists of more than one point, then ζ is a regular point for Ω. In particular, if Ω is simply connected, then every point of $\partial\Omega$ is a regular point.*

Proof. For $\varepsilon > 0$, let E_ε be the component of $\partial\Omega \cap \overline{B(\zeta, \varepsilon/2)}$ containing ζ. Then $E_\varepsilon \neq \{\zeta\}$. Let ψ_ε be a Riemann map of $\mathbb{C}^* \setminus E_\varepsilon$ onto $\mathbb{D}$ with $\psi_\varepsilon(\infty) = 0$. Then $|\psi_\varepsilon(z)| \to 1$ as $\mathbb{C}^* \setminus E_\varepsilon \ni z \to E_\varepsilon$ and

$$V_\varepsilon(z) \equiv \frac{\log \left| \frac{1}{\psi_\varepsilon(z)} \right|}{\inf_{\varepsilon < |z| < 1} \log \left| \frac{1}{\psi_\varepsilon(z)} \right|}$$

satisfies (6.2), (6.3), and (6.4) with $\rho = 1$. ∎

Theorem 6.3. *Assume $E = \mathbb{C}^* \setminus \Omega$ is a compact plane set with $\mathrm{Cap}(E) > 0$ and let μ_E be the equilibrium distribution for E. Then for $\zeta \in \partial\Omega$ the following conditions are equivalent:*

$$\zeta \text{ is a regular point,} \tag{6.6}$$

$$\lim_{\Omega \ni z \to \zeta} g_\Omega(z, \infty) = 0, \tag{6.7}$$

$$U_{\mu_E}(\zeta) = \gamma(E). \tag{6.8}$$

Proof. Take $R > 0$ such that $E \subset B(0, R)$. Then by Proposition 5.3, $g_\Omega(\cdot, \infty)$ is the solution to the Dirichlet problem in $\Omega^R = \Omega \cap B(0, R)$, for the boundary data

$$\begin{cases} 0 & \text{for } z \in \partial\Omega, \\ g_\Omega(z, \infty) & \text{for } |z| = R. \end{cases}$$

If $\zeta \in \partial\Omega$ is a regular point, then by Theorem 6.1, ζ is also a regular point for Ω^R and hence (6.7) holds.

Now suppose (6.7) holds. Then by (4.10)

$$U_{\mu_E}(z) = \gamma(E) - g_\Omega(z, \infty) \to \gamma(E), \qquad (6.9)$$

as $z \in \Omega \to E$. Because $\log 1/|z|$ is superharmonic, we have

$$U_{\mu_E}(\zeta) \geq \frac{1}{\pi r^2} \iint\limits_{B(\zeta,r)} U_{\mu_E}(z)dxdy,$$

while by Theorem 4.4,

$$U_{\mu_E}(z) \leq \gamma(E) \qquad (6.10)$$

for all z. Therefore

$$0 \leq \gamma(E) - U_{\mu_E}(\zeta) \leq \limsup_{r \to 0} \frac{1}{\pi r^2} \iint\limits_{B(\zeta,r) \cap \Omega} (\gamma(E) - U_{\mu_E}(z))dxdy$$

$$+ \limsup_{r \to 0} \frac{1}{\pi r^2} \iint\limits_{B(\zeta,r) \setminus \Omega} (\gamma(E) - U_{\mu_E}(z))dxdy.$$

By (6.9) and (6.10) the first limit is 0, while by the remark after Corollary 4.3 and Theorem 4.4,

$$\text{Area}(\{z \notin \Omega : U_{\mu_E}(z) \neq \gamma(E)\}) = 0,$$

so that the second limit is also 0. Hence (6.8) holds.

To show that (6.8) implies ζ is regular, set $E_\varepsilon = E \cap B(\zeta, \varepsilon^2/4)$. If ζ and $\{z : |z - \zeta| = \varepsilon^2/4\}$ lie in the same component of $E_\varepsilon \cup \{z : |z - \zeta| = \varepsilon^2/4\}$, then the complement in $\mathbb{C}^*$ of that component is a simply connected domain $\Omega_\varepsilon \supset \Omega \cap B(\zeta, \varepsilon^2/4)$, and $\zeta \in \partial\Omega_\varepsilon$. By Corollary 6.2, ζ is regular for Ω_ε, and then by Theorem 6.1, ζ is regular for Ω. Thus we may assume there is a (closed) C^2 Jordan curve $\sigma_\varepsilon \subset \Omega \cap B(\zeta, \varepsilon^2/4)$ such that ζ is in the bounded component W_ε of $\mathbb{C} \setminus \sigma_\varepsilon$.

Let μ_ε be the equilibrium distribution for E_ε, and let U_ε be its logarithmic potential. We will verify conditions (6.2), (6.3), and (6.4) for the functions

$$V_\varepsilon(z) = 1 - \frac{U_\varepsilon(z)}{\gamma(E_\varepsilon)}.$$

and this will show that ζ is regular. First note that $e^{-\gamma(E_\varepsilon)} = \text{Cap}(E_\varepsilon) > 0$, because if not, then by Corollary 4.3, $\mu_E(B(\zeta, \varepsilon^2/4)) = 0$. But then ζ would be in the unbounded component of $V = \mathbb{C} \setminus \text{supp}(\mu_E)$, where $\text{supp}(\mu_E)$ is the closed support of μ_E. Then (4.10) and the maximum principle would give $U_{\mu_E}(\zeta) < \gamma(E)$, contrary to the assumption (6.8).

By Theorem 4.4, (6.2) holds for V_ε. Note that if $|z - \zeta| > \varepsilon$, then

$$U_\varepsilon(z) = \int_{E_\varepsilon} \log \frac{1}{|z - w|} d\mu_{E_\varepsilon}(w) \leq \log \frac{2}{\varepsilon}.$$

Because $\gamma(E_\varepsilon) \geq \log \frac{4}{\varepsilon^2}$, we have $V_\varepsilon(z) > 1/2$ in $|z - \zeta| \geq \varepsilon$ and (6.4) holds for V_ε.

Choose $M_\varepsilon > 0$ so that when $z \in \sigma_\varepsilon$,

$$\gamma(E_\varepsilon) - U_\varepsilon(z) \leq M_\varepsilon \big(\gamma(E) - U_{\mu_E}(z)\big). \qquad (6.11)$$

By Proposition 5.3, each side of (6.11) is the solution to a Dirichlet problem in $W_\varepsilon \cap \Omega$ with boundary data 0 on $\partial\Omega \cap W_\varepsilon$, and hence the inequality (6.11) continues to hold in $W_\varepsilon \cap \Omega$. Sending $z \to \zeta$ then yields (6.3). Then ζ is a regular point by Theorem 6.1. ∎

Corollary 6.4 (Kellogg's theorem). *Let Ω be any domain in $\mathbb{C}^*$. Then*

$$\mathrm{Cap}\big(\{\zeta \in \partial\Omega : \zeta \text{ is irregular}\}\big) = 0.$$

Proof. We may assume $\infty \notin \Omega$. Then the corollary is immediate from Theorem 4.4 and Theorem 6.3. ∎

Corollary 6.5. *Let Ω be any domain in $\mathbb{C}^*$ with $\mathrm{Cap}(\mathbb{C}^* \setminus \Omega) > 0$. Then*

$$\omega(z, \{\zeta \in \partial\Omega : \zeta \text{ is irregular}\}, \Omega) = 0$$

for all $z \in \Omega$.

Proof. By (6.8), the set of irregular points is a Borel set, so the corollary follows from (5.4) and Corollary 6.4. ∎

7. Wiener Series

In a second paper [1924b], Wiener gave a geometric condition necessary and sufficient for $\zeta \in \partial\Omega$ to be a regular point for Ω. Let $A_n = A_n(\zeta)$ be the ring

$$A_n = \{z : 2^{-n} \leq |z - \zeta| \leq 2^{-n+1}\},$$

and recall the notation

$$\gamma(E) = \log\big(1/\mathrm{Cap}(E)\big).$$

Theorem 7.1 (Wiener). *Let Ω be a domain in $\mathbb{C}^*$ and let $\zeta \in \partial\Omega$. Then the following are equivalent:*

$$\zeta \text{ is a regular point,} \tag{7.1}$$

$$\sum_{n=1}^{\infty} \frac{n}{\gamma(A_n \setminus \Omega)} = \infty, \tag{7.2}$$

$$\sum_{n=1}^{\infty} \frac{n}{\gamma(A_n \cap \partial\Omega)} = \infty. \tag{7.3}$$

Proof. We can assume that $\zeta = 0$, and $\Omega \subset B(0, 1/2)$. It is clear that (7.3) implies (7.2) because the function $E \to \gamma(E)$ is decreasing. Now assume (7.2). By the definitions of regular point and capacity, we can replace Ω by a subdomain so that, for each n, $A_n \setminus \Omega$ is bounded by finitely many C^2 Jordan curves and so that the series (7.2) still diverges. Write $\gamma_n = \gamma(A_n \setminus \Omega)$, let μ_n be the equilibrium distribution for $A_n \setminus \Omega$, and let $u_n(z) = U_{\mu_n}(z)$. Fix $\varepsilon > 0$. Then because $\gamma_n \geq n \log 2$ there are integers $N_1 < N_2 < \ldots N_k \to \infty$ such that

$$1 \leq \varepsilon \sum_{N_j}^{N_{j+1}-1} \frac{n \log 2}{\gamma_n} \leq 1 + \varepsilon. \tag{7.4}$$

Set

$$W_j = \varepsilon \sum_{N_j}^{N_{j+1}-1} \frac{u_n}{\gamma_n}.$$

By our smoothness assumption on $A_n \setminus \Omega$, W_j is continuous on $\overline{B(0,1)}$, and W_j is harmonic and positive on $\Omega \cap \overline{B(0,1)}$. Also, by (7.4)

$$W_j(0) = \varepsilon \sum_{N_j}^{N_{j+1}-1} \frac{u_n(0)}{\gamma_n} \geq \varepsilon \sum_{N_j}^{N_{j+1}-1} \frac{n \log 2}{\gamma_n} \geq 1. \tag{7.5}$$

On the other hand, if $z \in A_k, k \geq 1$, then

$$\frac{u_n(z)}{\gamma_n} \leq \begin{cases} 1, & \text{if } |n - k| \leq 1 \\ \frac{(k+1)\log 2}{\gamma_n}, & \text{if } n > k + 1 \\ \frac{(n+1)\log 2}{\gamma_n}, & \text{if } n < k - 1. \end{cases}$$

Hence by (7.4),

$$W_j(z) \leq 3\varepsilon + \frac{N_j + 1}{N_j}(1 + \varepsilon) \leq 1 + 5\varepsilon \tag{7.6}$$

on $\overline{B(0,1)}$ provided N_j is sufficiently large. Also, for any $\delta > 0$,

$$\sup_{|z| \geq \delta} W_j(z) \leq \varepsilon \qquad (7.7)$$

if N_j is sufficiently large. Set $V_j = 1 + 5\varepsilon - W_j$. Then by (7.5), (7.6), and (7.7) the functions V_j satisfy the hypotheses of Theorem 6.1 (with 5ε), and $\zeta = 0$ is a regular point of $\partial\Omega$.

Now assume $\zeta = 0$ is a regular point of $\partial\Omega$ and assume the series (7.3) converges. Again we may replace Ω by a subdomain so that, for each n, $A_n \cap \partial\Omega$ consists of finitely many C^2 Jordan curves and so that (7.3) still converges. We may also assume $\Omega \subset B(0, 1/2)$. Write $\gamma_n = \gamma(A_n \cap \partial\Omega)$, let μ_n be the equilibrium distribution for $A_n \cap \partial\Omega$, and let $u_n = U_{\mu_n}$ be its potential.

Fix ε, $0 < \varepsilon < \frac{1}{3}$ and take $N \geq 3$ so that

$$\sum_{n=N}^{\infty} \frac{n}{\gamma_n} < \varepsilon. \qquad (7.8)$$

Let $F(z) \in C(\overline{\Omega})$ be harmonic on Ω with $0 \leq F \leq 1$, $F(0) = 1$, and $F = 0$ on $\partial\Omega \setminus B(0, 2^{-N+1})$. Since $\zeta = 0$ is a regular point there exists $s < 2^{-N}$ such that $F(z) > 1 - \varepsilon$ on $\overline{\Omega} \cap B(0, s)$. Let $0 < r < s$ be so small that if $U_r(z)$ is the equilibrium potential for $\partial\Omega \cap \overline{B(0, r)}$, then

$$V_r(z) = \frac{U_r(z)}{\gamma(\partial\Omega \cap \overline{B(0, r)})} \leq \varepsilon$$

on $\{|z| \geq s\}$. Let A be the annulus $\{r \leq |z| \leq 2^{-N+1}\}$. Let U denote the equilibrium potential for $\partial\Omega \cap A$ and take

$$V(z) = \frac{U(z)}{\gamma(\partial\Omega \cap A)}.$$

Then $V + V_r \geq F$ by Proposition 5.3. Consequently $V \geq 1 - 2\varepsilon$ on

$$\{|z| = s\} \cup \left(B(0, s) \cap \partial\Omega \cap A\right).$$

Hence by the maximum principle,

$$V(0) \geq 1 - 2\varepsilon. \qquad (7.9)$$

On the other hand, $V(z) \leq \frac{2}{N}$ on $|z| = \frac{1}{2}$, so that when $2^{-M} < r$ the maximum principle also gives

$$V(z) \leq \sum_{N}^{M} \frac{u_n(z)}{\gamma_n} + \frac{2}{N}.$$

Hence by (7.8) $V(0) \leq \varepsilon + \frac{2}{N}$, which contradicts (7.9). ∎

Theorem 7.1 has some important applications. The first shows there exist totally disconnected sets K such that each $\zeta \in K$ is a regular boundary point for $\Omega = \mathbb{C}^* \setminus K$.

Example 7.2. The Cantor set $K = \bigcap_{n=1}^{\infty} K_n$ of Example 4.7 has $\text{Cap}(K) > 0$. Suppose $\zeta \in K$ and suppose $\alpha \in K \cap A_j(\zeta)$. Then α belongs to an interval $I \subset K_j$ of length 3^{-j} and there is a second interval $J \subset K_j$ of length 3^{-j} with $\text{dist}(I, J) = 3^{-j}$. See Figure III.4. If j is such that $2^{-j-1} > 3^{-j+1}$ then one of the three rings $A_{j-1}(\zeta)$, $A_j(\zeta)$, or $A_{j+1}(\zeta)$ contains either I or J. Hence for some $n \in \{j-1, j, j+1\}$

$$\text{Cap}(A_n \cap K) \geq \text{Cap}(I \cap K) = \text{Cap}(J \cap K) = 3^{-j}\text{Cap}(K).$$

Therefore

$$\frac{n}{\gamma(A_n \cap K)} \geq C > 0$$

for some $C > 0$ and for infinitely many n and by (7.3) every ζ is a regular point for $\Omega = \mathbb{C}^* \setminus K$.

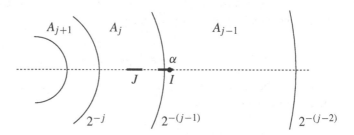

Figure III.4 Annuli and the Cantor set.

See Corollary 7.4 and Corollary D.3 of Appendix D for other proofs that every point of the Cantor set K is regular.

Corollary 7.3 (Beurling). *Let E be a compact set. Assume $0 \in E$ and set $E^* = \{|z| : z \in E\}$. If*

$$\int_{E^*} \frac{dr}{r} = \infty, \tag{7.9}$$

then 0 is a regular point for $\Omega = \mathbb{C}^* \setminus E$.

Proof. Set $E_n = E \cap A_n$ and $E_n^* = \{|z| : z \in E_n\}$. Then $\text{Cap}(E_n) \geq |E_n^*|/4$ by Theorem 4.5, and consequently we have $\sum 2^n \text{Cap}(E_n) = \infty$ by (7.9). But $2^n \text{Cap}(E_n) \leq Cn/\gamma(E_n)$, because $\text{Cap}(E_n) \leq 2^{-n+1}$ and the function

$x \log 1/x$ is increasing on $[0, 1/e]$. Therefore the series (7.3) also diverges and 0 is a regular point. ∎

Corollary 7.4. *Assume* $\zeta \in E = \partial\Omega$. *If*

$$\limsup_{r \to 0} \frac{\mathrm{Cap}\big(E \cap B(\zeta, r)\big)}{r} > 0, \qquad (7.11)$$

or if

$$\limsup_{r \to 0} \frac{\log(1/r)}{\gamma(E \cap B(\zeta, r))} > 0, \qquad (7.12)$$

then ζ *is a regular point of* $\partial\Omega$.

Proof. It is clear that (7.11) implies (7.12) and that (7.12) implies

$$\limsup_{n \to \infty} \frac{n}{\gamma(E \cap B(\zeta, 2^{-n}))} > 0.$$

Then by Lemma 7.5 below

$$\limsup_{n \to \infty} \sum_{n}^{\infty} \frac{k}{\gamma(E \cap A_k)} > 0,$$

and (7.3) again diverges. ∎

Lemma 7.5. *Suppose* $E_1, \dots, E_n$ *are disjoint Borel subsets of* $B(0, 1/2)$. *Then*

$$\frac{1}{\gamma(\bigcup E_k)} \le \sum_{k=1}^{n} \frac{1}{\gamma(E_k)}.$$

Proof. By the definition of capacity, we can suppose each E_k is bounded by finitely many pairwise disjoint C^2 Jordan curves. Let μ be the equilibrium distribution for $E = \bigcup E_k$, let μ_k be the equilibrium distribution for E_k, let $\sigma = \sum \mu_k / \gamma(E_k)$ and let U_μ, U_{μ_k}, and U_σ be their potentials. Then because $U_\mu = \gamma(E)$ on E and $\mathrm{supp}(\sigma) \subset E$, we have

$$\left(\sum \frac{1}{\gamma(E_k)} \right) \gamma(E) = \int U_\mu d\sigma = \int U_\sigma d\mu,$$

and because $U_{\mu_k} \ge 0$ on E and $U_{\mu_k} = \gamma(E_k)$ on E_k, we have $U_\sigma \ge 1$ on $\mathrm{supp}(\mu)$. Consequently

$$\int U_\sigma d\mu \ge 1$$

and the lemma follows. ∎

8. Polar Sets and Sets of Harmonic Measure Zero

In this section we prove a strong form of the maximum principle. It implies that if a compact set E is a polar set, i.e. $\mathrm{Cap}(E) = 0$, and if $u > 0$ is a harmonic function on $\mathbb{C}^* \setminus E$, then u is constant.

Theorem 8.1. *Suppose Ω is a domain such that $\mathrm{Cap}(\mathbb{C}^* \setminus \Omega) > 0$, and suppose A is a Borel subset of $\partial\Omega$ such that*

$$\omega(z_0, A, \Omega) = 0, \qquad (8.1)$$

holds for some $z_0 \in \Omega$. Let $u(z)$ be a subharmonic function in Ω. If u is bounded above in Ω and if

$$\limsup_{z \to \zeta} u(z) \le m \qquad (8.2)$$

for all $\zeta \in \partial\Omega \setminus A$, then

$$u(z) \le m$$

for all $z \in \Omega$.

The condition (8.1) is essential for Theorem 8.1. For example, if K is a compact subset of $\partial\Omega$ such that $0 < \omega(z_0, K) < 1$ for some $z_0 \in \Omega$, let $u(z) = \omega(z, K, \Omega)$. If $A = (\partial\Omega \setminus K) \cap \{\text{regular points}\}$ then $\omega(z_0, A, \Omega) > 0$ by Corollary 6.5. But $\lim_{z \to \zeta} u(z) = 0$ on A, while $u(z_0) > 0$.

Because a finite set has zero harmonic measure, Theorem 8.1 is a generalization of Lindelöf's maximum principle, Lemma I.1.1.

Proof. On $\overline{\Omega}$, define

$$V(\zeta) = \begin{cases} u(\zeta), & \zeta \in \Omega, \\[2mm] \displaystyle\limsup_{\Omega \ni z \to \zeta} u(z), & \zeta \in \partial\Omega. \end{cases}$$

Then by hypothesis, V is bounded above, and V is upper semicontinuous on $\overline{\Omega}$ because subharmonic functions are upper semicontinuous. By (8.1) and (8.2),

$$\int_{\partial\Omega} V(\zeta) d\omega(z, \zeta, \Omega) \le m \qquad (8.3)$$

for every $z \in \Omega$.

Let $\{\Omega_n\}$ be the subdomains from before. Then for any $z \in \Omega$,

$$\omega(z, \cdot, \Omega) = \lim_{n \to \infty} \omega(z, \cdot, \Omega_n),$$

with respect to weak-star convergence of measures on $\overline{\Omega}$. Let $V_k(\zeta)$ be a sequence of continuous functions that decreases to $V(\zeta)$ on $\overline{\Omega}$, and let $u_{k,n}(z)$ be the harmonic extension of V_k to Ω_n. Because u is upper semicontinuous on $\overline{\Omega_n} \subset \Omega$,

$$\limsup_{\Omega_n \ni z \to \zeta} \left(u(z) - u_{k,n}(z) \right) = u(\zeta) - V_k(\zeta) \le 0$$

at every $\alpha \in \partial \Omega_n$. Hence by the ordinary maximum principle,

$$u(z) \le u_{k,n}(z) = \int_{\partial \Omega_n} V_k(\zeta) d\omega(z, \zeta, \Omega_n)$$

for $z \in \Omega_n$. Then by the weak-star convergence

$$u(z) \le \int_{\partial \Omega} V_k(\zeta) d\omega(z, \zeta, \Omega), \tag{8.4}$$

so that by monotone convergence

$$u(z) \le \int_{\partial \Omega} V(\zeta) d\omega(z, \zeta, \Omega), \tag{8.5}$$

$z \in \Omega$. Then (8.3) and (8.5) imply that $u(z) \le m$ in $\Omega = \bigcup \Omega_n$. ∎

Theorem 8.2. *Suppose A is a Borel set. Then A is a polar set if and only if*

$$\omega(z, A \cap \partial\Omega, \Omega) = 0 \tag{8.6}$$

for every domain Ω such that $\mathrm{Cap}(\mathbb{C}^* \setminus \Omega) > 0$ and for every $z \in \Omega$.

Proof. By (4.8) and the regularity of harmonic measure, we may assume that the set A is compact. If $\mathrm{Cap}(A) = 0$, then (8.6) follows from (5.4). Conversely, if $\mathrm{Cap}(A) > 0$ let μ_A be the equilibrium distribution for A and let Ω be the unbounded component of $\mathbb{C} \setminus A$. Then $\mathrm{Cap}(\mathbb{C}^* \setminus \Omega) = \mathrm{Cap}(A) > 0$ and $\omega(\infty, A \cap \partial\Omega, \Omega) = \mu_A(A) = 1$. ∎

More quantitative forms of Theorem 8.2 will be given in the next section. Because of Theorem 8.2, Nevanlinna [1953] called a polar set a set of **absolute harmonic measure** 0.

Corollary 8.3. *Suppose Ω is a domain such that $\mathrm{Cap}(\mathbb{C}^* \setminus \Omega) > 0$, and suppose u is subharmonic and bounded above in Ω. If*

$$\limsup_{z \to \zeta} u(z) \le m$$

for all $\zeta \in \partial\Omega \setminus A$, where A is a Borel set and $\mathrm{Cap}(A) = 0$, then

$$\sup_{\Omega} u(z) \le m.$$

Corollary 8.3 is immediate from Theorems 8.1 and 8.2.

Corollary 8.4. *Suppose E is a compact polar set. If $u(z)$ is subharmonic and bounded above in $\mathbb{C}^* \setminus E$, then u is constant. If $\Omega \supset E$ is a Jordan domain then every bounded harmonic function on $\Omega \setminus E$ extends to be harmonic on E. Conversely, if $\mathrm{Cap}(E) > 0$, if Ω is the unbounded component of $\mathbb{C}^* \setminus E$, and if A is a compact subset of E such that $0 < \mu_E(A) < 1$, then $u(z) = \omega(z, A, \Omega)$ is a non-constant bounded harmonic function on Ω.*

If E is a compact polar set then by Corollary 8.4 every positive harmonic function in $\mathbb{C}^* \setminus E$ is constant and every bounded harmonic function in $\mathbb{C}^* \setminus E$ is constant.

Proof. Let $z_0 \in \mathbb{C}^* \setminus E$, $\delta < \mathrm{dist}(z_0, E)$, and $\Omega = \mathbb{C}^* \setminus (E \cup \overline{B}(z_0, \delta))$. Then $\mathrm{Cap}(\mathbb{C}^* \setminus \Omega) > 0$ and by Corollary 8.3,

$$\sup_{\Omega} u(z) \leq \sup_{\partial B(z_0, \delta)} u(z).$$

Letting $\delta \to 0$ we obtain

$$\sup_{\mathbb{C}^* \setminus E} u(z) \leq u(z_0).$$

Because z_0 is arbitrary in $\mathbb{C}^* \setminus E$, u is constant. To prove the second statement, we shrink Ω so that u is continous on $\partial\Omega$. Let v be the solution to the Dirichlet problem on Ω for the function u. Then by Corollary 8.3, $u = v$ and v is the harmonic extension of u to Ω. The converse assertion is clear. ∎

9. Estimates for Harmonic Measure

We use capacity and potentials to estimate $\omega(z, E, \mathbb{D} \setminus E)$ where E is a closed subset of the unit disc $\mathbb{D}$. By (5.3) we may assume $\partial E \subset \mathbb{D}$ is compact and consists of finitely many analytic curves. First consider the case when $\mathrm{dist}(E, \partial\mathbb{D})$ is large.

Theorem 9.1. *Suppose $E \subset \{z : 0 < \delta < |z| < r < 1\}$ and set*

$$\gamma(E) = -\log\left(\mathrm{Cap}(E)\right).$$

Then there are constants $c_1(r)$ and $c_2(\delta, r)$ such that

$$\frac{c_1(r)}{\gamma(E)} \leq \omega(0, E, \mathbb{D} \setminus E) \leq \frac{c_2(\delta, r)}{\gamma(E)}. \tag{9.1}$$

Proof. Let

$$g(z, \zeta) = \log \left| \frac{1 - \bar{z}\zeta}{z - \zeta} \right|$$

be Green's function on $\mathbb{D}$ and let μ_E be the equilibrium distribution for E. Then the Green potential

$$V(z) = \int_{\partial E} g(z, \zeta) d\mu_E(\zeta)$$

is continuous on $\overline{\mathbb{D} \setminus E}$ and harmonic on $\mathbb{D} \setminus E$. Also, $\log \frac{1}{r} \leq V(0) \leq \log \frac{1}{\delta}$, $V = 0$ on $\partial \mathbb{D}$, and for $z \in E$,

$$\gamma(E) + \log(1 - r^2) \leq V(z) \leq \gamma(E) + \log(1 + r^2)$$

by (2.3). Hence

$$\frac{\log \left(\frac{1}{r} \right)}{\gamma(E) + \log(1 + r^2)} \leq \omega(0, E, \mathbb{D} \setminus E) \leq \frac{\log \left(\frac{1}{\delta} \right)}{\gamma(E) + \log(1 - r^2)}, \qquad (9.2)$$

provided that $\gamma(E) + \log(1 - r^2) > 0$. This gives the left-hand inequality in (9.1) with $c_1(r) = (\log r)^2 / \log(r + 1/r)$, because $\gamma(E) \geq \log(1/r)$. If $\gamma(E) \geq 2 \log 1/(1 - r^2)$ the upper bound in (9.2) gives the upper bound in (9.1) with constant $2 \log(1/\delta)$, and if $\gamma(E) \leq 2 \log 1/(1 - r^2)$, the upper bound in (9.1) follows from the trivial estimate

$$\omega(0, E, \mathbb{D} \setminus E) \leq 1 \leq \frac{2 \log \frac{1}{1-r^2}}{\gamma(E)}. \qquad \blacksquare$$

The next result is called the **Beurling projection theorem**.

Theorem 9.2 (Beurling). *If $E \subset \overline{\mathbb{D}} \setminus \{0\}$, and $E^* = \{|z| : z \in E\}$ is the circular projection of E, then*

$$\omega(z, E, \mathbb{D} \setminus E) \geq \omega(-|z|, E^*, \mathbb{D} \setminus E^*).$$

Proof. By Green's theorem

$$\omega(z, E, \mathbb{D} \setminus E) = \omega(z) = \frac{1}{2\pi} \int_{\partial E} g(z, \zeta) \frac{\partial \omega(\zeta)}{\partial n_\zeta} ds$$

$$(9.3)$$

$$= \int_{\partial E} g(z, \zeta) d\sigma(\zeta),$$

in which the normal n_ζ points out of $\mathbb{D} \setminus E$, so that $\sigma \geq 0$. Consider the circular projection σ^* of σ,

$$\sigma^*(A) = \sigma(\{z \in E : |z| \in A\}),$$

and its Green potential

$$V(z) = \int_{(\partial E)^*} g(z, |\zeta|)d\sigma^*.$$

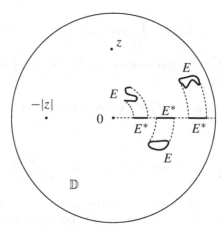

Figure III.5 Circular Projection in $\mathbb{D}$.

Because

$$g(-|z|, |\zeta|) \le g(z, \zeta) \le g(|z|, |\zeta|),$$

we have

$$V(-|z|) \le \omega(z, E, \mathbb{D} \setminus E) \le V(|z|).$$

Let $z_n \to z \in E$. By the right-hand inequality,

$$\liminf V(|z_n|) \ge 1,$$

so that by a comparison, $\omega(z, E^*, \mathbb{D} \setminus E^*) \le V(z)$. With the left-hand inequality that gives

$$\omega(-|z|, E^*, \mathbb{D} \setminus E^*) \le V(-|z|) \le \omega(z),$$

which proves the theorem.   ∎

Corollary 9.3. *Let Ω be a simply connected domain, let $z \in \Omega$ and let $\zeta \in \partial\Omega$. Then for r sufficiently small,*

$$\omega(z, B(\zeta, r) \cap \partial\Omega, \Omega) \le Cr^{\frac{1}{2}}.$$

Proof. We may assume $\zeta = 0$ and $z = -1$. We project $\partial\Omega \setminus B(0, r)$ to the positive real axis and apply Beurling's theorem in $\mathbb{C}^* \setminus B(0, r)$ to get

$$\omega(z, B(0, r) \cap \partial\Omega, \Omega) \le \omega\left(-1, B(0, r), \mathbb{C} \setminus (B(0, r) \cup [r, \infty))\right)$$
$$= \frac{2}{\pi} \arg\left(\frac{i - \sqrt{r}}{i + \sqrt{r}}\right) = \frac{4}{\pi} \tan^{-1} \sqrt{r}. \qquad \blacksquare$$

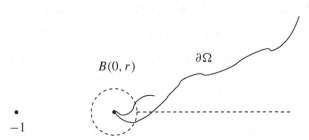

$B(0, r)$ $\partial\Omega$

-1

Figure III.6 The proof of Corollary 9.3.

The case when $\text{dist}(E, \partial\mathbb{D})$ is small leads to the connection between harmonic measure and Carleson measures. We work in the upper half-plane $\mathbb{H}$ and with dyadic Carleson boxes of the form

$$Q = \{j2^{-k} \le x \le (j + 1)2^{-k}; 0 < y \le 2^{-k}\},$$

and write $\ell(Q) = 2^{-k}$. Fix $z_0 = i$ and for convenience assume E is compact and

$$E \subset \{0 \le x \le 1, \ 0 < y \le 1/2\}. \tag{9.4}$$

Let $\mathcal{A} = \mathcal{A}_E$ be the set of positive measures σ on E such that

(i) σ is a Carleson measure with constant 1,

$$\sigma(Q) \le \ell(Q), \tag{9.5}$$

for every dyadic Carleson box Q (see Exercise I.19), and

(ii) the top half $T(Q) = Q \cap \{y \ge \ell(Q)/2\}$ of every dyadic Carleson box Q satisfies

$$\frac{\sigma(T(Q))}{\ell(Q)} \le \frac{1}{1 + \gamma(E \cap T(Q)) - \gamma(T(Q))}$$

$$\tag{9.6}$$

$$= \frac{1}{1 + \log\left(\frac{\text{Cap } T(Q)}{\text{Cap } E \cap T(Q)}\right)}.$$

Theorem 9.4. *Assume E satisfies (9.4). Then there are constants C_1 and C_2 such that*

$$C_1\omega(i, E, \mathbb{H} \setminus E) \le \sup\{\sigma(E) : \sigma \in \mathcal{A}_E\} \le C_2\omega(i, E, \mathbb{H} \setminus E). \quad (9.7)$$

In the special case $z = 0$ and $E \subset \{|\zeta| > 1/2\}$, Beurling's projection theorem (with a larger constant) is a consequence of Theorem 9.4. Suppose $\sigma_0(A) = \omega(0, A \cap E^*, \mathbb{D} \setminus E^*)$ and let σ be any positive measure on E that projects circularly onto σ_0. Then an elementary comparison gives

$$\sigma(Q) \le \sigma_0(Q^*) \le C\ell(Q^*)$$

for some constant C, so that $C^{-1}\sigma$ satisfies (9.5). To check (9.6) for $C\sigma$ notice that by (4.12) the right side of (9.6) becomes smaller when E is replaced by E^* and then apply the right-hand inequality in (9.2) with $\delta = 1 - \ell(Q)$ and $r = (1 + \delta)/2$.

When $E \subset \mathbb{H}$ we still write $E^* = \{|z| : z \in E\}$ for the circular projection of E, but for $z = x + iy \in \mathbb{H}$ we define $z^* = -|x| + iy$. The estimate

$$\omega(z, E, \mathbb{H} \setminus E) \ge \frac{2}{3}\omega(z^*, E^*, \mathbb{H})$$

from Hall [1937] is known as **Hall's lemma**. With a different constant, the most important special case, $E \subset \{\arg z < \pi/4\}$ and $z = z^*$ of Hall's lemma also follows easily Theorem 9.4. See Exercise 20.

Proof of Theorem 9.4. Write $\zeta = \xi + i\eta \in E$. Green's function for the half-plane $\mathbb{H}$ is

$$g(z, \zeta) = \log\left|\frac{z - \bar{\zeta}}{z - \zeta}\right|,$$

and $g(z, \zeta)$ satisfies

$$g(z, \zeta) = \log\frac{\eta}{|z - \zeta|} + O(1), \quad \text{if } \frac{|z - \zeta|}{|z - \bar{\zeta}|} \le c < 1, \quad (9.8)$$

and

$$g(z, \zeta) \sim \left|\frac{z - \bar{\zeta}}{z - \zeta}\right|^2 - 1 = \frac{4y\eta}{|z - \zeta|^2}, \quad \text{if } |z - \zeta| > \frac{\eta}{4}. \quad (9.9)$$

The **Green potential** of a measure μ on $\mathbb{H}$ is

$$G_\mu(z) = \int_{\mathbb{H}} g(z, \zeta)d\mu(\zeta).$$

Let μ_Q denote the (logarithmic) equilibrium distribution of $E \cap T(Q)$. Then for $z \in E \cap T(Q)$ we have

$$G_{\mu_Q}(z) = \int g(z, \zeta) d\mu_Q(\zeta) \sim 1 + \gamma(E \cap T(Q)) - \gamma(T(Q))$$

by (9.8). Thus if we define

$$\nu_Q = \frac{\mu_Q}{1 + \gamma(E \cap T(Q)) - \gamma(T(Q))},$$

then for some constant C

$$C^{-1} \le \int_{E \cap T(Q)} g(z, \zeta) d\nu_Q \le C \tag{9.10}$$

on $E \cap T(Q)$.

To prove the left half of (9.7) we may assume $E \subset \{y \ge 2^{-N}\}$. We inductively define a measure $\nu = \nu_N$ on E as follows: Set $\nu_0 = 0$. Assume that ν_k is defined and has support in $E \cap \{\zeta : \eta \le 2^{-N+k}\}$. Let

$$Q \in \mathcal{D}_k = \{Q \text{ dyadic} : \ell(Q) = 2^{-N+k+1} \text{ and } \mathrm{Cap}(E \cap T(Q)) > 0\}.$$

If $\inf_{E \cap T(Q)} G_{\nu_k} < 1$, take $\beta_Q > 0$ such that

$$\inf_{E \cap T(Q)} (G_{\nu_k}(z) + \beta_Q G_{\nu_Q}(z)) = 1. \tag{9.11}$$

If $\inf_{E \cap T(Q)} G_{\nu_k} \ge 1$, take $\beta_Q = 0$. Then define

$$\nu_{k+1} = \nu_k + \sum_{\mathcal{D}_{k+1}} \beta_Q \, \nu_Q.$$

Finally take $\nu = \nu_N$, and $\sigma = \eta \nu$. Then by construction $G_\nu \ge 1$ on E and $G_\nu = 0$ on $\mathbb{R}$. But by (9.4) and (9.9),

$$G_\nu(i) \le C \int_E \eta d\nu(\zeta) = C\sigma(E).$$

Therefore by the maximum principle

$$\omega(i, E, \mathbb{H} \setminus E) \le G_\nu(i) \le C\sigma(E),$$

and if we can show that $c\sigma \in \mathcal{A}_E$ for some constant c, then the left-hand inequality in (9.7) will hold with $C_1 = (cC)^{-1}$.

To verify (9.5) for $c\sigma$ let Q be a dyadic box of side 2^{-N+k+1} and first assume $\beta_Q > 0$. Then by (9.11) there exists $z \in E \cap T(Q)$ such that

$$1 = G_{\nu_k}(z) + \beta_Q G_{\nu_Q}(z)$$

$$\geq \int_Q g(z, \zeta) \frac{d\sigma}{\eta}$$

$$\geq C \frac{\sigma(T(Q))}{\ell(Q)} + C \int_{Q \setminus T(Q)} \frac{y}{|z - \zeta|^2} d\sigma$$

$$\geq C \frac{\sigma(Q)}{\ell(Q)},$$

and therefore (9.5) holds for Q. But if $\beta_Q = 0$ then $\sigma(T(Q)) = 0$ and

$$\sigma(Q) = \sum \{\sigma(Q') : Q' \subset Q, \ \beta_{Q'} > 0, \ Q' \ \text{maximal}\},$$

so that (9.5) also holds for Q.

To verify (9.6) for the dyadic box Q observe that

$$\frac{\sigma(T(Q))}{\ell(Q)} \leq \nu(T(Q))$$

$$\leq \frac{\beta_Q}{1 + \gamma(E \cap T(Q)) - \gamma(T(Q))}$$

$$\leq C$$

by (9.10) and (9.11).

To prove the right-hand part of (9.7) let $\sigma \in \mathcal{A}_E$ and set

$$\sigma' = \sum_{Q \text{ dyadic}} \sigma(T(Q)) \mu_Q \quad \text{and} \quad \nu = \frac{\sigma'}{\eta}.$$

Then G_ν is harmonic on $\mathbb{H} \setminus E$ and

$$G_\nu = 0 \tag{9.12}$$

on $\mathbb{R}$ since E is compact. By (9.4) and (9.9)

$$G_\nu(i) \sim \int_E \eta d\nu = \sigma(E).$$

Now let $z \in E$ and fix a dyadic box Q with $z \in T(Q)$. Then by (9.5) and (9.9)

$$\int_{|\zeta-z|\geq\ell(Q)/4} g(z,\zeta)d\nu(\zeta) \leq Cy \int_{|\zeta-z|\geq\ell(Q)/4} \frac{d\sigma'(\zeta)}{|z-\zeta|^2}$$

$$\leq Cy \sum_{n=0}^{\infty} \frac{\sigma'(\{|\zeta-z| \leq 2^{n-2}\ell(Q)\})}{2^{2n}\ell(Q)^2}$$

$$\leq C,$$

and by (9.6) and (9.8)

$$\int_{|\zeta-z|<\ell(Q)/4} g(z,\zeta)d\nu(\zeta) \leq C$$

for some constant C. Therefore $G_\nu(z) \leq C$ on E and by (9.12)

$$\omega(i, E, \mathbb{H} \setminus E) \geq C^{-1}G_\nu(i) \geq C\sigma(E).$$

That proves the right-hand inequality in (9.7). ∎

Every theorem in this section was proved using Green potentials and Green potentials have a theory parallel to the potential theory in Sections 2, 3, and 4. The **Green potential** of a signed measure σ supported on $E \subset \mathbb{D}$ is the function

$$G_\sigma(z) = \int_E g(z,\zeta)d\sigma(\zeta).$$

Assume E is compact and $\mathrm{Cap}(E) > 0$. Then by (9.3) and a limiting process there is a positive measure ν on E such that

$$\omega(z, E, \mathbb{D} \setminus E) = G_\nu(z). \qquad (9.13)$$

Moreover, there exists unique $\mu \in P(E)$ having minimal **Green energy**:

$$\int G_\mu d\mu = \inf\left\{\int G_\tau d\tau : \tau \in P(E)\right\} \equiv \gamma_G(E).$$

The minimum energy $\gamma_G(E)$ is called the **Green capacity** of E, and except for a set of capacity zero, $G_\mu = \gamma_G(E)$ on E. Then (9.13) becomes

$$\omega(z, E, \mathbb{D} \setminus E) = \frac{1}{\gamma_G(E)}G_\mu(z).$$

See Exercise 18 for the proofs of these results and for an application.

Notes

Sections 1 through 6 follow a UCLA lecture course given by Lennart Carleson. Most of Section 7 is in Wiener's original paper [1924b]. Actually, Wiener did not prove Theorem 7.1 in [1924b]. He proved the $\mathbb{R}^d$, $d \geq 3$, version of Theorem 7.1, but for $\mathbb{C} = \mathbb{R}^2$ he proved $\zeta \in \partial\Omega$ is regular if and only if

$$\sum_{n=1}^{\infty} \frac{2^n}{\gamma(J_n \setminus \Omega)} = \infty, \qquad \text{(N.1)}$$

where

$$J_n = \bigcup \{A_k : 2^{n-1} < k \leq 2^n\} = \{z : 2^{n-1} \leq \log_2 \frac{1}{|z - \zeta|} \leq 2^n\}.$$

In $\mathbb{R}^d$ Wiener's proof hinged on an elegant estimate of Green potentials (see [1924b], p. 133). In $\mathbb{R}^2$ the analogous estimate holds for the big rings J_n, but it is false for the rings A_n because $\log\frac{1}{|z|}$ decays more slowly in $\mathbb{R}^2$ than $\frac{1}{|X|}$ does in $\mathbb{R}^d$. The first proof of Theorem 7.1 was published by de la Vallée Poussin [1949]. See also Tsuji [1959]. Our proof follows Tsuji [1959] when the series diverges and copies Wiener [1924b] when it converges. See Exercise 9 for Wiener's proof of (N.1) and for the equivalence of (N.1) and (7.2). Corollary 7.3 is from Beurling's thesis [1933].

Theorem 9.4 is from Carleson [1982]. Potential theory is a vast subject and we have only touched on the two-dimensional theory. For the broader picture we refer to the books of Brelot [1965], Carleson [1967a], Landkof [1972], Ransford [1995], Tsuji [1959], and Wermer [1974]. See Doob [1984] for the connections with probability, and see the books of Heinonen, Kilpeläinen, and Martio [1993] and Adams and Hedberg [1996] for nonlinear potential theory.

Appendix E gives a geometric description of capacity by transfinite diameter and constructs Evans functions for sets of capacity zero.

Exercises and Further Results

1. Suppose $p(z) = a_k z^k + \ldots + a_0$ is a polynomial. Set $\Omega = \{|p| > \varepsilon\}$ and $\omega = \omega(\infty, ., \Omega)$. Then

   (i) $g_\Omega(z, \infty) = \frac{1}{k} \log \left| \frac{p(z)}{\varepsilon} \right|.$

   (ii) $\gamma(\partial\Omega) = \frac{1}{k} \log \left| \frac{a_k}{\varepsilon} \right|.$

   (iii) $d\omega = -\frac{\partial g}{\partial n} \frac{ds}{2\pi} = \frac{1}{k} \frac{p'(z)}{p(z)} \frac{dz}{2\pi i}.$

Exercises and Further Results 113

(iv) The harmonic measure at ∞ of a component J of $\partial\Omega$ is the number of zeros of p in the bounded component of $\mathbb{C} \setminus J$, divided by k.

(v) Formulate and prove similar results for rational functions.

2. Let Ω be a finitely connected Jordan domain such that $\infty \in \Omega$ and every component of $\partial\Omega$ is a C^2 curve. Let $G(z) = g_\Omega(z, \infty)$ and let $\{z_j\}$ be the (finite) set of critical points of G in Ω. Prove

$$\frac{1}{2\pi} \int_{\partial\Omega} \frac{\partial G}{\partial n} \log \left| \frac{\partial G}{\partial n} \right| ds = \gamma(\partial\Omega) + \sum G(z_j),$$

where $\gamma(\partial\Omega)$ is Robin's constant for $\partial\Omega$. This identity from Ahlfors [1947] will play a critical role in Chapter IX. Hint: Use Green's theorem and the Taylor expansion of $G_x - iG_y$ near each critical point z_j.

3. Let Ω be a bounded plane domain and let $v(z)$ be subharmonic in a neighborhood of $\overline{\Omega}$. Then there is a unique positive finite measure μ supported on Ω and a unique function $u(z)$ harmonic on Ω such that

$$v(z) = u(z) - U_\mu(z)$$

on Ω. Hint: If $v \in C^2(\overline{\Omega})$, take $d\mu = \frac{\Delta v}{2\pi} dxdy$. In the general case, replace v by $v * K_\varepsilon$ as in the proof of Theorem 3.1 and take a limit. This result is known as the F. Riesz decomposition theorem for subharmonic functions.

4. (a) Let g be a compactly supported real valued C^2 function on the plane. Then $g = \varphi_1 - \varphi_2$ where φ_1 and φ_2 are C^2 and subharmonic on an open set containing $\{z : g(z) \neq 0\}$.

(b) Let Ω be a bounded plane domain and let $\Omega_n \supset \overline{\Omega_{n-1}} \ldots$ be an increasing sequence of finitely connected subdomains of Ω such that $\Omega = \bigcup_n \Omega_n$ and $\partial\Omega_n$ consists of C^2 Jordan curves. Fix $z \in \Omega_1$ and consider the sequence μ_n of probability measures on $\overline{\Omega}$ defined by $\mu_n(E) = \omega(z, E \cap \partial\Omega_n, \Omega_n)$. Show that if φ is continuous on $\overline{\Omega}$ and subharmonic on Ω, then

$$\int \varphi d\mu_n \leq \int \varphi d\mu_{n+1}.$$

Using this result, part (a), and the boundedness of $\{\mu_n\}$, prove that for all $f \in C(\overline{\Omega})$,

$$\lim_n \int f d\mu_n \quad \text{exists.}$$

It follows that for bounded Ω the Wiener definition of harmonic measure does not depend on the approximating sequence Ω_n.

(c) In Theorem 4.1, if $\gamma = \lim \gamma_n = \infty$, then for any $\mu \in P(E)$ the sequence $\{E_n\}$ can be chosen so that μ is the weak-star limit of the sequence $\{\mu_n\}$.

5. A property is said to hold **p.p.** (for presque partout) on a set E if the subset of E where the property fails has capacity zero. Suppose E is compact and $\gamma(E) > 0$. Let $M(E) = \{v : v$ is a signed measure supported on $E\}$. Then

$$\frac{1}{\gamma(E)} = \inf\{v(E) : v \in M(E) \text{ and } U_v \geq 1 \text{ p.p. on } E\}$$

$$= \sup\{v(E) : v \in M(E) \text{ and } U_v \leq 1 \text{ p.p. on } E\}.$$

6. Let $\Omega = \mathbb{D} \setminus \{0\}$ and let $f \in C(\partial\Omega)$. Prove $\lim_{z \to 0} u_f(z) = f(0)$ if and only if $f(0) = \int_{\partial\mathbb{D}} f(e^{i\theta})\frac{d\theta}{2\pi}$.

7. Let

$$p(z) = z^d + a_{d-1}z^{d-1} + \ldots + a_0$$

be a monic polynomial of degree $d \geq 2$ and let $p^{(n)}$ be the n-th iterate

$$p^{(n)}(z) = p \circ p \circ \ldots \circ p(z)$$

and write

$$p^{(-n)} = \left(p^{(n)}\right)^{-1}.$$

The **Julia set** of $p(z)$ is

$$E = \{z : p^{(n)} \text{ is not a normal family in any neighborhood of } z\}.$$

Here "normal" includes convergence to the constant $f = \infty$. Thus, for example, if $p(z) = z^2$ then E is the unit circle $\{|z| = 1\}$. The **Fatou set** of $p(z)$ is the complement $C^* \setminus E$. The **basin of attraction** of ∞ for $p(z)$ is

$$\Omega = \{z : \lim_{n \to \infty} p^{(n)}(z) = \infty\}$$

and $\Omega \subset C^* \setminus E$. It is known that the Julia set E is a non-empty, compact, perfect set, that E is **completely invariant**, $E = p(E) = p^{-1}(E)$, and that $E = \partial\Omega$. See Chapter III of Carleson and Gamelin [1993]. This exercise gives Brolin's beautiful [1965] description of the harmonic measure $\omega(\infty, \cdot, \Omega)$.

(a) Let U_0 be a disc such that $\overline{U_0} \subset p(U_0)$, so that $C^* \setminus U_0 \subset \Omega$. The disc

$$U_0 = \{|z| < 1 + \sum_{j=0}^{d-1} |a_j|\}$$

is an example. If $U_n = p^{(-n)}(U_0)$ and $E_n = \overline{U_n}$ then $p(E_{n+1}) = E_n$, $p^{-1}(E_n) = E_{n+1}$, and $\bigcap E_n = E$. Let $g_n(z)$ denote Green's function for $C^* \setminus E_n$ with pole at ∞. Then

$$g_n(z) = \log|z| + \gamma_n + o(1),$$

where $\text{Cap}(E_n) = e^{-\gamma_n}$. Show that

$$g_{n+1}(z) = \frac{g_n(p(z))}{d},$$

by uniqueness. Then conclude that $\gamma_{n+1} = \gamma_n/d$ and $\text{Cap}(E) = 1$.

(b) Let μ_E be the equilibrium distribution for E. It is known that if V is open and $V \cap E \neq \emptyset$, then for some n, $p^{(n)}(V \cap E) = E$. Again see Carleson and Gamelin [1993]. Use this fact and Theorem 4.5 to show that μ_E has closed support E.

(c) If V is a neighborhood of $z \in E$ then by Montel's theorem $\bigcup_n p^{(n)}(V)$ can omit at most two values ζ_1 and ζ_2, and ζ_1 and ζ_2 do not depend on the point z. Fix any point $z_0 \notin \{\zeta_1, \zeta_2\}$ and take the point mass $\mu_0 = \delta_{z_0}$. Write $p^{(-n)}(z_0) = \{z_1, \ldots, z_{d^n}\}$, counting points with their multiplicities, and form the discrete measure

$$\mu_n = \frac{1}{d^n} \sum_j \delta_{z_j}.$$

Show $\mu_n(p(A)) = \mu_{n+1}(A)$ for any Borel set A. Show μ_n converges weak-star to μ_E. Hint: Use the identity

$$|p^{(n)}(z) - z_0| = \prod_j |z - z_j|$$

to estimate U_{μ_n}, and use Lemma 4.2 to show that any weak-star convergent subsequence of $\{\mu_n\}$ has limit μ_E.

(d) If $A \subset E$, then $\mu_E(p^{-1}(A)) = \mu_E(A)$. In other words, p is a **measure preserving transformation** on the measure space (E, μ_E). Moreover, p is **strongly mixing**: If $f, g \in L^2(E, \mu_E)$, then

$$\lim_{n \to \infty} \int f(p^{(n)}(z))g(z)d\mu_E(z) = \int f d\mu_E \int g d\mu_E.$$

8. (a) Prove ζ is a regular point for Ω if and only if $\lim_{z \to \zeta} g(z, z_0) = 0$. Give a proof with Wiener series and give a proof without Wiener series.

(b) Prove ζ is a regular point for Ω if and only if ζ is a regular point for $\Omega \cap B(\zeta, r), r > 0$. Give a proof with Wiener series and give a proof without Wiener series.

(c) Let $\Omega_j = \mathbb{C} \setminus E_j$, where E_1 and E_2 are disjoint compact sets and let $\Omega = \Omega_1 \cap \Omega_2$. Prove $\zeta \in \partial\Omega$ is regular for Ω if and only if ζ is regular for Ω_1 or ζ is regular for Ω_2.

9. Let $\zeta \in \partial\Omega$. (a) Let $\alpha < 1$ and set $G_n = \{z : \alpha^n \leq |z - \zeta| \leq \alpha^{n-1}\}$. Prove

ζ is a regular point if and only if

$$\sum_{n=1}^{\infty} \frac{n}{\gamma(G_n \setminus \Omega)} = \infty,$$

and if and only if

$$\sum_{n=1}^{\infty} \frac{n}{\gamma(G_n \cap \partial\Omega)} = \infty.$$

(b) Lemma 7.5 has a converse. Let $J_n = \bigcup\{A_k : 2^{n-1} < k \leq 2^n\}$. There is a universal constant $C < \infty$ so that if $\gamma(J_n \setminus \Omega) > C2^n$ and if $E_k \subset A_k$ for $2^{n-1} < k \leq 2^n$ then

$$\sum_{2^{n-1}+1}^{2^n} \frac{1}{\gamma(E_k)} \leq \frac{4}{\gamma(\cup E_k)}.$$

Hint: If $\mu = \sum \mu_k / \gamma(E_k)$ where μ_k is the equilibrium distribution for E_k, then $\|\mu\| = \sum 1/\gamma(E_k)$. By estimating $\int U_{\mu_k} d\mu_m$ as in the proof of Theorem 7.1, show that $I(\mu) \leq 3\|\mu\| + C_1 2^n \|\mu\|^2$. Now apply Theorem 4.1 and the hypothesis on $\gamma(J_n \setminus \Omega)$ above.

(c) Set $B_r = \{z : |z - \zeta| \leq r\}$. Prove that the following conditions are equivalent to the regularity of ζ:

(i) $\displaystyle\sum \frac{2^n}{\gamma(J_n \setminus \Omega)} = \infty,$

(ii) $\displaystyle\int_0^{\infty} \frac{1}{\gamma(B_{2^{-s}} \setminus \Omega)} ds = \infty,$

(iii) $\displaystyle\int_0^1 \frac{1}{\gamma(B_r \setminus \Omega)} \frac{dr}{r} = \infty,$

(iv) $\displaystyle\sum \frac{1}{\gamma(B_{2^{-n}} \setminus \Omega)} = \infty,$

(v) $\displaystyle\sum \frac{n}{\gamma(A_n \setminus \Omega)} = \infty.$

Hint: Note that the integrand in (ii) is decreasing. Bound the left side of each displayed formula by a multiple of the left side of the subsequent formula. If the first sum is infinite, then apply Theorem 7.1 to conclude ζ is regular, and if the first sum is finite, use part (b) to show the sum in (v) is finite.

(d) Wiener's paper [1924b] proved that (i) implies ζ is a regular point. Here is his argument. Suppose $\zeta = 0$ and suppose

$$\sum_{n=1}^{\infty} \frac{2^{2n}}{\gamma_{2n}} = \infty, \tag{E.1}$$

where $\gamma_n = \gamma(J_n \setminus \Omega)$. Set $r_n = 2^{-2^n}$. Then $J_n = \{z : r_n \le |z| \le r_{n-1}\}$ and $J_n \subset B_{n-1} = \{z : |z| \le r_{n-1}\}$. Define

$$W_n(z) = \int_{J_n \setminus \Omega} \log \left| \frac{r_{n-2} - z\bar{w}/r_{n-2}}{z - w} \right| \frac{d\mu_n(w)}{\gamma_n},$$

where μ_n is the equilibrium distribution for $J_n \setminus \Omega$. The function W_n is a Green potential because the kernel in its definition is Green's function for the disc B_{n-2}. Then

$$W_n(z) = 0 \quad \text{on } |z| = r_{n-2}, \tag{E.2}$$

$$W_n(z) \le 1 + \frac{1}{\gamma_n} \log \left| r_{n-2} + \frac{r_{n-1}^2}{r_{n-2}} \right| \le 1 \quad \text{on } J_n \cap \Omega, \tag{E.3}$$

for $n \ge 2$, and the critical

$$W_n(z) \ge \frac{1}{\gamma_n} \log \left| \frac{r_{n-2} - r_{n+1}r_{n-1}/r_{n-2}}{r_{n+1} + r_{n-1}} \right| \ge \frac{C2^n}{\gamma_n} \quad \text{on } |z| \le r_{n+1}. \tag{E.4}$$

Let $V_{n,m}$ be the Wiener solution to the Dirichlet problem on the domain $B_0 \setminus (\bigcup_{p=0}^{m} J_{2n+2p} \setminus \Omega)$ for the boundary data

$$V_{n,m}(z) = \begin{cases} 0, & \text{if } z \in \bigcup_{p=0}^{m} J_{2n+2p} \setminus \Omega, \\ 1, & \text{if } |z| = r_0. \end{cases}$$

We show by induction on m that

$$V_{n,m} \le \prod_{p=0}^{m} \left(1 - \frac{C2^{2n+2p}}{\gamma_{2n+2p}} \right) \quad \text{on } \{z : |z| < r_{2n+2m+1}\}. \tag{E.5}$$

To prove (E.5), we may assume $\bigcup_{p=0}^{m} J_{2n+2p} \setminus \Omega$ is bounded by disjoint C^2 Jordan curves. Then by the maximum principle, (E.2), (E.3), and (E.4),

$$V_{n,0} \le 1 - W_{2n} \le 1 - \frac{C''2^{2n}}{\gamma_{2n}}.$$

on B_{2n+1}. If (E.5) holds for $m = k - 1$, then by the maximum principle

$$V_{n,k}(z) \le \left(1 - W_{2n+2k}(z) \right) \prod_{p=0}^{k-1} \left(1 - \frac{C2^{2n+2p}}{\gamma_{2n+2p}} \right)$$

on $B_{2n+2k+1}$, so that by (E.4), we obtain (E.5) when $m = k$. Now by (E.1) and (E.5)

$$\lim_{m \to \infty} V_{n,m}(0) = 0.$$

For $|z| = r_{2n-2}$ and all m,

$$V_{n,m}(z) \geq \omega(z, B_0, B_0 \setminus D_n) \geq \varepsilon_n,$$

where $D_n = \{z : |z| \leq r_{2n-1}\}$ and it follows via Theorem 6.1 that 0 is a regular point. A similar argument works if $\sum \frac{2^{2n+1}}{\gamma_{2n+1}} = \infty$.

10. (a) Let $E \subset \{\frac{1}{2} \leq |z| \leq 1\}$ be a continuum with $E \cap \{z : |z| = \frac{1}{2}\} \neq \emptyset$ and $E \cap \{|z| = 1\} \neq \emptyset$. Then

$$\omega(0, E, \mathbb{D} \setminus E) \geq \frac{2}{\pi} \tan^{-1}\left(\frac{1}{\sqrt{8}}\right).$$

This follows from Beurling's projection theorem, but it also has an elementary proof, albeit with a smaller constant: We can assume $\frac{1}{2} \in E$. Set $\overline{E} = \{\overline{z} : z \in E\}$. Since E is a continuum, the component U of $\mathbb{D} \setminus (E \cup \overline{E})$ satisfying $0 \in U$ has $[\frac{1}{2}, 1] \cap U = \emptyset$ and $\partial U \subset \partial \mathbb{D} \cup E \cup \overline{E}$.

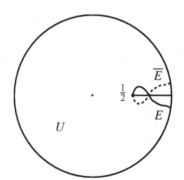

Figure III.7 Lower bound via reflection.

Then by the maximum principle,

$$\omega(z, E, \mathbb{D} \setminus E) + \omega(z, \overline{E}, \mathbb{D} \setminus \overline{E}) \geq \omega(z, (E \cup \overline{E}) \cap \partial U, U)$$

$$\geq \omega\left(z, [\frac{1}{2}, 1], \mathbb{D} \setminus [\frac{1}{2}, 1]\right).$$

Hence

$$2\omega(0, E, \mathbb{D} \setminus E) \geq \omega\left(0, [\frac{1}{2}, 1], \mathbb{D} \setminus [\frac{1}{2}, 1]\right) = \frac{2}{\pi} \tan^{-1}\left(\frac{1}{\sqrt{8}}\right).$$

See Milloux [1924] and Garnett [1986].

(b) Let $\varphi : \mathbb{D} \to \Omega$ be conformal, let I be an arc of $\partial\mathbb{D}$ with length $|I| < 1$ and center ζ_I and let $z_I = (1 - |I|)\zeta_I$. Let ω denote the harmonic measure on Ω for the point $\varphi(0)$. Show that if $C > 1$, then $\omega\big(B(\varphi(z_I), C\mathrm{dist}(\varphi(z_I), \partial\Omega))\big) \geq c|I|$, where c is a constant depending only on C.

11. Let K be a compact set and set $\Omega = \mathbb{C}^* \setminus K$. The set K is called **uniformly perfect** if there is a constant A such that

$$\mathrm{Cap}(K \cap B(z, \delta)) > A\delta \tag{E.6}$$

for all $z \in K$ and all $\delta < \mathrm{diam}(K)$.

(a) Condition (E.6) holds if and only if there is a constant $\eta > 0$ such that

$$\omega(z, B(z, 2d(z)), \Omega) \geq \eta$$

for all $z \in \Omega$, where $d(z) = \mathrm{dist}(z, K)$. Exercise IX.3 will give several other conditions equivalent to (E.6).

(b) Assume K satisfies (E.6), let U be an open subset of $\partial\Omega$ and let $u(z)$ be continuous on $\overline{\Omega}$ and harmonic on Ω. If $u = 0$ on U and if $J \subset U$ is compact, then

$$u(z) \leq C(\mathrm{dist}(z, J))^{\alpha}$$

for constants $C = C(J)$ and $\alpha = \alpha(A)$ satisfying $0 < \alpha \leq 1$. See Jerison and Kenig [1982a], Lemma 4.1. In particular, if K is uniformly perfect and $g(z) = g_{\Omega}(z, \infty)$, then on Ω,

$$g(z) \leq C(\mathrm{dist}(z, K))^{\alpha} \tag{E.7}$$

and

$$|g(z_1) - g(z_2)| \leq C'|z_1 - z_2|^{\alpha}.$$

(c) However, (E.7) does not imply (E.6). If

$$K = \{0\} \cup \bigcup_{n=1}^{\infty} \{|z| = 2^{-n^2}, |\arg z| \geq \varepsilon_n\}$$

for small ε_n, then (E.7) holds but (E.6) fails.

12. Øksendal [1972] showed that for any plane domain, harmonic measure is singular to area measure (although $\partial\Omega$ may have positive area). The following proof is from a lecture by Carleson: Let K be a compact subset of $\partial\Omega$ with $\mathrm{Area}(K) > 0$ and suppose ζ is a point of area density of K:

$$\frac{\mathrm{Area}(K \cap B(\zeta, \delta))}{\pi\delta^2} > 1 - \varepsilon$$

for small δ. Then there is $\frac{\delta}{2} < r(\delta) < \frac{3\delta}{4}$ so that

$$\frac{|\{|z - \zeta| = r(\delta)\} \setminus K|}{2\pi r(\delta)} < 6\varepsilon,$$

and by a comparison,

$$\omega(z, B(\zeta, \frac{\delta}{2}) \cap \partial\Omega, \Omega) < 42\varepsilon$$

on $\Omega \cap \{|z - \zeta| = \delta\}$. Taking $42\varepsilon < 1/4$ yields

$$2^{2n}\omega(z, B(\zeta, 2^{-n}) \cap \partial\Omega, \Omega) \longrightarrow 0,$$

and by the Radon–Nikodym theorem ω is singular to area measure.

13. Let Ω be a domain such that $\mathrm{Cap}(\partial\Omega) > 0$ and let $E \subset \partial\Omega$ be a Borel set such that $\omega(z, E, \Omega) > 0$ for some $z \in \Omega$. Prove

$$\sup_{\Omega} \omega(z, E, \Omega) = 1.$$

Find a proof without using the Perron solution.

14. Let Ω be a simply connected domain and let $E \subset \mathbb{R} \cap \partial\Omega$ have $|E| = 0$. Prove $\omega(z, E, \Omega) = 0$ for all $z \in \Omega$. Hint: The result is easy if $\Omega \subset \mathbb{H}$ or if $\Omega = \{\bar{z} : z \in \Omega\}$. In general, we can assume $\Omega_0 = \Omega \cap \mathbb{H} \neq \emptyset$. Then Ω_0 is simply connected. Write $\mathbb{R} \cap \Omega = \bigcup J_j$ where J_j is an open interval and let Ω_j be the component of $\Omega \cap \{\bar{z} : z \in \Omega\}$ with $J_j \subset \Omega_j$. Then Ω_j is also simply connected. Now if $z_0 \in \Omega_0$, then

$$\omega(z_0, E, \Omega) = \sum_j \int_{J_j} \omega(\zeta, E, \Omega) d\omega_{\Omega_0}(z_0, \zeta)$$

$$= \sum_j \int_{J_j} \int_{\partial\Omega_j \setminus \partial\Omega} \omega(z, E, \Omega) d\omega_{\Omega_j}(\zeta, z) d\omega_{\Omega_0}(z_0, \zeta),$$

but by symmetry $\omega_{\Omega_j}(\zeta, \partial\Omega_j \setminus \Omega, \Omega_j) = 1/2$ for $\zeta \in J_j$. The result now follows from Exercise 13. See Øksendal [1980].

15. (a) Let $0 \in \Omega$, let $\zeta \in \partial\Omega$ satisfy $|\zeta| > 1$, and let $0 < r < \frac{1}{2}$. Then

$$\omega(0, \partial\Omega \cap B(\zeta, r), \Omega) \leq C \sup_{|z - \zeta| < \frac{3r}{2}} g_\Omega(z, 0)$$

with constant C independent of Ω. Hint: Assume Ω is a finitely connected Jordan domain with smooth boundary. Take $\chi \in C^\infty(B(\zeta, \frac{3r}{2}))$ such that

$0 \le \chi \le 1$, $\chi(z) = 1$ on $B(\zeta, r)$, and $|\Delta \chi| \le \frac{C}{r^2}$. Write

$$\omega(0, \partial\Omega \cap B(\zeta, r), \Omega) \le \frac{-1}{2\pi} \int_{\partial\Omega} \chi \frac{\partial g(0, w)}{\partial n_w} ds(w)$$

and apply Green's theorem. See Jerison and Kenig [1982a].

(b) Let Ω be a simply connected domain containing the disc $\mathbb{D}$, let $z_1 \in \Omega$ satisfy $|z_1| > 1$, and set $g(z) = g_\Omega(z, z_1)$. Let Γ be the hyperbolic geodesic of Ω joining 0 to z_1 and for $0 \le t \le 1$ define

$$\delta(t) = \sup_{\Gamma \cap \{|z|=t\}} \text{dist}(z, \partial\Omega).$$

Then

$$g(0) \le C_0 \exp\left(-\frac{1}{2}\int_0^1 \frac{dt}{\delta(t)}\right).$$

Hint: Let $\rho(z, w)$ be the hyperbolic distance on Ω and for $0 \le t \le 1$ pick $z_t \in \Gamma$ with $|z_t| = t$. Then $g(z) \sim e^{-2\rho(z, z_1)}$ and by Koebe's estimate

$$\rho(0, z_1) \ge \int_\Gamma \frac{ds}{4 \, \text{dist}(z, \partial\Omega)} \ge \int_0^1 \frac{dt}{4\delta(t)}.$$

Carleson and Jones [1992].

(c) Let Ω be a simply connected domain containing the disc$\{|z| < \frac{1}{2}\}$, let $\zeta \in \partial\Omega$ satisfy $|\zeta| > 1$, and let

$$\delta(t) = \sup\{\text{dist}(z, \partial\Omega) : z \in \Omega, \ |z - \zeta| = t\}.$$

Then for $r < \frac{1}{4}$,

$$\omega(0, \partial\Omega \cap B(\zeta, r), \Omega) \le C \exp\left(-\frac{1}{2}\int_{\frac{3r}{2}}^1 \frac{dt}{\delta(t)}\right).$$

Hint: Use (a) and (b). See Jones and Makarov [1995].

(d) Part (c) is false with $\frac{3}{2}r$ replaced by r. Hint: $\zeta = 2$,

$$\Omega = 2\mathbb{D} \setminus \{|z - 2| \le r, \ |y| \ge \varepsilon\}.$$

(e) Let $\Omega = \{z : |\arg z| < \theta\}$ and $B_r = \{|z| \le r\}$. Use part (c) to show

$$\omega(1, B_r, \Omega \setminus B_r) \le Cr^{1/(2\sin\theta)}.$$

Also show by an explicit conformal map that

$$\omega(1, B_r, \Omega \setminus B_r) = Cr^{\pi/\theta}.$$

Theorem IV.6.2 gives the accurate upper bound $Cr^{\pi/\theta}$.

For related results see also Beurling [1933], Exercise 24, Theorem G.1, and the comments following the proof of Theorem G.1.

16. (a) The following result is from Gehring and Hayman [1962].

Let $\varphi : \mathbb{D} \to \Omega$ be univalent and assume φ extends continuously to the open arc $\sigma = \partial \mathbb{D} \cap \{\text{Im} z > 0\}$. Then

$$\ell\big(\varphi((-1, 1))\big) \leq K \ell\big(\varphi(\sigma)\big),$$

with constant K not depending on φ.

Set $I_j = [1 - 2^{-j}, 1 - 2^{-j-1}]$ and $\sigma_j = \sigma \cap (B(1, 2^{-j}) \setminus \overline{B(1, 2^{-j-1})})$, for $j = 0, 1, \ldots$. Then clearly

$$\inf_{z \in I_j} \omega(z, \sigma_j, \mathbb{D}) \geq c_1 > 0. \tag{E.8}$$

We may assume $\ell_j = \ell(\varphi(\sigma_j)) < \infty$. Suppose we have

$$\sup_{I_j} |\varphi'(z)| \leq c_2 2^j \ell_j \tag{E.9}$$

for all j. Then summing over j, and doing the same for $-I_j$ we get

$$\int_{-1}^1 |\varphi'(x)| dx \leq c_2 \ell(\varphi(\sigma)),$$

which gives the Gehring–Hayman theorem.

Now by Theorem I.4.3, (E.9) holds if and only if

$$\text{dist}(\varphi(z), \partial\Omega) \leq c_3 \ell_j \tag{E.10}$$

for some constant c_3 and for all $z \in I_j$. Let c_3 be large and let β_j be the arc length midpoint of the curve $\varphi(\sigma_j)$. Then

$$B_j = B(\beta_j, c_3 \ell_j) \supset \varphi(\sigma_j).$$

If (E.10) fails at z_j, then $\varphi(z_j) \notin B_j$. But using the inversion $Tz = \dfrac{\ell_j}{z - \beta_j}$ and the continuum $E = \mathbb{D} \cap T(\partial\Omega)$, we see by (E.8) and the Beurling projection theorem that (E.10) will hold at z_j if c_3 is sufficiently large.

(b) Here are two consequences. If there is an arc $\gamma \subset \mathbb{D}$, joining 0 to 1 such that

$$\ell(\varphi(\gamma)) = \int_\gamma |\varphi'(z)| ds = \ell < \infty,$$

then

$$\ell\big(\varphi((0, 1))\big) = \int_0^1 |\varphi'(x)| dx \leq K\ell. \tag{E.11}$$

Write $I_z = \partial\mathbb{D} \cap \{|\zeta - z| < 2(1 - |z|)\}$, for $z \in \mathbb{D}$. If φ extends continuously to I_z then $\ell(\varphi(I_z)) \geq K' \text{dist}(\varphi(z), \partial\Omega)$. Inequality (E.11) will be used in the next chapter.

See Heinonen and Näkki [1994], Heinonen and Rohde [1993], and Bonk, Koskela, and Rohde [1998] for extensions.

17. Suppose $E \subset \{\zeta : |\zeta| < \delta\}$ and suppose $\delta < r < 1$. Then there are constants $c_1(\delta, r)$ and $c_2(\delta, r)$ such that if $|z| = r$, then

$$\frac{c_1(\delta, r)}{\gamma(E)} \leq \omega(z, E, \mathbb{D} \setminus E) \leq \frac{c_2(\delta, r)}{\gamma(E)}.$$

This can be proved directly by mimicking the proof of Theorem 9.1, or derived from the statement of Theorem 9.1 by relating $\gamma(E)$ to $\gamma(TE)$ where T is a Möbius transformation of the disc.

18. Let $g(z, \zeta) = \log \left|\frac{1 - \bar\zeta z}{z - \zeta}\right|$ be Green's function for $\mathbb{D}$ and let σ be a compactly supported signed measure on $\mathbb{D}$.

(a) Prove that the Green potential $G_\sigma(z) = \int g(z, \zeta) d\sigma(\zeta)$ satisfies

$$J(\sigma) = \int G_\sigma d\sigma \geq 0$$

and $J(\sigma) = 0 \iff \sigma \equiv 0$. Hint: Follow the argument in Section 3.

(b) Let $E \subset \mathbb{D}$ be compact. Prove $\text{Cap}(E) > 0$ if and only if there exists $\mu \in P(E)$ such that $J(\mu) < \infty$. Assuming $\text{Cap}(E) > 0$, prove there is a unique $\mu \in P(E)$ such that

$$\int G_\mu d\mu = \inf \left\{\int G_\tau d\tau : \tau \in P(E)\right\} \equiv \gamma_G(E).$$

Also show that $G_\mu(z) = \gamma_G(E)$ for all $z \in E$ except for set of capacity zero and that on $\mathbb{D} \setminus E$

$$\omega(z, E, \mathbb{D} \setminus E) = \frac{1}{\gamma_G(E)} G_\mu(z).$$

(c) Let $E \subset [-1, 0)$ be compact and let $E^* = [-1, a]$ be a closed interval with the same logarithmic measure as E, $\int_E \frac{dt}{t} = \int_{E^*} \frac{dt}{t}$. Prove

$$\omega(0, E, \mathbb{D} \setminus E) > \omega(0, E^*, \mathbb{D} \setminus E^*).$$

This result, called the "Beurling shove theorem", is also from Beurling's thesis [1933]. See Essèn and Haliste [1989], Pruss [1999], and Betsakos and Solynin [2000] for refinements.

19. Write $\tilde z = x + i|y|$ when $z = x + iy$ and write $\tilde F = \{\tilde z : z \in F\}$.

(a) Let F be a compact subset of $\mathbb{D}$, and let $z \in \mathbb{D} \setminus F$. Prove

$$\omega(z, F, \mathbb{D} \setminus F) \geq \omega(\tilde{z}, \tilde{F}, \mathbb{D} \setminus \tilde{F}).$$

Hint: The left side is superharmonic in $\mathbb{D}$ and the right side is harmonic in $\mathbb{D} \setminus F$. Consequently we may assume $z \in \mathbb{D} \cap \mathbb{R}$. In fact, via a Möbius transformation of the disc, we can assume $z = 0$. Represent the right side as the Green potential of a positive measure μ on F. Write $F^+ = F \cap \{\mathrm{Im}\, z \geq 0\}$ and $F^- = F \cap \{\mathrm{Im}\, z < 0\}$. When $A \subset \tilde{F}$, define

$$v(A) = \mu(F^+ \cap A) + \mu(\overline{F^- \cap A})$$

and let $v(z)$ be the Green potential of v. Show $v(z) \geq \omega(z, \tilde{F}, \mathbb{D} \setminus \tilde{F})$.
(b) Let E be a compact plane set of positive capacity, let $\Omega = \mathbb{C} \setminus E$, and let $\Omega^* = \mathbb{C} \setminus \tilde{E}$. Prove

$$g_\Omega(x, w) \leq g_{\Omega^*}(x, \tilde{w})$$

for $x \in \mathbb{R} \cup \{\infty\}$ and $w \in \mathbb{C}$. Hint: We may take $x = \infty$. Now use the general and trivial fact

$$g_\Omega(\infty, w) = \lim_{R \to \infty} \log R\omega\big(w, \partial B(0, R), B(0, R) \setminus E\big).$$

(c) Let Ω be a simply connected domain with $0 \in \Omega$, let $a = \inf \Omega \cap \mathbb{R}$, $b = \sup \Omega \cap \mathbb{R}$, and

$$E = (-\infty, a) \cup \partial\Omega \cup (b, \infty),$$

and let Ω_1 be the component of $\mathbb{C} \setminus \tilde{E}$ containing the lower half-plane. Then Ω_1 is simply connected. Let $\psi : \Omega \to \mathbb{D}$ and $\psi_1 : \Omega_1 \to \mathbb{D}$ be conformal mappings with $\psi(0) = \psi_1(0) = 0$. Prove

$$|\psi'(x)| \leq 4|\psi_1'(x)|, \qquad x \in \mathbb{R}.$$

Consequently the Hayman–Wu theorem, Theorem I.5.1, can be reduced to the special case of $L \subset \Omega$. That case has been proved by B. Flinn [1983]. Parts (a), (b), and (c) are attributed to Baernstein in Fernández, Heinonen, and Martio [1989].

20. Hall's lemma asserts that if $E \subset \mathbb{H}$ is compact, $E^* = \{|z| : z \in E\}$, and $z^* = -|\mathrm{Re}\, z| + i\mathrm{Im}\, z$, then

$$\omega(z, E, \mathbb{H} \setminus E) \geq \frac{2}{3}\omega(z^*, E^*, \mathbb{H}). \qquad (E.12)$$

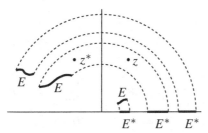

Figure III.8 Hall's lemma.

(a) Derive the special case $z = z^*$ and $E \subset \{\arg z < \frac{\pi}{4}\}$ of Hall's lemma from Theorem 9.4. Hint: For $0 < x < 1$, $x \leq 1/(1 + \log(1/x))$. Split E into $E \cap \mathbb{D}$ and $E \cap \mathbb{C} \setminus \mathbb{D}$, using a reflection to estimate the harmonic measure of the latter.

(b) The general case is also proved using Green potentials. See Hall [1937] or Duren [1970].

(c) Contrary to intuition, (E.12) is not true with the constant $\frac{2}{3}$ replaced by 1. See Hayman [1974] and Marshall and Sundberg [1989].

(d) (radial Hall's lemma) If $E \subset \mathbb{D}$ and if $\delta = \text{dist}(0, E) > 0$ then

$$\omega(0, E, \mathbb{D} \setminus E) \geq C\omega(0, E^*, \mathbb{D}),$$

where E^* is the radial projection of E onto $\partial \mathbb{D}$ and C is independent of E. Hint: Assume that E is the union of circular slits contained in $\{|z| > \frac{1}{2}\}$ such that no radius meets E more than once. Let V be the Green potential of $|d\zeta|/(|\zeta| \log(1/|\zeta|))$. Then $V(z) \leq c\omega(z, E, \mathbb{D} \setminus E)$ and $V(0) = |E^*|$ (Øksendal, private communication).

21. Let $Q = \{0 < x < 1, 0 < y < 1\}$ and let $\{Q_j\}$ be a sequence of pairwise disjoint dyadic squares

$$Q_j = \{k_j 2^{-n_j} < x < (k_j + 1)2^{-n_j}, 0 < y < 2^{-n_j}\}$$

with sidelength $\ell(Q_j) = 2^{n_j} \leq \frac{1}{4}$. Set $\Omega = Q \setminus \bigcup \overline{Q}_j$ and $z_0 = \frac{1}{2} + i\frac{3}{4}$. Prove that given $\varepsilon > 0$ there is $\delta > 0$ such that if $\omega(z_0, \bigcup \partial Q_j, \Omega) < \delta$ then $\sum \ell(Q_j) < \varepsilon$.

22. The following results are from Beurling's elegant paper [1940]. Also see Beurling's thesis [1933]. Let $u(z)$ be the Poisson integral of a real function $f \in L^2(\partial \mathbb{D})$.

(a) If f has Fourier series

$$\frac{a_0}{2} + \sum_{n=1}^{\infty} a_n \cos n\theta + b_n \sin n\theta,$$

then

$$S(f) \equiv D(u) = \iint_{\mathbb{D}} |\nabla u|^2 dxdy =$$

$$2\int_0^1 \int_0^{2\pi} |\frac{\partial u}{\partial r}|^2 r dr d\theta = \pi \sum n(a_n^2 + b_n^2).$$

(b) Assume $a_0 = u(0) = 0$, and assume $D(u) < \infty$. Set

$$U(Re^{i\theta}) = \int_0^R |\frac{\partial u}{\partial r}| dr.$$

Then

(i) $|u| \le U$,

(ii) U is subharmonic,

(iii) $F(\theta) = U(e^{i\theta}) \in L^2(\partial\mathbb{D})$,

and

$$\frac{1}{2\pi} \int_0^{2\pi} |F(\theta)|^2 d\theta \le \frac{1}{2} \sum n(a_n^2 + b_n^2).$$

(c) Prove

$$\int_0^{2\pi} \left(\frac{\partial U}{\partial \theta}(re^{i\theta})\right)^2 d\theta \le r^2 \int_0^{2\pi} \left(\frac{\partial U}{\partial r}(re^{i\theta})\right)^2 d\theta.$$

Hint: First suppose $U \in C^2$. Then by a partial integration (with respect to θ),

$$2\int_0^R \int_0^{2\pi} r^2 \frac{\partial U}{\partial r} \Delta U d\theta dr = \int_0^{2\pi} \left(r^2(\frac{\partial U}{\partial r})^2 - (\frac{\partial U}{\partial \theta})^2\right) d\theta.$$

(d) Because

$$D(U) = \iint \left((\frac{\partial U}{\partial r})^2 + \frac{1}{r^2}(\frac{\partial U}{\partial \theta})^2\right) d\theta,$$

part (c) implies that $D(U) \le D(u)$ and by the Dirichlet principle (see Appendix C), $S(F) \le S(f)$. In fact, $S(F) < S(f)$ because U is not harmonic.

(e) Let μ be a measure on $\partial\mathbb{D}$ and define

$$h_n = \int \cos n\theta d\mu(\theta),$$

$$k_n = \int \sin n\theta d\mu(\theta).$$

Then

$$U_\mu(re^{i\theta}) = \int \log\left|\frac{1}{re^{i\theta} - e^{it}}\right| d\mu(t) = \sum_0^\infty \frac{r^n}{n}(h_n \cos n\theta + k_n \sin n\theta).$$

(f) If $E \subset \partial\mathbb{D}$ is compact, if $\mathrm{Cap}(E) < \infty$, and if $\mu = \mu_E$, then

$$I(\mu) = \lim_{r\to 1}\int U_\mu(re^{i\theta})d\mu(\theta) = \sum \frac{h_n^2 + k_n^2}{n}.$$

If $S(f) < \infty$, then

$$\lim_{r\to 1}\int u(re^{i\theta})d\mu(\theta) = \sum(a_n h_n + b_n k_n) \le \sqrt{\frac{1}{\pi}S(f)}\sqrt{I(\mu)}.$$

(g) If $S(f) \le \pi$, then

$$E_\lambda = \{e^{i\theta} : \limsup_{r\to 1}|u(re^{i\theta})| > \lambda\}$$

satisfies

$$\mathrm{Cap}(E_\lambda) \le e^{-\lambda^2}.$$

(h) Therefore, if $S(f) < \infty$, then except on a set of capacity zero, $\lim_{r\to 1} u(re^{i\theta})$ exists, and in fact, $\int_0^1 |\frac{\partial u}{\partial r}(re^{i\theta})|dr < \infty$.
(i) Prove the theorem of Fejér that if $f(z) = \sum_0^\infty c_n z^n$, $\sum n|c_n|^2 < \infty$, and $\lim_{r\to 1} f(re^{i\theta}) = \alpha$, then $\sum c_n e^{in\theta} = \alpha$. Thus if $S(f) < \infty$, the Fourier series $\sum a_n \cos n\theta + b_n \sin n\theta$ converges except on a set of capacity zero.

23. (a) Let $\varphi(z)$ be a univalent function on the unit disc. Prove that φ has a nontangential limit and that $\int_0^1 |\varphi'(r\zeta)|dr < \infty$ for all $\zeta \in \partial\mathbb{D} \setminus E$, where $\mathrm{Cap}(E) = 0$. This is also from Beurling [1940].
(b) Let $f(z) = \sum c_n z^n$ be analytic on $\mathbb{D}$, and assume $\sum n|c_n|^2 \le 1$. Then

$$E_\lambda = \{\zeta \in \partial\mathbb{D} : \int_0^1 |f'(r\zeta)|dr > \lambda\}$$

has $\mathrm{Cap}(E_\lambda) < e^{-\lambda^2}$. Consequently, $\frac{1}{2\pi}|E_\lambda| < e^{1-\lambda^2}$. The second estimate is from Beurling [1933]. See also Marshall [1989].

24. (a) Now let Ω be a simply connected domain, and let $z, z_0 \in \Omega$. Consider the family Γ of rectifiable curves γ in Ω connecting z to z_0. With Beurling, define

$$\lambda(z_0, z) = \sup\{\inf_{\gamma\in\Gamma}\int_\gamma |\psi'(z)|ds : \psi \text{ univalent on } \Omega, \mathrm{Area}(\psi(\Omega)) \le \pi\}.$$

Prove

$$e^{-2g(z,z_0)} + e^{-\lambda^2} = 1.$$

(b) Now let $E \subset \partial\Omega$. Consider the family Γ of rectifiable curves γ in Ω having one endpoint z_0 and the other endpoint in E. With Beurling, define

$$L(z_0, E) = \sup\{\inf_{\gamma \in \Gamma} \int_\gamma |\psi'(z)|ds : \psi \text{ univalent on } \Omega, \text{ Area}(\psi(\Omega)) \leq \pi\}.$$

Prove

$$\omega(z_0, E, \Omega) \leq e^{1-L^2(z_0, E)}.$$

This estimate, from Beurling's thesis [1933], is a precursor of extremal length. Note that (b) follows from 23(b).

IV

Extremal Distance

Extremal distance is a conformally invariant version of distance. As such it is a powerful tool for estimating conformal invariants like harmonic measure in terms of more geometric quantities. We begin the chapter with the basic properties of extremal length and of extremal distance, which is the most important application of extremal length. We then characterize certain extremal distances in Jordan domains by means of conformal mappings from the domains to slit rectangles. This characterization leads to fundamental estimates for harmonic measure by extremal distance and the famous integral

$$\int \frac{dx}{\theta(x)}.$$

1. Definitions and Examples

By definition, a **path family** in a domain Ω is a non-empty set Γ of countable unions of rectifiable arcs in Ω. An element $\gamma \in \Gamma$ is called a **curve** even though γ may not be connected and may have many self-intersections. The **euclidean length** of the path family Γ is

$$\inf_{\gamma \in \Gamma} \int_\gamma ds, \tag{1.1}$$

where ds denotes arc length. The quantity (1.1) is not conformally invariant, but a conformally invariant version of (1.1) is

$$\sup_\varphi \left\{ \inf_{\gamma \in \Gamma} \int_\gamma |\varphi'(z)| ds \right\}, \tag{1.2}$$

where the supremum is taken over all conformal maps $\varphi : \Omega \to \varphi(\Omega)$ such that $\mathrm{Area}\big(\varphi(\Omega)\big) = 1$. See Exercise III.24 for an inequality about (1.2). However, the supremum in (1.2) is difficult to estimate geometrically and (1.2) becomes more manageable if we use functions more general than $|\varphi'(z)|$ in the arc length formula.

By definition, a **metric** is a non-negative Borel measurable function ρ on Ω such that the **area**,

$$A(\Omega, \rho) \equiv \int_{\Omega} \rho^2 dx dy,$$

satisfies $0 < A(\Omega, \rho) < \infty$. When ρ is a metric and Γ is a path family, we define the ρ-**length** of Γ by

$$L(\Gamma, \rho) = \inf_{\gamma \in \Gamma} \int_{\gamma} \rho |dz|,$$

and the **extremal length** of Γ by

$$\lambda_{\Omega}(\Gamma) = \sup_{\rho} \frac{L(\Gamma, \rho)^2}{A(\Omega, \rho)}. \tag{1.3}$$

A conformal mapping φ transforms the metric ρ on Ω into the metric $\rho(\varphi^{-1}(z))|(\varphi^{-1})'(z)|$ on $\varphi(\Omega)$. Therefore $\lambda_{\varphi(\Omega)}(\varphi(\Gamma)) = \lambda_{\Omega}(\Gamma)$ and the extremal length (1.3) is conformally invariant.

An **extremal metric** for Γ is a metric ρ which attains the supremum (1.3). In many important cases there exists an extremal metric of the form $|\varphi'(z)|$ for some conformal mapping φ, and in those cases (1.3) is just the square of (1.2). The advantage of the more general expression (1.3) is that every metric ρ provides a lower bound for $\lambda_{\Omega}(\Gamma)$. Indeed, one of the main results of the theory, Theorem 6.1 below, is proved using metrics not of the form $|\varphi'(z)|$.

Because of its homogeneity, the ratio (1.3) is unchanged if the metric ρ is multiplied by a positive constant. Thus we can normalize the metric to satisfy $A(\Omega, \rho) = 1$ so that

$$\lambda_{\Omega}(\Gamma) = \sup\{L^2(\Gamma, \rho) : A(\Omega, \rho) = 1\},$$

or we can normalize the metric by $L(\Gamma, \rho) = 1$ so that

$$\lambda_{\Omega}(\Gamma)^{-1} = \inf\{A(\Omega, \rho) : L(\Gamma, \rho) = 1\}.$$

The quantity $\lambda_{\Omega}(\Gamma)^{-1}$ is called the **modulus** or **module** of Γ. Some authors emphasize moduli instead of extremal lengths.

Extremal distance is the most useful instance of extremal length. When $E \subset \overline{\Omega}$ and $F \subset \overline{\Omega}$ the **extremal distance** $d_{\Omega}(E, F)$ from E to F is defined by

$$d_{\Omega}(E, F) = \lambda_{\Omega}(\Gamma),$$

where Γ is the family of connected arcs in Ω that join E and F. The **conjugate extremal distance**, $d_\Omega^*(E, F)$, is defined to be the extremal length of the family Γ^* of curves that separate E from F. To compute $d_\Omega^*(E, F)$ we allow a curve $\gamma^* \in \Gamma^*$ to be any finite union (not necessarily connected) of arcs and closed curves in Ω such that E and F lie in the boundaries of distinct components of $\Omega \setminus \gamma$. See Figure IV.1.

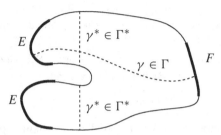

Figure IV.1 Connecting and separating curves.

Example 1.1 (The rectangle). *If* $R = \{(x, y) : 0 < x < \ell$ *and* $0 < y < h\}$ *is a rectangle of length ℓ and height h, then the extremal distance between the vertical sides E and F is*

$$d_R(E, F) = \frac{\ell}{h}. \qquad (1.4)$$

The rectangle example is the building block for many results about extremal length.

Proof. Let Γ be the family of connected arcs in R joining E and F and let ρ be a metric on R. Then

$$L^2(\Gamma, \rho) \le \left(\int_0^\ell \rho(x + iy)dx\right)^2 \le \ell \int_0^\ell \rho^2(x + iy)dx \qquad (1.5)$$

by the Cauchy–Schwarz inequality. Integrating (1.5) with respect to y gives

$$L^2(\Gamma, \rho)h \le \ell A(R, \rho)$$

and hence

$$d_R(E, F) = \sup_\rho \frac{L^2(\Gamma, \rho)}{A(R, \rho)} \le \frac{\ell}{h}. \qquad (1.6)$$

Equality holds when $\rho = 1$, and that proves (1.4). ∎

By the condition for equality in the Cauchy–Schwarz inequality, (1.5) is an equality if and only if ρ is constant almost everywhere $dxdy$. Therefore every extremal metric is constant almost everywhere. The conjugate extremal distance $d_R^*(E, F)$ is the extremal distance between the two horizontal sides of R, and thus $d_R^*(E, F) = \frac{h}{\ell}$.

More generally, let Ω be a Jordan domain, let $E = [\zeta_1, \zeta_2]$ and $F = [\zeta_3, \zeta_4]$ be two disjoint subarcs of $\partial\Omega$, where $\zeta_1, \zeta_2, \zeta_3$, and ζ_4 are listed in the counterclockwise ordering of $\partial\Omega$. There exist a unique $\ell > 0$ and a conformal mapping φ from Ω to a rectangle $R = \{(x, y) : 0 < x < \ell, \ 0 < y < 1\}$ so that $\zeta_1, \zeta_2, \zeta_3$, and ζ_4 are mapped respectively to the corners i, 0, ℓ, and $\ell + i$. To prove the rectangle exists, map Ω to a half-plane and apply a Schwarz–Christoffel map or use Theorem 4.1. To prove ℓ is unique use equality (1.4) and the conformal invariance of extremal lengths or apply Schwarz reflection to the map between two rectangles. By conformal invariance

$$\ell = d_\Omega(E, F)$$

and the extremal metrics for $d_\Omega(E, F)$ are the constant multiples of $|\varphi'|$. Thus (1.3) is the square of (1.2) in this case. For the rectangle we also have

$$d_\Omega(E, F) \, d_\Omega^*(E, F) = \ell \cdot \frac{1}{\ell} = 1. \tag{1.7}$$

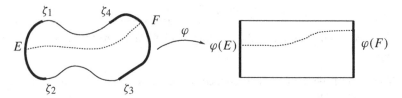

E ζ_1 ζ_4 F φ $\varphi(E)$ $\varphi(F)$ ζ_2 ζ_3

Figure IV.2 Extremal distance in a quadrilateral.

Recall that every metric gives a lower bound for $d_\Omega(E, F)$. Equality (1.7) is significant because every metric also gives a lower bound for $d_\Omega^*(E, F)$, and when (1.7) holds every lower bound for $d_\Omega^*(E, F)$ provides an upper bound for $d_\Omega(E, F)$. We shall see in Theorem 4.1 and Theorem H.1 in Appendix H that (1.7) holds in very general circumstances.

Example 1.2 (The annulus). *The extremal distance between the two boundary circles C_r and C_R of the annulus $A = \{z : r < |z| < R\}$ is*

$$d_A(C_r, C_R) = \frac{1}{2\pi} \log\left(\frac{R}{r}\right).$$

Proof. The reasoning is similar to the rectangle case. Since (1.3) is a supremum, we can assume $0 < r < R < \infty$. Let Γ be the family of connected arcs in A joining C_r to C_R and let ρ be a metric on A. Then again by Cauchy–Schwarz,

$$L^2(\Gamma, \rho) \leq \left(\int_r^R \rho(te^{i\theta}) dt \right)^2 \leq \log\left(\frac{R}{r}\right) \int_r^R \rho(te^{i\theta})^2 t \, dt. \qquad (1.8)$$

Integrating (1.8) with respect to θ gives

$$2\pi L^2(\Gamma, \rho) \leq \log\left(\frac{R}{r}\right) A(A, \rho)$$

and consequently

$$d_A(C_r, C_R) = \sup_\rho \frac{L^2(\Gamma, \rho)}{A(A, \rho)} \leq \frac{1}{2\pi} \log\left(\frac{R}{r}\right). \qquad (1.9)$$

We will get equality holds in (1.8) if and only if $\rho(z) = c/|z|$ a.e., where c is a positive constant, and for such ρ we also get equality in (1.9). Thus $d_A(C_r, C_R) = \frac{1}{2\pi} \log\left(\frac{R}{r}\right)$. ∎

The proof shows that the extremal metrics for $d_A(C_r, C_R)$ have the form $c/|z|$. The conjugate extremal distance, $d_A^*(C_r, C_R)$, is equal to the extremal length of the family Γ of closed curves in A separating the two boundary circles C_r and C_R. By a similar argument we also have

$$d_A^*(C_r, C_R) = \frac{2\pi}{\log\left(\frac{R}{r}\right)} = \frac{1}{d_A(C_r, C_R)}.$$

More generally, a **ring domain** is a doubly connected plane Ω. By Exercise 7 of Chapter II every ring domain is conformally equivalent to an annulus of the form $A = \{z : r < |z| < R\}$. Therefore the module of the family Γ of curves in Ω separating the two components of $\partial\Omega$ is $\lambda_\Omega(\Gamma)^{-1} = \frac{1}{2\pi} \log\left(\frac{R}{r}\right)$, and an extremal metric is $\frac{|\psi'(z)|}{|\psi(z)|}$ where $\psi : \Omega \to A$ is a conformal map. The quantity $\lambda_\Omega(\Gamma)^{-1}$ is also called the **module** of the ring domain Ω and denoted by

$$\lambda_\Omega(\Gamma)^{-1} = \text{mod}(\Omega) = \frac{1}{2\pi} \log\left(\frac{R}{r}\right).$$

2. Uniqueness of Extremal Metrics

It is an open problem to determine when a path family Γ has an extremal metric. However, if an extremal metric does exist, then up to a multiplicative constant it is unique area almost everywhere.

Theorem 2.1. *Let Γ be a path family on Ω and let ρ_1 and ρ_2 be metrics on Ω satisfying*

$$\frac{L^2(\Gamma, \rho_1)}{A(\Omega, \rho_1)} = \frac{L^2(\Gamma, \rho_2)}{A(\Omega, \rho_2)}.$$

Let $\rho_3 = c_1\rho_1 + c_2\rho_2$, where $c_j = \frac{1}{2}A(\Omega, \rho_j)^{-\frac{1}{2}}$. Then

$$\frac{L^2(\Gamma, \rho_3)}{A(\Omega, \rho_3)} \geq \frac{L^2(\Gamma, \rho_1)}{A(\Omega, \rho_1)}. \tag{2.1}$$

If equality holds in (2.1) then $c_1\rho_1 = c_2\rho_2$ a.e. $dxdy$. In particular, if ρ_1 and ρ_2 are extremal metrics for Γ, normalized by $A(\Omega, \rho_1) = A(\Omega, \rho_2)$, then $\rho_1 = \rho_2$ a.e. $dxdy$.

Proof. Without loss of generality we assume $A(\Omega, \rho_1) = A(\Omega, \rho_2) = 1$. Then $L(\Gamma, \rho_1) = L(\Gamma, \rho_2)$ and by the definition of $L(\Gamma, \rho)$,

$$L(\Gamma, \rho_3) \geq \frac{1}{2}\Big(L(\Gamma, \rho_1) + L(\Gamma, \rho_2)\Big) = L(\Gamma, \rho_1).$$

On the other hand, by the Cauchy–Schwarz inequality,

$$A(\Omega, \rho_3) = \frac{1}{4}A(\Omega, \rho_1) + \frac{1}{4}A(\Omega, \rho_2) + \frac{1}{2}\int_\Omega \rho_1\rho_2 dxdy \leq 1. \tag{2.2}$$

Therefore (2.1) holds. If equality occurs in (2.1), then $A(\Omega, \rho_3) = 1$ and equality occurs in (2.2), so that $\rho_1 = \rho_2$ almost everywhere $dxdy$, again by the condition for equality in the Cauchy–Schwarz inequality. In particular, if ρ_1 and ρ_2 are extremal metrics, then equality holds in (2.1) and $\rho_1 = \rho_2$ almost everywhere. ∎

3. Four Rules for Extremal Length

Extremal lengths obey four basic rules.

1. The extension rule. *Let $\Omega \subset \Omega'$ be domains and let Γ be a path family in Ω. Then*

$$\lambda_{\Omega'}(\Gamma) = \lambda_\Omega(\Gamma). \tag{3.1}$$

Moreover, if Γ' is a path family in Ω' such that every $\gamma' \in \Gamma'$ contains some $\gamma \in \Gamma$, then

$$\lambda_{\Omega'}(\Gamma') \geq \lambda_\Omega(\Gamma). \tag{3.2}$$

Equality (3.1) says that the extremal length depends on the path family Γ and not the domain Ω, and for this reason we often write $\lambda(\Gamma)$ for $\lambda_\Omega(\Gamma)$. The extension rule shows that $d_\Omega(E, F)$ is decreased if any of the sets E, F, or Ω is increased and $d_\Omega(E, F) = d_{\Omega \setminus (E \cup F)}(E, F)$ when E and F are closed.

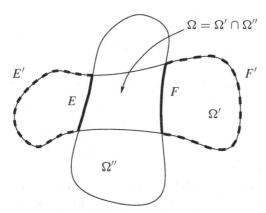

Figure IV.3 Extension rule.

In Figure IV.3, two applications of the extension rule give

$$d_{\Omega'}(E', F') \geq d_\Omega(E, F) \geq d_{\Omega''}(E, F).$$

Proof. The curves in the family Γ' are longer and fewer than the curves in the family Γ. Therefore $\lambda_{\Omega'}(\Gamma') \geq \lambda_{\Omega'}(\Gamma)$ and inequality (3.2) will follow from equality (3.1).

Let ρ' be any metric on Ω' and set $\rho = \rho'|_\Omega$. Then $L(\Gamma, \rho) = L(\Gamma, \rho')$ and $A(\Omega, \rho) \leq A(\Omega', \rho')$. Taking the supremum over ρ' yields $\lambda_{\Omega'}(\Gamma) \leq \lambda_\Omega(\Gamma)$. Now let ρ be any metric on Ω and define the metric ρ' on Ω' by $\rho' = \rho \chi_\Omega$. Then $L(\Gamma, \rho) = L(\Gamma, \rho')$ and $A(\Omega, \rho) = A(\Omega', \rho')$. Taking the supremum over ρ, we get $\lambda_{\Omega'}(\Gamma) \geq \lambda_\Omega(\Gamma)$, which proves (3.1). ∎

2. The serial rule. *Let Γ_1 and Γ_2 be path families contained in disjoint open sets Ω_1 and Ω_2 respectively, and let Γ be a path family contained in a domain $\Omega \supset \Omega_1 \cup \Omega_2$. If each $\gamma \in \Gamma$ contains some $\gamma_1 \in \Gamma_1$ and some $\gamma_2 \in \Gamma_2$, then*

$$\lambda(\Gamma) \geq \lambda(\Gamma_1) + \lambda(\Gamma_2).$$

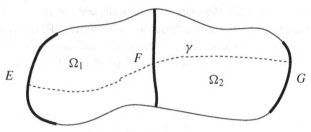

Figure IV.4 Serial rule.

In Figure IV.4, the crosscut F divides Ω into regions Ω_1 and Ω_2 with $E \subset \partial\Omega_1$ and $G \subset \partial\Omega_2$, and by the serial rule,

$$d_\Omega(E, G) \geq d_{\Omega_1}(E, F) + d_{\Omega_2}(F, G).$$

Proof. If either $\lambda(\Gamma_1)$ or $\lambda(\Gamma_2)$ is 0 or ∞, the result follows from the extension rule. Otherwise, choose metrics ρ_1 in Ω_1 and ρ_2 in Ω_2, normalized by the conditions $L(\Gamma_1, \rho_1) = A(\Omega_1, \rho_1)$ and $L(\Gamma_2, \rho_2) = A(\Omega_2, \rho_2)$. Then the metric $\rho = \rho_1 X_{\Omega_1} + \rho_2 X_{\Omega_2}$ on Ω satisfies $L(\Gamma, \rho) \geq L(\Gamma_1, \rho_1) + L(\Gamma_2, \rho_2)$ and $A(\Omega, \rho) = A(\Omega_1, \rho_1) + A(\Omega_2, \rho_2) = L(\Gamma_1, \rho_1) + L(\Gamma_2, \rho_2)$. Thus

$$\frac{L^2(\Gamma, \rho)}{A(\Omega, \rho)} \geq L(\Gamma_1, \rho_1) + L(\Gamma_2, \rho_2) = \frac{L^2(\Gamma_1, \rho_1)}{A(\Omega_1, \rho_1)} + \frac{L^2(\Gamma_2, \rho_2)}{A(\Omega_2, \rho_2)}.$$

Taking the supremum over all ρ_1 and ρ_2, we obtain $\lambda(\Gamma) \geq \lambda(\Gamma_1) + \lambda(\Gamma_2)$. ∎

3. The parallel rule. *Let Γ_1 and Γ_2 be path families contained in disjoint open sets Ω_1 and Ω_2 respectively. If Γ is a path family in $\Omega \supset \Omega_1 \cup \Omega_2$ such that every $\gamma \in \Gamma_1 \cup \Gamma_2$ contains some $\gamma' \in \Gamma$, then*

$$\frac{1}{\lambda(\Gamma)} \geq \frac{1}{\lambda(\Gamma_1)} + \frac{1}{\lambda(\Gamma_2)}. \tag{3.3}$$

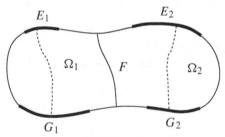

Figure IV.5 Parallel rule.

In Figure IV.5, F is a crosscut dividing Ω into regions Ω_1 and Ω_2 and $E_j = E \cap \partial\Omega_j$ and $G_j = G \cap \partial\Omega_j$ $j = 1, 2$. By the parallel rule,

$$\frac{1}{d_\Omega(E, G)} \geq \frac{1}{d_{\Omega_1}(E_1, G_1)} + \frac{1}{d_{\Omega_2}(E_2, G_2)}.$$

In situations where (1.7) holds, this inequality can also be obtained by applying the serial rule to the conjugate extremal distances.

Proof. If $\lambda(\Gamma) > 0$, let ρ be a metric on Ω, normalized by $L(\Gamma, \rho) = 1$. Then $L(\Gamma_1, \rho) \geq 1$ and $L(\Gamma_2, \rho) \geq 1$ and

$$A(\Omega, \rho) \geq A(\Omega_1, \rho) + A(\Omega_2, \rho) \geq \frac{1}{\lambda(\Gamma_1)} + \frac{1}{\lambda(\Gamma_2)}.$$

Taking the supremum (1.3) over all ρ, we obtain (2.3). ∎

4. The symmetry rule. *Let $T : \Omega \to \Omega$ satisfy $T \circ T(z) = z$ and suppose that either $T(z)$ or $\overline{T(z)}$ is analytic. If Γ is a path family in Ω such that $T(\Gamma) = \Gamma$, then*

$$\lambda(\Gamma) = \sup\left\{ \frac{L^2(\Gamma, \rho)}{A(\Omega, \rho)} : \rho = (\rho \circ T)\,|J_T| \right\}, \tag{3.4}$$

where $|J_T| = |T'|$ when T is analytic and $|J_T| = |\overline{T}\,'|$ when $\overline{T(z)}$ is analytic.

Proof. If $\rho_1 = (\rho \circ T)|J_T|$ then

$$\int_\gamma \rho|dz| = \int_{T^{-1}(\gamma)} \rho_1|dz|.$$

Because T maps Γ onto Γ, $L(\Gamma, \rho) = L(\Gamma, \rho_1)$, and because T is one-to-one, $A(\Omega, \rho) = A(\Omega, \rho_1)$. Then by Theorem 2.1, $\rho_2 = \frac{1}{2}(\rho + \rho_1)$ satisfies

$$\frac{L^2(\Gamma, \rho_2)}{A(\Omega, \rho_2)} \geq \frac{L^2(\Gamma, \rho)}{A(\Omega, \rho)}.$$

But since $T \circ T(z) = z$, we have $\rho_2 \circ T|J_T| = \rho_2$, and hence (3.4) holds. ∎

For example, let $T(z) = \bar{z}$, let $\Omega \subset \{z : \operatorname{Im} z > 0\}$, and let $\Omega_1 \supset \Omega \cup T(\Omega)$ be a domain such that $T(\Omega_1) = \Omega_1$. Let $E, F \subset \partial\Omega$ and set $E_1 = E \cup T(E)$ and $F_1 = F \cup T(F)$. See Figure IV.6. Then

$$d_{\Omega_1}(E_1, F_1) = \frac{1}{2}d_\Omega(E, F) \tag{3.5}$$

because to compute $d_{\Omega_1}(E_1, F_1)$ it suffices by the symmetry rule to consider only metrics ρ satisfying $\rho(z) = \rho(\bar{z})$ and $\rho = \infty$ on $\mathbb{R} \cap \Omega_1$. If γ is a curve connecting E_1 to F_1, then reflecting $\gamma \cap T(\Omega)$ yields a curve $\widetilde{\gamma} \subset \overline{\Omega} \cup \mathbb{R}$ that joins E to F and

$$\int_{\widetilde{\gamma}} \rho|dz| = \int_{\gamma} \rho|dz|.$$

Because ρ is a metric, we can find curves $\nu \subset \Omega$ with ρ-length arbitrarily close to the length of $\widetilde{\gamma}$. Then (3.5) follows since $A(\Omega_1, \rho) = 2A(\Omega, \rho)$. We will use this idea several times.

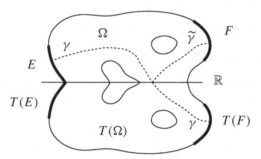

Figure IV.6 Symmetry rule.

There is an analogy between extremal distance and electrical resistance or resistivity. When the opposite ends of an electrical conductor are joined to the two terminals of a battery and the remaining sides of the conductor are insulated, there is a drop in potential energy from one end of the conductor to the other, and this drop is equal to the voltage of the battery. Suppose the conductor is a uniformly thin sheet of material in the shape of a Jordan domain Ω. Put a copper coating on two arcs of $\partial \Omega$ and attach each copper arc to a terminal of the battery. Assume the atmosphere acts as an insulator on the rest of the surface of the region. Then the electrical resistance of the conductor is the extremal distance between the two intervals, the magnitude of the electric field is the extremal metric, the "field lines" are the shortest curves in the extremal metric, and the "equipotential lines" are the shortest curves for the conjugate extremal distance. The extension rule says that the resistance $d_\Omega(E, F)$ increases if Ω or E or F is decreased. The serial and parallel rules correspond to the familiar rules for calculating resistance when circuits are attached in series or in parallel.

4. Extremal Metrics for Extremal Distance

Let Ω be a Jordan domain and let E and F be disjoint closed subsets of $\partial\Omega$, each consisting of a finite union of arcs. Recall that the extremal distance between E and F is the extremal length of the family of curves in Ω connecting E to F, and the conjugate extremal distance is the extremal length of the family of curves separating E from F.

In this section everything will hinge on the following observation: Let

$$R = \{(x, y) : 0 < x < 1, \ 0 < y < h\}$$

be a rectangle with sides parallel to the axes, let $\Omega = R \setminus \bigcup L_j$ where $\{L_j\}$ is a finite family of horizontal line segments in R, and let E and F be the vertical sides of the rectangle R. Then

$$d_\Omega(E, F) = d_R(E, F)$$

by the argument used in Example 1.1. Moreover, every extremal metric for $d_\Omega(E, F)$ is constant almost everywhere. The same holds for the conjugate extremal distance

$$d_\Omega^*(E, F) = d_R^*(E, F)$$

because by definition the "curves" that separate E from F need not be connected.

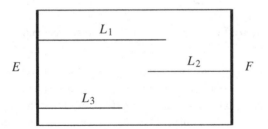

Figure IV.7 Slit rectangle.

Now suppose there exists a conformal mapping $\varphi : \Omega \to R \setminus \bigcup L_j$, where R and L_j are as in the preceding paragraph, such that ψ is continuous on $\overline{\Omega}$, $\varphi(E)$ is the left vertical side of R, and $\varphi(F)$ is the right vertical side of R. Then by the conformal invariance of extremal length and the preceding observation,

$$d_\Omega(E, F) = d_{\varphi(\Omega)}(\varphi(E), \varphi(F)) = 1/h = 1/d_\Omega^*(E, F),$$

and an extremal metric for $d_\Omega(E, F)$ is $\rho_0(z) = |\varphi'(z)|$. In the next theorem, we determine when Ω can be mapped to a slit rectangle R so that E and F correspond to the vertical edges of R.

Theorem 4.1. *Let Ω be a Jordan domain and let E and F be finite unions of closed subarcs of $\partial\Omega$. Assume $E \cap F = \emptyset$. Then there is a rectangle R having sides parallel to the axes and a conformal map φ of Ω onto the rectangle R with a finite number of horizontal line segments removed such that $\varphi \in C(\overline{\Omega})$ and $\varphi(E)$ and $\varphi(F)$ are the vertical sides of the rectangle if and only if there is an arc $\sigma \subset \partial\Omega$ such that*

$$E \subset \sigma \quad \text{and} \quad F \cap \sigma = \emptyset. \tag{4.1}$$

In this case, the extremal distance from E to F is the ratio of the length to height of this rectangle, the conjugate extremal distance satisfies

$$d^*_\Omega(E, F) = 1/d_\Omega(E, F),$$

and the extremal metrics on Ω for $d_\Omega(E, F)$ are the positive constant multiples of $\rho_0(z) = |\varphi'(z)|$.

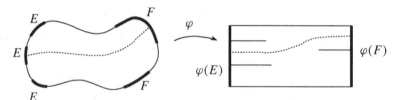

Figure IV.8 Extremal distance and slit rectangles.

Proof. Assume φ is a conformal map of Ω onto the rectangle R with a finite number of horizontal line segments removed such that $\varphi(E)$ and $\varphi(F)$ are the vertical sides of R. Let γ be a curve in $\varphi(\Omega)$ connecting the top and bottom edges of the rectangle. The curve γ divides $\varphi(\Omega)$ into two regions U_1 and U_2 such that $\varphi(E) \subset \partial U_1$ and $\varphi(F) \subset \partial U_2$. Then $\varphi^{-1}(\gamma)$ is a curve in Ω with two endpoints on $\partial\Omega$. Because Ω is a Jordan curve, these endpoints divide $\partial\Omega$ into two arcs σ_1 and σ_2 with $E \subset \sigma_1$ and $F \subset \sigma_2$.

Conversely, assume there is an arc σ such that (4.1) holds. By the conformal invariance of extremal distance we may suppose that Ω is the unit disc $\mathbb{D}$ and that $E \cup F \subset \partial\mathbb{D}$. Let $\Omega_d = \mathbb{C}^* \setminus (E \cup F)$ and let $\omega(z) = \omega(z, F, \Omega_d)$ be the harmonic measure of F in Ω_d. Let $\widetilde{\omega}(z)$ be the harmonic conjugate of ω in $\mathbb{D}$. We claim that

$$\varphi = \omega + i\widetilde{\omega}$$

is the conformal map promised by the theorem.

Because ω is harmonic in Ω_d, φ extends to be analytic across $\partial\mathbb{D} \setminus (E \cup F)$. By the symmetry of Ω_d and ω,

$$\frac{\partial\widetilde{\omega}}{\partial\theta} = \frac{\partial\omega}{\partial r} = 0 \tag{4.2}$$

on $\partial\mathbb{D} \setminus (E \cup F)$.

Let $\zeta \in E^\circ$, where E° is the relative interior of E in $\partial\mathbb{D}$. Because $\omega = 0$ on E, we can by the Schwarz reflection principle extend φ to be analytic in a neighborhood W of ζ. The extension satisfies $\text{Re}\varphi > 0$ on $W \cap \mathbb{D}$ and $\text{Re}\varphi < 0$ on $W \setminus \overline{\mathbb{D}}$. This implies $\varphi'(\zeta) \neq 0$ and $\partial\omega/\partial r \leq 0$ by Lemma II.2.4. Then because $\partial\omega/\partial\theta = 0$, we obtain

$$\frac{\partial\widetilde{\omega}}{\partial\theta} = \frac{\partial\omega}{\partial r} < 0 \tag{4.3}$$

at ζ, and hence on all of E°. Applying the same argument to $1 - \varphi$ on F° we obtain

$$\frac{\partial\widetilde{\omega}}{\partial\theta} = \frac{\partial\omega}{\partial r} > 0 \tag{4.4}$$

on F°.

Let ζ be an endpoint of $E \cup F$ and let W be a neighborhood of ζ so that $U = W \setminus \gamma$ is simply connected, where γ is the component of $E \cup F$ containing ζ. Then ω is continuous on W and $\varphi = \omega + i\widetilde{\omega}$ is analytic in U. By mapping $\mathbb{C}^* \setminus \gamma$ to the upper half-plane and using Schwarz reflection again, we see that φ extends to be continuous on $\mathbb{D} \cup \{\zeta\}$. We conclude that φ is continuous on $\overline{\mathbb{D}}$.

Now let us follow $\varphi(z)$ as z traverses $\partial\mathbb{D}$ counterclockwise. On each component of E, $\omega = 0$ and $\widetilde{\omega}$ is strictly decreasing, by (4.3). On each component of F, $\omega = 1$ and $\widetilde{\omega}$ is strictly increasing, by (4.4). On any component γ_k of $\partial\mathbb{D} \setminus (E \cup F)$, $0 < \omega < 1$ and $\widetilde{\omega}$ is constant, by (4.2). If both endpoints of γ_k are in E, then the curve $\varphi(\gamma_k)$ traces a horizontal line segment $L_k \subset \{z : 0 \leq \text{Re}\, z \leq 1\}$, beginning and ending on $\{z : \text{Re}\, z = 0\}$. If both endpoints are in F, then $\varphi(\gamma_k)$ traces a horizontal segment $L_k \subset \{z : 0 \leq \text{Re}\, z \leq 1\}$, beginning and ending on $\{z : \text{Re}\, z = 1\}$. There are only two components γ_k having endpoints on both E and F; on these $\widetilde{\omega}$ is constant and $\omega = \text{Re}\varphi$ goes from 0 to 1 or from 1 to 0. Thus $\varphi(\partial\mathbb{D})$ contains the boundary of a rectangle R and the contour $\varphi(\partial\mathbb{D})$ has winding number 1 about each point in $R \setminus \bigcup L_k$ and winding number 0 about each point in $\mathbb{C} \setminus \overline{R}$. The argument principle then shows that

$$\varphi(\mathbb{D}) = R \setminus \bigcup L_k$$

and that φ is one-to-one.

The statements about the extremal distance and the extremal metrics now follow from Theorem 2.1 and the observation made at the beginning of this section.   ∎

By the proof of Example 1.2, the extremal distance and conjugate extremal distance between the boundary circles of an annulus will also not change if radial slits are removed from the annulus. Thus we can replace the rectangle R by an annulus and obtain a ring domain version of Theorem 4.1.

Theorem 4.2. *Suppose Ω is bounded by two disjoint Jordan curves Γ_1, Γ_2 and suppose E and F are finite unions of closed arcs with $E \subset \Gamma_1$ and $F \subset \Gamma_2$. Then there is a conformal map ψ of Ω onto an annulus with finitely many radial slits removed such that ψ is continuous on $\overline{\Omega}$ and the images of E and F are the two boundary circles. The extremal metrics for the extremal distance between E and F are constant multiples of $|\psi'(z)|/|\psi(z)|$ and*

$$d_\Omega(E, F) = \frac{1}{d_\Omega^*(E, F)} = \frac{1}{2\pi} \log R,$$

where $R > 1$ is the ratio of the radii of the boundary circles.

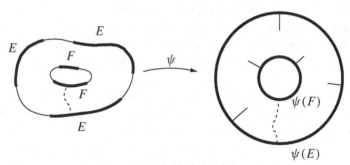

Figure IV.9 Extremal distance and slit annuli.

Proof. By conformal invariance, we may suppose that $\Gamma_2 = \partial \mathbb{D}$ and that $\Gamma_1 \subset \mathbb{D}$ is an analytic Jordan curve. If $\partial \Omega \setminus (E \cup F) \neq \emptyset$, take two copies of Ω and form the doubled Riemann surface Ω_d by attaching them along $\partial \Omega \setminus (E \cup F)$. Then Ω_d is a torus with finitely many arcs removed, where the arcs correspond to the intervals in $E \cup F$. If $E \cup F = \partial \Omega$, then set $\Omega_d = \Omega$. Let $\omega(z) = \omega(z, F_d, \Omega_d)$ be the harmonic measure of the double F_d of F in Ω_d. (The proof of Theorem B.4 in Appendix B explains the construction of Ω_d and the solution of the Dirichlet problem on Ω_d.) Let $\widetilde{\omega}$ be the (locally defined) harmonic conjugate of ω in Ω. Choose a Jordan curve $\gamma \subset \Omega$ homologous to $\partial \mathbb{D}$. If

$$\int_\gamma \frac{\partial \widetilde{\omega}}{\partial s} ds = 0, \tag{4.5}$$

then $\varphi = \omega + i\widetilde{\omega}$ is single valued and analytic in Ω and by the proof of Theorem 4.1 φ extends to be continuous on $\overline{\Omega}$. But then by (4.2) and (4.4), $\widetilde{\omega}$ increases as z traces $\partial \mathbb{D}$, in contradiction to (4.5). Thus there exists a non-zero constant c so that

$$c \int_\gamma \frac{\partial \widetilde{\omega}}{\partial s} ds = 2\pi.$$

Then $\psi = e^{c\varphi}$ is single valued and analytic on Ω, and ψ is continuous on $\partial \Omega$. As z traces $\partial \mathbb{D}$, $c\widetilde{\omega}$ increases by 2π and, as z traces Γ_1, $c\widetilde{\omega}$ decreases by 2π. Hence $\psi(\partial \Omega)$ contains the boundary of an annulus, and the remaining assertions of the theorem follow just as in Theorem 4.1. ∎

5. Extremal Distance and Harmonic Measure

We begin with the special case of a rectangle. Set

$$R_L = \{z : |\mathrm{Re}z| < L \text{ and } |\mathrm{Im}z| < 1\}.$$

Lemma 5.1. *If* $E_L = \{z \in \partial R_L : |\mathrm{Re}z| = L\}$ *is the union of the vertical edges of* R_L, *then*

$$e^{-\frac{\pi}{2}L} \le \omega(0, E_L, R_L) \le \frac{8}{\pi} e^{-\frac{\pi}{2}L}. \tag{5.1}$$

The estimates in (5.1) *are sharp, because*

$$\lim_{L \to \infty} \omega(0, E_L, R_L) e^{\frac{\pi}{2}L} = \frac{8}{\pi} \quad and \quad \lim_{L \to 0} \omega(0, E_L, R_L) e^{\frac{\pi}{2}L} = 1. \tag{5.2}$$

The lemma connects the harmonic measure of the two ends of R_L at its center point to the extremal distance L between the two ends. When $L \ge 1$, the lower bound in (5.1) can be improved to $(.93)(8/\pi)e^{-\frac{\pi}{2}L}$. When $L < 1$, better bounds can be obtained by applying (5.1) to $\omega(0, E_{1/L}, R_{1/L}) = 1 - \omega(0, E_L, R_L)$. See Exercise 10.

Proof. Let S_L be the infinite strip $\{z : \mathrm{Re}z > -L \text{ and } |\mathrm{Im}z| < 1\}$ and let $\omega_L(z)$ be the harmonic measure of the left edge $\partial S_L \cap \{\mathrm{Re}z = -L\}$ in S_L. Using the conformal map $z \to e^{\pi z/2}$ and some elementary geometry, we obtain

$$\omega_L(z) = \frac{2}{\pi} \arg \left(\frac{e^{\frac{\pi}{2} z} - (-ie^{-\frac{\pi}{2}L})}{e^{\frac{\pi}{2} z} - ie^{-\frac{\pi}{2}L}} \right). \tag{5.3}$$

In particular, $\omega_L(0) = \frac{4}{\pi} \tan^{-1}(e^{-\frac{\pi}{2}L})$. By (5.3) $\omega_L(z) \le \omega_L(L)$ on $\mathrm{Re}z = L$. Thus for $z \in \partial R_L$,

$$\omega(z, E, R_L) \le \omega_L(z) + \omega_L(-z) \le (1 + \omega_L(L))\omega(z, E, R_L).$$

By the maximum principle, this inequality persists at $z = 0$, and we conclude that

$$\frac{2\omega_L(0)}{1 + \omega_L(L)} \le \omega(0, E, R_L) \le 2\omega_L(0).$$

Because

$$\omega_L(L) = \omega_{2L}(0) = \frac{4}{\pi} \tan^{-1}(e^{-\pi L}),$$

we obtain

$$\frac{\frac{8}{\pi} \tan^{-1}(e^{-\frac{\pi}{2}L})}{1 + \frac{4}{\pi} \tan^{-1}(e^{-\pi L})} \le \omega(0, E, R_L) \le \frac{8}{\pi} \tan^{-1}(e^{-\frac{\pi}{2}L}). \tag{5.4}$$

The estimates (5.1) now follow from the elementary inequalities

$$\frac{\pi}{4} t \le \tan^{-1}(t) \le \min\left(t, \frac{\pi}{4}\right),$$

valid for $0 \le t \le 1$, and both equalities in (5.2) are immediate from (5.4). ∎

Now let Ω be a Jordan domain, let E be an arc on $\partial\Omega$ and let $z_0 \in \Omega$. Consider all Jordan arcs $\sigma \subset \Omega$ that join z_0 to $\partial\Omega \setminus E$, and define

$$\lambda(z_0, E) = \sup_{\sigma} d_{\Omega\setminus\sigma}(\sigma, E),$$

where the supremum is taken over all such Jordan arcs.

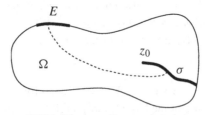

Figure IV.10 Distance from a point to an arc.

Both quantities $\lambda(z_0, E)$ and $\omega(z_0, E, \Omega)$ are conformally invariant but $\omega(z, E, \Omega)$ is strictly increasing in E, while $\lambda(z_0, E)$ is strictly decreasing in E. This means there is a function connecting λ to ω, and the next result examines that function.

Theorem 5.2. *Let Ω be a Jordan domain, let E be a subarc of $\partial\Omega$ and let $z_0 \in \Omega$. Then*

$$e^{-\pi\lambda(z_0, E)} \le \omega(z_0, E, \Omega) \le \frac{8}{\pi} e^{-\pi\lambda(z_0, E)}.$$

Moreover

$$\lim_{\lambda \to \infty} \omega(z_0, E, \Omega) e^{\pi\lambda(z_0, E)} = \frac{8}{\pi} \quad and \quad \lim_{\lambda \to 0} \omega(z_0, E, \Omega) e^{\pi\lambda(z_0, E)} = 1.$$

By Theorem 5.2 every choice of the arc σ and every choice of the metric ρ give an upper bound for $\omega(z_0, E, \Omega)$, because $\lambda(z_0, E)$ and the extremal distance $d_{\Omega\setminus\sigma}(\sigma, E)$ are both suprema. This idea of varying σ is from Beurling [1989].

Proof. By conformal invariance we may suppose that $\Omega = \mathbb{D}$, that $z_0 = 0$ and that E is an arc on $\partial\mathbb{D}$. Let E_1 and E_2 be the two disjoint arcs on $\partial\mathbb{D}$ given by $E_1 \cup E_2 = \{e^{i\theta} : e^{2i\theta} \in E\}$. Because the arcs E_1 and E_2 are symmetric about 0, there is a conformal map f of $\mathbb{D}$ onto a rectangle R such that R has center 0 and sides parallel to the axes, such that $f(0) = 0$ and such that $\widetilde{E}_j = f(E_j)$, $j = 1, 2$, are the vertical sides of R. Let σ be any arc in $\mathbb{D}$ connecting 0 to $\partial\mathbb{D} \setminus E$ and let $\widetilde{\sigma} = \{z \in R : (f^{-1}(z))^2 \in \sigma\}$. The map $\sqrt{z}$ is conformal on $\mathbb{D} \setminus \sigma$ and the two branches of $f(\sqrt{z})$ map $\mathbb{D} \setminus \sigma$ conformally onto the two components of $R \setminus \widetilde{\sigma}$. Hence by the serial rule and by conformal invariance,

$$d_R(\widetilde{E}_1, \widetilde{E}_2) \ge d(\widetilde{E}_1, \widetilde{\sigma}) + d(\widetilde{\sigma}, \widetilde{E}_2) = 2d_{\mathbb{D}}(\sigma, E).$$

Equality holds if $\widetilde{\sigma}_0$ is the vertical line segment in R through 0, so that $2\lambda(0, E) = d_R(\widetilde{E}_1, \widetilde{E}_2)$.

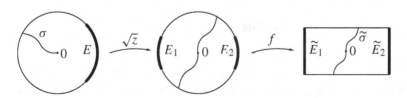

Figure IV.11 Proof of Theorem 5.2.

Since $\omega(z^2, E, \mathbb{D}) = \omega(z, E_1 \cup E_2, \mathbb{D})$, we have

$$\omega(0, E, \mathbb{D}) = \omega(0, E_1 \cup E_2, \mathbb{D}) = \omega(0, \widetilde{E}_1 \cup \widetilde{E}_2, R).$$

Then by Lemma 5.1,

$$e^{-\pi\lambda(0,E)} \leq \omega(0, E, \mathbb{D}) \leq \frac{8}{\pi} e^{-\pi d_{\mathbb{D}}(\sigma, E)}.$$

∎

The upper estimate is also valid when E is a finite union of arcs on $\partial\Omega$. In that case we consider all Jordan arcs σ in Ω connecting z_0 to $\partial\Omega \setminus E$, and define

$$\lambda(z_0, E) = \sup_{\sigma} d_{\Omega\setminus\sigma}(\sigma, E).$$

Theorem 5.3. *Let Ω be a Jordan domain, and let E be a finite union of arcs contained in $\partial\Omega$. Then*

$$\omega(z_0, E, \Omega) \leq \frac{8}{\pi} e^{-\pi\lambda(z_0, E)}. \tag{5.5}$$

Proof. We can assume $\Omega = \mathbb{D}$ and $z_0 = 0$. Let σ be an arc from 0 to $\partial\mathbb{D} \setminus E$ and set $\lambda = d_{\Omega\setminus\sigma}(\sigma, E)$. Write $\sigma_1 = \{z : z^2 \in \sigma\}$ and $\{e^{i\theta} : e^{2i\theta} \in E\} = E_1 \cup E_2$ where σ_1 separates E_1 from E_2 and $z^2(E_1) = z^2(E_2) = E$. By Theorem 4.1, there is a conformal map φ of $\mathbb{D}$ onto a rectangle R with horizontal slits removed so that $\widetilde{E}_j = \varphi(E_j)$ are the vertical ends of R. Then as in Theorem 5.2 we obtain $d_R(\widetilde{E}_1, \widetilde{E}_2) \geq 2d_{\mathbb{D}}(\sigma, E)$ and consequently

$$\omega(0, E, \mathbb{D}) \leq \frac{8}{\pi} e^{-\pi\lambda}.$$

Taking the supremum over σ then gives (5.5). ∎

There is no lower bound in Theorem 5.3 because the slits can consume nearly all of the harmonic measure. See Exercise 12.

6. The $\int \frac{dx}{\theta(x)}$ Estimate

Every lower bound for extremal distance yields an upper bound for harmonic measure, and by (1.3) every metric ρ yields a lower bound for extremal distance. Constructing good metrics is therefore an important method. Consider the special case of a **strip domain**:

$$\Omega = \{(x, y) : |y - m(x)| < \theta(x)/2, \ a < x < b\},$$

having varying width $\theta(x)$ and mid-line $y = m(x)$.

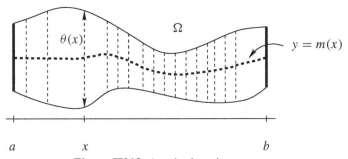

Figure IV.12 A strip domain.

Let $a = x_0 < x_1 < \ldots < x_n = b$ and set $\sigma_j = \Omega \cap \{z : \text{Re}z = x_j\}$, where $j = 0, \ldots, n$. By the serial rule,

$$d_\Omega(\sigma_0, \sigma_n) \geq \sum_{j=1}^{n} d(\sigma_{j-1}, \sigma_j).$$

If $\Delta x = x_j - x_{j-1}$ is small, the region between σ_{j-1} and σ_j is approximately a thin rectangle having Δx as base, $\theta(x_j)$ as height, and $x_j + im(x_j)$ as the midpoint of its right vertical side. Under the linear map

$$\frac{z - im(x_j)}{\theta(x_j)},$$

this rectangle is sent to a rectangle centered on $\mathbb{R}$ with height 1 and width $\Delta x/\theta(x_j)$, so that $d(\sigma_{j-1}, \sigma_j) \approx \Delta x/\theta(x_j)$. Consider the (non-analytic) map

$$\Phi : (x, y) \longrightarrow \left(\int_a^x \frac{dt}{\theta(t)}, \frac{y - m(x)}{\theta(x)} \right) \qquad (6.1)$$

from Ω onto a rectangle of height 1 and length $\int_a^b \frac{1}{\theta(t)} dt$. The extremal metric for $d_\Omega(\sigma_0, \sigma_n)$ is given by $\rho_0(z) = |\varphi'(z)| = |\nabla \text{Re}\varphi(z)|$, where φ is a conformal map of Ω onto some rectangle, but the key idea here is to replace the analytic φ by Φ and use as metric

$$\rho(z) = |\nabla \text{Re}\Phi| = \frac{1}{\theta(x)}, \qquad z = x + iy.$$

If γ is any curve connecting σ_0 to σ_n in Ω, then

$$\int_\gamma \rho(z)|dz| \geq \int_a^b \frac{1}{\theta(x)} dx.$$

Moreover

$$\iint_\Omega \rho^2 dy dx = \int_a^b \frac{1}{\theta(x)} dx,$$

so that

$$d_\Omega(\sigma_0, \sigma_n) \geq \int_a^b \frac{1}{\theta(x)} dx.$$

More generally, let Ω be a finitely connected Jordan domain and assume $E \subset \{z : \text{Re} z \leq a\} \cap \partial\Omega$ and $F \subset \{z : \text{Re} z \geq b\} \cap \partial\Omega$ are finite unions of arcs. Suppose $I_x \subset \{z : \text{Re} z = x\}$ separates E from F and let $\theta(x)$ be the length of I_x, $a < x < b$. We assume that $\theta(x)$ is measurable but we do not assume that $I_x = \Omega \cap \{z : \text{Re} z = x\}$ or that I_x is connected. Define the metric ρ_A by

$$\rho_A(x, y) = \begin{cases} \frac{1}{\theta(x)} & \text{if } (x, y) \in I_x, \\ 0 & \text{elsewhere in } \Omega. \end{cases}$$

Then by the above argument

$$d_\Omega(E, F) \geq \int_a^b \frac{1}{\theta(x)} dx. \tag{6.2}$$

Theorem 6.1. *Let Ω be a Jordan domain and let $z_0 \in \Omega$. Let $b > x_0 = \text{Re} z_0$ and suppose $F \subset \{z \in \partial\Omega : \text{Re} z \geq b\}$. Assume that for $x_0 < x < b$, there exists $I_x \subset \Omega \cap \{\text{Re} z = x\}$ separating z_0 from F and set $\theta(x) \equiv \ell(I_x)$. Then*

$$\omega(z_0, F) \leq \frac{8}{\pi} \exp\left(-\pi \int_{x_0}^b \frac{dx}{\theta(x)}\right). \tag{6.3}$$

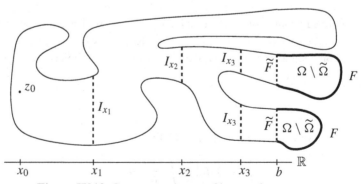

Figure IV.13 Crosscut estimate of harmonic measure.

For example, if $I_x = \{z \in \Omega : \mathrm{Re}\, z = x\}$, then $\theta(x)$ is lower semicontinuous and measurable.

Proof. Let $\widetilde{\Omega}$ denote the points of Ω which are not separated from z_0 by some I_x and let $\widetilde{F}$ denote the points of $\partial\widetilde{\Omega}$ which are separated from z_0 by all I_x, $x_0 < x < b$. We can suppose that $\widetilde{F}$ is a finite union of arcs on $\partial\widetilde{\Omega}$ by replacing Ω with $\varphi^{-1}(|z| < r)$ where φ is a conformal map of Ω onto $\mathbb{D}$. Let $\sigma \subset \{\mathrm{Re}\, z = x_0\}$ be a curve connecting z_0 to $\partial\widetilde{\Omega}$. By the maximum principle,

$$\omega(z_0, F, \Omega) \le \omega(z_0, \widetilde{F}, \widetilde{\Omega}),$$

and then Theorem 5.3 and inequality (6.2) give (6.3). ∎

Theorem 6.1 remains true if $\theta(x)$ is replaced by $\ell(\{z \in \Omega : \mathrm{Re}\, z = x\})$, because that change only increases the right side of (6.3). When $\{z \in \Omega : \mathrm{Re}\, z = x\}$ contains several crosscuts that separate z_0 from F, sharper versions of (6.3) are available. See Appendix G. Theorem H.8 in Appendix H extends Theorem 6.1 to finitely connected domains.

Theorem 6.1 is often used in polar coordinates, as in the following theorem.

Theorem 6.2. *Let Ω be a Jordan domain and let $E \subset \overline{\Omega} \cap \{|z| \ge R\}$. If $z_0 \in \Omega$ satisfies $r_0 = \max(|z_0|, \mathrm{dist}(0, \partial\Omega)) < R$. Suppose that $J_r \subset \{z \in \Omega : |z| = r\}$ separates z_0 from E, $r_0 < r < R$, and let $r\Theta(r)$ be the length of J_r. Then*

$$\omega(z_0, E) \le \frac{8}{\pi} \exp\left(-\pi \int_{r_0}^{R} \frac{dr}{r\Theta(r)}\right), \qquad (6.4)$$

if $\Theta(r)$ is measurable.

Proof. Define the metric

$$\rho_A(z) = \begin{cases} \frac{1}{r\Theta(r)} & \text{in } J_r, \\ 0 & \text{in } \Omega \setminus \bigcup\{J_r : r_0 < r < R\}, \end{cases}$$

for $z \in \Omega$ and $r = |z|$ and repeat the proof of Theorem 6.1. ∎

Notes

Much of this chapter is based on Beurling's Mittag–Leffler Institut lectures, published in [1989], and on Ahlfors [1973]. In [1973] Ahlfors says that Beurling invented extremal length in 1943 or 1944, but waited until 1946 to announce his results. The application of the Cauchy–Schwarz inequality in the proof of (1.4) is often referred to as the **length–area principle**; it goes back to early work by Grötzsch [1928] and others. Both Beurling [1989] and Ahlfors [1973] obtained the extremal metric $|\varphi'(z)| = |\nabla \mathrm{Re}\varphi(z)|$ of Section 4 by solving a mixed Dirichlet–Neumann problem for $\mathrm{Re}\varphi(z)$. Lemma 5.1 is from Marshall and Sundberg [1989]. Theorem 5.3 is in Beurling [1989], with different constants. The first form of Theorem 6.1 was proved by Ahlfors [1930], but we have followed Beurling [1989]. Fuchs [1967], Hersch [1955], and Ohtsuka [1970] have other presentations. Lord Rayleigh [1871], [1876] also derived the estimate (6.2) in 3 dimensions for the resistance of a conductor using the serial rule and proved that the conjugate resistance (extremal distance) is the reciprocal of the resistance for a quadrilateral. Baernstein [1988] says Carleman was the first to use integrals of the form $\int \frac{dx}{\theta(x)}$ to measure a domain and calls it "surely one of the most brilliant ideas in the history of complex function theory". Carleman's original [1933] estimate of harmonic measure by the above integral, proved with a differential inequality, will be given in Appendix G. In Appendix H, extremal distances on finitely connected domains will be studied by embedding the domains in their Riemann surface doubles.

Exercises and Further Results

1. Let Ω be a Jordan domain and let A_1, B_1, A_2, and B_2 be arcs of $\partial\Omega$ listed in counterclockwise order. Assume these arcs have disjoint interiors and assume $A_1 \cup A_2 \cup B_1 \cup B_2 = \partial\Omega$. Prove

$$\mathrm{Area}(\Omega) \geq \mathrm{dist}_{\mathbb{C}}(A_1, A_2)\mathrm{dist}_{\mathbb{C}}(B_1, B_2),$$

where $\mathrm{dist}_{\mathbb{C}}$ denotes euclidean distance. Prove equality holds only if A_1, B_1, A_2, and B_2 are the four sides of a rectangle.

2. Suppose two annuli A_1 and A_2 are each bounded by concentric circles. Prove that if A_1 and A_2 are conformally equivalent, then the ratios of the radii of their bounding circles are the same. Find one proof with extremal lengths and find one proof without extremal lengths.

3. Let S be a one-to-one analytic or conjugate analytic map from a domain Ω_1 onto a domain Ω_2 such that $\Omega_1 \cap \Omega_2 = \emptyset$. Let Γ_1 be a path family in Ω_1, set $\Gamma_2 = S(\Gamma_1)$ and suppose Γ is a path family in $\Omega_1 \cup \Omega_2$.
(a) If $\gamma_1 \cup S(\gamma_1) \in \Gamma$ for every $\gamma_1 \in \Gamma_1$, and if every $\gamma \in \Gamma$ contains some $\gamma_1 \in \Gamma_1$ and some $\gamma_2 \in \Gamma_2$, then $\lambda(\Gamma) = \lambda(\Gamma_1) + \lambda(\Gamma_2) = 2\lambda(\Gamma_1)$. In other words, equality holds in the serial rule. Moreover, $\lambda(\Gamma^*) = \lambda(\Gamma)$ where $\Gamma^* = \{\gamma_1 \cup S(\gamma_1) : \gamma_1 \in \Gamma_1\}$. For example, if $S(z) = \bar{z}$ and $S(\Omega) = \Omega$, set

$\Omega_1 = \Omega \cap \{z : \operatorname{Im} z > 0\}$. Then if $E \subset \partial\Omega \cap \{z : \operatorname{Im} z > 0\}$, we obtain

$$d_\Omega(E, S(E)) = 2d_{\Omega_1}(E, \mathbb{R} \cap \Omega).$$

(b) If every $\gamma_1 \in \Gamma_1$ and every $\gamma_2 \in \Gamma_2$ contains some $\gamma \in \Gamma$ and if whenever $\gamma \in \Gamma$, $\gamma \cup (S^{-1}\gamma \cap \Omega_1)$ contains some $\gamma_1 \in \Gamma_1$, then

$$\frac{1}{\lambda(\Gamma)} = \frac{1}{\lambda(\Gamma_1)} + \frac{1}{\lambda(\Gamma_2)} = \frac{2}{\lambda(\Gamma_1)}.$$

In other words, equality holds in the parallel rule. For example, if $S(z) = \bar{z}$ and $S(\Omega) = \Omega$, and if $E_1, F_1 \subset \partial\Omega \cap \{z : \operatorname{Im} z > 0\}$, then for $E_2 = S(E_1)$, $F_2 = S(F_1)$, and $\Omega_1 = \Omega \cap \{z : \operatorname{Im} z > 0\}$,

$$\frac{1}{d_\Omega(E_1 \cup E_2, F_1 \cup F_2)} = \frac{1}{d_{\Omega_1}(E_1, F_1)} + \frac{1}{d_{\Omega_2}(E_2, F_2)} = \frac{2}{d_{\Omega_1}(E_1, F_1)}.$$

4. Let $a < b < c$ be real, let Γ be the family of arcs in $\mathbb{C}$ joining $[-\infty, a]$ to $[b, c]$, and let Γ^+ be the family of arcs in the upper half-plane joining $[-\infty, a]$ to $[b, c]$.
(a) Show that

$$2\lambda(\Gamma) = \lambda(\Gamma^+) = f\left(\frac{b - a}{c - b}\right),$$

where $f(t)$ is strictly increasing, $f(0) = 0$, and $f(\infty) = \infty$.
(b) Write the function f as an elliptic integral, using the Schwarz–Christoffel formula.
(c) If C is the circle centered on $\mathbb{R}$ through a and c, let ζ be the intersection of $\{\operatorname{Re} z = b\}$ and C in $\{\operatorname{Im} z > 0\}$. Show that

$$\omega(\zeta, (-\infty, a] \cup [b, c], \mathbb{H}) = \frac{2}{\pi} \tan^{-1} \sqrt{\frac{c - b}{b - a}} \sim \frac{8}{\pi} e^{-\pi\lambda(\Gamma^+)/2}.$$

5. A ring domain Ω **separates** the two sets E and F if E and F are contained in different components of $\mathbb{C} \setminus \Omega$.
(a) (Grötzsch [1928]) If Ω is a ring domain separating $\{0, r\}$ from the unit circle $\{\zeta : |\zeta| = 1\}$, then

$$\operatorname{mod}(\Omega) \le \operatorname{mod}(\mathbb{D} \setminus [0, r]), \tag{E.1}$$

where $\operatorname{mod}(\Omega)$ denotes the modulus of Ω. Hint: Apply the extension rule for the (larger) family of curves in $\mathbb{D}$ separating $\{0, r\}$ from $\partial\mathbb{D}$. Then use the symmetry rule, and consider curves in $\mathbb{D} \cap \{\operatorname{Im} z > 0\}$ connecting $[-1, 0]$ to $[r, 1]$.

(b) (Teichmüller [1938]) If z_1, z_2, and 0 are three distinct points and if Ω is a ring domain separating $\{0, z_1\}$ from $\{z_2, \infty\}$, then

$$\text{mod}(\Omega) \leq \text{mod}\big(\mathbb{C}^* \setminus ([-|z_1|, 0] \cup [|z_2|, \infty])\big). \tag{E.2}$$

Many authors call (E.2) Teichmüller's module theorem, but in [1938] Teichmüller called a weak version of Theorem 4.1 of Chapter V "Der Modulsatz".
(c) If $\Omega = \mathbb{C}^* \setminus ([0, 1 - \varepsilon] \cup [1, \infty])$ then

$$\text{mod}(\Omega) \geq \frac{C}{\log \frac{1}{\varepsilon}}.$$

Hint: See Exercise 4 and (1.7) or Lehto and Virtanen [1973], page 61.
(d) If $\partial\Omega = \Gamma_1 \cup \Gamma_2$ where Γ_1 is a rectifiable Jordan curve with $\ell(\Gamma_1) \leq 1$ and $\text{dist}(\Gamma_1, \Gamma_2) > \varepsilon$, then

$$\text{mod}(\Omega) \geq C\varepsilon.$$

Hint: If $E = \{z : \text{dist}(z, \Gamma_1) < \varepsilon\}$ and $\rho = \chi_E$, then $\iint \rho^2 dxdy \leq C\varepsilon$.
(e) Given $R > 0$, there exists $A > 0$ such that every ring domain Ω with $\text{mod}(\Omega) > A$ contains an annulus $\{r_1 < |z - z_0| < r_2\}$ with $r_2/r_1 > R$.

6. (a) Let A and B be disjoint arcs in $\mathbb{C}$. Then

$$d_{\mathbb{C}}(A, B) \geq \frac{1}{\pi}\left(\frac{\text{dist}(A, B)}{\text{diam}(A) + \text{dist}(A, B)}\right)^2.$$

Hint: Take $z_0 \in A$ such that $\text{dist}(z_0, B) = \text{dist}(A, B)$ and define

$$\rho(z) = \begin{cases} \frac{1}{\text{dist}(A,B)} & \text{if } |z - z_0| < \text{dist}(A, B) + \text{diam}(A), \\ 0 & \text{if } |z - z_0| \geq \text{dist}(A, B) + \text{diam}(A). \end{cases}$$

(b) Suppose Ω is a ring domain having boundary components Γ_1 and Γ_2 such that with respect to the spherical metric $\rho(z) = (1 + |z|^2)^{-1}$, $\text{diam}(\Gamma_j) \geq \delta$. Prove

$$\text{mod}(\Omega) \leq \frac{\pi}{4\delta^2}.$$

If also $\text{dist}(\Gamma_1, \Gamma_2) \leq \varepsilon \leq \delta$, prove

$$\frac{1}{\text{mod}(\Omega)} \geq \frac{2}{\pi} \log\left(\frac{\tan(\frac{\delta}{2})}{\tan(\frac{\varepsilon}{2})}\right).$$

See Lehto and Virtanen [1973], page 34.

7. (Beurling's criterion for extremal metrics) Let Γ be a path family in a domain Ω and let ρ_0 be a metric on Ω. Suppose Γ contains a subfamily Γ_0 such that
(i) $\int_\gamma \rho_0|dz| = L(\Gamma, \rho_0)$ for all $\gamma \in \Gamma_0$, and

(ii) If h is real valued and Borel measurable in Ω and if $\int_\gamma h|dz| \geq 0$ for all $\gamma \in \Gamma_0$, then $\int_\Omega h\rho_0 dxdy \geq 0$.

Then ρ_0 is an extremal metric for Γ. It follows that all other extremal metrics are of the form $c\rho_0$, where $c > 0$ is a constant. See Ahlfors [1973].

8. Let $\varphi : \mathbb{D} \to \Omega$ be conformal, let I be an arc of $\partial\mathbb{D}$ with length $|I| < \pi$ and center ζ_I and let $z_I = (1 - |I|)\zeta_I$ and let σ be any arc in $\mathbb{D}$ that connects the endpoints of I. Prove $\text{dist}(\varphi(z_I), \partial\Omega) \leq C\text{diam}\varphi(\sigma)$.

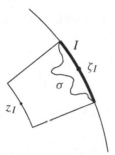

Figure IV.14

9. Let $\Omega = (0, 1) \times (0, 1)$ be the unit square, let $F = \{(1, y) : 0 \leq y \leq 1\}$, and let $E = \{(0, y) : 0 \leq y \leq 1\} \cup \{(1/2, 1)\}$.
(a) Prove $d_\Omega(E, F) = 1$. Hint: let

$$\rho_\varepsilon(z) = 1 + \frac{\chi_{B_\varepsilon \cap \Omega}(z)}{|z - \zeta| \log 1/|z - \zeta|},$$

where B_ε is a ball of radius ε centered at $\zeta = (1/2, 1)$.
(b) Let E_δ consists of the set E together with an interval centered at $(1/2, 1)$ of length δ, and let $\rho_\delta = |\varphi'_\delta|$ be the extremal metric for $d_\Omega(E_\delta, F)$. Prove $\lim_{\delta \to 0} \rho_\delta$ exists and is the constant metric, but the limit is not an extremal metric for $d_\Omega(E_0, F)$.
(c) Prove there is no extremal metric for $d_\Omega(E_0, F)$. See Ohtsuka [1970].

10. Let $R_1 = \{z : |\text{Re}z| < L, |\text{Im}z| < 1\}$, $R_2 = \{z : |\text{Im}z| \leq 1\}$, where $L > 2$. Let Ω be a Jordan domain with $R_1 \subset \Omega \subset R_2$. Let $J = \{(0, y) : |y| \leq 1\}$, $E \subset \{z \in R_2 : \text{Re}z \leq -L\}$ and $F \subset \{z \in R_2 : \text{Re}z \geq L\}$ be such that E and F are finite unions of closed arcs in $\partial\Omega$. Then there is a constant C such that

$$d_\Omega(E, F) \leq d_\Omega(E, J) + d_\Omega(J, F) + ce^{-\pi L/2}. \tag{E.3}$$

In other words, if L is large then equality almost holds in the serial rule. This result will be used in Chapter VIII. The proof of Theorem V.5.7 will include a simple special case of (E.3).

Hints: By the extension rule, $d_\Omega(E, F) \geq L$. By Theorem 4.1 there exists a conformal map φ from Ω to a rectangle R with a finite number of horizontal slits removed. We may assume $R = \{z : |\mathrm{Re}z| < d_\Omega(E, F), |\mathrm{Im}z| < 1\}$, $\mathrm{Im}(\varphi(z) - z) = 0$ on the top and bottom edges of R_1, $\mathrm{Re}\varphi = -d_\Omega(E, F)$ on E, and $\mathrm{Re}\varphi = d_\Omega(E, F)$ on F. Since $|\mathrm{Im}(\varphi(z) - z)| \leq 2$ on the vertical edges of R_1, $|\mathrm{Im}(\varphi(z) - z)| \leq C_1 e^{-\pi L/2}$ on $R_1 \cap \{|\mathrm{Re}z| \leq 2\}$ by Lemma 5.1. By Schwarz reflection, φ extends to a conformal mapping from some domain containing the rectangle $\{z : |\mathrm{Re}z| < L, |\mathrm{Im}z| < 3\}$ into the rectangle $\{z : |\mathrm{Re}z| < d_\Omega(E, F), |\mathrm{Im}z| < 3\}$, and $|\mathrm{Im}(\varphi(z) - z)| \leq C_1 e^{-\pi L/2}$ on $\mathbb{D}_2 = \{|z| < 2\}$. Differentiating the Herglotz integral formula on $\partial \mathbb{D}_2$ shows that there is a constant C such that

$$|\varphi'(z) - 1| < Ce^{-\pi L/2},$$

for $z \in J$. It follows that

$$\sup_J \mathrm{Re}\varphi - \inf_J \mathrm{Re}\varphi \leq 2\sup_J |\mathrm{Im}\varphi'(z)| \leq 2Ce^{-\pi L/2}.$$

Therefore (E.3) holds because $d_\Omega(E, J) \geq \frac{1}{2}(d_\Omega(E, F) + \inf_J \mathrm{Re}\varphi)$ and $d_\Omega(J, F) \geq \frac{1}{2}(d_\Omega(E, F) - \sup_J \mathrm{Re}\varphi)$.

11. Let $R_L = \{|\mathrm{Re}z| < L, |\mathrm{Im}z| < 1\}$ and let $E_L = \partial R_L \cap \{|\mathrm{Re}z| = L\}$ be as in Lemma 5.1.

(a) Prove that $\omega(0, E_L, R_L) \geq (.93)\frac{8}{\pi}e^{-\frac{\pi}{2}L}$ when $L \geq 1$. See Marshall and Sundberg [1989].

(b) If $L < 1$, prove that $1 - \frac{8}{\pi}e^{-\frac{\pi}{2L}} \leq \omega(0, E_L, R_L) \leq 1 - (.93)\frac{8}{\pi}e^{-\frac{\pi}{2L}}$.

(c) Let $S = \{|\mathrm{Im}z| < 1\}$ and let $F_L = \partial S \cap \{|\mathrm{Re}z| < L\}$. Prove

$$\omega(z, E_L, R_L) = \sum_{n=-\infty}^{+\infty} 2\omega(z + (4n + 2)L, F_L, S).$$

(d) Using a conformal map, find explicit formulae for $\omega(z, F_L, S)$, and $\omega(0, E_L, R_L)$.

12. Let R_L and E_L be as in Exercise 10.

(a) Set

$$x_n = \left(\sin\frac{\pi}{2}\omega(0, E_{2^n L}, R_{2^n L})\right)^{-\frac{1}{2}}.$$

Prove $x_{n-1}^2 = \frac{1}{2}(x_n + 1/x_n)$. Hint: Use the Schwarz reflection principle to relate the conformal map of $\mathbb{D}$ onto R_L to the conformal map of $\mathbb{D}$ onto R_{2L}.

(b) Given n and L, set

$$y_0 = \frac{1}{2} e^{\pi L 2^{n-1}},$$

$$y_k = \sqrt{\frac{1}{2}(y_{k-1} + 1/y_{k-1})}, \quad 1 \le k \le n,$$

$$\omega = \frac{1}{\pi} \tan^{-1}\left(\frac{2}{y_n - 1/y_n}\right).$$

Prove

$$|\omega(0, E_L, R_L) - \omega| \le C e^{-\pi L 2^{n-1}}.$$

When $n = 3$ and $L \ge 1$ the computation of ω is accurate to 16 decimal places. In general this method gives quadratic convergence to $\omega(0, E_L, R_L)$, which is considerably faster than other methods such as the one given in Exercise 11.

(c) Use (a) to find a quadratically convergent algorithm to compute the conformal map of D onto R_L. Hint: For large N, approximate $R_{2^n L}$ by a semi-infinite strip. See Marshall and Sundberg [1989].

13. (a) Given $\varepsilon > 0$, construct a set $E \subset \partial \mathbb{D}$ consisting of finitely many closed arcs so that $\omega(0, E, \mathbb{D}) < \varepsilon$ but $\lambda(0, E) = \sup_\sigma d_{\mathbb{D} \setminus \sigma}(\sigma, E) = 1$ where the supremum is taken over all arcs σ connecting 0 to $\partial \mathbb{D}$. Hint: Remove horizontal slits from a rectangle.

(b) Show that for each set $E \subset \partial \mathbb{D}$ that consists of n arcs, there are n possible rectangles that occur in the proof of Theorem 5.3. Also show that the vertical line through 0 can meet the excised slits in the proof of Theorem 5.3. Is there a curve σ so that $\lambda(z_0, E) = d_{\mathbb{D}}(\sigma, E)$?

14. Let Ω be simply connected, let $z_0, z_1 \in \Omega$, and let γ_j be an arc in Ω joining z_j to $\partial \Omega$. Then

$$\mathrm{Min}\big(g(z_0, z_1), 1\big) \le C \exp\left(-\pi d_{\Omega \setminus \gamma_1 \cup \gamma_2}(\gamma_1, \gamma_2)\right)$$

for some absolute constant C. See Beurling [1989].

15. Let $L_n = \{re^{i\theta_n} : 0 \le r \le 1/2, \theta_n = 2^{-n}\pi\}$ and

$$\Omega = \{z : |z| < 1, \mathrm{Im}\, z > 0, |z - i/4| > 1/4\} \setminus \bigcup_{n=1}^{\infty} L_n.$$

Let $E = \{e^{i\theta} : 0 \le \theta \le \pi/6\}$ and $F = \{-r : 0 \le r \le 1\}$. If Γ denotes the curves in Ω connecting E to F and Γ_n denotes the curves in Ω connecting E to $F_n = F \cup B(0, 2^{-n})$, then for $\rho \equiv 1$,

$$\inf_{\gamma \in \Gamma_n} \int_\gamma \rho |dz| \le 1$$

and

$$\inf_{\gamma \in \Gamma} \int_\gamma \rho |dz| = \frac{\sqrt{3}}{2} + \frac{\pi}{8} + \frac{1}{4} > 1.$$

The approximation does not work here because the prime end corresponding to 0 includes the interval [0, 1/2]. See Ohtsuka [1970].

16. (a) Let $\gamma_n \subset \Omega$ have endpoints $\gamma_n(0) \to a \in \partial\Omega$ and $\gamma_n(1) \to b \in \partial\Omega$. Moreover suppose that there are arcs $I_k \subset \Omega \cap \{|z - a| = \delta_k\}$ with $\delta_k \to 0$ so that I_k separates all but finitely many $\{\gamma_n(0)\}$ from some fixed point $z_0 \in \Omega$. Similarly suppose there are arcs $J_k \subset \{|z - b| = \delta_k\}$ so that J_k separates all but finitely many $\{\gamma_n(1)\}$ from z_0. Let ρ be a metric on Ω with $A(\Omega, \rho) < \infty$. Prove that for all $\varepsilon > 0$ there exists a curve $\sigma_\varepsilon \subset \Omega$ connecting a to b so that

$$\int_{\sigma_\varepsilon} \rho |dz| \le \liminf \int_{\gamma_n} \rho |dz| + \varepsilon.$$

Hint: Use the length–area principle. See Ohtsuka [1970].
(b) Suppose Ω is a finitely connected Jordan domain and suppose E_n and F_n are each finite unions of pairwise disjoint closed arcs in $\partial\Omega$ with $E_n \supset E_{n+1}$ and $F_n \supset F_{n+1}$. Let $E = \bigcap E_n$ and $F = \bigcap F_n$. Use (a) to prove that

$$\lim d_\Omega(E_n, F_n) = d_\Omega(E, F).$$

V

Applications and Reverse Inequalities

In Section 1 we present Ahlfors' solution of the Denjoy problem, historically the first application of an estimate like (IV.6.4). In Section 2 we give Beurling's converse to Theorem IV.6.1. It is proved by constructing a new metric. In Section 3 we discuss reduced extremal distance and derive a recent sharp theorem of Balogh and Bonk about the lengths of conformal images of radii. It is also proved by constructing a new metric. In Section 4 we prove Teichmüller's theorem on the addition of moduli of rings. In Sections 5 and 6 we use extremal distances to to obtain two different, definitive results on the existence of angular derivatives.

1. Asymptotic Values of Entire Functions

An entire function $f(z)$ has **asymptotic value** a if there is a Jordan arc Γ tending to ∞ such that

$$\lim_{\Gamma \ni z \to \infty} f(z) = a.$$

For example

$$f(z) = \int_0^z e^{-w^n} dw$$

has n distinct asymptotic values, one along each of the curves $\arg z = 2\pi j / n$, $j = 1, \ldots, n$. Write

$$M(r) = \sup_{|z|=r} |f(z)|$$

for the maximum of $|f(z)|$ on $\{|z| = r\}$.

Theorem 1.1. *If the non-constant entire function $f(z)$ has n distinct finite*

157

asymptotic values, then

$$\liminf_{r \to \infty} \left(\frac{\log M(r)}{r^{n/2}} \right) > 0. \tag{1.1}$$

Proof. Assume f has n distinct asymptotic values a_j taken along n distinct curves Γ_j. Moving each Γ_j slightly, we may assume that Γ_j is a simple polygonal arc whose vertices tend only to ∞. Because the asymptotic values a_j are distinct, we may further move the Γ_j so that $\Gamma_j \cap \Gamma_k = \{0\}$ for $j \neq k$, and so that $0 \in \Gamma_1$ when $n = 1$. Then the Γ_j divide the plane into n Jordan domains $G_1, G_2, \ldots, G_n$, and when $n > 1$ each G_j is bounded by two distinct Γ_k.

On each domain G_j the function $f(z)$ must be unbounded. When $n = 1$, this holds because f is not constant. When $n > 1$, it follows from Lindelöf's theorem, Exercise II.3(d), applied to $f \circ \varphi$ where φ is a conformal map from $\mathbb{D}$ onto G_j.

We can assume $|f(z)| \leq 1$ on $\bigcup \Gamma_j$, and by the maximum principle we can choose R_0 large so that for each $j = 1, \ldots, n$ there is a $z_j \in G_j$ with $|z_j| = R_0$ and $|f(z_j)| > 1$. Let $R > R_0$. Write

$$\omega_j(R) = \omega(z_j, \{|z| = R\}, G_j \cap \{|z| < R\}),$$

and write $\Theta_j(r)$ for the angular measure of $G_j \cap \{|z| = r\}$. Then by the maximum principle

$$\log|f(z_j)| \leq \omega_j(R)\log M(R)$$

and hence by (IV.6.4),

$$\log|f(z_j)| \leq \frac{8}{\pi} \exp\left(-\pi \int_{R_0}^{R} \frac{dr}{r\Theta_j(r)} \right) \log M(R).$$

(When $n = 1$, G_1 can be mapped to a Jordan domain using $\sqrt{z}$ and (IV.6.4) can then be applied.) By our assumptions on the Γ_j,

$$\sum_{j=1}^{n} \Theta_j(r) = 2\pi,$$

and by the Cauchy–Schwarz inequality,

$$\left(\sum_{j=1}^{n} \frac{1}{\Theta_j(r)} \right) \left(\sum_{j=1}^{n} \Theta_j(r) \right) \geq n^2,$$

so that

$$\sum_{j=1}^{n} \frac{1}{\Theta_j(r)} \geq \frac{n^2}{2\pi}.$$

Hence

$$\frac{1}{n}\sum_{j=1}^{n}\int_{R_0}^{R}\frac{dr}{r\Theta_j(r)} \geq \frac{n}{2\pi}\log\left(\frac{R}{R_0}\right),$$

and there is $j = j(R)$ such that

$$\exp\left(-\pi\int_{R_0}^{R}\frac{dr}{r\Theta_j(r)}\right) \leq \left(\frac{R_0}{R}\right)^{n/2}.$$

We conclude that

$$0 < \min_{j}\log|f(z_j)| \leq \frac{8}{\pi}R_0^{\frac{n}{2}}\left(\frac{\log M(R)}{R^{\frac{n}{2}}}\right),$$

from which (1.1) follows. ∎

2. Lower Bounds

For a domain of the form

$$\Omega = \{(x, y) : |y - m(x)| < \theta(x)/2, \ a < x < b\},$$

the inequality (IV.6.3) of Theorem IV.6.1 is not sharp, because the mapping Φ defined in (IV.6.1) is not conformal. However, minor smoothness assumptions on $m(x)$ and $\theta(x)$ will yield a lower bound for $\omega(z_0, F)$ that complements (IV.6.3). Lower bounds on harmonic measure come from upper bounds on extremal distances or, equivalently, from lower bounds on conjugate extremal distances. In view of (IV.6.1) a natural metric for studying the conjugate extremal distance in Ω is

$$\rho_B(x, y) = \left|\nabla\left(\frac{y - m(x)}{\theta(x)}\right)\right| = \sqrt{\frac{1}{\theta^2(x)} + \left(\frac{(y - m)\theta' + \theta m'}{\theta^2}\right)^2}. \quad (2.1)$$

This metric was invented by Beurling [1989].

Let $E = \partial\Omega \cap \{x = a\}$ and $F = \partial\Omega \cap \{x = b\}$. If γ is any curve in Ω connecting the curve $y = m(x) + \theta(x)/2$ to the curve $y = m(x) - \theta(x)/2$, then

$$\int_{\gamma}\rho_B|dz| \geq \left|\int_{\gamma}\nabla\left(\frac{y - m}{\theta}\right)\cdot dz\right| = 1.$$

Furthermore,

$$A(\Omega, \rho_B) = \int_a^b \int_{m-\frac{\theta}{2}}^{m+\frac{\theta}{2}} \left\{ \frac{1}{\theta^2} + \left[\frac{(y-m)\theta' + \theta m'}{\theta^2} \right]^2 \right\} dy dx$$

$$= \int_a^b \frac{dx}{\theta(x)} + \int_a^b \frac{m'(x)^2 + \frac{1}{12}\theta'(x)^2}{\theta(x)} dx.$$

Thus

$$d_\Omega(E, F) = \frac{1}{d_\Omega^*(E, F)} \leq \int_a^b \frac{dx}{\theta(x)} + \int_a^b \frac{m'(x)^2 + \frac{1}{12}\theta'(x)^2}{\theta(x)} dx. \quad (2.2)$$

Theorem 2.1. *Suppose* $\Omega = \{(x, y) : |y - m(x)| < \theta(x)/2, a < x < b\}$ *is a Jordan domain and suppose* θ *and* m *are absolutely continuous on* (a, b). *Suppose further that*

$$\{(x, y) : |x - x_0| < \delta, \ |y - y_0| < \delta\} \subset \Omega,$$

and set $z_0 = x_0 + iy_0$. *If* $F = \partial\Omega \cap \{(x, y) : x = b\}$, *then*

$$\omega(z_0, F) \geq C(\delta) \exp \left\{ -\pi \int_{x_0}^b \frac{1 + m'(x)^2 + \frac{1}{12}\theta'(x)^2}{\theta(x)} dx \right\}, \quad (2.3)$$

where

$$C(\delta) = \exp \left(-\frac{\pi}{\delta^2} \int_{x_0 - \delta}^{x_0 + \delta} \theta(x) dx \right)$$

depends only on δ *and* $\theta(x)$ *for* $x \in (x_0 - \delta, x_0 + \delta)$.

Proof. Define the metric $\tilde{\rho}$ for $(x, y) \in \Omega$ by

$$\tilde{\rho}(x, y) = \begin{cases} \rho_B(x, y) & \text{if } x > x_0 + \delta, \\ \sqrt{\rho_B^2(x, y) + \frac{1}{\delta^2}} & \text{if } x_0 < x < x_0 + \delta, \\ \frac{1}{\delta} & \text{if } x_0 - \delta < x < x_0, \\ 0 & \text{elsewhere in } \Omega, \end{cases}$$

where ρ_B is the Beurling metric (2.1).

Let σ be any curve connecting z_0 to $\partial\Omega$. If γ is any curve in $\Omega \setminus \sigma$ that separates σ from F, then $\int_\gamma \tilde{\rho}|dz| \geq 1$. Indeed suppose $\gamma \subset \{z : \text{Re}\, z > x_0\}$. Then $\int_\gamma \tilde{\rho}|dz| \geq \int_\gamma \rho_B|dz| \geq 1$. On the other hand, if $\gamma \not\subset \{z : \text{Re}\, z > x_0\}$, then the euclidean length of $\gamma_\delta = \gamma \cap \{z : x_0 - \delta < \text{Re} z < x_0 + \delta\}$ is at least δ, because γ separates σ from F. Hence $\int_\gamma \tilde{\rho}|dz| \geq \int_{\gamma_\delta} \frac{1}{\delta}|dz| \geq 1$.

Also,

$$\iint_\Omega \tilde{\rho}^2 dx dy = \int_{x_0}^b \int_{m-\theta/2}^{m+\theta/2} \rho_B^2(x, y) dy dx + \frac{1}{\delta^2}\int_{x_0-\delta}^{x_0+\delta} \theta(x) dx.$$

Thus,

$$d_\Omega(\sigma, F) = \frac{1}{d_\Omega^*(\sigma, F)} \leq \int_{x_0}^b \frac{1}{\theta(x)} dx$$
$$+ \int_{x_0}^b \frac{m'(x)^2 + \frac{1}{12}\theta'(x)^2}{\theta(x)} dx + \frac{1}{\delta^2}\int_{x_0-\delta}^{x_0+\delta} \theta(x) dx$$

by (2.2). Because the crosscut σ is arbitrary, the result now follows from Theorem IV.5.2. ∎

Theorem 2.1 implies that Theorem IV.6.1 is the best result possible for domains with smooth boundaries.

Corollary 2.2. *Suppose* $\Omega = \{(x, y) : |y - m(x)| < \frac{\theta(x)}{2}, \; x > a\}$ *is a Jordan domain and suppose*

$$\{(x, y) : |x - x_0| < \delta, |y - y_0| < \delta\} \subset \Omega.$$

Set $z_0 = x_0 + i y_0$. *If*

$$\int_a^\infty \left(\frac{m'(x)^2 + \theta'(x)^2}{\theta(x)}\right) dx = A < \infty,$$

then whenever $b > x_0$ *and* $F = \{z \in \overline{\Omega} : \text{Re} z \geq b\}$,

$$C \exp\left(-\pi \int_{x_0}^b \frac{dx}{\theta(x)}\right) \leq \omega(z_0, F) \leq \frac{8}{\pi} \exp\left(-\pi \int_{x_0}^b \frac{dx}{\theta(x)}\right),$$

where C *is a constant depending only on* A *and* $\theta(x)$ *for* $x \in (x_0 - \delta, x_0 + \delta)$.

The lower bound (2.3) and the upper bound (IV.6.3) were both obtained by studying metrics of the form $\rho = |\nabla u|$. Metrics of this form have the obvious advantage that lengths can be estimated by

$$\int_\gamma |\nabla u||dz| \geq \left|\int_\gamma \nabla u \cdot dz\right|,$$

which is the change in u along γ. Other metrics of this form will appear in Section 6.

3. Reduced Extremal Distance

Let Ω be a finitely connected Jordan domain, let $z_0 \in \Omega$, and let E be a finite union of closed subarcs of $\partial\Omega$. The reduced extremal distance $\delta_\Omega(z_0, E)$ is a conformally invariant version of the distance

$$\inf\left\{ \int_\gamma ds : \gamma \subset \Omega, \ \gamma \text{ joins } z_0 \text{ to } E \right\}$$

in Ω from z_0 to E.

Let us try to measure the distance from z_0 to E by deleting a small disc, $B_\varepsilon = B_\varepsilon(z_0) = \{z : |z - z_0| < \varepsilon\}$ from Ω and computing the extremal distance $d_{\Omega \setminus B_\varepsilon}(\partial B_\varepsilon, E)$. If $\beta < \varepsilon$, then by the serial rule and Example IV.1.2,

$$
\begin{aligned}
d_{\Omega \setminus B_\beta}(\partial B_\beta, E) &\geq d_{\Omega \setminus B_\varepsilon}(\partial B_\varepsilon, E) + d_{B_\varepsilon \setminus B_\beta}(\partial B_\beta, \partial B_\varepsilon) \\
&= d_{\Omega \setminus B_\varepsilon}(\partial B_\varepsilon, E) + \frac{1}{2\pi}\log(\varepsilon/\beta).
\end{aligned}
\tag{3.1}
$$

Hence

$$h(\varepsilon) = d_{\Omega \setminus B_\varepsilon}(\partial B_\varepsilon, E) + \frac{1}{2\pi}\log\varepsilon$$

is a decreasing function of ε. We will show in a moment that $h(\varepsilon)$ is bounded above. It will then follow that the three limits

$$\lim_{\varepsilon \to 0} d_{\Omega \setminus B_\varepsilon}(\partial B_\varepsilon, E) + \frac{1}{2\pi}\log\varepsilon,$$

$$\lim_{\varepsilon \to 0} d_{\Omega \setminus B_\varepsilon}(\partial B_\varepsilon, \partial\Omega) + \frac{1}{2\pi}\log\varepsilon,$$

and

$$\delta(z_0, E) = \delta_\Omega(z_0, E) = \lim_{\varepsilon \to 0} \left(d_{\Omega \setminus B_\varepsilon}(\partial B_\varepsilon, E) - d_{\Omega \setminus B_\varepsilon}(\partial B_\varepsilon, \partial\Omega) \right)$$

all exist and are finite. The **reduced extremal distance** is defined to be the third limit $\delta(z_0, E)$.

To see that $h(\varepsilon)$ is bounded above, choose a component E_1 of E and let ψ be a conformal map of $\mathbb{C}^* \setminus E_1$ to $\mathbb{D}$ such that $\psi(z_0) = 0$. Then $\psi(B_\varepsilon) \supset B(0, a\varepsilon)$ for some $a > 0$, so that by conformal invariance and the extension rule,

$$
\begin{aligned}
h(\varepsilon) &\leq d_{\Omega \setminus B_\varepsilon}(\partial B_\varepsilon, E_1) + \frac{1}{2\pi}\log\varepsilon \\
&\leq d_{\mathbb{D} \setminus B(0, a\varepsilon)}(\partial B(0, a\varepsilon), \partial\mathbb{D}) + \frac{1}{2\pi}\log\varepsilon = \frac{1}{2\pi}\log\frac{1}{a},
\end{aligned}
$$

and $h(\varepsilon)$ is bounded.

Lemma 3.1. *The reduced extremal distance* $\delta(z_0, E)$ *is conformally invariant.*

Proof. Suppose φ is a conformal map defined on Ω. We may suppose that $z_0 = 0$ and $\varphi(z_0) = 0$. Because φ is conformal at 0, there exist $\varepsilon_0 > 0$ and $K > 0$ such that for $\alpha = \varepsilon|\varphi'(0)| - K\varepsilon^2$ and $\beta = \varepsilon|\varphi'(0)| + K\varepsilon^2$,

$$B_\alpha \subset \varphi(B_\varepsilon) \subset B_\beta$$

when $\varepsilon < \varepsilon_0$. Because $k(\alpha) = d_{\varphi(\Omega)\setminus B_\alpha}(\partial B_\alpha, \varphi(E)) + \frac{1}{2\pi}\log\alpha$ is decreasing and bounded above and because $\lim_{\varepsilon\to 0}\log(\beta/\alpha) = 0$, we have

$$\lim_{\varepsilon\to 0}\big(d_{\varphi(\Omega)\setminus B_\alpha}(\partial B_\alpha, \varphi(E)) - d_{\varphi(\Omega)\setminus B_\beta}(\partial B_\beta, \varphi(E))\big) = \lim_{\varepsilon\to 0}(k(\alpha) - k(\beta)) = 0.$$

We conclude that

$$\begin{aligned}
\delta(0, E) &= \lim_{\varepsilon\to 0}\big(d_\Omega(\partial B_\varepsilon, E) - d_\Omega(\partial B_\varepsilon, \partial\Omega)\big)\\
&= \lim_{\varepsilon\to 0}\big(d_{\varphi(\Omega)}(\varphi(\partial B_\varepsilon), \varphi(E)) - d_{\varphi(\Omega)}(\varphi(\partial B_\varepsilon), \partial\varphi(\Omega))\big)\\
&= \lim_{\varepsilon\to 0}\big(d_{\varphi(\Omega)}(\partial B_\alpha, \varphi(E)) - d_{\varphi(\Omega)}(\partial B_\alpha, \partial\varphi(\Omega))\big)\\
&= \delta(0, \varphi(E)),
\end{aligned}$$

and $\delta(z_0, E)$ is a conformal invariant. ∎

Recall that when E is a compact plane set, the logarithmic capacity of E is $e^{-\gamma_E}$, where $g(z, \infty) = \log|z| + \gamma_E + o(1)$ (as $z \to \infty$), is Green's function for $\mathbb{C}^* \setminus E$, and γ_E is Robin's constant for E In [1950] and [1952] Ahlfors and Beurling obtained the following connection between capacity and reduced extremal distance.

Theorem 3.2 (Ahlfors–Beurling). *If E is a finite union of closed arcs in $\partial\mathbb{D}$, then*

$$\delta_\mathbb{D}(0, E) = \delta_{\overline{\mathbb{D}}^c}(\infty, E) = \gamma_E/\pi,$$

where γ_E is Robin's constant for E.

Proof. Set $\Omega_d = \mathbb{C}^* \setminus E$ and $G(z) = g(z) + g(1/\overline{z})$, where $g(z)$ is Green's function in Ω_d with pole at ∞. Note that $g(z) - g(1/\overline{z}) - \log|z|$ is harmonic in Ω_d and zero on $\partial\Omega_d$ and hence $g(1/\overline{z}) = g(z) - \log|z|$ in Ω_d. Thus $G(z) = 2g(z) - \log|z|$ and $g(0) = \lim_{|z|\to\infty} g(z) - \log|z| = \gamma_E$. Since G is symmetric about the unit circle, $\partial G/\partial r = 0$ on $\partial\mathbb{D}\setminus E$ and $G = 0$ on E. If $\widetilde{g}$ is a harmonic conjugate of g in $\mathbb{D}$, then $f(z) = ze^{-2(g+i\widetilde{g})}$ is analytic in $\mathbb{D}$ and $-\log|f| = G$. By repeating the proof of Theorem IV.4.1, but working with the symmetric function G instead of with ω, we see that f extends to be continuous on $\overline{D}$, that $|f| = 1$ on E, and that $\partial(\arg f)/\partial\theta = 0$ on $\partial\mathbb{D} \setminus E$. The argument principle shows that f is a conformal map of $\mathbb{D}$ onto the disc with radial slits removed, because f has exactly one zero in $\mathbb{D}$. Furthermore, $\partial\mathbb{D} = f(E)$ and $f(0) = 0$. The extremal distance from $F_\varepsilon \equiv \{z : |f(z)| = \varepsilon\}$ to E is then $\frac{1}{2\pi}\log(1/\varepsilon)$. Because $g(0) = \gamma_E$, we have $\varepsilon = |z|e^{-2g(z)} \sim |z|e^{-2\gamma_E}$ on F_ε and hence F_ε is close to the circle of radius $\varepsilon e^{2\gamma_E}$. But then the proof of Lemma 3.1 yields

$$\delta_{\mathbb{D}}(0, E) = \lim_{\varepsilon \to 0} d_{\mathbb{D}}(F_\varepsilon, E) - d_{\mathbb{D}}(F_\varepsilon, \partial\mathbb{D})$$

$$= \lim_{\varepsilon \to 0}\left(\frac{1}{2\pi}\log\frac{1}{\varepsilon} - \frac{1}{2\pi}\log\frac{1}{\varepsilon e^{2\gamma_E}}\right)$$

$$= \gamma_E/\pi. \qquad\blacksquare$$

The proof of Theorem 3.2 was a reprise of the proof of Theorem IV.4.1, with harmonic measure replaced by Green's function and for that reason reduced extremal distance leads to estimates in which harmonic measures are replaced by capacities. Theorem 3.2 includes an estimate for the harmonic measure of $E \subset \partial\mathbb{D}$, because by Example III.1.2,

$$e^{-\gamma_E} \geq \sin\frac{|E|}{4} \geq \frac{|E|}{2\pi} = \omega(0, E). \qquad (3.2)$$

When E is a single arc, equality holds in the first inequality in (3.2). Thus we have the following corollary.

Corollary 3.3. *If Ω is a Jordan domain and if E is a finite union of closed arcs on $\partial\Omega$, then*

$$\omega(z_0, E) \leq e^{-\pi\delta(z_0, E)}.$$

If E is a single arc, then

$$\omega(z_0, E) = \frac{2}{\pi}\sin^{-1}(e^{-\pi\delta(z_0, E)}) \geq \frac{2}{\pi}e^{-\pi\delta(z_0, E)}.$$

Pfluger [1955] gives a quantitative version of Theorem 3.2.

Corollary 3.4 (Pfluger). *If $\Omega = \mathbb{D}$, if E is a finite union of closed arcs in $\partial\mathbb{D}$, and if ∂B_ε is the circle centered at 0 of radius ε, then*

$$\frac{\gamma_E}{\pi} + \frac{\log 1/\varepsilon}{2\pi} + \frac{\log(1-\varepsilon)}{\pi} \leq d_{\mathbb{D}}(\partial B_\varepsilon, E) \leq \frac{\gamma_E}{\pi} + \frac{\log 1/\varepsilon}{2\pi}.$$

Proof. By (3.1) the function

$$h(\varepsilon) = d_{\mathbb{D}}(\partial B_\varepsilon, E) - d_{\mathbb{D}}(\partial B_\varepsilon, \partial\mathbb{D}) = d_{\mathbb{D}}(\partial B_\varepsilon, E) + \frac{1}{2\pi}\log\varepsilon$$

increases to γ_E / π as $\varepsilon \to 0$. This proves the right-hand inequality. By Theorem III.2.2 , we have

$$\gamma_E - g(z) = U_\mu(z) = \int_E \log \frac{1}{|z - \zeta|} d\mu(\zeta),$$

where μ is a probability measure and $g(z) = g_\Omega(z, \infty)$. If $f = z e^{-2(g + i\tilde{g})}$ as in the proof of Theorem 3.2, then for $z \in \partial B_\varepsilon$

$$\log |f| = -2g(z) + \log \varepsilon \le -2\gamma_E - 2\log(1 - \varepsilon) + \log \varepsilon.$$

Thus if $\log 1/r = 2\gamma_E + 2\log(1 - \varepsilon) - \log \varepsilon$,

$$d_{\mathbb{D}}(\partial B_\varepsilon, E) \ge d_{f(\mathbb{D})}(\partial B_r, \partial \mathbb{D}) = d_{\mathbb{D}}(\partial B_r, \partial \mathbb{D})$$

$$= \frac{1}{2\pi}(2\gamma_E + 2\log(1 - \varepsilon) - \log \varepsilon).$$

That proves the left-hand inequality. ∎

See Exercise 2 for a proof of Theorem 3.2 using logarithmic potentials only. The next result introduces a metric found in Bonk, Koskela, and Rohde [1998] and Balogh and Bonk [1999].

Fix $z_0 \in \Omega$ and let

$$\text{dist}_\Omega(z_0, w) = \inf \left\{ \int_\gamma ds : \gamma \text{ is a curve in } \Omega \text{ and } z_0, w \in \overline{\gamma} \right\}$$

be the **euclidean distance in** Ω from z_0 to w.

Theorem 3.5. *Let $\psi : \mathbb{D} \to \Omega$ be a normalized univalent function, $\psi(0) = 0$ and $\psi'(0) = 1$, and let $E \subset \{\zeta \in \partial \Omega : \text{dist}_\Omega(0, \zeta) \ge R\}$. Then*

$$\delta_\Omega(0, E) \ge \frac{1}{2\pi} \log R. \tag{3.3}$$

Equality holds in (3.3) if Ω is a disc of radius R, with radial slits removed.

Proof. We may assume that $\partial \Omega$ is an analytic curve. Since $\psi(0) = 0$ and $|\psi'(0)| = 1$,

$$\delta_\Omega(0, E) = \lim_{\varepsilon \to 0} \left(d_\Omega(\partial B_\varepsilon, E) - \frac{1}{2\pi} \log \frac{1}{\varepsilon} \right).$$

By the extension rule, we may suppose that $E = \{z \in \Omega : \text{dist}_\Omega(0, z) \ge R\}$ and by another application of the extension rule, we may suppose that $\text{dist}_\Omega(0, \zeta) \le R$ for all $\zeta \in \partial \Omega$ and $E = \{\zeta \in \partial \Omega : \text{dist}_\Omega(0, \zeta) = R\}$. Define a metric ρ on Ω by

$$\rho(z) = \frac{1}{\text{dist}_\Omega(0, z)}.$$

If $\gamma : [0, L) \to \Omega$ is a curve parameterized by arc length with $\gamma(0) = 0$, then

$$|\gamma(s)| \leq \text{dist}_\Omega(0, \gamma(s)) \leq s.$$

Any curve σ connecting ∂B_ε to E in $\Omega \setminus B_\varepsilon$ can be extended to a curve γ connecting 0 to E by adding a line segment of length ε. Thus

$$\int_\sigma \rho(z)ds = \int_{|\gamma| \geq \varepsilon} \rho(z)ds \geq \int_\varepsilon^R \frac{1}{s}ds = \log \frac{R}{\varepsilon}$$

and

$$\iint_{\Omega \setminus B_\varepsilon} \rho^2(z)dxdy \leq \iint_{\varepsilon < |w| < R} \frac{1}{|w|^2}dA(w) = 2\pi \log \frac{R}{\varepsilon},$$

where $B_\varepsilon = \{z : |z| < \varepsilon\}$. Therefore

$$d_\Omega(\partial B_\varepsilon, E) = d_{\Omega \setminus B_\varepsilon}(\partial B_\varepsilon, E) \geq \frac{1}{2\pi} \log \frac{R}{\varepsilon}$$

and

$$\delta(0, E) = \lim_{\varepsilon \to 0}\left(d_\Omega(\partial B_\varepsilon, E) - \frac{1}{2\pi} \log \frac{1}{\varepsilon}\right) \geq \frac{1}{2\pi} \log R. \qquad (3.4)$$

By Theorem IV.4.2, equality holds in (3.4) when E is the circle of radius R and Ω is the disc of radius R with radial slits removed. ∎

Corollary 3.6 (Balogh–Bonk). *Let $\psi(z)$ be a normalized univalent function, $\psi(0) = 0$ and $\psi'(0) = 1$, in $\mathbb{D}$ and let $R > 0$. Then there is a constant C, independent of ψ and R, such that*

$$\text{Cap}\left\{\zeta \in \partial\mathbb{D} : \int_0^1 |\psi'(r\zeta)|dr > \lambda\right\} \leq C\lambda^{-1/2}.$$

Proof. Set $\Omega = \psi(\mathbb{D})$. By an approximation, we may suppose that $\partial\Omega$ is an analytic Jordan curve. By the Gehring–Hayman inequality, Exercise III.16,

$$\int_0^1 |\psi'(r\zeta)|dr \leq K\text{dist}_\Omega(0, \psi(\zeta))$$

for some constant K independent of ψ and ζ. Corollary 3.6 now follows from Theorems 3.2 and 3.5 with $R = \lambda/K$ and $C = \sqrt{K}$. ∎

For a set $E \subset \partial\mathbb{D}$, (3.2) shows that capacity can be replaced by length in the statement of Corollary 3.6. See Exercises III.22 and III.23 for precursors of Corollary 3.6.

4. Teichmüller's Modulsatz

Let $\Omega = \{R_1 < |z| < R_2\}$ and let $J \subset \Omega$ be a continuum that separates Ω into disjoint ring domains Ω_1, Ω_2 with $\{|z| = R_1\} \subset \partial\Omega_1$ and $\{|z| = R_2\} \subset \partial\Omega_2$.

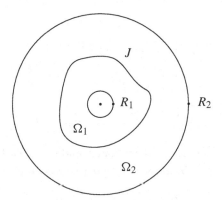

Figure V.1 Teichmüller's Modulsatz.

By the serial rule

$$\mathrm{mod}(\Omega_1) + \mathrm{mod}(\Omega_2) \le \mathrm{mod}(\Omega) = \frac{1}{2\pi} \log\left(\frac{R_2}{R_1}\right), \qquad (4.1)$$

and equality holds in (4.1) if J is a circle $\{|z| = r\}$. Theorem 4.1 says that if equality approximately holds in (4.1) then J is approximately a circle centered at 0.

Theorem 4.1 (Teichmüller). *If*

$$\mathrm{mod}(\Omega_1) + \mathrm{mod}(\Omega_2) \ge \frac{1}{2\pi} \log\left(\frac{R_2}{R_1}\right) - \delta \qquad (4.2)$$

and if $\delta > 0$ is sufficiently small, then there is a $C < \infty$ independent of Ω and δ such that

$$\frac{\sup_J |z|}{\inf_J |z|} \le 1 + C\sqrt{\delta \log \frac{1}{\delta}}. \qquad (4.3)$$

In [1938] Teichmüller derived his Modulsatz, Theorem 4.1, from his special Modulsatz, Theorem 4.2 below, and we do the same.

Let G be a simply connected domain such that $0 \in G$. For small ρ define

$$M_\rho = M_\rho(G) = \mathrm{mod}(G \setminus \overline{B(0, \rho)}).$$

As in (3.1),

$$M_\rho + \frac{\log \rho}{2\pi} \leq M_{\rho'} + \frac{\log \rho'}{2\pi}, \qquad (4.4)$$

if $\rho' < \rho$. Let $\varphi : \mathbb{D} \to G$ be a conformal mapping with $\varphi(0) = 0$. Then for small ρ

$$M_\rho(G) = \mathrm{mod}\big(\mathbb{D} \setminus \varphi^{-1}(\overline{B(0, \rho)})\big),$$

so that by the proof of Lemma 3.1,

$$\lim_{\rho \to 0} M_\rho + \frac{\log \rho}{2\pi} = \frac{1}{2\pi} \log |\varphi'(0)|.$$

We call $M(G) = \frac{1}{2\pi} \log |\varphi'(0)|$ the **reduced modulus** of G. If $G \subset C^*$ is simply connected and $\infty \in G$, the reduced modulus of G is defined by

$$M(G) = \frac{1}{2\pi} \log |(\frac{1}{\psi})'(0)|,$$

where $\psi : \mathbb{D} \to G$ is a conformal map with $\psi(0) = \infty$. It follows from Example III.1.1 that $2\pi M(G) = \gamma_{\partial G}$ when $\infty \in G$. If J is a continuum and if G_1 and G_2 are disjoint components of $\mathbb{C}^* \setminus J$ such that $0 \in G_1$ and $\infty \in G_2$, then by (4.1)

$$M(G_1) + M(G_2) \leq 0,$$

and equality holds if J is a circle $\{|z| = r\}$.

Theorem 4.2. *If*

$$M(G_1) + M(G_2) \geq -\delta \qquad (4.5)$$

and if $\delta > 0$ is sufficiently small, then there is a $C < \infty$ independent of J and δ such that

$$\frac{\sup_J |z|}{\inf_J |z|} \leq 1 + C\sqrt{\delta \log \frac{1}{\delta}}. \qquad (4.6)$$

By inequality (4.4), Theorem 4.1 is a corollary of Theorem 4.2.

Proof. We follow Pommerenke [1991]. We assume J is not a circle. Let

$$\varphi(z) = \sum_{n=1}^{\infty} a_n z^n$$

be a conformal map of $\mathbb{D}$ onto G_1. Also let

$$\psi(z) = \frac{b_{-1}}{z} + \sum_{n=0}^{\infty} b_n z^n$$

be a conformal map of $\mathbb{D}$ onto G_2. Then by (4.5)

$$\log \left| \frac{b_{-1}}{a_1} \right| \le 2\pi \delta. \tag{4.7}$$

Inequality (4.7) is the main character in the proof of Theorem 4.2.

By the proof of the area theorem, Lemma I.4.2, we have

$$\pi \sum_{n=1}^{\infty} n|a_n|^2 = \text{Area}(G_1) \le \pi \left(|b_{-1}|^2 - \sum_{n=1}^{\infty} n|b_n|^2 \right).$$

Therefore

$$|b_{-1}|^2 - |a_1|^2 \ge |b_1|^2 + \sum_{n=2}^{\infty} n \left(|b_n|^2 + |a_n|^2 \right).$$

In particular, $|a_1| < |b_{-1}|$, because J is not a circle. Set $|z| = r = |a_1/b_{-1}|$. Then by (4.7), $e^{-2\pi\delta} \le r < 1$ and

$$(1 - r^2) \log(1/(1 - r^2)) \le C_1 \delta \log(1/\delta),$$

so that by the Cauchy–Schwarz inequality

$$|\varphi(z) - a_1 z|^2 \le \left(\sum_{n=2}^{\infty} n|a_n|^2 \right) \sum_{n=1}^{\infty} \frac{r^{2n}}{n}$$

$$\le \left(|b_{-1}|^2 - |a_1|^2 \right) \log \frac{1}{1 - r^2} \le C_1 |b_{-1}|^2 \delta \log \frac{1}{\delta}.$$

Thus

$$\inf_{z \in J} |z| \ge \inf_{|z|=r} |\varphi(z)| \ge |b_{-1}| \left(1 - C_2 \sqrt{\delta \log \frac{1}{\delta}} \right) \equiv A(\delta) \tag{4.8}$$

when δ is small, because by (4.7)

$$\frac{|a_1|}{|b_{-1}|} \ge 1 - 2\pi\delta.$$

Similarly, on $|z| = r$

$$\left| \psi(z) - \left(\frac{b_{-1}}{z} + b_0 \right) \right|^2 \le \left(\sum_{n=1}^{\infty} n|b_n|^2 \right) \log \frac{1}{1 - r^2} \le C_1 |b_{-1}|^2 \delta \log \frac{1}{\delta}.$$

But then by (4.7),

$$\sup_{z \in J} |z - b_0| \le \sup_{|z|=r} |\psi(z) - b_0|$$

$$\le |b_{-1}| \left(1 + \sqrt{C_1} \sqrt{\delta \log \tfrac{1}{\delta}} \right) \equiv B(\delta), \tag{4.9}$$

when δ is small. Therefore J lies inside the circle centered at b_0 of radius $B(\delta)$ and outside the circle centered at 0 of radius $A(\delta)$. Consequently

$$|b_0| \le B(\delta) - A(\delta) = C_3 |b_{-1}| \sqrt{\delta \log \tfrac{1}{\delta}},$$

and with (4.8) and (4.9) this gives (4.6).   ■

Corollary 4.3. *Let $R > 1$, let $\Omega = \{1 < |z| < R\}$ be the annulus with boundary curves $\Gamma_1 = \{|z| = 1\}$ and $\Gamma_2 = \{|z| = R\}$, let σ be a curve in Ω joining $1 \in \Gamma_1$ to Γ_2, and assume*

$$\sup_{\sigma} |\arg z| = \varepsilon. \tag{4.10}$$

If $\varepsilon < \varepsilon_0 < \pi$, there is a constant $C = C(\varepsilon_0)$ such that if $1 < R \le e$ then

$$d_{\Omega \setminus \sigma}(\Gamma_1, \Gamma_2) \ge \frac{1}{2\pi} \log R + \frac{C\varepsilon^2}{\log \tfrac{1}{\varepsilon}}, \tag{4.11}$$

while if $R > e$ then

$$d_{\Omega \setminus \sigma}(\Gamma_1, \Gamma_2) \ge \frac{1}{2\pi} \log R + \frac{C}{\log R} \frac{\varepsilon^2}{\left(\log \log R + \log \tfrac{1}{\varepsilon}\right)}. \tag{4.12}$$

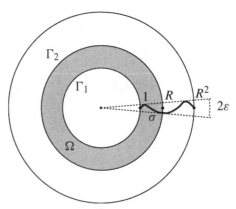

Figure V.2 Ω, σ and their reflection about $|z| = R$.

By the extension rule,

$$d_{\Omega \setminus \sigma}(\Gamma_1, \Gamma_2) \geq \frac{1}{2\pi} \log R \qquad (4.13)$$

and equality holds in (4.13) if σ is the radial slit $[1, R]$. Corollary 4.3 says that if equality almost holds in (4.13) then σ is almost a radial segment.

Proof. We first prove (4.11) in the case $R \leq e$ and then derive (4.12) for $R > e$ from (4.11).

Write $\lambda = d_{\Omega \setminus \sigma}(\Gamma_1, \Gamma_2)$ and assume

$$\lambda < \frac{\log R}{2\pi} + \delta. \qquad (4.14)$$

Let $\widetilde{\Omega}$ and $\widetilde{\sigma}$ be the extensions of Ω and σ by reflection through $\{|z| = R\}$, so that $\widetilde{\sigma}$ joins $z = 1$ to $z = R^2$ and $\mathrm{mod}(\widetilde{\Omega}) = 2\mathrm{mod}(\Omega)$. Then by the symmetry rule (see Exercise IV.3(a)) the family $\widetilde{\Gamma}$ of symmetric curves in $\widetilde{\Omega} \setminus \widetilde{\sigma}$ joining a point $e^{i\theta} \in \Gamma_1$ to $R^2 e^{i\theta} \in \partial\widetilde{\Omega}$ satisfies $\lambda(\widetilde{\Gamma}) = 2\lambda$. Set

$$B = e^{2\pi^2/\log R}.$$

The map

$$z \to \exp\left(\frac{-\pi i \log z}{\log R}\right)$$

sends $\widetilde{\Omega} \setminus \widetilde{\sigma}$ to the slit ring $\Omega_1 \setminus [1, B]$ bounded by Σ and $B\Sigma = \{Bz : z \in \Sigma\}$ and cut along the positive real axis. The map also sends Γ_1 and $\{|z| = R^2\}$ to the slit $[1, B]$ from Σ to $B\Sigma$ and sends $\widetilde{\Gamma}$ to the family of closed curves in Ω_1 separating Σ from $B\Sigma$. Therefore for δ sufficiently small,

$$\mathrm{mod}(\Omega_1) \geq \frac{1}{2\lambda} \geq \frac{\pi}{\log R} - 2\left(\frac{\pi}{\log R}\right)^2 \delta + \dots$$

$$\geq \frac{\log B}{2\pi} - C\delta.$$

Let $\varphi(z)$ be a conformal map from $\mathbb{D}$ to the domain inside of Σ such that $\varphi(0) = 0$. Let $r > 0$ be so small that $\Omega_2 = B\varphi(\{r < |z| < 1\}) \supset \Omega_1$ and set $\Omega_3 = \Omega_2 \setminus (\Omega_1 \cup \Sigma)$. Then

$$\mathrm{mod}(\Omega_2) = \frac{1}{2\pi} \log\left(\frac{1}{r}\right),$$

while

$$\mathrm{mod}(\Omega_3) \geq \frac{1}{2\pi} \log\left(\frac{1}{Br}\right) - O(r),$$

because

$$\varphi^{-1}(\Omega_3) = \mathbb{D} \setminus \varphi^{-1}\big(B\varphi(\{|z| \le r\})\big) \supset \mathbb{D} \setminus \{|z| \le Br - O(r^2)\}.$$

Therefore for r small,

$$\mathrm{mod}(\Omega_1) + \mathrm{mod}(\Omega_3) \ge \mathrm{mod}(\Omega_2) - C\delta. \qquad (4.15)$$

Let $J = (B\varphi)^{-1}(\Sigma) = \varphi^{-1}(\frac{1}{B}\Sigma)$. Then by (4.15) and Theorem 4.1

$$\sup_J |z| / \inf_J |z| \le 1 + C\sqrt{\delta \log \frac{1}{\delta}}. \qquad (4.16)$$

By (4.10) we have

$$\Sigma \subset \{e^{\frac{-\pi \varepsilon_0}{\log R}} \le |z| \le e^{\frac{\pi \varepsilon_0}{\log R}}\},$$

$$B\Sigma \subset \{Be^{\frac{-\pi \varepsilon_0}{\log R}} \le |z| \le Be^{\frac{\pi \varepsilon_0}{\log R}}\},$$

and

$$e^{\frac{\pi \varepsilon_0}{\log R}} \le c(\varepsilon_0) B e^{\frac{-\pi \varepsilon_0}{\log R}}$$

with $c(\varepsilon_0) < 1$ because $R \le e$. By Koebe's theorem $J \subset \{|z| \le c'(\varepsilon_0) < 1\}$ so that by (4.16) and the Schwarz lemma

$$\sup_\Sigma |z| / \inf_\Sigma |z| \le 1 + C'\sqrt{\delta \log \frac{1}{\delta}}.$$

Therefore by definition of Σ

$$\varepsilon \le C\sqrt{\delta \log \frac{1}{\delta}}$$

and

$$\delta \ge C \frac{\varepsilon^2}{\log \frac{1}{\varepsilon}}.$$

Now assume $R > e$. The map $z \to \log z$ sends Ω to a region $\log \Omega$ bounded by $\{\mathrm{Re}\, z = 0\}$, $\{\mathrm{Re}\, z = \log R\}$, a curve γ from 0 to $\{\mathrm{Re}\, z = \log R\}$, and $\gamma + 2\pi i$. Let n be the integer such that $n - 1 < \log R \le n$ and let Ω_n be the region bounded by $\{\mathrm{Re}\, z = 0\}$, $\{\mathrm{Re}\, z = \log R\}$, γ, and $\gamma + 2\pi n i$. Let $E_n = \partial \Omega_n \cap \{\mathrm{Re}\, z = 0\}$ and let $F_n = \partial \Omega_n \cap \{\mathrm{Re}\, z = \log R\}$. Then by the parallel rule and conformal invariance

$$\frac{1}{d_{\Omega_n}(E_n, F_n)} \ge \frac{n}{d_{\Omega_1}(E_1, F_1)} = \frac{n}{d_{\Omega \setminus \sigma}(\Gamma_1, \Gamma_2)}.$$

By the $R \leq e$ case of (4.11) and a rescaling, we have

$$d_{\Omega_n}(E_n, F_n) \geq \frac{1}{2\pi n} \log R + C \frac{(\varepsilon/n)^2}{(\log n/\varepsilon)^3}$$

and so

$$d_{\Omega \setminus \sigma}(\Gamma_1, \Gamma_2) \geq \frac{1}{2\pi} \log R + \frac{C}{\log R} \frac{\varepsilon^2}{\left(\log \log R + \log \frac{1}{\varepsilon}\right)}. \qquad \blacksquare$$

Slightly weaker forms of (4.3) and (4.11) are easier to prove and sufficient for most applications. See Exercise 4.

Teichmüller's [1938] paper actually shows that the inequality (4.3) is sharp, except for the choice of constant. For his example Teichmüller formed a Riemann surface (which is $\mathbb{C}^*$) by fixing $q > 1$, taking two domains $D_1 = \mathbb{D}$ and $D_2 = \{1 < |z| \leq \infty\}$, and making the identifications:

(i) $e^{i\theta} \in \partial \mathbb{D}_1$ with $e^{iq\theta} \in \partial \mathbb{D}_2$, for $0 \leq |\theta| < \pi/q$,
(ii) $e^{i\theta} \in \partial \mathbb{D}_1$ with $e^{-i\theta} \subset \partial \mathbb{D}_1$, for $\pi/q \leq |\theta| < \pi$, and
(iii) $e^{i\pi/q}$ with $e^{-i\pi/q}$ and with $e^{i\pi}$.

By the uniformization theorem this gives two domains G_1 and G_2 such that $J = \mathbb{C}^* \setminus (G_1 \cup G_2)$ is a slit cardioid. Teichmüller showed (4.6) is sharp by varying the parameter q. It follows that (4.3) is also sharp.

5. Boundary Conformality and Angular Derivatives

Let Ω be a simply connected domain, let $\zeta \in \partial \Omega$, and let $F : \Omega \to \Omega'$ be a conformal mapping. In this section we study two kinds of local boundary behavior:

(a) The conformality of F at ζ, and
(b) The existence of a non-zero angular derivative $F'(\zeta)$.

We seek geometric conditions on Ω that are necessary and sufficient for either (a) or (b) to hold.

Let $\zeta \in \mathbb{C}$. For $\theta \in [0, 2\pi)$, $\beta \in (0, \pi/2)$, and $\varepsilon > 0$, define the **truncated cone**

$$\Gamma_\beta^\varepsilon(\zeta, \theta) = \{z : |\arg(z - \zeta) - \theta| < \beta, \ 0 < |z - \zeta| < \varepsilon\}.$$

Definition 5.1. *We say that $\partial \Omega$ has an **inner tangent** with inner normal $e^{i\theta}$ at $\zeta \in \partial \Omega$ if for every $\beta \in (0, \pi/2)$ there is an $\varepsilon = \varepsilon(\beta) > 0$ so that*

$$\Gamma_\beta^\varepsilon(\zeta, \theta) \subset \Omega. \qquad (5.1)$$

When $\theta = \pi/2$ we say $\partial\Omega$ has [vertical] **vertical inner normal**.

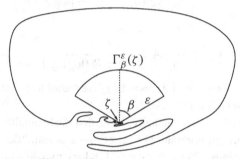

Figure V.3 Truncated cone at ζ.

When $\partial\Omega$ has an inner tangent, the inner normal $e^{i\theta}$ may not be unique. For example, at each point on a slit there are two inner tangents with opposite normal directions and the Jordan domain $\mathbb{D} \cap \{0 < \arg z < 3\pi/2\}$ has inner tangent at 0 with inner normal $e^{i\theta}$ for all $\theta \in [\pi/2, \pi]$. Nevertheless, when $\partial\Omega$ has an inner tangent, we fix one $e^{i\theta}$ for which (5.1) holds and write

$$\Gamma_{\beta}^{\varepsilon}(\zeta) = \Gamma_{\beta}^{\varepsilon}(\zeta, \theta).$$

If $\partial\Omega$ has an inner tangent at ζ and if

$$\lim_{\Gamma_{\beta}^{\varepsilon}(\zeta) \ni z \to \zeta} f(z) = A,$$

we say the function f has **nontangential limit** A at ζ.

Definition 5.2. *Suppose $F : \Omega \to \Omega'$ is a conformal mapping and suppose $\partial\Omega$ has an inner tangent at $\zeta \in \partial\Omega$. We say F is* **conformal at** ζ *if F has a non-tangential limit*

$$F(\zeta) \equiv \lim_{\Gamma_{\beta}^{\varepsilon}(\zeta) \ni z \to \zeta} F(z)$$

and if the limit

$$A_{\zeta} = \lim_{\Gamma_{\beta}^{\varepsilon}(\zeta) \ni z \to \zeta} \arg \frac{F(z) - F(\zeta)}{z - \zeta} \tag{5.2}$$

exists for every $\beta \in (0, \pi/2)$.

If F is conformal at ζ and if $|\alpha - \theta| < \pi/2$, then as $z \to \zeta$ the image arc $F(\{\arg(z - \zeta) = \alpha\})$ is asymptotic to the ray $\{\arg(w - F(\zeta)) = \alpha + A_{\zeta}\}$. In other words, F preserves angles between nontangential rays. If F is conformal at ζ, then $\partial \big(F(\Omega) \big)$ has an inner tangent at $F(\zeta)$ and F^{-1} is conformal at $F(\zeta)$. If Ω is a Jordan domain, if $F : \mathbb{D} \to \Omega$ is a conformal mapping, and if $\partial \Omega$ has a tangent at $F(\zeta)$, then by Theorem II.4.2, F is conformal at ζ. Moreover, in this case the convergence $z \to \zeta$ need not be restricted to cones $\Gamma_\beta(\zeta)$. In general, F can be conformal at $\zeta \in \partial \mathbb{D}$ even when Γ does not have a tangent at $F(\zeta)$. See Exercise VII.12 or Theorem 5.5 below.

Definition 5.3. *Suppose $F : \Omega \to \Omega'$ is a conformal map defined on Ω and suppose $\partial \Omega$ has an inner tangent at $\zeta \in \partial \Omega$. We say F has* **angular derivative** *$F'(\zeta)$ if for all $\beta \in (0, \pi/2)$ the two limits*

$$F(\zeta) \equiv \lim_{\Gamma_\beta^\varepsilon(\zeta) \ni z \to \zeta} F(z)$$

and

$$F'(\zeta) \equiv \lim_{\Gamma_\beta^\varepsilon(\zeta) \ni z \to \zeta} \frac{F(z) - F(\zeta)}{z - \zeta} \tag{5.3}$$

exist and are finite.

Lemma 5.4. *Suppose $F : \Omega \to \Omega'$ is a conformal map defined on Ω and suppose $\partial \Omega$ has an inner tangent at $\zeta \in \partial \Omega$. Then F has an angular derivative at ζ if and only if F' has a finite nontangential limit at ζ.*

Proof. If F' has nontangential limit L, then by integration F has finite nontangential limit $F(\zeta)$ and by the fundamental theorem of calculus (5.3) holds with $F'(\zeta) = L$. Conversely, if F has an angular derivative at ζ and if $z_n \to \zeta$ nontangentially, then uniformly in some disc $\{w : |w| < \delta\}$,

$$g_n(w) = \frac{F(z_n + (\zeta - z_n)w) - F(\zeta)}{\zeta - z_n} \to (w - 1)F'(\zeta).$$

Therefore

$$\lim F'(z_n) = \lim g_n'(0) = F'(\zeta)$$

and F' has nontangential limit $F'(\zeta)$ at ζ. $\blacksquare$

It is clear that if F has a *non-zero* angular derivative at ζ then F is conformal at ζ and $A_\zeta = \arg F'(\zeta)$. Furthermore, F^{-1} also has a non-zero angular derivative at $F(\zeta)$. On the other hand, F can be conformal at $\zeta \in \partial \Omega$ without having an angular derivative at ζ. See Exercise VI.8.

However, the conformality of F at ζ does yield some information about the difference quotients (5.3). Suppose $\partial \Omega$ has an inner tangent with vertical inner normal at $0 \in \partial\Omega$ and suppose F is conformal at 0. Form the functions

$$f_\varepsilon(z) = \log\left(\frac{F(\varepsilon z) - F(0)}{\varepsilon z} \cdot \frac{\varepsilon i}{F(\varepsilon i) - F(0)}\right).$$

Then $\operatorname{Im} f_\varepsilon$ converges to 0, as $\varepsilon \to 0$, uniformly on compact subsets of the half annulus

$$A = \{z : 1/2 < |z| < 2\} \cap \mathbb{H}.$$

Since $\operatorname{Re} f_\varepsilon(i) = 0$, it follows that $\operatorname{Re} f_\varepsilon$ converges to 0 uniformly on compact subsets of A. Therefore

$$\lim_{\varepsilon \to 0} \frac{F(\varepsilon z) - F(0)}{\varepsilon z} \cdot \frac{\varepsilon i}{F(\varepsilon i) - F(0)} = \lim_{\varepsilon \to 0} e^{f_\varepsilon} = 1$$

uniformly on compact subsets of A. Consequently F has an angular derivative at 0 if and only if the vertical limit

$$\lim_{\varepsilon \to 0} \frac{F(\varepsilon i) - F(0)}{\varepsilon i}$$

exists. This fact also follows from Lindelöf's theorem, Exercise II.3(d).

For the rest of this section we assume

$$\varphi : \mathbb{H} \to \Omega$$

is a conformal mapping from upper half-plane $\mathbb{H}$ onto a domain Ω such that $0 \in \partial\Omega$ and we assume φ has nontangential limit

$$\lim_{\Gamma_\alpha(0)\ni z \to 0} \varphi(z) = 0. \tag{5.4}$$

It will often be convenient to transform both $\mathbb{H}$ and $\Omega \equiv \varphi(\mathbb{H})$ to "strips." The "standard strip"

$$\mathbb{S} = \{z : |\operatorname{Im} z| < \pi/2\}$$

is mapped by the function $\tau(z) = ie^{-z}$ onto $\mathbb{H}$ so that $\tau(+\infty) = 0$ and $\tau(-\infty) = \infty$. Then

$$\psi(z) = \tau^{-1} \circ \varphi \circ \tau(z) = i\pi/2 - \log\varphi(ie^{-z}) \tag{5.5}$$

is a conformal map of $\mathbb{S}$ onto the region $\widetilde{\Omega}$ defined by

$$\widetilde{\Omega} = \{i\pi/2 - \log w : w \in \Omega\} \quad \text{or} \quad \Omega = \{ie^{-z} : z \in \widetilde{\Omega}\}.$$

With this correspondence the cones $\Gamma^1_\beta(0)$ in $\mathbb{H}$ or in Ω are transformed to half-strips

$$S_\delta = \{z : |\operatorname{Im} z| < \pi/2 - \delta \text{ and } \operatorname{Re} z > 0\}, \quad \delta \in (0, \pi/2)$$

in $\mathbb{S}$ or in $\tilde{\Omega}$. With this transformation (5.4) holds if and only if

$$\lim_{S_\delta \ni z \to +\infty} \text{Re}\psi(z) = +\infty$$

for all $\delta \in (0, \pi/2)$, and condition (5.2) holds if and only if

$$\lim_{S_\delta \ni z \to +\infty} \left(\text{Im}\psi(z) - \text{Im}z \right) = -A_\zeta$$

for all $\delta \in (0, \pi/2)$. Similarly, $\partial\Omega$ has an inner tangent at 0 with a vertical normal if and only if for each $\delta \in (0, \pi/2)$ there is $x_\delta > 0$ such that

$$x_\delta + S_\delta \subset \tilde{\Omega}. \tag{5.6}$$

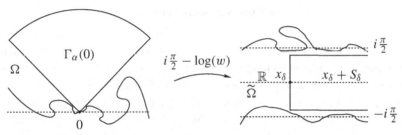

Figure V.4 Transforming the half-plane version to the strip version.

The first theorem gives local geometric conditions on a region $\Omega = \varphi(\mathbb{H})$ that are necessary and sufficient for φ to be conformal $\zeta \in \mathbb{R}$. The theorem is from Ostrowski [1937]. For convenience we treat only the special case $\zeta = 0$, $\varphi(0) = 0$, and $A_\zeta = 0$ in (5.2); the other cases reduce to this case by translations and rotations.

Theorem 5.5 (Ostrowski). *Suppose Ω is a simply connected domain in $\mathbb{C}$ and suppose $0 \in \partial\Omega$. Let $\varphi : \mathbb{H} \to \Omega$ be a conformal map having nontangential limit $\varphi(0) = 0$. If φ is conformal at 0 and if (5.2) holds with $A_0 = 0$, then $\partial\Omega$ has an inner tangent at 0 with a vertical inner normal and*

$$\lim_{\mathbb{R}\ni x \to 0} \frac{\text{dist}(x, \partial\Omega)}{x} = 0. \tag{5.7}$$

Conversely, if $\partial\Omega$ has an inner tangent at 0 with a vertical inner normal and if (5.7) holds, then there exists a conformal map φ of $\mathbb{H}$ onto Ω having nontangential limit $\varphi(0) = 0$ such that φ is conformal at 0 and the limit A_0 in (5.2) is 0.

The comb in Figure V.5 has tines $[x_n - i, x_n]$ with $x_n = -x_{-n} \to 0$. If φ is a conformal map of the upper half-plane onto the complement of the comb (including ∞) and if $\varphi(0) = 0$, then by Ostrowski's theorem φ is conformal at 0 if and only if $\lim_{n\to\infty} x_{n+1}/x_n = 1$.

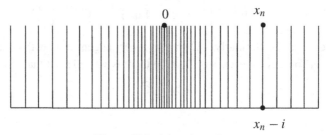

Figure V.5 A comb region.

Proof. If $\varphi(0) = 0$ and if the limit in (5.2) is 0, then for all $\theta \in (0, \pi)$ the ray $\{re^{i\theta} : r > 0\}$ and its image are asymptotic and therefore $\partial\Omega$ has an inner tangent at 0 with a vertical inner normal. If (5.7) fails, there is $\varepsilon > 0$ and a sequence $x_n \to 0$ such that $\text{dist}(x_n, \partial\Omega) \geq \varepsilon x_n$ for all n. We may suppose $x_n > 0$ and $x_n \in \Omega$ by (5.1). The sequence of harmonic functions

$$u_n(z) = \arg \frac{x_n z}{\varphi^{-1}(x_n z)}$$

is then a normal family on

$$A_\varepsilon = B(1, \varepsilon) \cup \{z \in \mathbb{H} : 1 - \varepsilon < |z| < 1 + \varepsilon\},$$

and for $0 < \delta < \pi/2$, $u_n(z)$ converges to 0 on $A_\varepsilon \cap \{z : \delta < \arg z < \pi - \delta\}$. Fix $z = 1 - i\frac{\varepsilon}{2} \in A_\varepsilon$. Then $u_n(z)$ converges to 0, so that $\arg \varphi^{-1}(x_n z) < 0$ for large n. Because $x_n z \in \Omega$ and $\varphi^{-1}(\Omega) = \mathbb{H}$, this is a contradiction and (5.7) is established.

Conversely, assume $\partial\Omega$ has an inner tangent at 0 with a vertical inner normal and assume (5.7) holds. Let E_r be the component of $\Omega \cap \{z : |z| = r\}$ such that $ir \in E_r$ and let Φ be a conformal map of Ω onto $\mathbb{D}$. By the proof of Carathéodory's Theorem I.3.1, applied to Φ (instead of Φ^{-1}), we conclude that

$$\lim_{\substack{z \in E_r \\ r \to 0}} \Phi(z)$$

exists in $\partial\mathbb{D}$. Composing Φ^{-1} with a linear fractional transformation, we obtain a conformal map φ of $\mathbb{H}$ onto Ω such that

$$\lim_{\substack{z \in E_r \\ r \to 0}} \varphi^{-1}(z) = 0. \tag{5.8}$$

Because conformality is a local property of $\partial\Omega$, we may suppose Ω contains $i\mathbb{R}^+ = \{iy : y > 0\}$ and $\lim_{\Omega \ni z \to \infty} |\varphi^{-1}(z)| = +\infty$. In terms of the strip regions this means that $\mathbb{R} \subset \widetilde{\Omega}$ and that if ψ is given by (5.5) then $\psi^{-1}(\mathbb{R})$ is a curve in $\mathbb{S}$ along which $\mathrm{Re}\,\psi^{-1}(x) \to +\infty$ as $x \to +\infty$ and $\mathrm{Re}\,\psi^{-1}(x) \to -\infty$ as $x \to -\infty$. Condition (5.7) becomes

$$\lim_{\substack{z \in \partial \widetilde{S} \\ x \to +\infty}} \mathrm{dist}(z, \partial\widetilde{\Omega}) = 0, \tag{5.9}$$

the set E_r is replaced by the component F_x of $\{z : \mathrm{Re}\,z = x\} \cap \widetilde{\Omega}$ such that $F_x \cap \mathbb{R} \neq \emptyset$, and (5.8) translates to

$$\lim_{\substack{z \in F_x \\ x \to +\infty}} \mathrm{Re}\,\psi^{-1}(z) = +\infty. \tag{5.10}$$

Take $\zeta \in \partial\mathbb{S}$ with $\mathrm{Im}\,\zeta = \pi/2$. For $\delta < 1$, set $B_\delta = \{z : |z - \zeta| < \delta\}$, and let U_δ be the component of $B_\delta \cap \widetilde{\Omega}$ satisfying $U_\delta \cap F_{\mathrm{Re}\,\zeta} \neq \emptyset$. We claim there is an absolute constant C so that when $U_\delta \neq \emptyset$

$$\mathrm{diam}\,\psi^{-1}(U_\delta) \leq C \left(\log \frac{1}{\delta} \right)^{-\frac{1}{2}}. \tag{5.11}$$

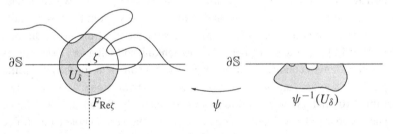

Figure V.6 Distortion near the ends of vertical crosscuts.

Indeed, by the extension rule and the annulus example

$$d_{\widetilde{\Omega}}(U_\delta, \mathbb{R}) \geq d_{\mathbb{C}}(B_\delta, \mathbb{R}) \geq C + \frac{1}{2\pi} \log \frac{1}{\delta}. \tag{5.12}$$

By conformal invariance and the extension rule,

$$d_\Omega(U_\delta, \mathbb{R}) = d_{\mathbb{S}}(\psi^{-1}(U_\delta), \psi^{-1}(\mathbb{R})) \leq d_{\mathbb{S}}(\psi^{-1}(U_\delta), L), \tag{5.13}$$

where $L = -\pi i/2 + \mathbb{R}$ is the component of $\partial\mathbb{S}$ such that $\zeta \notin L$. Let Γ be the family of curves in $\mathbb{S}$ separating the connected set $\psi^{-1}(U_\delta)$ from L. Using the spherical metric, as in Exercise IV.6, we obtain

$$\frac{1}{d_{\mathbb{S}}(\psi^{-1}(U_\delta), L)} = \lambda_{\mathbb{S}}(\Gamma) \geq C(\mathrm{diam}\,\psi^{-1}(U_\delta))^2. \tag{5.14}$$

Now (5.11) follows from (5.12), (5.13), and (5.14). The same estimate (5.11) also holds if $\zeta \in \partial \mathbb{S}$ and $\operatorname{Im}\zeta = -\pi i/2$.

Finally, if $\delta > 0$ and if $z \in \partial S_\delta$ then $\operatorname{dist}(z, \partial\Omega) < 2\delta$ for $\operatorname{Re}z$ is sufficiently large by (5.9). Thus by (5.11)

$$\operatorname{dist}(\psi^{-1}(z), \partial\mathbb{S}) < C(\log\frac{1}{\delta})^{-\frac{1}{2}}$$

and

$$|\operatorname{Im}z - \operatorname{Im}\psi^{-1}(z)| \le C(\log\frac{1}{\delta})^{-\frac{1}{2}}. \tag{5.15}$$

In other words $\gamma = \psi^{-1}(\partial S_\delta) \cap \{\operatorname{Re}z > x(\delta)\}$ consists of two curves situated outside S_ε, where $\varepsilon = 2C(\log 1/\delta)^{-\frac{1}{2}}$. Moreover, by (5.10) $\gamma \to +\infty$. By the maximum principle and (5.15),

$$|\operatorname{Im}\psi(w) - \operatorname{Im}w| \le C(\log\frac{1}{\delta})^{-\frac{1}{2}}$$

when $w \in M + S_\varepsilon$ if $M > 0$ is sufficiently large. Because $\delta > 0$ was arbitrary, this proves that φ has nontangential limit 0 at 0 and that (5.2) holds with limit $A_0 = 0$. ∎

Ostrowski's theorem gave a geometric condition at $p \in \partial\Omega$ which is necessary and sufficient for the conformal map $\varphi : \mathbb{H} \to \Omega$ from the disc or half-plane to Ω to be conformal at $\varphi^{-1}(p)$. The next theorem, due to Rodin and Warschawski [1976] and independently to Jenkins and Oikawa [1977], gives an extremal length condition that is necessary and sufficient for the conformal map to have non-zero angular derivative at $\varphi^{-1}(p)$. In Section 6 we will reinterpret this extremal length condition more geometrically. Because the existence of a non-zero angular derivative depends on the domain Ω and not on the choice of the conformal map φ, we make the following definition:

Definition 5.6. *If Ω is simply connected and if $0 \in \partial\Omega$, we say Ω has a **positive angular derivative** at $0 \in \partial\Omega$ if there is a conformal map φ of $\mathbb{H}$ onto Ω which has nontangential limit $\varphi(0) = 0$ and which has angular derivative $\varphi'(0)$ such that $0 < \varphi'(0) < \infty$.*

Theorem 5.7 (Jenkins–Oikawa, Rodin–Warschawski). *Suppose Ω is a simply connected domain such that $0 \in \partial\Omega$ and $i\mathbb{R}^+ = \{iy : y > 0\} \subset \Omega$. For $r > 0$, let E_r denote the component of $\Omega \cap \{z : |z| = r\}$ containing the point ir. Then Ω has positive angular derivative at 0 if and only if*
(a) Ω has an inner tangent at 0 with a vertical inner normal, and
(b) $\lim_{s<r\to 0} d_\Omega(E_r, E_s) - \frac{1}{\pi}\log\frac{r}{s} = 0$. $\tag{5.16}$

If the Ostrowski condition (5.7) holds, then Theorem 5.7 is still true if condition (b) is replaced by the weaker

(b') $\lim_{n > m \to +\infty} d_\Omega(E_{2^{-m}}, E_{2^{-n}}) - \frac{1}{\pi} \log 2^{n-m} = 0.$

See Exercise 5.

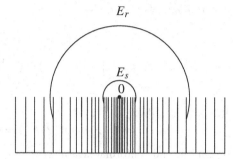

Figure V.7 Positive angular derivative at a tine.

Proof. First assume that Ω has a positive angular derivative at 0. Then (a) holds by Theorem 5.5. To establish (b), we transform the problem using (5.5). The conformal map ψ of the standard strip $\mathbb{S}$ onto $\widetilde{\Omega} = \{i\pi/2 - \log w : w \in \Omega\}$ satisfies

$$\lim_{\substack{\mathrm{Re} z \to +\infty \\ z \in S_\delta}} \psi(z) - z = C$$

for each $\delta \in (0, \pi/2)$, where C is a (finite) real number; or equivalently,

$$\lim_{\substack{\mathrm{Re} w \to +\infty \\ w \in S_\delta}} w - \psi^{-1}(w) = C \qquad (5.17)$$

for each $\delta \in (0, \pi/2)$. Let F_x denote the component of $\widetilde{\Omega} \cap \{z : \mathrm{Re} z = x\}$ such that $x \in F_x$. Then (b) will hold if and only if

$$\pi d_{\widetilde{\Omega}}(F_s, F_t) - |s - t| \to 0 \qquad (5.18)$$

as $s, t \to +\infty$. By translating $\widetilde{\Omega}$, we may suppose that the constant C in (5.17) is zero. That will not affect (5.18).

With $\varepsilon > 0$ and C as in (5.11) choose δ so that $C(\log 1/\delta)^{-1/2} < \varepsilon$. By Theorem 5.5, we can choose x_ε so that $x_\varepsilon + S_\delta \subset \Omega$ and $B(\zeta, \delta) \cap \partial\Omega \neq \emptyset$ if $\zeta \in \partial\mathbb{S}$ satisfies $\mathrm{Re}\zeta > x_\varepsilon$. Let σ_ζ be the component of $\{z : |z - \zeta| = \delta\} \cap \Omega$ such that $\sigma_\zeta \cap F_{\mathrm{Re}\zeta} \neq \emptyset$. Then if U_ζ is the component of $\Omega \setminus \sigma_\zeta$ that meets

$B(\zeta, \delta)$, we have by (5.11)

$$\text{diam}\,\psi^{-1}(U_\zeta) = \text{diam}\,\psi^{-1}(\sigma_\zeta) \le C(\log \frac{1}{\delta})^{-\frac{1}{2}} < \varepsilon.$$

By (5.17), we may also suppose that $|\text{Re}\,\psi^{-1}(z) - x| < \varepsilon$ for all $z \in S_\delta \cap F_x$ and all $x > x_\varepsilon$. Consequently,

$$\sup_{z \in F_x} |\text{Re}\,\psi^{-1}(z) - x| < \varepsilon, \quad x > x_\varepsilon.$$

Thus for $t > s > x_\varepsilon$,

$$\{z : s - \varepsilon < \text{Re}z < t + \varepsilon, \ |\text{Im}z| < \pi/2\} \supset \psi^{-1}(F_s) \cup \psi^{-1}(F_t), \quad (5.19)$$

whereas

$$\begin{aligned} \{z : s + \varepsilon < \text{Re}z < t - \varepsilon, \ |\text{Im}z| < \pi/2\} \cap \\ (\psi^{-1}(F_s) \cup \psi^{-1}(F_t)) = \emptyset. \end{aligned} \quad (5.20)$$

Then by the extension rule and conformal invariance

$$t - s - 2\varepsilon < \pi d_{\mathbb{S}}(\psi^{-1}(F_s), \psi^{-1}(F_t)) = \pi d_{\widetilde{\Omega}}(F_s, F_t) < t - s + 2\varepsilon$$

and (5.18) follows.

Conversely, assume (a) and (b) hold. We first establish (5.7). If (5.7) fails, then as in the proof of Theorem 5.5 we may suppose there is a region

$$A = \{z : \text{Im}z > 0, \ 1 - \varepsilon < |z| < 1 + \varepsilon\} \cup B(1, \varepsilon)$$

and a sequence $x_n \to 0$, $x_n > 0$ such that $x_n A \subset \Omega$. Let $r_n = (1 + \varepsilon)x_n$ and $s_n = (1 - \varepsilon)x_n$. Then by the extension rule and conformal invariance,

$$\begin{aligned} d_\Omega(E_{r_n}, E_{s_n}) &\le d_A(\{|z| = 1 + \varepsilon\}, \{|z| = 1 - \varepsilon\}) \\ &= -C(\varepsilon) + \frac{1}{\pi} \log \frac{1 + \varepsilon}{1 - \varepsilon} = -C(\varepsilon) + \frac{1}{\pi} \log \frac{r_n}{s_n}, \end{aligned}$$

for some $C(\varepsilon) > 0$. Letting $x_n \to 0$ then contradicts (b), and therefore (5.7) holds.

Now choose φ as in Theorem 5.5, define ψ by (5.5), and recall that F_x is that component of $\widetilde{\Omega} \cap \{\text{Re}z = x\}$ containing x. We claim that the variation of $\text{Re}z$ along the image of F_x satisfies

$$V_x = \sup_{u, v \in F_x} |\text{Re}\,\psi^{-1}(u) - \text{Re}\,\psi^{-1}(v)| \to 0 \quad (5.21)$$

as $x \to +\infty$. If (5.21) is true, then by a comparison with rectangles as in (5.19) and (5.20) we obtain via (5.18) that

$$\lim_{\substack{\text{Re}w \to +\infty \\ w \in S_\delta}} \text{Re}w - \text{Re}\,\psi^{-1}(w)$$

exists for all $\delta \in (0, \pi/2)$. Then by Theorem 5.5

$$\lim_{\substack{\text{Re}w \to +\infty \\ w \in S_\delta}} \text{Im}w - \text{Im}\psi^{-1}(w) = 0,$$

and (5.17) follows. But (5.17) clearly implies that φ has a positive angular derivative. Thus to complete the proof of Theorem 5.7, we need only prove (5.21).

To prove (5.21) note that by (5.18) there exists $x_\delta > 0$ so that for all r, s, t with $x_\delta < r < s < t < \infty$ we have approximate equality in the serial rule:

$$d_{\widetilde{\Omega}}(F_r, F_s) + d_{\widetilde{\Omega}}(F_s, F_t) \geq d_{\widetilde{\Omega}}(F_r, F_t) - \delta.$$

Let f be the conformal map of the subregion of $\mathbb{S}$ between $\psi^{-1}(F_r)$ and $\psi^{-1}(F_t)$ onto the rectangle $R = \{z : 0 < \text{Im}z < \pi, \ 0 < \text{Re}z < \pi d_{\widetilde{\Omega}}(F_r, F_t)\}$, so that $f \circ \psi^{-1}(F_r)$ is the left vertical end I_r of R and $f \circ \psi^{-1}(F_t)$ is the right vertical end I_t of R. Then $I_s = f \circ \psi^{-1}(F_s)$ divides R into two regions U_1 and U_2, and

$$d_{U_1}(I_r, I_s) + d_{U_2}(I_s, I_t) \geq d_R(I_r, I_t) - \delta. \tag{5.22}$$

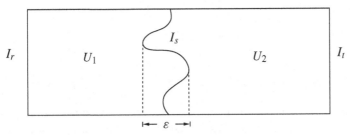

Figure V.8 Image of a vertical crosscut.

Reflect R through its top edge, obtaining a rectangle $\widetilde{R}$ with ends $\widetilde{I}_r$ and $\widetilde{I}_t$ and $\widetilde{R}$ contains regions $\widetilde{U}_1$ and $\widetilde{U}_2$ separated by a curve $\widetilde{I}_s$ which connects a_s to $a_s + 2\pi i$. By the symmetry rule,

$$d_{\widetilde{U}_1}(\widetilde{I}_r, \widetilde{I}_s) + d_{\widetilde{U}_2}(\widetilde{I}_s, \widetilde{I}_t) \geq d_R(\widetilde{I}_r, \widetilde{I}_t) - \delta/2. \tag{5.23}$$

Because $\widetilde{I}_s$ connects $a_s \in \mathbb{R}$ to $a_s + 2\pi i$, Theorem 4.1 can now be applied to $\{e^z : z \in \widetilde{R}\}$ to derive (5.21) from (5.23). ∎

Exercises 4 and 5 include proofs of (5.21) without Teichmüller's Modulsatz.

When $\partial\Omega$ is smooth, Theorem 5.7 and the estimates in Sections 2 and IV.6 yield necessary and sufficient geometric conditions for the existence of a positive angular derivative.

Corollary 5.8. *Suppose Ω is a simply connected domain given by*

$$\Omega = \{re^{i\alpha} : \alpha_1(r) < \alpha < \alpha_2(r); \; r > 0\},$$

where α_j are absolutely continuous in $(0, 1)$, $\alpha_1(r) < \pi/2 < \alpha_2(r)$, $\alpha_1(0) = 0$, $\alpha_2(0) = \pi$, and

$$\int_0^1 \alpha_j'(r)^2 r\, dr < \infty,$$

for $j = 1, 2$. Let $\Theta(r) = \alpha_2(r) - \alpha_1(r)$. Then Ω has a positive angular derivative at 0 if and only if

$$\int_0^1 \left(\frac{1}{r\Theta(r)} - \frac{1}{\pi r} \right) dr \tag{5.24}$$

exists.

A related result can be found in Rodin and Warschawski [1976].

Proof. Since $\alpha_1(0) = 0$ and $\alpha_2(0) = \pi$, Ω has an inner tangent with a vertical inner normal. Let

$$m(x) = \frac{\pi}{2} - \frac{\alpha_1(e^{-x}) + \alpha_2(e^{-x})}{2},$$

$$\theta(x) = \alpha_2(e^{-x}) - \alpha_1(e^{-x}),$$

and

$$\widetilde{\Omega} = \{x + iy : m(x) - \theta(x)/2 < y < m(x) + \theta(x)/2\}.$$

Then $\varphi(z) = ie^{-z}$ is one-to-one from $\widetilde{\Omega}$ onto Ω. Let $F_x = \widetilde{\Omega} \cap \{z : \operatorname{Re} z = x\}$. For s, t sufficiently large, by (IV.6.2) and (2.2) we have the estimates

$$\int_s^t \frac{1}{\theta(x)} dx \le d_{\widetilde{\Omega}}(F_r, F_s) \le \int_s^t \frac{1}{\theta(x)} dx + \int_s^t \frac{m'(x)^2 + \frac{1}{12}\theta'(x)^2}{\theta(x)} dx.$$

By our assumptions on α_j,

$$\lim_{s,t\to+\infty} \int_s^t \frac{m'(x)^2 + \frac{1}{12}\theta'(x)^2}{\theta(x)} dx = 0. \tag{5.25}$$

Thus by Theorem 5.7, Ω has a positive angular derivative at 0 if and only if

$$\lim_{s,t\to+\infty} \int_s^t \left(\frac{1}{\theta(x)} - \frac{1}{\pi} \right) dx = 0, \tag{5.26}$$

and by a change of variables, (5.26) is equivalent to (5.24). ∎

6. Conditions More Geometric

In this section we seek more geometric conditions on a simply connected domain Ω that are equivalent to the existence of a non-zero angular derivative. We concentrate on the normalized domain

$$\widetilde{\Omega} = \{i\pi/2 - \log w : w \in \Omega\}$$

and on the point $z = +\infty \in \partial\widetilde{\Omega}$. From this perspective, Ω has an inner tangent and vertical inner normal if and only if (5.6) holds for $\widetilde{\Omega}$, and Ω has a positive angular derivative if and only if (5.6) and (5.18) hold for $\widetilde{\Omega}$. If $\widetilde{\Omega}$ is a simply connected domain satisfying (5.6) and (5.18), we say that $\widetilde{\Omega}$ has an **angular derivative at** $+\infty$. We do not require that the map $z \to e^{-z}$ be one-to-one on $\widetilde{\Omega}$.

We begin by observing that under mild assumptions, condition (5.26) equivalent to a statement about areas. Set $\Omega_{s,t} = \widetilde{\Omega} \cap \{z : s < \mathrm{Re}\,z < t\}$ and $\mathbb{S}_{s,t} = \{z : s < \mathrm{Re}\,z < t, \ |\mathrm{Im}\,z| < \pi/2\}$ and note that then

$$\int_s^t (\pi - \theta)dx = \mathrm{Area}(\mathbb{S}_{s,t} \setminus \Omega_{s,t}) - \mathrm{Area}(\Omega_{s,t} \setminus \mathbb{S}_{s,t}).$$

Now if $\partial\Omega$ has an inner tangent at 0 with a vertical inner normal, then for x sufficiently large

$$\frac{\pi}{2} \le \theta(x) \le 2\pi.$$

Hence if $\pi - \theta$ has constant sign on (s, t), then

$$\frac{1}{2\pi^2} \le \frac{\left| \int_s^t \left(\frac{1}{\theta} - \frac{1}{\pi} \right) dx \right|}{\left| \mathrm{Area}(\mathbb{S}_{s,t} \setminus \Omega_{s,t}) - \mathrm{Area}(\Omega_{s,t} \setminus \mathbb{S}_{s,t}) \right|} \le \frac{2}{\pi^2}. \tag{6.1}$$

In particular, if

$$\mathrm{Area}(\mathbb{S} \setminus \widetilde{\Omega}) + \mathrm{Area}(\widetilde{\Omega} \setminus \mathbb{S}) < \infty$$

then

$$\lim_{s,t \to \infty} \int_s^t \left| \frac{1}{\theta(x)} - \frac{1}{\pi} \right| dx = 0$$

and (5.26) holds.

When (5.25) fails, upper and lower estimates can still be found in terms of areas if $\widetilde{\Omega}$ is replaced by a smaller region with a Lipschitz boundary. Let $M > 0$. Recall that an M-Lipschitz graph is the graph of a function h for which

$$|h(s) - h(t)| \le M|s - t|$$

for all $s, t \in \mathbb{R}$. Suppose $\tilde{\Omega}$ is a region such that $\mathbb{R} \subset \tilde{\Omega}$. Let $\mathcal{T}_M$ denote the collection of isosceles triangles T such that

(i) $T \subset \Omega$,

(ii) T has base on $\mathbb{R}$, and

(iii) T has sides with slope $\pm M$,

and define

$$\tilde{\Omega}_M = \bigcup \{T : T \in \mathcal{T}_M\}. \tag{6.2}$$

Then $\tilde{\Omega}_M$ is the largest domain such that $\mathbb{R} \subset \tilde{\Omega}_M \subset \tilde{\Omega}$ and $\partial \tilde{\Omega}_M$ consists of two M-Lipschitz graphs.

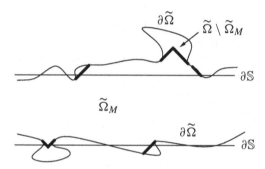

Figure V.9 M-Lipschitz subregion.

Theorem 6.1. *Suppose* $\tilde{\Omega} \supset \mathbb{R}$ *is a simply connected domain such that*

$$\partial \tilde{\Omega} \cap \{z : \mathrm{Re} z < 0\} = \partial \mathbb{S} \cap \{z : \mathrm{Re} z < 0\}. \tag{6.3}$$

Let $\tilde{\Omega}_M \subset \tilde{\Omega}$ *be the region defined by (6.2) above.*

(a) *Suppose* $\mathrm{Area}(\mathbb{S} \setminus \tilde{\Omega}_M) < \infty$. *Then* $\tilde{\Omega}$ *has an angular derivative at* $+\infty$ *if and only if*

$$\mathrm{Area}(\tilde{\Omega}_M \setminus \mathbb{S}) < \infty.$$

(b) *Suppose* $\mathrm{Area}(\tilde{\Omega}_M \setminus \mathbb{S}) < \infty$. *Then* $\tilde{\Omega}$ *has an angular derivative at* $+\infty$ *if and only if*

$$\mathrm{Area}(\mathbb{S} \setminus \tilde{\Omega}_M) < \infty,$$

provided that $M > 8\pi$.

The hypothesis (6.3) is not crucial because the existence of an angular derivative at $+\infty$ is a local property of $\partial \Omega$ near $+\infty$. See Marshall [1995] for an example where M is small and (b) fails.

Proof. If Area$(\mathbb{S} \setminus \widetilde{\Omega}_M) < \infty$ and Area$(\widetilde{\Omega}_M \setminus \mathbb{S}) < \infty$, then

$$\lim_{\substack{\mathrm{Re}z \to +\infty \\ z \in \partial\widetilde{\Omega}_M}} \mathrm{dist}(z, \partial\mathbb{S}) = 0 \tag{6.4}$$

and (5.9) holds. Conversely, if $\widetilde{\Omega}$ has an angular derivative at $+\infty$, then by Ostrowski's theorem, (6.4) holds. Thus we may assume (6.4) during the remainder of the proof. If F_s denotes the component of $\widetilde{\Omega} \cap \{z : \mathrm{Re}z = s\}$ such that $s \in F_s$, then $\widetilde{\Omega}$ has an angular derivative at $+\infty$ if and only if

$$\lim_{\substack{s,t \to \infty \\ s<t}} d_{\widetilde{\Omega}}(F_s, F_t) - \frac{t-s}{\pi} = 0. \tag{6.5}$$

To prove Theorem 6.1 we will obtain upper and lower bounds for the left side of (6.5). The upper bound is given by Lemma 6.2, and the lower bound by Lemma 6.4.

Now suppose that $\partial\widetilde{\Omega}_M$ is given by the two curves $y = h_1(x) + \pi/2$ and $y = -h_2(x) - \pi/2$, $-\infty < x < +\infty$, where $h_1 > -\pi/2$ and $h_2 > -\pi/2$. By (6.4), $\lim_{x \to +\infty} h_j(x) = 0$. As before, for any region U, let

$$U_{s,t} = U \cap \{z : s < \mathrm{Re}z < t\}.$$

Lemma 6.2 (Sastry). *Suppose $|h_j(x)| \leq \pi M^2$ for $s < x < t$. Then*

$$d_{\widetilde{\Omega}}(F_s, F_t) - \frac{t-s}{\pi} \leq \frac{(4M^2+2)\mathrm{Area}(\mathbb{S} \setminus \widetilde{\Omega}_M)_{s,t}}{\pi^2} - \frac{\mathrm{Area}(\widetilde{\Omega}_M \setminus \mathbb{S})_{s,t}}{(4M^2+2)\pi^2}.$$

Sastry [1995] proved this lemma with different constants and used a piecewise constant metric. We will use a metric of the form $\rho = |\nabla u|$, where it is easier to estimate lengths.

Proof. To give an upper bound for extremal distance, we construct a metric that gives a lower bound for the conjugate extremal distance. The idea is that the conformal map of $\widetilde{\Omega}$ onto $\mathbb{S}$ is nearly the identity in $\mathbb{S}_\delta$ and it compresses the regions near $\partial\widetilde{\Omega} \setminus \mathbb{S}$ and expands near $\partial\widetilde{\Omega} \cap \mathbb{S}$. The expansion and compression will be accomplished by imitating Beurling's metric from Section 2 in those regions.

Because

$$d_{\widetilde{\Omega}}(F_s, F_t) \leq d_{\widetilde{\Omega}_M}(F_s, F_t) = d_{(\widetilde{\Omega}_M)_{s,t}}(F_s, F_t),$$

without loss of generality, we may suppose (for notational convenience) that $\widetilde{\Omega} = (\widetilde{\Omega}_M)_{s,t}$. Define

$$U_1 = \{(x, y + \pi/2) \in \widetilde{\Omega} : h_1(x) > 0 \text{ and } -h_1/(2M^2) < y < h_1\},$$
$$U_2 = \{(x, y + \pi/2) \in \widetilde{\Omega} : h_1(x) < 0 \text{ and } 2h_1 < y < h_1\},$$
$$U_3 = \{(x, y - \pi/2) \in \widetilde{\Omega} : h_2(x) > 0 \text{ and } -h_2 < y < h_2/(2M^2)\},$$
$$U_4 = \{(x, y - \pi/2) \in \widetilde{\Omega} : h_2(x) < 0 \text{ and } -h_2 < y < -2h_2\},$$

and

$$U^* = \widetilde{\Omega} \setminus \bigcup_{j=1}^{4} U_j.$$

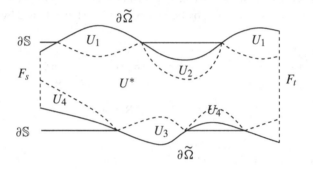

Figure V.10 Proof of Sastry's lemma.

Consider the continuous map $(x, y) \to (x, u(x, y))$ of $\widetilde{\Omega}$ onto $\mathbb{S}$ which fixes U^* and is linear in y on $\widetilde{\Omega} \setminus U^*$. Thus

$$u(x, y) = \begin{cases} y & \text{for } (x, y) \in U^*, \\ (y - (h_1 + \pi/2))/(2M^2 + 1) + \pi/2 & \text{for } (x, y) \in U_1, \\ 2(y - (h_1 + \pi/2)) + \pi/2 & \text{for } (x, y) \in U_2, \\ (y + h_2 + \pi/2)/(2M^2 + 1) - \pi/2 & \text{for } (x, y) \in U_3, \\ 2(y + h_2 + \pi/2) - \pi/2 & \text{for } (x, y) \in U_4. \end{cases}$$

Set $\rho = |\nabla u|$ on $\widetilde{\Omega}$. Note that

$$|\nabla u|^2 \leq \begin{cases} 1 & \text{on } U^*, \\ 4(1 + M^2) & \text{on } U_2 \cup U_4, \\ (1 + M^2)/(2M^2 + 1)^2 & \text{on } U_1 \cup U_3. \end{cases}$$

Thus,

$$\int_{\widetilde{\Omega}} |\nabla u|^2 dy dx - \int_{\mathbb{S}_{s,t}} dy dx$$

$$\leq 4(1 + M^2)\text{Area}(U_2 \cup U_4) + \frac{1 + M^2}{(2M^2 + 1)^2}\text{Area}(U_1 \cup U_3)$$

$$- \frac{1}{2M^2}\text{Area}(\widetilde{\Omega} \setminus \mathbb{S}_{s,t}) - 2\text{Area}(\mathbb{S}_{s,t} \setminus \widetilde{\Omega})$$

$$= (4M^2 + 2)\text{Area}(\mathbb{S}_{s,t} \setminus \widetilde{\Omega}) - \frac{1}{4M^2 + 2}\text{Area}(\widetilde{\Omega} \setminus \mathbb{S}_{s,t}).$$

If γ is a curve connecting $\{y = h_1(x) - \pi/2\}$ to $\{y = -h_2(x) - \pi/2\}$ and contained in $\widetilde{\Omega}$, then

$$\int_{\gamma} |\nabla u||dz| \geq \left|\int_{\gamma} \nabla u \cdot dz\right| = \pi.$$

Thus

$$d_{\widetilde{\Omega}}(F_s, F_t) \leq \frac{\int_{\widetilde{\Omega}} |\nabla u|^2 dy dx}{(\inf_{\gamma} \int_{\gamma} |\nabla u||dz|)^2}$$

$$\leq \frac{1}{\pi^2}\left(\pi(t - s) + (4M^2 + 2)\text{Area}(\mathbb{S}_{s,t} \setminus \widetilde{\Omega})\right.$$

$$\left. - \frac{1}{4M^2 + 2}\text{Area}(\widetilde{\Omega} \setminus \mathbb{S}_{s,t})\right). \qquad \blacksquare$$

The next step is to derive a lower bound for the extremal distance in (6.5). Write $\partial\widetilde{\Omega}_M \setminus \partial\widetilde{\Omega} = \cup_j \sigma_j$ where each σ_j is an open arc with endpoints $z_j^l, z_j^r \in \partial\widetilde{\Omega}$ where $\text{Re}z_j^l < \text{Re}z_j^r$. For $v \in \mathbb{C} \setminus \mathbb{R}$, let T_v denote the triangle in T_M with vertex v. There is a unique v_j with $\text{Re}z_j^l \leq \text{Re}v_j \leq \text{Re}z_j^r$ so that $z_j^l, z_j^r \in \partial T_{v_j}$. Then $\sigma_j \subset \partial T_{v_j}$ and σ_j consists of at most two line segments. If $|\text{Im}z_j^l| \leq |\text{Im}z_j^r|$, let $B_j = \{z : |z - z_j^l| < \text{Re}z_j^r - \text{Re}z_j^l\}$. Otherwise let $B_j = \{z : |z - z_j^r| < \text{Re}z_j^r - \text{Re}z_j^l\}$.

Lemma 6.3.

$$\text{Area}(B_j) \leq \frac{8\pi}{M}\int_{z \in \sigma_j} \left||\text{Im}z| - \frac{\pi}{2}\right| dx, \quad z = x + iy.$$

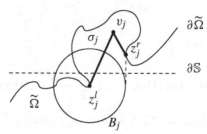

Figure V.11 Proof of Lemma 6.3.

Proof. Suppose f is defined and continuous on $[0, 1]$ with

$$f'(x) = \begin{cases} 1 & 0 \leq x \leq a, \\ -1 & a \leq x \leq 1. \end{cases}$$

Then by elementary geometry or calculus

$$\int_0^1 |f(x)| dx$$

is minimal when $a = \frac{1}{2}$ and $f(0) = -\frac{1}{4}$. Thus

$$\int_0^1 |f(x)| dx \geq \frac{1}{8}.$$

If $\text{Im} z > 0$ on σ_j, the map

$$(x, y) \rightarrow \left(\frac{x - \text{Re}z_j^l}{\text{Re}z_j^r - \text{Re}z_j^l}, \frac{y - \pi/2}{M(\text{Re}z_j^r - \text{Re}z_j^l)} \right)$$

transforms σ_j into the graph of one such f. Thus

$$\int_{z \in \sigma_j} \left| |\text{Im} z| - \frac{\pi}{2} \right| dx \geq \frac{M(\text{Re}z_j^r - \text{Re}z_j^l)^2}{8} = \frac{M}{8\pi} \text{Area}(B_j).$$

The inequality of the lemma for $\text{Im} z < 0$ follows by a reflection about $\mathbb{R}$. ∎

The quantity

$$\int_{z \in \sigma_j} \left| |\text{Im} z| - \frac{\pi}{2} \right| dx$$

is the total area "between" σ_j and $\partial\mathbb{S}$.

The next lemma gives a lower bound for the extremal distance in $\widetilde{\Omega}$ using the geometry of $\widetilde{\Omega}_M$. When $\widetilde{\Omega} = \widetilde{\Omega}_M$, it follows immediately from the discussion at the end of Section 5.

Lemma 6.4. *If $\varepsilon > 0$ there is an $s_0 < \infty$ so that for $s_0 < s < t < \infty$,*

$$d_{\widetilde{\Omega}}(F_s, F_t) - \frac{t-s}{\pi} \geq \frac{(M - 8\pi)\text{Area}(\mathbb{S} \setminus \widetilde{\Omega}_M)_{s,t}}{M\pi^2}$$
$$- \frac{(M + 8\pi)\text{Area}(\widetilde{\Omega}_M \setminus \mathbb{S})_{s,t}}{M\pi^2} - \varepsilon.$$

Proof. Suppose $0 \leq t - s \leq \varepsilon$. Then as $s, t \to \infty$, $\text{Area}(\widetilde{\Omega}_M \setminus \mathbb{S})_{s,t} \to 0$ and $\text{Area}(\mathbb{S} \setminus \widetilde{\Omega}_M)_{s,t} \to 0$ as $s, t \to \infty$ by (6.4) and the inequality follows. Now suppose that $t - s \geq \varepsilon$. Let ρ be the metric on $\widetilde{\Omega}$ given by

$$\rho = \begin{cases} 1 & \text{for } z \in \widetilde{\Omega}_M \cup \bigcup_j B_j, \\ 0 & \text{elsewhere on } \widetilde{\Omega}. \end{cases}$$

The metric ρ will provide the lower bound for the extremal distance. First we compute the ρ-area of $\widetilde{\Omega}_{s,t}$. By Lemma 6.3

$$\int_{\widetilde{\Omega}_{s,t}} \rho^2 dx dy - \int_{\mathbb{S}_{s,t}} dx dy$$

$$\leq \text{Area}(\widetilde{\Omega}_M \setminus \mathbb{S})_{s,t} - \text{Area}(\mathbb{S} \setminus \widetilde{\Omega}_M)_{s,t} + \sum_j \text{Area}(B_j) \qquad (6.6)$$

$$\leq (1 + \frac{8\pi}{M})\text{Area}(\widetilde{\Omega}_M \setminus \mathbb{S})_{s,t} - (1 - \frac{8\pi}{M})\text{Area}(\mathbb{S} \setminus \widetilde{\Omega}_M)_{s,t}.$$

Now suppose γ is a curve in $\widetilde{\Omega}$ connecting F_s to F_t, with $s < t$. Note that

$$\text{Re}z_j^r - \text{Re}z_j^l \leq \frac{4}{M} \max_{z \in \sigma_j} |\,|\text{Im}z| - \pi/2| \to 0 \qquad (6.7)$$

as $\sigma_j \to +\infty$, by (6.4). Thus by deleting an initial and terminal portion of γ, if necessary, we may suppose that γ begins and ends in $\widetilde{\Omega}_M \cup \bigcup_j B_j$ and that if γ meets σ_j, then $s < \text{Re}z_j^l < \text{Re}z_j^r < t$.

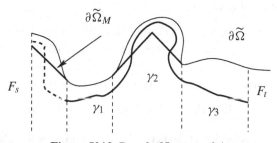

Figure V.12 Proof of Lemma 6.4.

Thus we can write $\gamma = \bigcup_k \gamma_k$ where $\{\gamma_k\}$ are disjoint subarcs of γ such that either

(i) $\gamma_k \subset \widetilde{\Omega}_M \cup \bigcup_j B_j$, or

(ii) γ_k meets some σ_j and connects $F_{\mathrm{Re}z_j^l}$ to $F_{\mathrm{Re}z_j^r}$ with $s < \mathrm{Re}z_j^l < \mathrm{Re}z_j^r < t$.

If (i) holds, then $\int_{\gamma_k} \rho|dz|$ is at least equal to the change in $\mathrm{Re}z$ along γ_k. If (ii) holds, then

$$\mathrm{length}(\gamma_k \cap (\widetilde{\Omega}_M \cup B_j)) \geq \mathrm{Re}z_j^r - \mathrm{Re}z_j^l.$$

Indeed, if z_j^l is the center of B_j and if $\zeta \in \gamma_k \cap \sigma_j$, then the shortest curve from $F_{\mathrm{Re}z_j^l}$ to ζ in $\widetilde{\Omega}$ is the straight line from z_j^l to ζ, since $|\mathrm{Im}\zeta| \geq |\mathrm{Im}z_j^l|$. Since γ_k must intersect ∂B_j, it must have length at least the radius of B_j, namely $\mathrm{Re}z_j^r - \mathrm{Re}z_j^l$. A similar argument works if z_j^r is the center of B_j.

Together with (6.7) this implies, for s, t sufficiently large, that

$$\int_\gamma \rho|dz| \geq t - s - \varepsilon.$$

We conclude by (6.6) that for $t - s \geq \varepsilon$,

$$d_{\widetilde{\Omega}}(F_s, F_t)$$
$$\geq \frac{(t - s - \varepsilon)^2}{\pi(t - s) + (1 + \frac{8\pi}{M})\mathrm{Area}(\widetilde{\Omega}_M \setminus \mathbb{S})_{s,t} - (1 - \frac{8\pi}{M})\mathrm{Area}(\mathbb{S} \setminus \widetilde{\Omega}_M)_{s,t}}.$$

Because $\partial\widetilde{\Omega}_M$ is Lipschitz and (6.4) holds, we have, as $s, t \to \infty$, both $\mathrm{Area}(\widetilde{\Omega}_M \setminus \mathbb{S})_{s,t}/(t - s) \to 0$ and $\mathrm{Area}(\mathbb{S} \setminus \widetilde{\Omega}_M)_{s,t}/(t - s) \to 0$. Hence for s, t sufficiently large with $t - s \geq \varepsilon$,

$$d_{\widetilde{\Omega}}(F_s, F_t) - \frac{t - s}{\pi} \geq \frac{(M - 8\pi)\mathrm{Area}(\mathbb{S} \setminus \widetilde{\Omega}_M)_{s,t}}{M\pi^2}$$
$$- \frac{(M + 8\pi)\mathrm{Area}(\widetilde{\Omega}_M \setminus \mathbb{S})_{s,t}}{M\pi^2} - \varepsilon.$$

concluding the proof of Lemma 6.4. ∎

We now complete the proof of Theorem 6.1. We have $\mathrm{Area}(\mathbb{S} \setminus \widetilde{\Omega}_M) < \infty$ if and only if $\mathrm{Area}(\mathbb{S} \setminus \widetilde{\Omega}_M)_{s,t} \to 0$ as $s, t \to +\infty$. Likewise, the condition $\mathrm{Area}(\widetilde{\Omega}_M \setminus \mathbb{S}) < \infty$ is equivalent to $\mathrm{Area}(\widetilde{\Omega}_M \setminus \mathbb{S})_{s,t} \to 0$ as $s, t \to +\infty$. Thus if $\mathrm{Area}(\mathbb{S} \setminus \widetilde{\Omega}_M) < \infty$ and $\mathrm{Area}(\widetilde{\Omega}_M \setminus \mathbb{S}) < \infty$, by Lemmas 6.2 and 6.4,

$$\lim_{s,t\to\infty} d_{\widetilde{\Omega}}(F_s, F_t) - (t - s)/\pi = 0,$$

and hence by the discussion at the beginning of the proof of the Theorem 6.1, $\widetilde{\Omega}$ has an angular derivative at $+\infty$.

If Area$(\mathbb{S} \setminus \widetilde{\Omega}_M) < \infty$ and if $\widetilde{\Omega}$ has an angular derivative at $+\infty$, then by (6.5) and Lemma 6.2,

$$\text{Area}(\widetilde{\Omega}_M \setminus \mathbb{S})_{s,t} \to 0$$

as $s, t \to +\infty$. That proves part (a) of Theorem 6.1. If $M > 8\pi$ and if Area$(\widetilde{\Omega}_M \setminus \mathbb{S}) < \infty$ then by (6.5) and Lemma 6.4,

$$\text{Area}(\mathbb{S} \setminus \widetilde{\Omega}_M)_{s,t} \to 0$$

as $s, t \to +\infty$, and that proves part (b) of Theorem 6.1. ∎

There exist regions that do not satisfy the hypotheses of Theorem 6.1, but have angular derivative at $+\infty$. See Exercise 8.

Notes

Theorem 1.1 was conjectured by Denjoy in [1907] and proved by Ahlfors in [1930]. Beurling [1933] and Carleman [1933] give different proofs. Ahlfors' paper, which predated the theory of extremal length, uses the length–area method. He obtained a version of (IV.6.4), with a slightly different $\Theta(r)$ and a constant larger than $\frac{8}{\pi}$, but his version was good enough for (1.1). See also Warschawski [1942] for another variant. We have followed Ahlfors' original argument, but starting from (IV.6.4). Carleman's proof is in Appendix G, and Beurling's proof is in Exercise 1. Lord Rayleigh [1871] gave the three-dimensional version of Beurling's upper estimate (2.2) for the resistance in a conductor using the parallel rule, assuming the conductor is a solid of revolution about an axis ($m'(x) \equiv 0$ in the two dimensional version) with smooth boundary. Maxwell [1891] gave an account of Rayleigh's method with a slight improvement of the upper bound.

Reduced extremal distance is close to a method used by Beurling in his thesis and explained in Exercise III.23. Theorem 3.2 is from Ahlfors and Beurling [1952], but Corollary 3.6 is a recent result by Balogh and Bonk [1999]. An extension of Theorem 3.2 to finitely connected domains is given in Theorem H.9 of Appendix H. Jenkins [1970] has a slightly weaker but easier analogue of Theorem 4.1. See Exercise 4.

Bertilsson [1999] proved (4.11) directly and derived Teichmüller's Modulsatz as a corollary. He also showed that the estimate $\varepsilon^2/(\log 1/\varepsilon)$ in (4.11) is sharp. See Exercise 10.

The proof of Ostrowski's theorem given here is based on Warschawski [1967]. The estimate (5.11) is due to J. Wolff [1935]. The problem of finding

geometric conditions equivalent to the existence of a non-zero angular derivative is very old, dating back at least to Ahlfors [1930]. Warschawski [1967],
Rodin and Warschawski [1977], Baernstein [1988], and their references provide
a history of the problem. The proof of Theorem 6.1 is from Marshall [1995].
Theorem 6.1(a) was proved by Burdzy [1986], using Brownian excursions, in
the half-plane form given in Exercise 9, which is the context of the original
angular derivative problem. Half of Theorem 6.1(a), in the "strip" form given
in the text, was proved by Rodin-Warschawski [1986] and the other half was
proved by Sastry [1995]. For other proofs of the half-plane version of Theorem
6.1(a), see Carroll [1988] and Gardiner [1991]. Theorem 6.1(b) was proved by
Marshall [1995] and answers a question of Burdzy [1986].

Exercises and Further Results

1. In [1933] Beurling derived Theorem 1.1 from his own inequality, proved in
Exercise III.22 above. With the notation of the proof of Theorem 1.1, take
$D_j = G_j \cap \{|z| > 1\}$, assume $|f| \le 1$ on ∂D_j, but take $z_j \in D_j$ such that
$\log |f(z_j)| \ge e$. For $R > \text{Max}|z_j|$ set

$$M_j(R) = \sup_{D_j \cap \{|z|=R\}} |f(z)|.$$

Beurling shows there is R_0 such that if

$$\log M_j(R) = \left(\frac{R}{R_0}\right)^{\sigma_j(R)}$$

then for all $R > R_0$,

$$\sum_{j=1}^{n} \frac{1}{\sigma_j(R)} < 2. \tag{E.1}$$

and (E.1) clearly implies (1.1).

To prove (E.1), write $\omega_j(R) = \omega(z_j, |z| = R, D_j \cap \{|z| < R\})$. Then

$$e \le \omega_j(R) \log M_j(R).$$

Now let C_j be the family of curves in D_j connecting z_j to $\partial D_j \cap \{|z| = R\}$
and define

$$L_j = \sup_{\gamma \in C_j} \inf \left\{\ell(\psi(\gamma)) : \psi \text{ is conformal on } D_j \text{ and } \text{Area}\,\psi(D_j) = \pi\right\}.$$

Beurling's inequality from Exercise III.24(b) gives

$$\omega_j \le e^{1-L_j^2},$$

so that

$$\log M_j(R) \ge e^{L_j^2}. \tag{E.2}$$

Under the logarithm D_j is mapped to a region D_j^* with area A_j and

$$\sum A_j \le 2\pi \log R. \tag{E.3}$$

Using

$$\psi(z) = \left(\frac{\pi}{A_j}\right)^{\frac{1}{2}} \log z$$

gives

$$L_j^2 \ge \frac{\pi}{A_j}\left(\log \frac{R}{|z_j|}\right)^2 \ge \frac{\pi}{A_j} \log R \log\left(\frac{R}{R_0}\right) \tag{E.4}$$

if $R_0 \ge \text{Max}|z_j|^2$, and together (E.2), (E.3), and (E.4) imply (E.1).

2. This exercise outlines a proof of Theorem 3.2 that does not use Green's functions.

(a) Let μ be the equilibrium distribution for $E \subset \partial\mathbb{D}$ and let U_μ be its logarithmic potential. Set $V(z) = \frac{1}{2}\left(U_\mu(z) + U_\mu(1/\bar{z})\right)$ and prove

$$\log\frac{1}{1+|z|} \le V(z) + \frac{1}{2}\log\frac{1}{|z|} \le \log\frac{1}{1-|z|}.$$

(b) Use Green's theorem to prove

$$D_{\Omega_\varepsilon}(V) \equiv \int_{\Omega_\varepsilon} |\nabla V|^2 dxdy = \pi\left(\gamma_E + \frac{1}{2}\log\frac{1}{\varepsilon}\right),$$

where $\Omega_\varepsilon = \{z \in \mathbb{D} : V(z) > (\log\varepsilon)/2\}$.

(c) Set

$$u_\varepsilon = \frac{V + \frac{1}{2}\log\frac{1}{\varepsilon}}{\gamma_E + \frac{1}{2}\log\frac{1}{\varepsilon}}.$$

Then $\rho = |\nabla u_\varepsilon|$ as the extremal metric for $d(\{z : V = (\log\varepsilon)/2\}, E)$.

(d) Deduce Theorem 3.2 from (c) and (a).

3. Compare the estimates of harmonic measure given by Theorem IV.5.2, Corollary 3.3, and Exercise III.24(b) when E is an arc on the boundary of a Jordan domain.

4. The following variations on Theorem 4.1 and Corollary 4.3 are from Jenkins [1970].

(a) Let Ω be the annulus bounded by $\Gamma_1 = \{|z| = 1\}$ and $\Gamma_2 = \{|z| = R\}$ where $R > 1$, and let σ be a curve joining Γ_1 to Γ_2 such that

$$0 = \inf_{\sigma} \arg z < \sup_{\sigma} \arg z = \varepsilon.$$

Prove

$$d_{\Omega \setminus \sigma}(\Gamma_1, \Gamma_2) \ge \frac{\log R}{2\pi} + \frac{1}{108} \frac{\varepsilon^3}{\pi^2 \log R} + O(\varepsilon^5).$$

Hint: If a curve $\gamma \subset \Omega \setminus \sigma$ joins Γ_1 to Γ_2 and enters $\{\frac{\varepsilon}{3} < \arg z < \frac{2\varepsilon}{3}\}$, then

$$\int_{\gamma} \frac{ds}{|z|} \ge \sqrt{(\log R)^2 + \frac{\varepsilon^2}{9}}.$$

(b) Let R be the rectangle $\{0 < x < L; 0 < y < \pi\}$ with left vertical side E_1 and right vertical side E_2. Suppose that T is a curve connecting the horizontal sides of R such that

$$a = \inf_{T} \mathrm{Re}\, z < \sup_{T} \mathrm{Re}\, z = a + \varepsilon.$$

Then T divides R into two components Ω_1 and Ω_2 such that $E_j \subset \partial \Omega_j$. Prove

$$d_{\Omega_1}(E_1, T) + d_{\Omega_2}(T, E_2) \le \frac{L}{\pi} - \frac{\varepsilon^3}{27\pi^3} + O(\varepsilon^5).$$

Hint: Bound $d_{\Omega_j}(E_j, T)^{-1}$ from below using the metric

$$\rho(z) = \begin{cases} \dfrac{1}{\sqrt{\pi^2 + \varepsilon^2/9}}, & \text{if } a + \varepsilon/3 < x < a + 2\varepsilon/3; \\[2mm] \dfrac{1}{\pi}, & \text{otherwise.} \end{cases}$$

(c) Use (b) and the symmetry rule to prove Theorem 4.1 with the weaker estimate $\delta^{\frac{1}{3}}$ in (4.3).

5. Suppose Ω is a simply connected domain such that $\{iy : y > 0\} \subset \Omega$ and $0 \in \partial\Omega$. For $r > 0$ let E_r be the arc of $\Omega \cap \{|z| = r\}$ such that $ir \in E_r$. Assume Ω has inner tangent at 0 with vertical inner normal and satisfies the Ostrowski condition

$$\lim_{\mathbb{R} \ni x \to 0} \frac{\mathrm{dist}(x, \partial\Omega)}{x} = 0.$$

(a) Prove that

$$\lim d_\Omega(E_r, E_s) - \frac{1}{\pi} \log \frac{r}{s} = 0,$$

where the limit is restricted to $1 < r/s < C$ and $r, s \to 0$. Hint: Use estimate (5.11) together with a normal families argument on $\{z : 1 < |z| < C\} \cap \mathbb{H}\}$ as given in the comments after Definition 5.3.

(b) Prove

$$\lim_{s<r\to 0} d_\Omega(E_r, E_s) - \frac{1}{\pi} \log \frac{r}{s} = 0$$

if and only if

$$\lim_{n>m\to+\infty} d_\Omega(E_{2^{-m}}, E_{2^{-n}}) - \frac{1}{\pi} \log 2^{n-m} = 0.$$

(c) Use part (a) to prove (5.21).

6. (a) Show that if

$$\int_{x_0}^{\infty} \frac{\theta'(x)^2}{\theta(x)} dx < \infty$$

and

$$\lim_{s,t\to+\infty} \int_s^t \left(\frac{1}{\theta(x)} - \frac{1}{\pi} \right) dx = 0$$

then

$$\lim_{x\to+\infty} \theta(x) = \pi.$$

(b) Show that if $\lim_{x\to+\infty} \theta(x) = \pi$ and if $\widetilde{\Omega} \subset \mathbb{S}$ then

$$\lim_{s,t\to\infty} \frac{\pi^2 \left| \int_s^t \left(\frac{1}{\theta} - \frac{1}{\pi} \right) dx \right|}{\text{Area}(\mathbb{S}_{s,t} \setminus \Omega_{s,t})} = 1,$$

and prove a similar limit holds if $\widetilde{\Omega} \supset \mathbb{S}$ using $\text{Area}(\Omega_{s,t} \setminus \mathbb{S}_{s,t})$.

7. Let $\Omega = \{(x, y) : y > \min(|x|/\log(1/|x|), 1)\}$. Then $\Gamma = \partial\Omega$ has a tangent at 0.

(a) If φ_+ is a conformal map of $\mathbb{H}$ onto Ω with $\varphi(0) = 0$, then φ_+ has infinite angular derivative at 0.

(b) If φ_- is a conformal map of $\mathbb{H}$ onto $\mathbb{C} \setminus \overline{\Omega}$ with $\varphi(0) = 0$, then φ_- has zero angular derivative at 0.

(c) Prove that if Ω is replaced by $\Omega_1 = \{(x, y) : y > \min(|x|/(\log|x|)^2, 1)\}$, then the conformal maps φ_+ and φ_- have non-zero angular derivatives at 1.

8. Set $\widetilde{\Omega} = \{(x, y) : |y - m(x)| < \frac{1}{2}\}$.

(a) If $|m'(x)| \leq 1$, $\int_0^\infty |m'(x)|^2 dx < \infty$, and $\int_0^\infty |m(x)| dx = \infty$, then by (2.2) and Theorem 5.7 $\widetilde{\Omega}$ has an angular derivative at $+\infty$, but Theorem 6.1 does not apply.

(b) If $|m'(x)| \leq 1$, $\int_0^\infty |m(x)| dx < \infty$, and $\int_0^\infty |m'(x)|^2 dx = \infty$, then by Theorem 6.1 $\widetilde{\Omega}$ has an angular derivative at $+\infty$, but this fact can not be deduced from (2.2).

9. Let Ω be a simply connected domain containing the positive imaginary axis and suppose $0 \in \partial\Omega$. Let h_M denote the smallest M-Lipschitz function whose graph is contained in Ω.

(a) (Burdzy [1986]) Suppose

$$\int_{-1}^1 \chi_{h_M > 0} \frac{h_M(x)}{x^2} dx < \infty. \tag{E.5}$$

Then Ω has a positive angular derivative at 0 if and only if

$$\int_{-1}^1 \chi_{h_M < 0} \frac{h_M(x)}{x^2} dx > -\infty. \tag{E.6}$$

For example if $\mathbb{H} \subset \Omega$ then Ω has a positive angular derivative at 0 if and only if (E.6) holds.

(b) (Marshall [1995]) There is an $M_0 < \infty$ so that if $M > M_0$ and if h_M satisfies (E.6) then Ω has a positive angular derivative at 0 if and only if h_M satisfies (E.5). For example, if $\Omega \subset \mathbb{H}$, then Ω has a positive angular derivative at 0 if and only if (E.5) holds.

(c) If $\tau(z) = ie^{-z}$ is one-to-one on $\widetilde{\Omega}$ show that Theorem 6.1 can be derived from parts (a) and (b), but with 8π replaced by a larger constant. See Marshall [1995].

10. Let

$$\Omega = \{(x, y) : 0 < x < 1, \ h(x) < y < h(x) + 1\},$$

where $h(x) = \varepsilon \log(1 + x/\varepsilon)/\log(1 + 1/\varepsilon)$ and set $E = \{\text{Re} z = 0\} \cap \partial\Omega$ and $F = \{\text{Re} z = 1\} \cap \partial\Omega$. Let φ be a conformal map of Ω onto a rectangle

$$R = \{(x, y) : 0 < x < 1, \ 0 < y < H\},$$

with $\text{Re}\varphi = 0$ on E and $\text{Re}\varphi = 1$ on F. Then

$$d_\Omega(E, F) = \frac{1}{H} = \frac{1}{\int_\Omega |\nabla \text{Re}\varphi|^2 dA}.$$

Show that $\int_\Omega |\nabla \text{Re}\varphi|^2 dA \geq 1 - C\frac{\varepsilon^2}{\log 1/\varepsilon}$ and hence

$$d_\Omega(E, F) \leq 1 + C\frac{\varepsilon^2}{\log \frac{1}{\varepsilon}}.$$

Hint: Write $\text{Re}\varphi = x + u(x, y)$. By Green's theorem, it is enough to show

$$\left| \int_{\partial\Omega} u\, dy \right| \leq C\frac{\varepsilon}{\sqrt{\log \frac{1}{\varepsilon}}} \left(\int_\Omega |\nabla u|^2 dx dy \right)^{\frac{1}{2}}.$$

Since $|h'|$ is small, it is enough to prove this for $v(x, y) \equiv u(x, h(x) + y)$ defined on the unit square S. Relate v on ∂S to $|\nabla v|$ on S by integrating the gradient along lines with slope -1. See Bertilsson [1999].

VI

Simply Connected Domains, Part One

The results of this chapter are largely elementary and independent of the three preceding chapters. First we prove the classical [1916] F. and M. Riesz theorem. Next we prove three "almost everywhere" theorems, due to Privalov, Plessner, and McMillan. Their proofs all rely on elementary measure theory and a cone construction (see Figure VI.2) which is a geometric variation on the proof of Lusin's theorem. We then give Makarov's elegant proof that on every simply connected domain

$$\omega \perp \Lambda_\alpha, \quad \alpha > 1,$$

and the extension by Pommerenke that almost everywhere identifies a non-zero angular derivative at ζ with the existence of a cone at $\varphi(\zeta)$.

1. The F. and M. Riesz Theorem

By definition an analytic function $f(z)$ on $\mathbb{D}$ is in the **Hardy space** H^p for $0 < p < \infty$, if

$$\sup_{0<r<1} \int |f(re^{i\theta})|^p d\theta = ||f||_{H^p}^p < \infty.$$

See Appendix A for a brief introduction to the Hardy spaces H^p.

Theorem 1.1. *Let Ω be a domain such that $\partial\Omega = \Gamma$ is a Jordan curve and let φ be a conformal map from $\mathbb{D}$ onto Ω. Then the curve Γ is rectifiable if and only if $\varphi' \in H^1$. If $\varphi' \in H^1$, then*

$$||\varphi'||_{H^1} = \ell(\Gamma) = \Lambda_1(\Gamma). \tag{1.1}$$

Proof. By Carathéodory's theorem, φ extends to a homeomorphism from $\overline{\mathbb{D}}$ to $\overline{\Omega}$. First assume $\varphi' \in H^1$. Let $\{0 = \theta_0 < \theta_1 < \ldots < \theta_n = 2\pi\}$ be any partition of $[0, 2\pi]$. Then

$$\sum_{j=1}^{n} \left| \varphi(e^{i\theta_j}) - \varphi(e^{i\theta_{j-1}}) \right| = \lim_{r \to 1} \sum_{j=1}^{n} \left| \varphi(re^{i\theta_j}) - \varphi(re^{i\theta_{j-1}}) \right|$$

$$= \lim_{r \to 1} \sum_{j=1}^{n} \left| \int_{\theta_{j-1}}^{\theta_j} \varphi'(re^{i\theta}) i re^{i\theta} \, d\theta \right| \qquad (1.2)$$

$$\leq \|\varphi'\|_{H^1}.$$

But $\ell(\Gamma)$ is the supremum, over all partitions, of the left side of (1.2). Therefore Γ is rectifiable and

$$\ell(\Gamma) \leq \|\varphi'\|_{H^1}.$$

Conversely, assume Γ is rectifiable. Then given $r < 1$, choose a partition $\{\theta_0 < \theta_1 < \ldots < \theta_n\}$ of $[0, 2\pi]$, so that

$$\sum_{j=1}^{n} \left| \varphi(re^{i\theta_j}) - \varphi(re^{i\theta_{j-1}}) \right| \geq \ell(\Gamma_r) - \varepsilon,$$

where $\Gamma_r = \varphi(\{|z| = r\})$. Write

$$\psi(z) = \sum_{j=1}^{n} \left| \varphi(ze^{i\theta_j}) - \varphi(ze^{i\theta_{j-1}}) \right|.$$

Then ψ is subharmonic on $\mathbb{D}$ and by Carathéodory's theorem ψ is continuous on $\overline{\mathbb{D}}$, so that

$$\sup_{\mathbb{D}} \psi(z) = \sup_{\theta} \psi(e^{i\theta}) \leq \ell(\Gamma).$$

Thus

$$\int |\varphi'(re^{i\theta})| d\theta = \ell(\Gamma_r) \leq \psi(r) + \varepsilon \leq \ell(\Gamma) + \varepsilon.$$

Therefore $\varphi' \in H^1$ and the equality (1.1) holds. ∎

Theorem 1.1 is equivalent to a theorem about harmonic measure. Assume that $\Gamma = \partial\Omega$ is a rectifiable Jordan curve and let $A \subset \Gamma$ be a Borel set. Write $A = \varphi(E)$ and $z_0 = \varphi(0)$. Then by Carathéodory's theorem

$$\omega(z_0, A, \Omega) = \frac{1}{2\pi} |E|.$$

When A is an arc, Theorem A.1 from Appendix A and the proof of (1.1) yield

$$\Lambda_1(A) = \Lambda_1(\varphi(E)) = \lim_{r \to 1} \int_E |\varphi'(re^{i\theta})| d\theta = \int_E |\varphi'(e^{i\theta})| d\theta, \qquad (1.3)$$

because $\varphi' \in H^1$, and if (1.3) holds for arcs then (1.3) also holds for all Borel sets $A \subset \partial\Omega$. Consequently

$$\omega(A) = 0 \Longrightarrow \Lambda_1(A) = 0.$$

Conversely,

$$\Lambda_1(A) = 0 \Longrightarrow \omega(A) = 0,$$

because by Corollary A.2,

$$\left|\{\theta : |\varphi'(\theta)| = 0\}\right| = 0.$$

Thus when Γ is rectifiable, harmonic measure for Ω and linear measure on Γ are mutually absolutely continuous,

$$\omega \ll \Lambda_1 \ll \omega.$$

This argument proves the following:

Theorem 1.2 (F. and M. Riesz). *Let Ω be a simply connected plane domain such that $\Gamma = \partial\Omega$ is a rectifiable Jordan curve and let $\varphi : \mathbb{D} \to \Omega$ be conformal. Then $\varphi' \in L^1(\partial\mathbb{D})$. For any Borel set $E \subset \partial\mathbb{D}$,*

$$\Lambda_1(\varphi(E)) = \int_E |\varphi'| d\theta,$$

and for any Borel set $A \subset \partial\Omega$,

$$\omega(A) = 0 \Longleftrightarrow \Lambda_1(A) = 0. \tag{1.4}$$

Theorem 1.2 is equivalent to Theorem 1.1 because (1.3) holds only if $\varphi' \in H^1$.

In [1936] Lavrentiev gave a quantitative version of (1.4). See Exercise 2 and (5.1) below. See Exercise 4 for an example of a Jordan domain Ω and a set $E \subset \mathbb{R} \cap \partial\Omega$ such that $\Lambda_1(E) > 0$ but $\omega(E) = 0$.

Theorem 1.1 has another important geometric consequence. Let φ be a conformal map from $\mathbb{D}$ to a domain Ω such that $\partial\Omega = \Gamma$ is a rectifiable Jordan curve. Then by Fatou's theorem, Theorem 1.1, and Corollary A.2, φ' has a non-zero nontangential limit $\varphi'(\zeta)$ at almost all $\zeta \in \partial\mathbb{D}$. But by calculus, φ has a non-zero angular derivative at $\zeta \in \partial\mathbb{D}$ whenever φ' has a non-zero nontangential limit at ζ. Therefore φ has a non-zero angular derivative $\varphi'(\zeta)$ at Λ_1 almost every $\varphi(\zeta) \in \Gamma$. In this way Theorem 1.1 is an improvement, almost everywhere, on the pointwise theorems II.4.2 and V.5.7.

Corollary 1.3. *Let Ω be a domain such that $\partial\Omega = \Gamma$ is a rectifiable Jordan curve and let φ be a conformal map from $\mathbb{D}$ onto Ω. Then φ has a non-zero* *angular derivative at ζ and Γ has a tangent at $\varphi(\zeta)$, both for Lebesgue almost every $\zeta \in \partial\mathbb{D}$ and for Λ_1 almost every $\varphi(\zeta) \in \Gamma$.*

In general, φ can have a non-zero angular derivative at a point ζ, so that $\partial\Omega$ then has an inner tangent at $\varphi(\zeta)$ even though $\partial\Omega$ does not have a tangent at $\varphi(\zeta)$. Bishop [1987] exhibited a domain bounded by a non-rectifiable Jordan curve Γ such that for almost every $\zeta \in \partial\mathbb{D}$, φ is conformal at ζ but Γ has no tangent at $\varphi(\zeta)$. The domain is $\{-f(x) < y < f(x),\ 0 < x < 1\}$ where $f(x)$ is continuous, $f(0) = f(1)$, $f(x) > 0$ on $(0, 1)$, and f has no finite or infinite derivative at any point. See Exercise 9.

If $\partial\Omega$ is not a Jordan curve, then φ will not be one-to-one on $\partial\mathbb{D}$. However, the nontangential limit $\varphi(\zeta)$ can still exist, and in that case the existence of an inner tangent at $\varphi(\zeta)$ depends on the point $\zeta \in \partial\mathbb{D}$ and not just on the point $\varphi(\zeta) \in \partial\Omega$. But since φ is one-to-one on $\mathbb{D}$, there can be at most two points $\zeta \in \varphi^{-1}(w)$ at which φ is conformal.

2. Privalov's Theorem and Plessner's Theorem

Let $\zeta \in \partial\mathbb{D}$, let $\alpha > 1$ and let $0 < h < 1$. The **truncated cone** Γ_α^h is

$$\Gamma_\alpha^h(\zeta) = \{z : |z - \zeta| < \alpha(1 - |z|) < \alpha h\}.$$

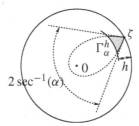

Figure VI.1 The truncated cone Γ_α^h.

Let $u(z)$ be a harmonic or meromorphic function on $\mathbb{D}$.

Definition 2.1. *The function u is **nontangentially bounded** at ζ if there exist $M < \infty, \alpha > 1$, and $0 < h < 1$ such that*

$$|u(z)| \le M \text{ on } \Gamma_\alpha^h(\zeta).$$

Notice that the definition of nontangential boundedness only requires that $u(z)$ be bounded on a single truncated cone, whereas the definition of nontangential convergence requires that $u(z)$ has a limit through every cone $\Gamma_\alpha(\zeta)$.

Theorem 2.2 (Privalov). *Suppose $E \subset \partial\mathbb{D}$ and suppose $u(z)$ is nontangentially bounded at each $\zeta \in E$. Then $u(z)$ has a nontangential limit at almost every $\zeta \in E$.*

Proof. Let $\varepsilon > 0$. By a little measure theory there exists a compact set $K \subset E$ with $|E \setminus K| < \varepsilon$ and there exist constants α, h, and M such that $|u| < M$ on

$$U = \bigcup_{\zeta \in K} \Gamma_\alpha^h(\zeta).$$

To prove the theorem it is enough to show u has a nontangential limit almost everywhere on K. Let U_i be one component of U and let φ_i be a conformal map of $\mathbb{D}$ onto U_i. Then $u \circ \varphi_i$ is a bounded harmonic function on $\mathbb{D}$ and by Fatou's theorem, $u \circ \varphi_i$ has a finite nontangential limit almost everywhere on $\partial\mathbb{D}$. Because α and h are fixed, ∂U_i is a rectifiable Jordan curve. See Figure VI.2. For Lebesgue almost every $\zeta \in K \cap \partial U_i$, φ_i is conformal at $\varphi_i^{-1}(\zeta)$ and $u \circ \varphi_i$ has a finite nontangential limit at $\varphi_i^{-1}(\zeta)$, both by the F. and M. Riesz theorem. But when φ_i is conformal at $\varphi_i^{-1}(\zeta)$, $u \circ \varphi_i$ has a nontangential limit at $\varphi_i^{-1}(\zeta)$ if and only if u has a nontangential limit at ζ. Since $K = \bigcup_i K \cap \partial U_i$, we conclude that u has a finite nontangential limit at almost every $\zeta \in K$.   ∎

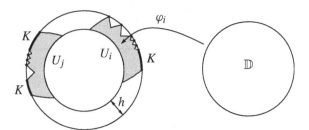

Figure VI.2 Cone domains U_i.

Many proofs in this book will be simple variations on Figure VI.2 and the previous argument. Define a **cone domain** to be a domain of the form

$$U = \bigcup_{\zeta \in K} \Gamma_\alpha(\zeta) \quad \text{or} \quad U = \bigcup_{\zeta \in K} \Gamma_\alpha^h(\zeta),$$

where $K \subset \partial\mathbb{D}$ is compact, $\alpha > 1$, and $0 < h < 1$. Cone domains will be recurrent in this chapter and again in Chapter X.

Theorem 2.3. *Let $f(z)$ be a meromorphic function on $\mathbb{D}$ and let $E \subset \partial\mathbb{D}$ have $|E| > 0$. If for each $\zeta \in E$ there is $\alpha = \alpha(\zeta) > 1$ such that*

$$\lim_{\Gamma_\alpha(\zeta) \ni z \to \zeta} f(z) = 0,$$

then $f(z) \equiv 0$ *in* $\mathbb{D}$.

Proof. Let $K \subset E$ be compact with $|K| > 0$ and take the cone domains U_i and the maps φ_i as in the proof of Privalov's theorem. Then $f \circ \varphi_i$ is bounded and analytic on $\mathbb{D}$, and by the F. and M. Riesz theorem, $f \circ \varphi_i$ has nontangential limit 0 on a set of positive measure. Then by (1.4), $f \circ \varphi_i \equiv 0$ so that $f \equiv 0$. ∎

Nontangential approach regions are necessary in Theorems 2.2 and 2.3. Bagemihl and Seidel [1954] showed that there is a non-constant analytic function on $\mathbb{D}$ having radial limit 0 almost everywhere on $\partial\mathbb{D}$. See Exercise 5.

Definition 2.4. *Let* $f(z)$ *be a meromorphic function on* $\mathbb{D}$. *A point* $\zeta \in \partial\mathbb{D}$ *is a* **Plessner point** *for* f *if for all* $\alpha > 1$ *and all* $0 < h < 1$, $f\left(\Gamma_\alpha^h(\zeta)\right)$ *is dense in* $\mathbb{C}$.

Theorem 2.5 (Plessner). *Let* $f(z)$ *be a nonconstant meromorphic function on* $\mathbb{D}$. *Then there are pairwise disjoint Borel subsets* N, G, *and* P *of* $\partial\mathbb{D}$ *such that* $\partial\mathbb{D} = N \cup G \cup P$ *and*
(a) $|N| = 0$,
(b) *At each* $\zeta \in G$, f *has a finite nontangential limit* $f(\zeta)$ *and* $f(\zeta) \neq 0$, *and*
(c) *Each* $\zeta \in P$ *is a Plessner point for* f.

Proof. Let P be the set of Plessner points for f and let $E = \partial\mathbb{D} \setminus P$. Then P and E are Borel sets, and the theorem is equivalent to the assertion that f has a finite non-zero nontangential limit at almost every $\zeta \in E$. Let $\{w_n\}$ be a countable dense subset of $\mathbb{C}$ and set

$$E_n = \Big\{ \zeta : \text{there exist } \alpha > 1,\ 0 < h < 1,$$
$$\text{and } \varepsilon > 0 \text{ such that } \big| f - w_n \big| \geq \varepsilon \text{ on } \Gamma_\alpha^h(\zeta) \Big\}.$$

Then $E = \bigcup_n E_n$. By Privalov's theorem $1/(f(z) - w_n)$ has a nontangential limit almost everywhere on E_n and hence f has a nontangential limit at almost every $\zeta \in E$. By Theorem 2.3, applied to f and to $1/f$, the limit is finite and non-zero almost everywhere. ∎

3. Accessible Points

Now let Ω be a simply connected domain and let $\varphi : \mathbb{D} \to \Omega$ be a conformal mapping. We do not assume $\partial\Omega$ is a Jordan curve. Nevertheless, φ has a nontangential limit $\varphi(\zeta)$ at almost every $\zeta \in \partial\mathbb{D}$, because the function

$$\frac{z}{\varphi(z) - \varphi(0)}$$

is bounded and analytic on $\mathbb{D}$. When φ has a finite nontangential limit at $\zeta \in \partial\mathbb{D}$ we *define* $\varphi(\zeta)$ to be that nontangential limit. Thus we have defined $\varphi(\zeta)$ at almost every $\zeta \in \partial\mathbb{D}$. Better yet, by Corollary V.3.6,

$$\int_0^1 |\varphi'(r\zeta)|dr < \infty \tag{3.1}$$

except on a capacity zero subset of $\partial\mathbb{D}$. If (3.1) holds then φ has a radial limit at ζ, and it follows from Lindelöf's theorem, Exercise 3(d) of Chapter II, that φ has a nontangential limit at ζ. Therefore the set where $\varphi(\zeta)$ is not defined is smaller than many sets of linear measure zero; it has capacity zero and Hausdorff dimension zero. We say $w \in \partial\Omega$ is an **accessible point** of $\partial\Omega$ if w is the endpoint of an open arc $\sigma \subset \Omega$.

Lemma 3.1. *If φ has a nontangential limit $\varphi(\zeta)$ at $\zeta \in \partial\mathbb{D}$, then $\varphi(\zeta) \in \partial\Omega$ and $\varphi(\zeta)$ is an accessible point of $\partial\Omega$. Conversely, every accessible $w \in \partial\Omega$ is a nontangential limit $w = \varphi(\zeta)$.*

Proof. Suppose φ has nontangential limit at $\zeta \in \partial\mathbb{D}$. Then clearly $\varphi(\zeta) \in \overline{\Omega}$. But if $\varphi(\zeta) \in \Omega$ then there is $z \in \mathbb{D}$ with $\varphi(\zeta) = \varphi(z)$, and that means that φ^{-1} is not single valued in some neighborhood of $\varphi(z)$. Therefore $\varphi(\zeta) \in \partial\Omega$ and clearly $\varphi(\zeta)$ is an accessible point.

For the converse, suppose $w \in \partial\Omega$ is the endpoint of an arc $\sigma \subset \Omega$. The preceding argument above shows that $\lim_{\sigma \ni z \to w} |\varphi^{-1}(z)| = 1$. Because φ has nontangential limits a.e. and $\varphi^{-1}(\sigma)$ is an arc, Theorem 2.3, applied to $\varphi(z) - w$, shows that $\lim_{\sigma \ni z \to w} \varphi^{-1}(z) = \zeta \in \partial\mathbb{D}$ exists. It then follows from Lindelöf's theorem, Exercise II.3(d), that φ has nontangential limit w at ζ. ∎

Write

$$A = \{\text{accessible points of } \partial\Omega\}.$$

Then A is the range of the boundary function $\varphi(\zeta)$, defined except on a set of capacity zero, and A is dense in $\partial\Omega$.

Theorem 3.2. *Let Ω be a simply connected domain. Then the set A of accessible points is an ω-measurable subset of $\partial\Omega$ and for all $z \in \Omega$,*

$$\omega(z, A, \Omega) = 1. \tag{3.2}$$

If $E \subset A$, then

$$\omega(z, E, \Omega) = \omega(\varphi^{-1}(z), \varphi^{-1}(E), \mathbb{D}). \tag{3.3}$$

In particular,

$$\omega(z, E, \Omega) = 0 \iff |\varphi^{-1}(E)| = 0.$$

Proof. By Egoroff's theorem and Lemma 3.1, there are closed sets $K_n \subset \partial \mathbb{D}$ such that $|\partial \mathbb{D} \setminus K_n| < 1/n$ and such that φ is continuous on the compact set $\cup_{K_n} \overline{\Gamma}_{\pi/2}(e^{i\theta})$. Hence $\varphi(K_n) \subset A$ is compact and $\omega(\cup_n \varphi(K_n)) = 1$. Thus A is ω-measurable and (3.2) holds. We may assume $z = \varphi(0)$. Let $r_n \uparrow 1$, and set $\Omega_n = \varphi(\{z : |z| < r_n\})$, and let $f \in C(\overline{\Omega})$. Then by the definition of harmonic measure, by Lemma 3.1, and by dominated convergence,

$$\int_{\partial \Omega} f(\zeta) d\omega(\zeta) = \lim_{n \to \infty} \frac{1}{2\pi} \int f(\varphi(r_n e^{i\theta})) d\theta = \frac{1}{2\pi} \int f(\varphi(e^{i\theta})) d\theta. \quad (3.4)$$

Each side of (3.3) defines a Borel measure and since Borel measures are determined by their actions on continuous functions, (3.4) also implies (3.3). ∎

4. Cone Points and McMillan's Theorem

Recall from Chapter V that if φ has a non-zero angular derivative at ζ, then φ is conformal at ζ. Using conjugate functions or Ostrowski's theorem and Corollary V.5.8 it is easy to make an example where φ has no non-zero angular derivative at ζ but φ is conformal at ζ. See Exercise 8 and Exercise V.7. On the other hand, Theorem 6.1 below will show that if φ is conformal at all $\zeta \in E \subset \partial \mathbb{D}$, then φ has a non-zero angular derivative at almost every $\zeta \in E$.

We permanently write

$$G = \left\{ \zeta \in \partial \mathbb{D} : \varphi \text{ has non-zero angular derivative at } \zeta \right\}$$

and

$$B = \Big\{ \zeta \in \partial \mathbb{D} : \varphi \text{ has a nontangential limit at } \zeta, \text{ and}$$
$$\liminf_{\Gamma_\alpha(\zeta) \ni z \to \zeta} |\varphi'(z)| = 0, \text{ for all } \alpha > 1 \Big\}. \quad (4.1)$$

Theorem 4.1. *Let φ be a univalent function on $\mathbb{D}$. Then $G \cap B = \emptyset$ and*

$$|G \cup B| = 2\pi.$$

Proof. Because φ has a nontangential limit almost everywhere, this theorem is nothing but Theorem 2.3 and Plessner's theorem, applied to the function φ'. ∎

We call a point $w \in \partial \Omega$ a **cone point** of $\partial \Omega$ if w is the vertex of an open isosceles triangle $T \subset \Omega$. Every cone point is an accessible boundary point. In review, we know that the following implications hold pointwise:

$$\varphi \text{ has non-zero angular derivative at } \zeta$$

$$\Downarrow$$

$$\varphi \text{ is conformal at } \zeta$$

$$\Downarrow$$

$$\partial\Omega \text{ has an inner tangent at } \varphi(\zeta)$$

$$\Downarrow$$

$$\varphi(\zeta) \text{ is a cone point of } \partial\Omega,$$

but that all three converse implications fail pointwise. However, Theorem 6.1 will show that these four conditions are almost everywhere equivalent. Write

$$K = K(\Omega) = \{\text{cone points for } \Omega\}.$$

Then it is clear that $\varphi(G) \subset K$.

Theorem 4.2 (McMillan). *Let Ω be a bounded simply connected domain. Then K is a Borel set, with σ-finite Λ_1 measure, and when $E \subset K$,*

$$\omega(E) = 0 \iff \Lambda_1(E) = 0. \tag{4.2}$$

Moreover, at almost every $w \in K$, $\partial\Omega$ has an inner tangent.

Proof. Let $\{L_n\}$ be the countable set of lines having rational slope and rational y intercept. For each n, put $w \in K_n$ if $w \in \partial\Omega \setminus L_n$ and w is the vertex of an open isosceles triangle $T_n(w) \subset \Omega$ such that

(a) $T_n(w)$ has base on the line segment $L_n \cap \{|z| \le n\}$,
(b) $T_n(w)$ has vertex angle $\frac{\pi}{n}$, and
(c) $T_n(w)$ has height $h_n(w)$ satisfying $\frac{1}{n} \le h_n(w) \le n$.

Then $K = \bigcup_n K_n$, where K_n is compact, so that K is an F_σ set. Note that $\text{dist}(K_n, L_n) \ge \frac{1}{n}$. Let $\Omega_n = \bigcup_{K_n} T_n(w)$. Then $\Omega_n \subset \Omega$, and by (a), (b), and (c), Ω_n has finitely many components $\Omega_{n,j}$. Each component $\Omega_{n,j}$ is bounded by a Jordan curve $\Gamma_{n,j}$, and except for endpoints on L_n, these Jordan curves are pairwise disjoint. We have $K \subset \bigcup(\Gamma_{n,j} \setminus L_n)$ and K has σ-finite Λ_1 measure. If we assume the line L_n is horizontal, then $\Gamma_{n,j}$ is the union of an arc of L_n and the graph of a Lipschitz function defined on that arc. In particular $\Gamma_{n,j}$ is rectifiable, and Ω has an inner tangent at Λ_1 almost every point of $\Gamma_{n,j}$, and thus at Λ_1 almost every point of K.

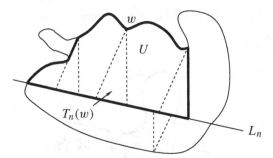

Figure VI.3 *n*-Lipschitz subregion.

Then to prove (4.2) we can assume $E \subset \Gamma_{n,j} \setminus L_n$. Suppose $\Lambda_1(E) > 0$. Then by the F. and M. Riesz theorem, $\omega(z, E, \Omega_{n,j}) > 0$ for all $z \in \Omega_{n,j}$, and therefore $\omega(z, E, \Omega) > 0$.

Conversely, suppose $E \subset \Gamma_{n,j} \setminus L_n$ is a compact set such that $\Lambda_1(E) = 0$. We may suppose L_n is the real axis. Set

$$U = \bigcup_E T_n(w).$$

Translate L_n by $\frac{1}{10n}$ units in the direction of $\Omega_{n,j}$ to a parallel line L'_n. When $w \in E$, let $T'_n(w) \subset T_n(w)$ be the isosceles triangle having vertex w, base on L'_n, and vertex angle $\frac{\pi}{n+1}$. Set

$$V = \bigcup_E T'_n(w).$$

Translate L_n by $\frac{1}{5n}$ units in the direction of $\Omega_{n,j}$ to a parallel line L''_n. Repeat this construction with L''_n and with triangles $T''_n(w)$ now having bases on L''_n and vertex angle $\frac{\pi}{n+2}$. Set

$$W = \bigcup_E T''_n(w).$$

Then W has finitely many components. We may assume E has no isolated points so that W is not a triangle. We can also assume W, and hence U and V, is connected by replacing E by a subset if necessary. Then

$$W \subset V \subset U \subset \Omega_{n,j} \subset \Omega$$

and like $\Gamma_{n,j}$, ∂U, ∂V, and ∂W consist of a segment of L_n, L'_n, L''_n, respectively, and a Lipschitz graph over that segment. Each component τ_k of $\partial V \setminus E$ is an arc having endpoints in E, and these endpoints are also the endpoints of a component σ_k of $\partial W \setminus E$ and a component γ_k of $\partial U \setminus E$. Each of the arcs

$\tau_k, \sigma_k, \gamma_k$ is a polygonal arc having at most three sides and angles bounded below.

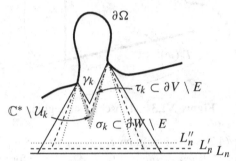

Figure VI.4 Nested regions in McMillan's theorem.

Let $\mathcal{U}_k$ be the unbounded component of $\mathbb{C} \setminus (\tau_k \cup \sigma_k)$. Only one σ_k satisfies $\sigma_k \cap L'_n \neq \emptyset$. Then for each E

$$\inf_{\gamma_k} \omega(z, \sigma_k, \mathcal{U}_k) = \alpha > 0,$$

uniformly in k, by a normal families argument for instance. Therefore

$$\inf_{\partial U \setminus E} \omega(z, \partial\Omega \setminus E, \Omega \setminus \overline{V}) \geq \alpha > 0 \qquad (4.3)$$

and by the maximum principle

$$\inf_{\partial U \setminus E} \omega(z, \partial\Omega \setminus E, \Omega) \geq \alpha > 0.$$

But ∂U is rectifiable and $\Lambda_1(E) = 0$, so by the F. and M. Riesz theorem, $\omega(z, E, U) = 0$. Thus $\omega(z, \partial\Omega \setminus E, \Omega) \geq \alpha$ on $U \supset \partial V \setminus E$ and

$$\beta = \inf_{\partial V \setminus E} \omega(z, \partial\Omega \setminus E, \Omega) \geq \alpha.$$

Therefore

$$\omega(z, \partial\Omega \setminus E, \Omega) = \omega(z, \partial\Omega \setminus E, \Omega \setminus \overline{V}) + \int_{\partial V} \omega(\zeta, \partial\Omega \setminus E, \Omega) d\omega_{\Omega \setminus \overline{V}}(z, \zeta)$$

$$\geq \omega(z, \partial\Omega \setminus E, \Omega \setminus \overline{V}) + \beta\omega(z, \partial V, \Omega \setminus \overline{V})$$

$$= (1 - \beta)\omega(z, \partial\Omega \setminus E, \Omega \setminus \overline{V}) + \beta,$$

for all $z \in \Omega \setminus \overline{V}$. But taking $z \in \partial U \setminus E$ then gives

$$\beta \geq \alpha \geq (1 - \beta)\alpha + \beta,$$

so that $\beta = 1$ and $\omega(E) = 0$.   ■

In [1969] McMillan proved Theorem 4.2 and also the following striking result, known as the McMillan twist point theorem: At almost every $\zeta \in \partial\mathbb{D} \setminus G$, $\arg(\varphi(z) - \varphi(\zeta))$ *is unbounded above and unbounded below* on every curve $\gamma \subset \mathbb{D}$ having one endpoint ζ. When $\arg(\varphi(z) - \varphi(\zeta))$ is so unbounded, $\varphi(\zeta)$ is called a **twist point** of $\partial\Omega$. In particular, for almost every $\zeta \in \partial\mathbb{D} \setminus G$, the radial image $\{\varphi(r\zeta) : 0 < r < 1\}$ is a rectifiable arc in Ω with endpoint $\varphi(\zeta)$, but $\arg(w - \varphi(\zeta))$ has no upper or lower bound along this arc. McMillan's twist point theorem will be proved in Appendix I. Example 4.3 below will show that we can have $|\partial\mathbb{D} \setminus G| = 2\pi$.

When $\partial\Omega$ is a Jordan curve, φ is one-to-one and the notions of twist point or inner tangent depend only on $\partial\Omega$ near the point $w = \varphi(\zeta) \in \partial\Omega$. But suppose $\partial\Omega$ is not a Jordan curve and let $w \in \partial\Omega$. If there exists $\zeta_0 \in \varphi^{-1}(w)$ such that on some curve $\gamma \subset \mathbb{D}$ with endpoint ζ_0,

$$\arg(\varphi(z) - \varphi(\zeta_0)) \text{ is unbounded above and below,} \qquad (4.4)$$

then (4.4) holds for every $\zeta \in \varphi^{-1}(w)$ and for every curve $\gamma \in \mathbb{D}$ ending at ζ_0.

Now assume Γ is a Jordan curve and let Ω_1 and Ω_2 be the two simply connected components of $\mathbb{C}^* \setminus \Gamma$. Fix $p_j \in \Omega_j$, let $\varphi_j : \mathbb{D} \to \Omega_j$ be a conformal mapping with $\varphi_j(0) = p_j$ and write $\omega_j = \omega(p_j, \cdot, \Omega_j)$. Set

$$\mathrm{Tn}(\Gamma) = \big\{w \in \Gamma : \Gamma \text{ has a tangent at } w\big\}.$$

If $w \in \mathrm{Tn}(\Gamma)$ then by Theorem II.4.2, φ_j is conformal at ${\varphi_j}^{-1}(w)$ for $j = 1, 2$. Consequently, by Theorem 4.2 the three measures ω_1, ω_2, and Λ_1 are in the same measure class on $\mathrm{Tn} = \mathrm{Tn}(\Gamma)$,

$$\chi_{\mathrm{Tn}}\omega_1 \ll \chi_{\mathrm{Tn}}\Lambda_1 \ll \chi_{\mathrm{Tn}}\omega_2 \ll \chi_{\mathrm{Tn}}\omega_1.$$

Theorem 6.3 below will show, without using the twist point theorem, that on $\Gamma \setminus \mathrm{Tn}(\Gamma)$, $\omega_1 \perp \omega_2$. Write $\mathrm{Twi}(\Gamma)$ for the set of twist points of Γ. In [1987], C. J. Bishop also made an example of a Jordan curve for which

$$\omega_1(\mathrm{Twi}(\Gamma)) = 1,$$

but

$$\omega_2\big(\mathrm{Tn}(\Gamma) \cup \mathrm{Twi}(\Gamma)\big) = 0.$$

Bishop's construction is outlined in Exercise 10.

Example 4.3 (The von Koch snowflake). Let $\Omega_0 \subset \Omega_1 \ldots$ be the increasing sequence of Jordan domains defined as follows: $\partial\Omega_0 = \Gamma_0$ is the equilateral triangle with vertices $\{0, 1, \frac{1}{2} + i\frac{\sqrt{3}}{2}\}$; $\Gamma_n = \partial\Omega_n$ consists of $3 \cdot 4^n$ segments $J_{n,j}$ of length 3^{-n}; and Γ_{n+1} is obtained from Γ_n by replacing the middle third

$I_{n,j}$ of each $J_{n,j}$ by the two remaining sides of the unique equilateral triangle $T_{n,j}$ satisfying $I_{n,j} = J_{n,j} \cap T_{n,j}$ and $T_{n,j} \cap \Omega_n = \emptyset$. Clearly $\Omega_{n+1} \supset \Omega_n$.

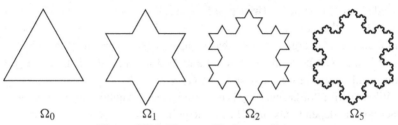

Ω_0 Ω_1 Ω_2 Ω_5

Figure VI.5 The von Koch snowflake.

Then $\Omega = \bigcup \Omega_n$ is a Jordan domain and $\partial \Omega = \Gamma = \lim \Gamma_n$, where the limit is taken in the Hausdorff metric. The Jordan curve $\Gamma = \partial \Omega$ is called the **von Koch snowflake**. The Hausdorff dimension of Γ is $\frac{\log 4}{\log 3}$ and $0 < \Lambda_{\frac{\log 4}{\log 3}}(\Gamma) < \infty$. See Appendix D. Let φ be the conformal map of $\mathbb{D}$ onto Ω. If Γ has an inner tangent at $\varphi(\zeta)$, then for any $\varepsilon > 0$, Ω contains a truncated cone with vertex $\varphi(\zeta)$ and with aperture $\pi - \varepsilon$. Consequently $\varphi(\zeta) \in \bigcup_n (\Gamma \cap \Gamma_n)$. But $E = \bigcup_n (\Gamma \cap \Gamma_n)$ is the union of a sequence of linear images of the Cantor set and $\dim E = \frac{\log 2}{\log 3}$. Therefore the set of points of Γ with an inner tangent in Ω has linear measure 0 and by Theorem 4.2, harmonic measure 0 with respect to Ω. If Γ' is a copy of Γ rotated by 90 degrees and scaled by $1/\sqrt{3}$, then six copies of Γ' in a hexagonal pattern fit exactly along the outer boundary of Γ. Thus the conformal maps φ and $\widetilde{\varphi}$ of $\mathbb{D}$ onto Ω and $\widetilde{\Omega}$ respectively have angular derivatives only on a set of measure 0 by Theorem 4.2. The theorems in Section 6 below imply that the three measures Λ_1, $\omega_1 = \omega_\Omega$, and $\omega_2 = \omega_{\widetilde{\Omega}}$ are pairwise mutually singular on Γ.

5. Compression and Expansion

Let Ω be a simply connected domain, let ω be harmonic measure for some fixed $z_0 \in \Omega$ and let $\varphi : \mathbb{D} \to \Omega$ be a conformal mapping. If $\Lambda_\alpha(\partial \Omega) > 0$ and $\omega \perp \Lambda_\alpha$, then the conformal map φ **compresses** a set $E \subset \partial \mathbb{D}$ having full harmonic measure into a set $\varphi(E)$ having $\Lambda_\alpha(\varphi(E)) = 0$, and φ **expands** a set $E \subset \partial \Omega$ with $\omega(E) = 0$ into a set $\varphi(E)$ with $\Lambda_\alpha(\varphi(E)) > 0$. On the other hand, if $\omega \ll \Lambda_\alpha$, then φ cannot compress a set E with $|E| > 0$ into a set with $\Lambda_\alpha(\varphi(E)) = 0$.

If $\partial\Omega$ is a rectifiable Jordan curve then by the F. and M. Riesz theorem, $\omega \ll \Lambda_1 \ll \omega$. In [1936] Lavrentiev made the more precise lower estimate

$$\omega(E) \le \frac{C \log \Lambda_1(\partial\Omega)}{1 + |\log \Lambda_1(E)|}, \tag{5.1}$$

when $\partial\Omega$ is rectifiable, $E \subset \partial\Omega$, and $\operatorname{dist}(z_0, \partial\Omega) \ge 1$. See Exercise 2 for the proof.

If $\partial\Omega$ is not rectifiable, no inequality like (5.1) can hold. Lavrentiev [1936] exhibited a Jordan domain for which $\omega \not\ll \Lambda_1$. McMillan and Piranian [1973] constructed a Jordan domain such that

$$\omega \perp \Lambda_1.$$

Then Kaufman and Wu [1982] built, for any measure function of the form

$$h(t) = t\exp(|\log t|^\alpha), \quad 0 < \alpha < \frac{1}{2},$$

a Jordan domain for which $\omega \perp \Lambda_h$. An even sharper result will be proved in Chapter VIII. These examples show φ can compress a set E with $|E| = 2\pi$ into $\varphi(E)$ where $\Lambda_h(\varphi(E)) = 0$.

On the other hand, φ cannot compress harmonic measure very much. By Beurling's projection theorem (see Corollary III.9.3) $\omega \ll \Lambda_{\frac{1}{2}}$. In the very influential paper [1973], Carleson improved Beurling's estimate to $\omega \ll \Lambda_\beta$ for some unknown β, $\frac{1}{2} < \beta < 1$. Then, in the paper that inspired much of this book, Makarov [1985] proved the following theorem.

Theorem 5.1 (Makarov). *Let ω be harmonic measure for a simply connected domain Ω and let $0 < \alpha < 1$. Then*

$$\omega \ll \Lambda_\alpha.$$

However, φ cannot expand harmonic measure very much either. In [1972] and [1981] Øksendal proved $\omega \perp \Lambda_2$ for plane domains of any connectivity, and he conjectured that $\omega \perp \Lambda_\alpha$ for all $\alpha > 1$ and all domains. See Exercise III.12 for a proof of Øksendal's result. Makarov, also in [1985], established the Øksendal conjecture for simply connected domains:

Theorem 5.2 (Makarov). *Let ω be harmonic measure for a simply connected domain Ω and let $h(t)$ be a measure function such that*

$$\lim_{t \to 0} \frac{h(t)}{t} = 0.$$

Then ω is singular to Λ_h, $\omega \perp \Lambda_h$.

In particular, $\omega \perp \Lambda_\alpha$ if $\alpha > 1$. Together Theorems 5.1 and 5.2 say that on a simply connected domain, harmonic measure has **dimension** 1 and φ can neither expand nor compress harmonic measure very much. Later Jones and Wolff [1988] proved that for every plane domain, $\omega \perp \Lambda_\alpha$ for all $\alpha > 1$. The Jones–Wolff theorem will be proved in Chapter IX.

Theorem 5.1 depends on Makarov's [1985] law of the iterated logarithm. It will be proved in Chapter VIII. Here we give Makarov's elegant proof of Theorem 5.2. We need a simple lemma.

Let $E \subset \partial \mathbb{D}$ and let $S = \{z_k\}$ be a sequence in $\mathbb{D}$. We say S is **nontangentially dense a.e.** on E if there is $\alpha > 1$ such that for almost every $\zeta \in E$,

$$\zeta \in \overline{S \cap \Gamma_\alpha(\zeta)}.$$

Lemma 5.3. *Assume $\{z_k\}$ is nontangentially dense a.e. on $E \subset \partial \mathbb{D}$ and let φ be the conformal map from $\mathbb{D}$ onto a simply connected domain Ω. Then set $w_k = \varphi(z_k), r_k = \text{dist}(w_k, \partial \Omega), B_k = B(w_k, 2r_k),$ and*

$$V = \partial \Omega \cap \left(\bigcup B_k \right) \cap A,$$

where A is the set of accessible boundary points. Then

$$|E \setminus \varphi^{-1}(V)| = 0. \tag{5.2}$$

Proof. Suppose (5.2) is false and let W_k be the component of $B_k \cap \Omega$ such that $w_k \in W_k$.

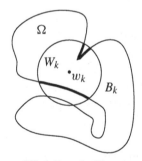

Figure VI.6 Proof of Lemma 5.3.

Since $\partial \Omega$ is connected,

$$\omega(w_k, V, \Omega) \geq \omega(w_k, \partial \Omega \cap B_k, W_k) \geq c,$$

by the Beurling projection theorem; see Exercise 10 of Chapter III. Consequently $u(z) = \omega(\varphi(z), V, \Omega)$ satisfies $u(z_k) \geq c$. But u has nontangential limits a.e. and almost every $\zeta \in E$ is the nontangential limit of a subsequence of

$\{z_k\}$, so that

$$\lim_{\Gamma_\alpha(\zeta)\ni z\to\zeta} u(z) \geq c$$

almost everywhere on E. On the other hand, by Theorem 3.2

$$u(z) = \omega(z, \varphi^{-1}(V), \mathbb{D}),$$

so that $u(z)$ has nontangential limit $\chi_{\varphi^{-1}(V)}$ almost everywhere on $\partial\mathbb{D}$. Consequently

$$\chi_{\varphi^{-1}(V)} \geq c$$

almost everywhere on E, and (5.2) follows. ∎

Proof of Theorem 5.2. Fix $\varepsilon > 0$. When $z \in \mathbb{D}$ and $|z| > \frac{1}{2}$, define

$$I(z) = \left\{ \zeta \in \partial\mathbb{D} : |z - \zeta| < 2(1 - |z|) \right\}.$$

Then $1 - |z|^2 \leq c|I(z)|$ and $\zeta \in I(z)$ if and only if $z \in \Gamma_2(\zeta)$.

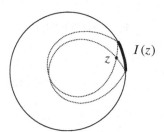

Figure VI.7 $I(z)$ consists of the base points of all cones Γ_2 containing z.

Let φ be the conformal mapping from $\mathbb{D}$ to Ω, and set $\alpha = 2$. By Theorem 2.3,

$$\liminf_{\Gamma_\alpha(\zeta)\ni z\to\zeta} |\varphi'(z)| < \infty$$

for almost every $\zeta \in \partial\mathbb{D}$, and the sets

$$E_n = \left\{ \zeta \in \partial\mathbb{D} : \liminf_{\Gamma_\alpha(\zeta)\ni z\to\zeta} |\varphi'(z)| < n \right\}$$

satisfy $E_n \subset E_{n+1}$ and $|\bigcup E_n| = 2\pi$. Every $\zeta \in E_n$ is covered by arbitrarily small arcs $I(z)$ such that

$$|\varphi'(z)| < n, \tag{5.3}$$

and

$$1 - |z|^2 < \delta_n, \tag{5.4}$$

where $\delta_n < \varepsilon/n$ is so small that

$$\frac{h(t)}{t} < \frac{\varepsilon}{n2^{n+2}} \tag{5.5}$$

whenever $t < 4n\delta_n$. By the Vitali covering lemma, Exercise I.9, there is a sequence $\{z_{n,j}\}$ satisfying (5.3) and (5.4) such that

$$\left| E_n \setminus \bigcup_j I(z_{n,j}) \right| = 0 \tag{5.6}$$

and

$$\sum_j |I(z_{n,j})| \le 2\pi.$$

Then by (5.3), (5.4), and (5.5),

$$\sum_j h\big(2|\varphi'(z_{n,j})|(1 - |z_{n,j}|^2)\big) \le \frac{C\varepsilon}{2^n}. \tag{5.7}$$

With ε fixed, we take $\{z_k\} = \bigcup_n \{z_{n,j}\}$, $w_k = \varphi(z_k)$, $r_k = \mathrm{dist}(w_k, \partial\Omega)$, $B_k = B(w_k, 2r_k)$, and

$$V_\varepsilon = \partial\Omega \cap (\cup B_k)).$$

Then by (5.4) and Theorem I.4.3,

$$2r_k \le 2|\varphi'(z_k)|(1 - |z_k|^2) \le 4\varepsilon,$$

so that (5.7) yields

$$\sum_k h(2r_k) \le C\varepsilon.$$

Consequently, $V = \bigcap_m^\infty V_{\frac{1}{m}}$ satisfies $\Lambda_h(V) = 0$. But by (5.6), $\{z_k\}$ is nontangentially dense a.e. in $\partial\mathbb{D}$. Therefore

$$\omega(\,\cdot\,, V_{\frac{1}{m}}, \Omega) = 1$$

by Lemma 5.3 and $\omega(\,\cdot\,, V, \Omega) = 1$. ∎

6. Pommerenke's Extension

In [1986a], Pommerenke extended Makarov's argument to obtain the following improvement of Theorem 5.2. We recall the sets

$$G = \Big\{ \zeta \in \partial\mathbb{D} : \varphi \text{ has non} - \text{zero angular derivative at } \zeta \Big\},$$

$$B = \Big\{ \zeta \in \partial \mathbb{D} : \varphi \text{ has a nontangential limit at } \zeta, \text{ but}$$

$$\liminf_{\Gamma_\alpha(\zeta) \ni z \to \zeta} |\varphi'(z)| = 0, \text{ for all } \alpha > 1 \Big\},$$

and

$$K = \Big\{ \text{cone points for } \Omega \Big\}.$$

Then we know $G \cap B = \emptyset$, $|G \cup B| = 2\pi$ and $\varphi(G) \subset K$.

Theorem 6.1 (Pommerenke). *Let Ω be a simply connected domain and let $\varphi : \mathbb{D} \to \Omega$. Then there is a subset $S \subset \varphi(B) \setminus K$ such that*

$$\Lambda_1(S) = 0. \tag{6.1}$$

and

$$\omega(S \cup \varphi(G)) = 1. \tag{6.2}$$

In particular, the set $P = \varphi(G) \cup S$ has σ−finite Λ_1 measure and $\omega(P) = 1$. Consequently,

$$\Lambda_1(K \setminus \varphi(G)) = \omega(K \setminus \varphi(G)) = 0.$$

Proof. We follow the proof of Theorem 5.2. Again by the Vitali covering lemma, there are $\{z_{n,j}\}$ such that

$$|\varphi'(z_{n,j})| < 2^{-n-3}, \tag{6.3}$$

$$\Big| B \setminus \bigcup_j I(z_{n,j}) \Big| = 0,$$

and

$$\sum_j |I(z_{n,j})| \leq 2\pi.$$

Now take $w_{n,j} = \varphi(z_{n,j})$, $r_{n,j} = \mathrm{dist}(w_{n,j}, \partial\Omega)$, $B_{n,j} = B(w_{n,j}, 2r_{n,j})$, and

$$V_n = \partial\Omega \cap \Big(\bigcup_j B_{n,j} \Big).$$

Then as before, (6.3) yields

$$\sum_j 2r_{n,j} \leq \sum_j 2|\varphi'(z_{n,j})|(1 - |z_{n,j}|^2) \leq c\pi 2^{-n},$$

and

$$V = \bigcap_k \bigcup_{n \geq k} V_n$$

has $\Lambda_1(V) = 0$. Set

$$S = V \cap \varphi(B).$$

Then (6.1) holds for S. By Beurling's projection theorem, $\omega(w_{n,j}, V_n, \Omega) \geq c$, and

$$\bigcup_{n \geq N} \{z_{n,j}\}$$

is nontangentially dense on B. Therefore by Lemma 5.3

$$\left| B \setminus \bigcup_{n \geq k} \varphi^{-1}(A \cap V_n) \right| = 0$$

for any k, where A is the set of accessible points. Hence

$$\left| B \setminus \varphi^{-1}(S) \right| = 0,$$

and (6.2) then follows from Theorem 4.1. Since $\Lambda_1(S) = 0$, if we replace S by $S \setminus K$ then (6.1) and (6.2) still hold by Theorem 4.2 and $\omega(K \setminus \varphi(G)) = 0$. By Theorem 4.2, $\Lambda_1(K \setminus \varphi(G)) = 0$. ∎

By McMillian's twist point theorem, Appendix I, ω almost every $w \in S$ is a twist point of $\partial\Omega$. From Theorems 6.1 and 4.2 we have the following corollary.

Corollary 6.2. *Let Ω be a simply connected domain and let $S \subset \partial\Omega$ be the set given in Theorem* 6.1. *Then*

$$\omega \ll \Lambda_1 \iff \omega(S) = 0 \iff |G| = 2\pi,$$

and

$$\omega \perp \Lambda_1 \iff |G| = 0.$$

Theorem 6.3. *Suppose Γ is a Jordan curve, and let Ω_1 and Ω_2 be the two components of the complement $\mathbb{C}^* \setminus \Gamma$. Let E be a Borel subset of Γ such that $\omega(z_j, E, \Omega) > 0$ for $z_j \in \Omega_j$, $j = 1, 2$, and let $\omega_j|_E$ be the restriction of $\omega_j = \omega(z_j, \cdot, \Omega_j)$ to E. Then*

$$\omega_1|_E \perp \omega_2|_E$$

if and only if

$$\Lambda_1(\mathrm{Tn}(\Gamma) \cap E) = 0,$$

where $\mathrm{Tn}(\Gamma)$ *is the set of tangent points of* Γ.

Browder and Wermer [1963] gave the first example of a Jordan domain for which $\Lambda_1(\mathrm{Tn}(\Gamma)) = 0$. See also Gamelin and Garnett [1971] and McMillan and Piranian [1973].

Proof. Assume $\Lambda_1(\mathrm{Tn}(\Gamma) \cap E) > 0$. On $\mathrm{Tn}(\Gamma)$, $\omega_j \ll \Lambda_1 \ll \omega_j$ by Theorem 4.2, so that for $j = 1, 2$, $\omega_j|_E(\mathrm{Tn}(\Gamma)) > 0$ and

$$\omega_1|_E \not\perp \omega_2|_E.$$

Now assume $\omega_1|_E \not\perp \omega_2|_E$. The argument is exactly like the proof of Theorem V.1.1. We may assume $\mathrm{dist}(z_j, \Gamma) \geq 1$. Let $w \in \Gamma$. For $0 < t < 1$, let $J_j(t)$ be any arc in $\Omega_j \cap \{z : |z - w| = t\}$ such that J_j separates z_j from w, and write

$$t\theta_j(t) = \ell(J_j(t)).$$

Then by Theorem IV.6.2,

$$\omega_j(B(w, r)) \leq \frac{8}{\pi} \exp\left(-\pi \int_r^1 \frac{dt}{t\theta_j(t)}\right).$$

Because $\left(\frac{1}{\theta_1} + \frac{1}{\theta_2}\right)(\theta_1 + \theta_2) \geq 4$, (by Cauchy–Schwarz) and $\theta_1 + \theta_2 \leq 2\pi$, we have

$$\frac{1}{\theta_1} + \frac{1}{\theta_2} \geq \frac{2}{\pi}.$$

Therefore

$$\omega_1(B(w, r))\omega_2(B(w, r)) \leq \frac{64}{\pi^2} \exp\left(-\int_r^1 \frac{2dt}{t}\right) = \frac{64}{\pi^2} r^2. \qquad (6.4)$$

By assumption, there is a compact set $E_N \subset E$ such that $\omega_j(E_N) > 0$ and

$$\frac{\omega_1(S)}{N} \leq \omega_2(S) \leq N\omega_1(S)$$

for all Borel $S \subset E_N$. Cover $S \subset E_N$ by balls $B(w_k, r_k)$, $w_k \in \Gamma$, and $r_k < 1$. Then by (6.4),

$$\sum r_k \geq \frac{\pi}{8\sqrt{N}} \omega_j(S),$$

and hence

$$\Lambda_1(S) \geq \frac{\pi}{4\sqrt{N}} \omega_j(S). \qquad (6.5)$$

Let φ_j be a conformal map of $\mathbb{D}$ onto Ω_j, let $G_j \subset \partial\mathbb{D}$ be the set where φ_j has non-zero angular derivative and let $S = S_j$ be the set given by Theorem 6.1

that satisfies (6.1) and (6.2) for Ω_j. Then $\omega_1(S \cap E_N) = 0$ by (6.1) and (6.5). Therefore

$$\omega_1\big(\varphi_1(G_1) \cap \varphi_2(G_2) \cap E_N\big) > 0$$

so that by Theorem 4.2,

$$\Lambda_1\big(\varphi_1(G_1) \cap \varphi_2(G_2) \cap E_N\big) > 0,$$

which proves the theorem because

$$\varphi_1(G_1) \cap \varphi_2(G_2) \subset \mathrm{Tn}(\Gamma). \qquad\blacksquare$$

The proof of Theorem 6.3 yields a purely geometric result about the tangent points of a Jordan curve Γ. Let $w \in \Gamma$, take $\theta_j(t)$ as in that proof, and set

$$\varepsilon(w, t) = \max\{|\pi - \theta_j(t)| : j = 1, 2\}.$$

Then by Taylor series

$$\frac{1}{\theta_1} + \frac{1}{\theta_2} \geq \frac{2}{\pi} + \frac{2}{\pi}\left(\frac{\varepsilon(w, t)}{\pi}\right)^2,$$

and the proof of (6.4) yields

$$\frac{\omega_1(B(w, r))}{r} \frac{\omega_2(B(w, r))}{r} \leq \frac{64}{\pi^2} \exp\left(-\frac{2}{\pi^2} \int_r^1 \varepsilon^2(w, t)\frac{dt}{t}\right). \qquad (6.6)$$

But by Theorem 4.2 the left side of (6.6) is bounded below at Λ_1 (or ω_j) almost every point of $\mathrm{Tn}(\Gamma)$. Consequently we have the following corollary.

Corollary 6.4. *If Γ is a Jordan curve, then at Λ_1 almost every tangent point w of Γ,*

$$\int_0^1 \varepsilon^2(w, t)\frac{dt}{t} < \infty. \qquad (6.7)$$

Carleson has conjectured that conversely, Γ has a tangent at almost every point where (6.7) holds. Bishop and Jones [1994] has a result slightly weaker than this conjecture. See the remarks following Theorem X.2.5 below.

We continue to assume Ω_1 and Ω_2 are the two complementary components of a Jordan curve Γ. Let K_j be the cone points for Ω_j, let G_j be the set where the conformal map φ_j of $\mathbb{D}$ onto Ω_j has non-zero angular derivative, and let $S_j \subset \partial\Omega_j \setminus K_j$ be a set satisfying (6.1) and (6.2) with respect to Ω_j. Note that

$\varphi_j(G_j) \cap S_j = \emptyset$. Consider the set

$$
\begin{aligned}
E = \varphi_1(G_1) \cup S_1 \cup \varphi_2(G_2) \cup S_2 &= \left(\varphi_1(G_1) \cap \varphi_2(G_2) \right) \\
&\cup \left(S_1 \cap S_2 \right) \\
&\cup \left(\varphi_1(G_1) \setminus \left(\varphi_2(G_2) \cup S_2 \right) \right) \cup \left(\varphi_2(G_2) \setminus \left(\varphi_1(G_1) \cup S_1 \right) \right) \\
&\cup \left(S_1 \setminus S_2 \right) \cup \left(S_2 \setminus S_1 \right).
\end{aligned}
$$

By (6.2), $\omega_1(E) = \omega_2(E) = 1$. By Theorems 6.3 and 6.1 we have the following table.

Subsets of Γ	Measures
$\varphi_1(G_1) \cap \varphi_2(G_2) \subset \mathrm{Tn}(\Gamma)$	$\omega_1 \ll \Lambda_1 \ll \omega_2 \ll \omega_1$
$S_1 \cap S_2$	$\omega_1 \perp \omega_2$, while $\Lambda_1 = 0$
$\psi_j(G_j) \setminus \left(\varphi_k(G_k) \cup S_k \right)$	$\omega_k = 0$, but $\omega_j \ll \Lambda_1 \ll \omega_j$
$S_j \setminus S_k$	$\omega_k = 0$ and $\Lambda_1 = 0$
$K_1 \cup K_2 \setminus \left(\varphi_1(G_1) \cup \varphi_2(G_2) \right)$	$\omega_1 = \omega_2 = \Lambda_1 = 0$

Examples that limit these results can be found in Exercises 8, 9 and 10.

Notes

Sections 1 through 4 are classical. Sections 5 is from Makarov [1985]. See Pommerenke [1986a] and Bishop, Carleson, Garnett and Jones [1989] for the results of Section 6.

Exercises and Further Results

1. (a) Let φ be the conformal mapping from $\mathbb{D}$ to a simple connected domain Ω, and assume $\varphi' \in H^1$. Prove that $\partial\Omega$ is a rectifiable curve, although not necessarily a Jordan curve.

(b) Let K be a compact connected subset of $\mathbb{C}$ such that $\Lambda_1(K) < \infty$. Let Ω_j be the components of $\mathbb{C}^* \setminus K$, and let φ_j be a conformal mapping from $\mathbb{D}$ onto Ω_j. If Ω_j is bounded, then $\varphi'_j \in H^1$, while if $\infty \in \Omega_j$ and $\varphi(0) = \infty$, then $z^2 \varphi_j'(z) \in H^1$. Furthermore,

$$\sum_j \int_0^{2\pi} \left| \varphi_j' \right| d\theta = 2\Lambda_1(K).$$

See Alexander [1989].

2. Suppose Ω is a bounded Jordan domain with rectifiable boundary, and suppose $z_0 \in \Omega$ is such that $\text{dist}(z_0, \partial\Omega) \geq 1$. Then for $E \subset \partial\Omega$,

$$\omega(z_0, E) \leq \frac{C \log \Lambda_1(\partial\Omega)}{1 + \left| \log \Lambda_1(E) \right|}.$$

See Lavrentiev [1936]. Hint: Take $z_0 = \varphi(0)$. Then $\ell(\partial\Omega) = \int |\varphi'| ds$ and $|\varphi'(0)| \geq 1$. By subharmonicity,

$$\int \log |\varphi'| d\theta \geq 0$$

so that by Jensen's inequality

$$\begin{aligned}
\left\| \log |\varphi'| \right\|_1 &\leq 2 \left\| \log^+ |\varphi'| \right\|_1 \\
&\leq 4\pi \log \left\| \varphi' \right\|_1 \\
&= 4\pi \log \Lambda_1(\partial\Omega).
\end{aligned}$$

Write $E = \varphi(S)$. We may assume $2\pi\omega(E) = |S|$ and $\Lambda_1(E) = \int_S |\varphi'| d\theta$ are small. Again by Jensen's inequality,

$$\frac{1}{|S|} \int_S \log |\varphi'| d\theta \leq \log \frac{\Lambda_1(E)}{|S|},$$

so that

$$|S| \log \frac{|S|}{\Lambda_1(E)} \leq \int_{\partial\mathbb{D}\setminus S} \log |\varphi'| d\theta \leq 4\pi \log \Lambda_1(\partial\Omega).$$

3. We recall the Hayman–Wu theorem: There exists a constant C such that if $\varphi : \mathbb{D} \to \Omega$ is univalent and if L is a line, then

$$\text{length} \left(\varphi^{-1}(L) \right) \leq C.$$

In Section 5 of Chapter I, we showed that $C \leq 4\pi$. Here we give in outline Øyma's [1993] example that $C \geq \pi^2$.

(a) If I is an interval in $\mathbb{R}$ and if $0 < \varepsilon < 1$, let $S_{I,\varepsilon}$ be the longer circular arc in $\mathbb{H}$ having the same endpoints as I and meeting $\mathbb{R}$ at angles ε and $\pi - \varepsilon$. Let $\Omega_{I,\varepsilon}$ be the unbounded component of $\mathbb{C}^* \setminus \left(S_{I,\varepsilon} \cup S_{I,\varepsilon/2} \right)$. Then for $z \in I$

$$\omega(z, S_{I,\varepsilon}, \Omega_{I,\varepsilon}) = \frac{(\pi + \varepsilon/2)}{(2\pi - \varepsilon/2)} < \frac{1}{2} + \varepsilon.$$

(b) Given $\delta > 0$, there is $b > 0$ and $\varepsilon > 0$ such that if I is a subarc of $\partial \mathbb{D}$ with $|I| < b$ and J is a crosscut in $\mathbb{D}$ joining the endpoints of I with

$$\sup_J \omega(z, I, \mathbb{D}) < \frac{1}{2} + \varepsilon,$$

then

$$\text{length}(J) > (1 - \delta)\frac{\pi}{2}|I|.$$

Hint: The level set $\Gamma = \{z : \omega(z, I) = \frac{1}{2} + \varepsilon\}$ is a circular arc, and if b and ε are small enough, then $\text{length}(J) \geq \text{length}(\Gamma) > (1 - \delta)\frac{\pi}{2}|I|$.

(c) Fix $\delta > 0$, and let ε and b be from part (b). Construct open intervals $I_n \subset \mathbb{R}$ of bounded length such that for $n \neq m$,

$$I_n \cap I_m = \emptyset,$$

$$S_{I_n, \varepsilon/2} \cap S_{I_m, \varepsilon/2} = \emptyset,$$

and

$$\left| \mathbb{R} \setminus \bigcup I_n \right| = 0.$$

Write $S_n = S_{I_n, \varepsilon}$, let V_n be the Jordan domain with $\partial V_n = I_n \cup S_n$, and set

$$\Omega = \{y < 0\} \cup \bigcup (V_n \cup I_n).$$

Then $\partial \Omega$ is locally rectifiable, so that if $\varphi : \mathbb{D} \to \Omega$ is conformal, then by the F. and M. Riesz theorem, $\left| \varphi^{-1}(\mathbb{R} \setminus \bigcup I_n) \right| = 0$. Hence

$$\sum \left| \varphi^{-1}(S_n) \right| = 2\pi.$$

Taking φ with $|\text{Im}\varphi(0)|$ large gives $|\varphi^{-1}(S_n)| < b$ for all n. Then by (a) and (b),

$$\text{length}\left(\varphi^{-1}(I_n) \right) > (1 - \delta)\frac{\pi}{2}|\varphi^{-1}(S_n)|$$

and thus

$$\text{length}\left(\varphi^{-1}(\mathbb{R}) \right) = \sum \text{length}\left(\varphi^{-1}(I_n) \right) \geq (1 - \delta)\pi^2.$$

4. There exists a simply connected Jordan domain Ω and a set $E \subset \mathbb{R} \cap \partial\Omega$ such that $\Lambda_1(E) = |E| > 0$ but $\omega(z, E, \Omega) = 0$. Let $\Omega_0 = (0, 1) \times (0, 1)$,

and $E_0 = [0, 1]$ and form closed sets $E_{n+1} \subset E_n$ and domains $\Omega_{n+1} \subset \Omega_n$ thusly: E_n consists of 2^n intervals $I_{n,j}$ with $|I_{n,j}| = a_n$, $2^{n+1}a_{n+1} < 2^n a^n$, but $\lim 2^n a_n > 0$. Let $I_{n,j} \setminus E_{n+1}$ be the middle open subinterval $J_{n,j} \subset I_{n,j}$ with $|J_{n,j}| = a_n - 2a_{n+1}$. Let $b_n \downarrow 0$ satisfy $\lim 2^n e^{-\pi \frac{b_n}{a_n}} = 0$, and take

$$\Omega_{n+1} = \Omega_n \setminus \bigcup_j (\bar{J}_{n,j} \times [0, b_{n+1}]).$$

Then $E = \bigcap E_n$ has $|E| > 0$ and $\Omega = \bigcap \Omega_n$ is a Jordan domain such that $E \subset \partial\Omega$, and for $z \in \Omega$,

$$\omega(z, E, \Omega) \leq \lim_n \omega(z, E_n, \Omega_n) = 0$$

by Theorem IV.6.1 or Theorem VI.6.1. See Lohwater and Seidel [1948].

5. (a) Let $\gamma \subset \partial\mathbb{D}$ be an arc with non-empty interior and let $f(z)$ be a meromorphic function on $\mathbb{D}$ such that

$$\lim_{r \to 1} f(r\zeta) = 0$$

for all $\zeta \in \gamma$. Prove $f \equiv 0$. Hint: Use Baire category and the proof of Privalov's theorem.

   (b) If $E \subset \partial\mathbb{D}$ is an F_σ set of the first Baire category in $\partial\mathbb{D}$, and if $g(z)$ is any continuous function on the open disc $\mathbb{D}$, then there is a function $f(z)$ analytic on $\mathbb{D}$ such that

$$\lim_{r \to 1} (f(r\zeta) - g(r\zeta)) = 0$$

for all $\zeta \in E$.

   (c) There is a non-constant analytic function $f(z)$ on $\mathbb{D}$ having radial limit 0 almost everywhere on $\partial\mathbb{D}$. See Bagemihl and Seidel [1954].

6. Let $f(z)$ be an analytic function on $\mathbb{D}$ and let $E \subset \partial\mathbb{D}$. If $\text{Re} f$ is nontangentially bounded on E, then f has a nontangential limit almost everywhere on E.

7. Let Ω be a simply connected domain, let $\alpha \subset \partial\Omega$ be a compact set, let σ be an arc in Ω connecting $z_0 \in \Omega$ to some $\zeta \in \partial\Omega \setminus \alpha$, and let Γ be the set of curves in Ω joining σ to α. Prove

$$\omega(z_0, \alpha, \Omega) \leq \frac{8}{\pi} e^{-\pi\lambda(\Gamma)}.$$

Hint: Use Theorems 3.2 and IV.5.3.

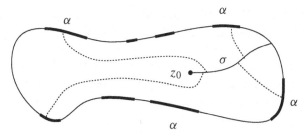

Figure VI.8 Extremal distance estimate of harmonic measure.

8. (a) Let $u(z)$ be real valued and continuous on $\overline{\mathbb{D}}$, and harmonic on $\mathbb{D}$, and assume $\sup_{z \in \mathbb{D}} |u(z)| < 1$ but $\lim_{r \to 1} \tilde{u}(r) = -\infty$. Then

$$\varphi(z) = \int_0^z e^{i(u+i\tilde{u})(\zeta)} d\zeta$$

is a conformal mapping from $\mathbb{D}$ to a simply connected Jordan domain, φ is conformal at $\zeta = 1$, but φ does not have a non-zero angular derivative at $\zeta = 1$.

(b) Find a simply connected domain Ω such that for some $\zeta \in \partial\mathbb{D}$, the conformal map $\varphi : \mathbb{D} \to \Omega$ is not conformal at ζ even though $\partial\Omega$ has an inner tangent at $\varphi(\zeta)$.

9. Suppose the Jordan curve Γ is a finite union of linear images $w = Az + B$ of graphs

$$\Gamma_k = \{y = f_k(x)\}.$$

Let Ω_1 and Ω_2 be the components of $\mathbb{C}^* \setminus \Gamma$ and let ω_j be harmonic measure for some point in Ω_j.

(a) Prove that $\omega_j \ll \Lambda_1$ on Γ.   Hint: Use the twist point theorem from Appendix I or use Lemmas I.2 and I.3 directly.

(b) Suppose f_k has no finite or infinite derivative at any point. For examples, see Hobson [1926]. Then $\omega_j \ll \Lambda_1$ but $\omega_1 \perp \omega_2$. Consequently if φ_j is the conformal map from $\mathbb{D}$ to Ω_j then at almost every $\zeta \in \partial\mathbb{D}$, φ is conformal but Γ has no tangent at $\varphi(\zeta)$.

(c) Suppose $f_k(x) = \sum 2^{-n} \cos 2^n x$. Then f_k is nowhere differentiable, but $f_k \in Z^*$. See Zygmund [1959], page 47. It follows that Γ has tangents ω_j almost everywhere.

(d) If f_k is nowhere differentiable, but $f_k \in Z^*$, then the graph $\{y = f_k(x)\}$ has a vertical tangent on a set of positive linear measure.

See Bishop [1987] for (a)–(d).

10. From the interval $E_0 = [0, 1]$ remove the middle open interval $I_{1,1}$ of length $|I_{1,1}| = 1/8$, obtaining a set E_1. Inductively, obtain E_n by removing the middle interval $I_{n,j}$ of length $|I_{n,j}| = 2^{-2n-1}$ from each of the 2^{n-1} components of E_{n-1}. That leaves $E = \bigcap E_n$ with $|E| = 3/4$. Each deleted interval $I_{n,j}$ is the base of a rectangle $R_{n,j}$ having dimensions $2^{-2n-1} \times 2^{-n-2}$. Let the "tower" $T_{n,j}$ consist of the three sides of $\partial R_{n,j} \setminus I_{n,j}$. Then

$$\sigma_1 = E \cup \bigcup_{n,j} T_{n,j}$$

is a Jordan arc connection 0 to 1. Note that σ_1 falls inside the triangle with vertices $\{0, 1, 1/2 + i/6\}$. On each side of the unit square $[0, 1] \times [0, 1]$ put an isometric copy of σ, with the towers pointing out from $[0, 1] \times [0, 1]$, to obtain a Jordan curve Γ_1. Let Ω_1 denote the bounded domain with $\partial \Omega_1 = \Gamma_1$.

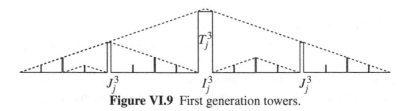

Figure VI.9 First generation towers.

Partition each tower $T_{n,j}$ into $2^n + 1$ segments of length $|I_{n,j}| = 2^{-2n-1}$, and repeat the tower construction on each segment using the same length ratios as before. That gives a second Jordan curve Γ_2 bounding a domain $\Omega_2 \supset \Omega_1$. Continuing, we get a Jordan curve $\Gamma = \lim \Gamma_n$ bounding a domain $\Omega = \bigcup_n \Omega_n$. Let $\widetilde{\Omega}$ be the exterior $\mathbb{C}^* \setminus (\Gamma \cup \Omega)$. Write $\omega(E) = \omega(E, \Omega)$ and $\widetilde{\omega} = \omega(E, \widetilde{\Omega})$. Prove $\omega \ll \Lambda_1 \ll \omega$ but $\widetilde{\omega} \perp \Lambda_1$ (Bishop [1987]).

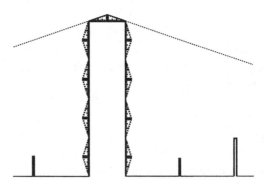

Figure VI.10 Second generation towers.

11. Let Γ be the von Koch snowflake curve. Prove

$$0 < \Lambda_{\frac{\log 4}{\log 3}}(\Gamma) < \infty,$$

and consequently that $\dim(\Gamma) = \frac{\log 4}{\log 3}$. (See Appendix D for the definition of Λ_α.)

12. Let $S = \{z_k\}$ be a discrete sequence in $\mathbb{D}$. Then the following are equivalent:

   (i) S is a nontangentially dense a.e. on $\partial\mathbb{D}$.

   (ii) For a.e. $\zeta \in \partial\mathbb{D}$ there is $\alpha > 0$ such that $\zeta \in \overline{S \cap \Gamma_\alpha(\zeta)}$.

   (iii) For every bounded harmonic function $u(z)$ on $\mathbb{D}$,

$$\sup_{\mathbb{D}} |u(z)| = \sup_{S} |u(z_k)|.$$

   (iv) There is $M < \infty$ such that for every bounded analytic function $f(z)$ on $\mathbb{D}$,

$$\sup_{\mathbb{D}} |f(z)| \leq M \sup_{S} |f(z_k)|.$$

   (v) For each $z \in \mathbb{D}$, there exist positive weights λ_k such that for every bounded harmonic function u on $\mathbb{D}$,

$$u(z) = \sum \lambda_k u(z_k).$$

   (vi) For some $z_0 \in \mathbb{D} \setminus S$, there are complex weights μ_k with $\sum |\mu_k| < \infty$ such that for every bounded analytic function f on $\mathbb{D}$,

$$f(z) = \sum \mu_k f(z_k).$$

See Brown, Shields, and Zeller [1960], and Hoffman and Rossi [1967].

13. Let Ω be a simply connected domain and fix a point $z_0 \in \Omega$. For $w \in \partial\Omega$, set $\omega(\delta) = \omega(w, \delta) = \omega(z_0, B(w, \delta) \cap \partial\Omega, \Omega)$. Then at ω almost every cone point, the limit

$$\lim_{\delta \to 0} \frac{\omega(\delta)}{\delta} = L$$

exists and $0 < L < \infty$. Let S be the set given by Theorem 6.1. Then ω almost everywhere on S

$$\limsup_{\delta \to 0} \frac{\omega(\delta)}{\delta} = \infty.$$

Hint: Use the results or proofs in Section 5. Bishop [1994] conjectured that

$$\liminf_{\delta \to 0} \frac{\omega(\delta)}{\delta} = 0$$

almost everywhere on S. This has been proved recently by S. Choi [2004].

14. Let Ω be a simply connected domain, and let $\varphi : \mathbb{D} \to \Omega$ be conformal. Fix $w_0 \in \Omega$. For accessible $\zeta \in \partial\mathbb{D}$ and $a = \varphi(\zeta)$, let $\gamma(a, r)$ be an arc of the circle $\{|w - a| = r\}$ such that $\gamma(a, r)$ is a crosscut of Ω separating w_0 from $\{\varphi(r\zeta) : r_0 < r < 1\}$ for some $r_0 < 1$. (Thus $\gamma(a, r)$ actually depends on ζ.) Write

$$L(a, r) = \ell(\gamma(a, r)),$$

and

$$A(a, r) = \int_0^r L(a, s)\,ds.$$

McMillan [1970] proved these results:

(a) Almost everywhere with respect to ω,

$$\limsup_{r \to 0} \frac{A(a, r)}{\pi r^2} \geq \frac{1}{2}.$$

(b) Consequently, ω almost everywhere,

$$\limsup_{r \to 0} \frac{L(a, r)}{2\pi r} \geq \frac{1}{2}.$$

(c) There is an example for which, ω almost everywhere,

$$\liminf_{r \to 0} \frac{A(a, r)}{\pi r^2} = 0, \text{ and } \limsup_{r \to 0} \frac{A(a, r)}{\pi r^2} = 1.$$

See McMillan [1970] and [1969] or O'Neill [1999] for the proof of (a), and note that (b) also follows from Theorems 5.2 and IV.6.2. In [1970] had McMillan conjectured that ω almost everywhere,

$$\liminf_{r \to 0} \frac{A(a, r)}{\pi r^2} \leq \frac{1}{2}$$

and

$$\liminf_{r \to 0} \frac{L(a, r)}{2\pi r} \leq \frac{1}{2}.$$

These conjectures were recently established by O'Neill and Thurman [2000] and [2001].

VII

Bloch Functions and Quasicircles

1. Bloch Functions

An analytic function $g(z)$ on $\mathbb{D}$ is a **Bloch function** if

$$||g||_{\mathcal{B}} = \sup_{z \in \mathbb{D}} |g'(z)|(1 - |z|^2) < \infty. \tag{1.1}$$

We write $\mathcal{B}$ for the space of Bloch functions and we call $||g||_{\mathcal{B}}$ the **Bloch norm**[1] of g. Because

$$\frac{d}{dz} g\left(\frac{z + z_0}{1 + \bar{z}_0 z}\right)\bigg|_{z=0} = g'(z_0)(1 - |z_0|^2), \tag{1.2}$$

we have

$$||g||_{\mathcal{B}} = \sup_{T \in \mathcal{M}} |(g \circ T)'(0)|, \tag{1.3}$$

where $\mathcal{M}$ is the set of conformal self maps of $\mathbb{D}$:

$$T(z) = \lambda \frac{z + a}{1 + \bar{a} z},$$

with $a \in \mathbb{D}$ and $|\lambda| = 1$. Thus the Bloch space $\mathcal{B}$ is invariant under $\mathcal{M}$:

$$||g \circ T||_{\mathcal{B}} = ||g||_{\mathcal{B}}, \quad T \in \mathcal{M}. \tag{1.4}$$

If g is bounded, then $g \in \mathcal{B}$ by (1.2). In fact by (II.3.4) and (II.3.5), $g \in \mathcal{B}$ if only $\operatorname{Re} g$ is bounded. The function

$$g(z) = \sum_{n=1}^{\infty} z^{2^n} \tag{1.5}$$

---

[1] Actually $||g||_{\mathcal{B}}$ is only a semi-norm, but $||g||_{\mathcal{B}} = 0$ only if g is constant.

is a good example of a Bloch function in no H^p space. See Theorem 1.3 below.

A function $g(z)$ on $\mathbb{D}$ is **normal** if $\{g \circ T : T \in \mathcal{M}\}$ is a normal family, in the extended sense that $f \equiv \infty$ is a permitted limit. By Marty's theorem (Ahlfors [1979]) $g(z)$ is normal if and only if

$$\sup_{T \in \mathcal{M}} \frac{|(g \circ T)'(z)|}{1 + |g \circ T(z)|^2}$$

is bounded on each compact subset of $\mathbb{D}$. Since $\mathcal{M}$ is transitive, it follows from (1.2) that g is normal if and only if

$$\sup_{\mathbb{D}} \frac{|g'(z)|(1 - |z|^2)}{1 + |g(z)|^2} < \infty.$$

Hence (1.1) yields:

Theorem 1.1. *Every Bloch function is a normal function.*

Recall that the hyperbolic distance from $z_1 \in \mathbb{D}$ to $z_2 \in \mathbb{D}$ is

$$\rho(z_1, z_2) = \inf \int_{z_1}^{z_2} \frac{dz}{1 - |z|^2},$$

where the infimum is taken over all arcs in $\mathbb{D}$ connecting z_1 to z_2.

Theorem 1.2. *Let $g(z)$ be analytic on $\mathbb{D}$. Then $g \in \mathcal{B}$ if and only if g is Lipschitz continuous as a map from the hyperbolic metric on $\mathbb{D}$ to the euclidian metric on $\mathbb{C}$. Furthermore,*

$$\|g\|_{\mathcal{B}} = \sup_{z,w \in \mathbb{D}} \frac{|g(z) - g(w)|}{\rho(z, w)}.$$

Proof. If

$$|g(z_1) - g(z_2)| \leq B \int_{z_1}^{z_2} \frac{1}{1 - |z|^2} |dz|,$$

then a differentiation shows $\|g\|_{\mathcal{B}} \leq B$.

Conversely, if $g \in \mathcal{B}$ then we get

$$|g(z_1) - g(z_2)| \leq \|g\|_{\mathcal{B}} \, \rho(z_1, z_2) \tag{1.6}$$

by integrating along a hyperbolic geodesic. ∎

Theorem 1.3. *Let $g(z)$ be analytic on $\mathbb{D}$ and let $G(z)$ be the primitive*

$$G(z) = \int_0^z g(w) dw.$$

Then $g \in \mathcal{B}$ if and only if G is continuous on $\overline{\mathbb{D}}$ and $G(e^{i\theta})$ is a Zygmund function, $G(e^{i\theta}) \in Z^*$. When $g \in \mathcal{B}$, there is a constant C independent of g such that

$$\frac{1}{C}(\|g\|_{\mathcal{B}} + |g(0)|) \leq \|G\|_{Z^*} \leq C(\|g\|_{\mathcal{B}} + |g(0)|). \tag{1.7}$$

The Bloch function g has nontangential limit at $e^{i\theta_0} \in \partial\mathbb{D}$ if and only if its primitive $G(e^{i\theta})$ is differentiable at θ_0.

Proof. Suppose $g \in \mathcal{B}$. Then (1.6) gives, for $0 < r < s < 1$,

$$\left|G(se^{i\theta}) - G(re^{i\theta})\right| = \left|\int_r^s g(te^{i\theta})e^{i\theta}dt\right| \leq |g(re^{i\theta})|(s-r) \\ + \frac{\|g\|_{\mathcal{B}}}{2}\int_r^s \log\left(\frac{1+t}{1-t}\right)dt, \tag{1.8}$$

and hence G extends continuously to $\overline{\mathbb{D}}$. Then (1.1) and Theorem II.3.4 give (1.7).

The proof of (1.8) also gives the estimate

$$\int_r^1 |g(se^{i\theta}) - g(re^{i\theta})|ds \leq \|g\|_{\mathcal{B}}\int_r^1 \rho(r,s)ds \\ = \|g\|_{\mathcal{B}}\log\left(\frac{2}{1+r}\right) \tag{1.9} \\ \leq \|g\|_{\mathcal{B}}(1-r).$$

Now suppose $\|g\|_{\mathcal{B}} \leq 1$ and suppose g has nontangential limit 0 at $e^{i\theta}$. Then (1.9) implies that $\frac{dG}{d\theta} = 0$. Indeed, let ε be small and take $\alpha > 1$ so that

$$(1 - \varepsilon t)e^{i(\theta+t)} \in \Gamma_\alpha(e^{i\theta}) \tag{1.10}$$

for all small t.

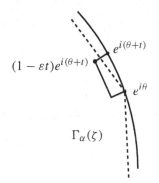

Figure VII.1

Set $r = 1 - \varepsilon t$. Then

$$
\begin{aligned}
G(e^{i(\theta+t)}) - G(e^{i\theta}) &= \lim_{s \to 1} \int_{se^{i\theta}}^{se^{i(\theta+t)}} g(z)dz \\
&= \int_{re^{i\theta}}^{re^{i(\theta+t)}} g(z)dz - e^{i\theta} \int_r^1 g(se^{i\theta})ds \\
&\quad + (1-r)e^{i(\theta+t)}g(re^{i(\theta+t)}) \\
&\quad + e^{i(\theta+t)} \int_r^1 \Big(g(se^{i(\theta+t)}) - g(re^{i(\theta+t)}) \Big)ds.
\end{aligned}
\tag{1.11}
$$

By (1.10) the first three terms on the far right-hand side of (1.11) are each o(t), because $g(z) \to 0$ when $\Gamma_\alpha(e^{i\theta}) \ni z \to e^{i\theta}$, and by (1.9) the last term in (1.11) is bounded by

$$
\|g\|_B \log\Big(\frac{2}{1+r}\Big) = \mathrm{O}(\varepsilon t).
$$

Hence $G(e^{i(\theta+t)}) - G(e^{i\theta}) = \mathrm{o}(t)$ and $G'(e^{i\theta}) = 0$.

The converse assertion, that G' has a nontangential limit wherever $\frac{dG}{d\theta}$ exists, is a simple property of the Poisson kernel. See Exercise I.8. ∎

Theorem 1.3 shows that the Bloch function (1.5) has a nontangential limit at no point, because the Weierstrass function

$$
\mathrm{Re}\, G(z) = \sum 2^{-n} \cos(2^n\theta)
$$

is differentiable at no point. See Katznelson [1968], page 106.

2. Bloch Functions and Univalent Functions

We fix a pair of functions $\varphi(z)$ and $g(z)$ analytic on $\mathbb{D}$ such that

$$
g(z) = \log(\varphi'(z)), \qquad \varphi(0) = 0,
\tag{2.1}
$$

or equivalently

$$
\varphi(z) = \int_0^z e^{g(w)}dw.
$$

Theorem 2.1. *If φ is univalent on $\mathbb{D}$, then $g \in B$ and $\|g\|_B \le 6$. Conversely, there is $\beta < 1$ such that if $\|g\|_B < \beta$, then φ is univalent and $\varphi(\mathbb{D})$ is bounded by a Jordan curve.*

Proof. Fix $z \in \mathbb{D}$ and write

$$\psi(w) = \frac{\varphi\left(\frac{w+z}{1+\bar{z}w}\right) - \varphi(z)}{\varphi'(z)(1 - |z|^2)}.$$

Then ψ is a normalized univalent function, $\psi(0) = 0$ and $\psi'(0) = 1$, and by Theorem I.4.1, $|\psi''(0)| \leq 4$. Then because

$$\psi''(0) = \frac{\varphi''(z)}{\varphi'(z)}(1 - |z|^2) - 2\bar{z} = g'(z)(1 - |z|^2) - 2\bar{z},$$

it follows that

$$|g'(z)|(1 - |z|^2) \leq 6.$$

To prove the converse, let I be an arc on $\partial\mathbb{D}$ with $|I| \leq \pi$. Let C_I be the circle that intersects $\partial\mathbb{D}$ orthogonally at the endpoints of I, and let z_I be the point on C_I closest to 0.

Lemma 2.2. *If $\|g\|_B \leq \beta < 2$, then at each $\zeta \in \partial\mathbb{D}$, $\varphi(\zeta) = \lim_{\mathbb{D}\ni z\to\zeta} \varphi(z)$ exists and*

$$\left|\varphi(\zeta) - \varphi(z_I) - (\zeta - z_I)\varphi'(z_I)\right| \leq \frac{2\beta}{2 - \beta}|\varphi'(z_I)||I| \qquad (2.2)$$

for all $\zeta \in I$. Furthermore, φ extends continuously to $\overline{\mathbb{D}}$ and $\varphi \in C^{1-\beta/2}(\overline{\mathbb{D}})$.

Proof. Let $J = \partial\mathbb{D} \cap \{\mathrm{Re}\,w \geq 0\}$ and let $T \in \mathcal{M}$ satisfy $T(J) = I$. Then

$$T(w) = \frac{z_I}{|z_I|} \frac{w + |z_I|}{1 + |z_I|w}.$$

Note that when $\mathrm{Re}\,w \geq 0$,

$$|T'(w)| \leq 1 - |z_I|^2 \leq |I| \qquad (2.3)$$

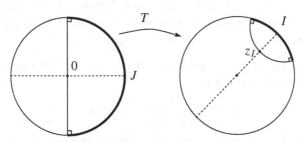

Figure VII.2

Suppose for a moment that the limit $\varphi(\zeta)$ exists. Then

$$\varphi(\zeta) - \varphi(z_I) - \varphi'(z_I)(\zeta - z_I) = \varphi'(z_I) \int_{z_I}^{\zeta} \left(e^{g(z) - g(z_I)} - 1\right) dz$$

$$= \varphi'(z_I) \int_{0}^{T^{-1}(\zeta)} \left(e^{g(T(w)) - g(T(0))} - 1\right) T'(w) dw.$$

The inequality $|e^z - 1| \le e^{|z|} - 1$ is trivial by Taylor series. With (2.3) and the hypothesis $\|g\|_B \le \beta$, it yields

$$\left|\varphi(\zeta) - \varphi(z_I) - \varphi'(z_I)(\zeta - z_I)\right| \le |\varphi'(z_I)||I| \int_{0}^{T^{-1}(\zeta)} \left(\left(\frac{1 + |w|}{1 - |w|}\right)^{\frac{\beta}{2}} - 1\right) d|w|$$

$$\le \frac{2\beta}{2 - \beta} |\varphi'(z_I)||I|,$$

which is (2.2). Since (1.6) and (2.1) give

$$|\varphi'(z_I)| \le e^{|g(0)|} \left(\frac{2}{1 - |z_I|}\right)^{\beta/2},$$

Theorem II.3.2 shows that φ extends continuously to $\overline{\mathbb{D}}$, $\varphi \in C^{1 - \beta/2}(\overline{\mathbb{D}})$, and

$$|\varphi(\zeta_1) - \varphi(\zeta_2)| \le \frac{C}{2 - \beta} |z_1 - z_2|^{1 - \beta/2}. \qquad \blacksquare$$

To conclude the proof of Theorem 2.1, let $\zeta_1, \zeta_2 \in \partial \mathbb{D}$ and let I be the shorter arc of ∂D having endpoints ζ_1 and ζ_2. Then by (2.2) and triangle inequalities,

$$|\varphi(\zeta_1) - \varphi(\zeta_2)| \ge |\varphi'(z_I)||\zeta_1 - \zeta_2| - \frac{4\beta}{2 - \beta} |\varphi'(z_I)||I|$$

$$\ge \left(1 - \frac{2\pi\beta}{2 - \beta}\right) |\varphi'(z_I)||\zeta_1 - \zeta_2|, \qquad (2.4)$$

because $|I| \le \frac{\pi}{2} |\zeta_1 - \zeta_2|$. Thus when $\frac{2\pi\beta}{2-\beta} < 1$, φ is a one-to-one map from $\partial \mathbb{D}$ onto a Jordan curve, and it then follows by the argument principle that φ is univalent on $\mathbb{D}$. $\qquad \blacksquare$

Lemma 2.2 proves more than it says. A Jordan curve Γ is a **quasicircle** if there is a constant A such that

$$\frac{|w_1 - w| + |w - w_2|}{|w_1 - w_2|} \le A, \qquad (2.5)$$

whenever $w \in \Gamma$ lies on that subarc of Γ having endpoints w_1 and w_2 and smaller diameter. See Ahlfors [1963] and [1966]. We call (2.5) the **Ahlfors condition.** Any C^1 curve or Lipschitz curve is a quasicircle and the von Koch snowflake is a nonrectifiable quasicircle. A **quasidisc** is a domain bounded by a quasicircle.

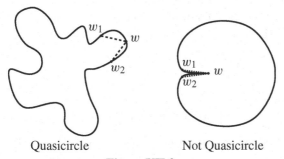

Quasicircle   Not Quasicircle

Figure VII.3

Theorem 2.3. *There is $\beta_0 < 1$ such that if $\|g\|_B < \beta < \beta_0$, then $\varphi(\partial \mathbb{D})$ is a quasicircle with constant*

$$A \leq A(\beta) \leq \sqrt{2}(1 + C_1 \beta). \tag{2.6}$$

Proof. Take $w_j = \varphi(\zeta_j)$ and let $w = \varphi(\zeta)$ lie on the smaller diameter subarc $\varphi(I)$ of $\partial \mathbb{D}$ with endpoints ζ_1 and ζ_2. Replacing φ by $\varphi \circ T$ for some $T \in \mathcal{M}$ if necessary, we can assume that $|I| \leq \pi$. Then by (2.2),

$$\left| \varphi(\zeta_j) - \varphi(\zeta) - (\zeta_j - \zeta)\varphi'(z_I) \right| \leq \frac{4\beta}{2 - \beta} |\varphi'(z_I)||I|,$$

so that

$$|w_1 - w| + |w - w_2| \leq |\varphi'(z_I)|(|\zeta_1 - \zeta| + |\zeta - \zeta_2|) + \frac{8\beta}{2 - \beta}|\varphi'(z_I)||I|$$

$$\leq \left(1 + \frac{4\pi\beta}{2 - \beta}\right)|\varphi'(z_I)|(|\zeta_1 - \zeta| + |\zeta - \zeta_2|).$$

Then by (2.4) we obtain (2.5) with the estimate

$$A \leq A(\beta) = \left(\frac{2 + (4\pi - 1)\beta}{2 - (2\pi + 1)\beta}\right)\sqrt{2}.$$

Therefore $\varphi(\partial \mathbb{D})$ is a quasicircle and (2.6) holds provided

$$\beta < \frac{2}{(2\pi + 1)}. \qquad \blacksquare$$

The constants β and β_0 in Theorems 2.1 and 2.3 are not sharp. Becker [1972] showed that φ is univalent if $\|g\|_B \leq 1$, and that $\varphi(\partial \mathbb{D})$ is a quasicircle if $\|g\|_B < 1$. The constant 1 is sharp in both of Becker's results, but when $\|g\|_B = 1$, $\varphi(\partial\mathbb{D})$ is still a Jordan curve. See Becker and Pommerenke [1984]. The constant $\sqrt{2}$ in (2.6) is needed because $A = \sqrt{2}$ for a circle.

Later we will need the following geometric result, which can be viewed as a converse of Theorem 2.3. For $\delta > 0$ write

$$D_\delta = \{z : |z| < \frac{1}{\delta} \text{ and } \operatorname{Im} z > \delta\},$$

and

$$L_\delta = \{x \in \mathbb{R} : x + i\delta \in \partial D_\delta\}.$$

When $w \in \Omega$ choose $w_0 \in \partial\Omega$ to satisfy $|w - w_0| = \operatorname{dist}(w, \partial\Omega)$, and write $\tau(z) = \tau_w(z) = w_0 - i(w - w_0)z$, $D_\delta(w) = \tau(D_\delta)$, and $L_\delta(w) = \tau(L_\delta)$.

Theorem 2.4. *Let φ be a conformal map from $\mathbb{D}$ onto a simply connected domain Ω, and let $g(z) = \log(\varphi'(z))$. Then the following conditions are equivalent:*

(a) *There are $\eta > 0$ and $M < \infty$ such that*

$$\sup_{\{z:\rho(z,z_0)<M\}} \left(1 - |z|^2\right)\left|g'(z)\right| \geq \eta, \tag{2.7}$$

for all $z_0 \in \mathbb{D}$.

(b) *The family of maps*

$$\psi_\zeta(z) = \frac{\varphi\left(\frac{z+\zeta}{1-\bar{\zeta}z}\right) - \varphi(\zeta)}{\varphi'(\zeta)\left(1 - |\zeta|^2\right)}, \quad \zeta \in \mathbb{D},$$

contains no sequence converging, uniformly on compact subsets of $\mathbb{D}$, to a map of the form

$$\psi^{(\lambda)}(z) = \frac{z}{1 + \bar{\lambda}z}, \quad |\lambda| = 1.$$

(c) *There is $\delta > 0$ such that for all $w \in \Omega$, either*

$$D_\delta(w) \cap \partial\Omega \neq \emptyset,$$

or there exists $w_1 \in L_\delta(w)$ so that

$$B(w_1, \delta|w - w_0|) \cap \partial\Omega = \emptyset.$$

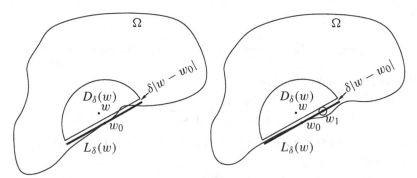

Figure VII.4 The two possibilities in Condition (c).

Proof. Suppose (a) is false. Then there exists $\{z_n\} \subset \mathbb{D}$, with $z_n \to \lambda \in \partial\mathbb{D}$ such that

$$\lim_{n\to\infty} \sup_{\{z:\rho(z,z_n)\le M\}} (1 - |z|^2)|g'(z)| = 0,$$

for all $M < \infty$. Then, uniformly on compact subsets of $\mathbb{D}$,

$$g\left(\frac{z + z_n}{1 + \bar{z}_n z}\right) - g(z_n) \to 0,$$

and

$$\frac{\varphi'\left(\frac{z+z_n}{1+\bar{z}_n z}\right)}{\varphi'(z_n)} \to 1,$$

so that

$$\psi'_{z_n}(z) = \frac{\varphi'\left(\frac{z+z_n}{1+\bar{z}_n z}\right)}{\varphi'(z_n)(1 + \bar{z}_n z)^2} \to \frac{1}{(1 + \bar{\lambda}z)^2}. \tag{2.8}$$

Because $\psi_{z_n}(0) = \psi^{(\lambda)}(0) = 0$, it follows that

$$\psi_{z_n}(z) \to \psi^{(\lambda)}(z)$$

and that (b) is false.

Now suppose (b) fails. Then there is a sequence $\{z_n\} \to \mu \in \partial\mathbb{D}$ such that $\psi_{z_n}(z) \to \psi^{(\lambda)}(z)$, uniformly on compact subsets of $\mathbb{D}$. If $\mu = \lambda$, then we can reverse the previous argument to conclude that (a) does not hold. If $\mu \ne \lambda$ then by (2.8)

$$\frac{\varphi'\left(\frac{z+z_n}{1+\bar{z}_n z}\right)}{\varphi'(z_n)} \to \left(\frac{1 + \bar{\mu}z}{1 + \bar{\lambda}z}\right)^2. \tag{2.9}$$

Fix $|\zeta| < 1$ and $|z| < 1$. Set

$$\alpha_n = \frac{\zeta + z_n}{1 + \bar{z}_n \zeta}, \quad \beta_n = \frac{1 + \bar{z}_n \zeta}{1 + z_n \bar{\zeta}}, \quad \beta = \frac{1 + \bar{\mu}\zeta}{1 + \mu\bar{\zeta}}, \text{ and } w_n = \frac{\beta_n z + \zeta}{1 + \bar{\zeta}\beta_n z}.$$

Then $\alpha_n \to \mu$ and

$$\frac{w_n + z_n}{1 + \bar{z}_n w_n} = \frac{z + \alpha_n}{1 + \bar{\alpha}_n z}.$$

Because $w_n \to (\beta z + \zeta)/(1 + \bar{\zeta}\beta z) \in \mathbb{D}$ and (2.9) holds uniformly on compact subsets of $\mathbb{D}$

$$\frac{\varphi'\left(\frac{z+\alpha_n}{1+\bar{\alpha}_n z}\right)}{\varphi'(\alpha_n)} = \frac{\varphi'\left(\frac{w_n+z_n}{1+\bar{z}_n w_n}\right)}{\varphi'(z_n)} \cdot \frac{\varphi'(z_n)}{\varphi'\left(\frac{\zeta+z_n}{1+\bar{z}_n \zeta}\right)}$$

$$\to \left(\frac{1 + \bar{\mu}\left(\frac{\beta z+\zeta}{1+\bar{\zeta}\beta z}\right)}{1 + \bar{\lambda}\left(\frac{\beta z+\zeta}{1+\bar{\zeta}\beta z}\right)}\right)^2 \left(\frac{1 + \bar{\lambda}\zeta}{1 + \bar{\mu}\zeta}\right)^2$$

$$= \left(\frac{1 + \bar{\mu}z}{1 + \bar{\sigma}z}\right)^2,$$

where

$$\sigma = \lambda \frac{1 + \bar{\lambda}\zeta}{1 + \bar{\mu}\zeta} \cdot \frac{1 + \mu\bar{\zeta}}{1 + \lambda\bar{\zeta}}. \tag{2.10}$$

When $|\sigma| = 1$ but $\sigma \neq \mu$, we can solve (2.10) for ζ with $|\zeta| < 1$. Take $\sigma_k \to \mu$ and $\alpha_{n,k}$ so that $\lim_{n\to\infty} \alpha_{n,k} = \mu$ and $\lim_{n\to\infty} \psi_{\alpha_{n,k}} = \psi^{(\sigma_k)}$. Then a diagonalization argument will give (2.9) with $\lambda = \mu$. Hence (a) implies (b).

The equivalence of (b) and (c) is nothing but the Carathéodory convergence theorem; see Tsuji [1959], page 381.   ∎

We remark that if (2.7) fails for $M = 1$ then it fails for all $M < \infty$ by Theorems 1.6 and 2.1.

Theorem 2.4 is from Jones [1989]. Exercise 2 gives a result from Pommerenke [1978] that complements Theorems 2.3 and 2.4.

Together Theorems 1.3 and 2.1 set up a one-to-one correspondence

$$\log \varphi' = (u_h + i\tilde{u}_h)', \tag{2.11}$$

between univalent functions φ and small norm Zygmund functions h. This correspondence was first noticed by Duren, Shapiro, and Shields [1966]. We give two applications.

Example 2.5. *There exists a Jordan domain Ω such that the conformal mapping $\varphi : \mathbb{D} \to \Omega$ has a non-zero angular derivative at no point.*

Proof. Actually we will give two examples. For the first example, let

$$\log(\varphi') = \varepsilon \sum z^{2^n}.$$

If ε is small, then φ maps $\mathbb{D}$ to a quasidisc; and by the remark following Theorem 1.3, φ has an angular derivative at no $\zeta \in \partial\mathbb{D}$.

The second example is the von Koch snowflake, Example VI.4.3. The conformal map from $\mathbb{D}$ to the interior domain Ω has a non-zero angular derivative at no point of $\partial\Omega$. Indeed if there is an inner tangent at ζ, then ζ belongs to Γ_n for some n, as shown in Example VI.4.3. If ζ is a vertex of Γ_n, then Ω either contains a truncated cone with opening $4\pi/3$ or contains no truncated cone with opening greater than $\pi/3$ at ζ. If ζ is not a vertex then Ω contains the union of a half disc and an equilateral triangle where the disc, centered at ζ, can be arbitrarily small and the size of the equilateral triangle is comparable to the diameter of the disc. See Figure VII.5. By the easy half of Ostrowski's Theorem V.5.5, Ω does not have a nonzero angular derivative at ζ. In fact the conformal map of the unit disc onto Ω is not conformal at any point of $\partial\mathbb{D}$. ∎

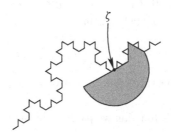

Figure VII.5 Union of a half disc and triangle.

Duren, Shapiro, and Shields [1966] introduced (2.11) in order to construct a Jordan domain that is not a Smirnov domain. If Ω is a Jordan domain with rectifiable boundary Γ the conformal map $\varphi : \mathbb{D} \to \Omega$ satisfies $\varphi' \in H^1$ and

$$\log |\varphi'(z)| = \int_{\partial\mathbb{D}} P_z(e^{i\theta}) \log |\varphi'(e^{i\theta})| d\theta - \int_{\partial\mathbb{D}} P_z(e^{i\theta}) d\mu_s(\theta),$$

where the singular measure μ_s is the weak-star limit

$$\mu_s = \lim_{r \to 1} \left(\log |\varphi'(e^{i\theta})| d\theta - \log |\varphi'(re^{i\theta})| d\theta \right).$$

Because $\varphi' \in H^1$,

$$\log |\varphi'(z)| \leq \int P_z(e^{i\theta}) \log |\varphi'(e^{i\theta})| d\theta$$

by Theorem A.1 and it follows that $\mu_s \geq 0$. The domain Ω is called a **Smirnov domain** if $\mu_s = 0$, or equivalently if

$$\log |\varphi'(z)| = \int \log |\varphi'(e^{i\theta})| P_z(e^{i\theta}) d\theta \qquad (2.12)$$

for all (or one) $z \in \mathbb{D}$. Smirnov domains are important in approximation theory because Ω is a Smirnov domain if and only if, for $1 \leq p < \infty$, the analytic polynomials

$$a_0 + a_1 z + a_2 x^2 + \dots$$

are $L^p(\Gamma, ds)$ dense in the space, called $H^p(\Omega)$, of analytic functions satisfying $(f \circ \varphi)(\varphi')^{1/p} \in H^p(\mathbb{D})$. See Duren [1970], p. 173.

Example 2.6. *There exists a Jordan domain Ω with rectifiable boundary such that Ω is not a Smirnov domain.*

Proof. Kahane [1969] exhibited an increasing function $h \in Z^*$ such that $h' = 0$ almost everywhere. A different construction appears in Piranian [1966]. We may assume $\|h\|_{Z^*}$ is small. Then the map φ defined by (2.11) has image $\varphi(\mathbb{D})$ bounded by a rectifiable Jordan curve and

$$\log |\varphi'(z)| = \int_{\partial \mathbb{D}} P_z(\theta) d\mu(\theta),$$

where μ is the singular measure having primitive h. Hence $\varphi(\mathbb{D})$ is not a Smirnov domain.

To construct Kahane's function we partition the unit interval into 4^n subintervals,

$$I_{n,j} = [j4^{-n}, (j+1)4^{-n}),$$

and define a sequence μ_n of probability measures. Let $d\mu_0 = dx$. When $n \geq 0$, assume $d\mu_n$ is absolutely continuous with constant density $d_{n,j} \geq 0$ on each $I_{n,j}$. Write

$$I_{n,j} = I_{n+1,4j} \cup I_{n+1,4j+1} \cup I_{n+1,4j+2} \cup I_{n+1,4j+3}.$$

If $d_{n,j} > 0$, define

$$d_{n+1,k} = \begin{cases} d_{n,j} - 1, & \text{if } k = 4j, \ 4j+3; \\ d_{n,j} + 1, & \text{if } k = 4j+1, \ 4j+2. \end{cases}$$

If $d_{n,j} = 0$, define $d_{n+1,k} = 0$ for all $I_{n+1,k} \subset I_{n,j}$. Then the sequence μ_n converges weak-star, because $\mu_{n+1}(I_{n,j}) = \mu_n(I_n)$. Take $\mu = \lim \mu_n$ and

$$h(\theta) = \mu([0, \frac{\theta}{2\pi})).$$

If S_n is the closed support of μ_n, then $S_{n+1} \subset S_n$ and $\left| \bigcap S_n \right| = 0$ because an unbiased random walk on the integers which begins at 1 almost surely hits 0. See for example Feller [1968]. Therefore h is singular.

To see that $h \in Z^*$, note that whenever I and J are adjacent intervals of the same length,

$$|\mu(I) - \mu(J)| \leq C|I|. \tag{2.13}$$

By construction, (2.13) holds whenever $I = I_{n,j}$ and $J = I_{n,j+1}$, even though $I_{n,j}$ and $I_{n,j+1}$ may fall in different $I_{n-1,k}$. In general, let $4^{-n} \leq |I|/2 < 4^{-n+1}$, and take $I_{n,j} \subset I$ and $I_{n,k} \subset J$. Then by the 4-adic case of (2.13),

$$|\mu(I_{n,j}) - \mu(I_{n,k})| \leq 2C4^{-n},$$

$\mu(I) - \mu(I_{n,j}) \leq 4C4^{-n}$, and $\mu(J) - \mu(I_{n,k}) \leq 4C4^{-n}$. That establishes (2.13) for general I and J and hence $h \in Z^*$. ∎

3. Quasicircles

We first describe without proof some of the magical properties of quasiconformal mappings and quasicircles. The proofs can be found in the delightful books of Ahlfors [1966] and Lehto and Virtanen [1973], and some are outlined in the exercises. We end the section with the Jerison and Kenig characterization of quasicircles in terms of harmonic measure.

Let Ω and Ω' be domains in the extended plane $\mathbb{C}^*$, let $f : \Omega \to \Omega'$ be an orientation preserving homeomorphism, and let $K \geq 1$. Then we say f is a **K-quasiconformal mapping** if

(a) f is absolutely continuous on almost every horizontal or vertical line segment in Ω, and

(b) the (area almost everywhere defined) derivatives

$$f_z = \frac{f_x - if_y}{2} \quad \text{and} \quad f_{\bar{z}} = \frac{f_x + if_y}{2}$$

satisfy

$$|f_{\bar{z}}| \leq \frac{K-1}{K+1}|f_z| \tag{3.1}$$

almost everywhere on Ω.

When condition (a) holds we say f is ACL, for **absolutely continuous on lines.** The smallest constant K for which (3.1) holds is called the **dilatation** of f on Ω.

Let $f : \Omega \to \Omega'$ be K-quasiconformal, let $z \in \Omega \setminus \{\infty, f^{-1}(\infty)\}$, and suppose $B(z, r) \subset \Omega$. Define

$$m(z, r) = \inf_{|w-z|=r} |f(w) - f(z)|,$$

$$M(z, r) = \sup_{|w-z|=r} |f(w) - f(z)|,$$

and

$$H(z) = \limsup_{r \to 0} \frac{M(z, r)}{m(z, r)}.$$

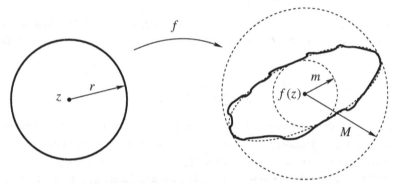

Figure VII.6 The quasiconformal image of a small circle.

When f is ACL it follows from (3.1) that

$$H(z) \le K \tag{3.2}$$

almost everywhere on Ω. Conversely, f is K-quasiconformal if f is an orientation preserving ACL homeomorphism for which (3.2) holds almost everywhere. The equivalence of (3.1) and (3.2) is easy to prove when f is an orientation preserving diffeomorphism, but the general case requires the differentiation theorem of Gehring and Lehto [1959]. Surprisingly, Heinonen and Koskela [1995] derived (3.1) from a weaker form of (3.2).

If f is a K-quasiconformal, then f^{-1} is also K-quasiconformal, and f is a 1-quasiconformal mapping if and only if f is a conformal mapping.

Let f be a K-quasiconformal on a domain Ω, let Γ be a path family in Ω, and write $f(\Gamma) = \{f(\gamma) : \gamma \in \Gamma\}$. Then (3.1) implies

$$\frac{1}{K}\lambda_\Omega(\Gamma) \le \lambda_{f(\Omega)}(f(\Gamma)) \le K\lambda_\Omega(\Gamma). \tag{3.3}$$

Therefore a K-quasiconformal mapping can only increase or decrease an extremal distance by the factor K,

$$\frac{1}{K}d_\Omega(E, F) \le d_{f(\Omega)}(f(E), f(F)) \le K d_\Omega(E, F),$$

whenever f is K-quasiconformal on a neighborhood of $\overline{\Omega}$. It follows that a K-quasiconformal mapping can only increase or decrease the module of a ring domain $W \subset \overline{\Omega}$ by the same factor K,

$$\frac{1}{K}\text{mod}(f(W)) \le \text{mod}(W) \le K\text{mod}(f(W)). \tag{3.4}$$

Conversely, if f is an orientation preserving homeomorphism defined on an open set U and if (3.4) holds for every ring $W \subset U$, then f is K-quasiconformal on U. Again it is not hard to prove that an orientation preserving diffeomorphism is quasiconformal if and only if it satisfies (3.4), but in the general case the argument is much deeper.

One consequence of (3.3) is the large-scale version of (3.2):

Proposition 3.1. *If $f : \mathbb{C} \to \mathbb{C}$ is K-quasiconformal, then for all $z \in \mathbb{C}$ and all $r > 0$,*

$$M(z, r) \le e^{8K}m(z, r).$$

In other words, if $|z_1 - z| \le |z_2 - z|$, then

$$|f(z_1) - f(z)| \le e^{8K}|f(z_2) - f(z)|.$$

The proof of Proposition 3.1 is outlined in Exercise 3.

If $f : \mathbb{C} \to \mathbb{C}$ is K-quasiconformal and if $f(\partial\mathbb{D}) = \partial\mathbb{D}$, then (3.4) implies there is a constant $M = M(K)$ such that

$$\frac{1}{M} \le \frac{|f(e^{i(\theta+t)}) - f(e^{i\theta})|}{|f(e^{i\theta}) - f(e^{i(\theta-t)})|} \le M, \tag{3.5}$$

for all θ and all $t < \pi$. The converse is also true. See Beurling and Ahlfors [1956], Ahlfors [1966], or the sketch in Exercise 4. Homeomorphisms of $\partial\mathbb{D}$ satisfying (3.5) are called **quasisymmetric functions**.

Theorem 3.2. *Let f be an orientation preserving homeomorphism of $\partial\mathbb{D}$ to $\partial\mathbb{D}$. Then f is the restriction to $\partial\mathbb{D}$ of a quasiconformal homeomorphism of $\mathbb{C}$ to $\mathbb{C}$ if and only if (3.5) holds. If (3.5) holds with constant M then the extension satisfies (3.1) with $K \le K(M)$, and if f is K-quasiconformal and $f(\partial\mathbb{D}) = \partial\mathbb{D}$, then (3.5) holds with $M \le M(K)$.*

Let $f : \mathbb{C} \to \mathbb{C}$ be K-quasiconformal. It then follows easily from (3.4) that $\Gamma = f(\partial\mathbb{D})$ is a quasicircle with constant $A \le A(K)$. The converse holds also.

Theorem 3.3. *Let Γ be a Jordan curve in the plane. Then Γ satisfies the Ahlfors condition (2.5) if and only if*

$$\Gamma = f(\partial \mathbb{D}), \text{ where } f : \mathbb{C} \to \mathbb{C} \text{ is } K\text{-quasiconformal} . \qquad (3.6)$$

If (2.5) holds with constant A, then (3.6) holds with $K \leq K(A)$, and if (3.6) holds with constant K, then (2.5) holds with $A \leq A(K)$.

See Exercise 5 for a proof of Theorem 3.3.

Corollary 3.4. *Let Ω be a quasidisc and let $\varphi : \mathbb{D} \to \Omega$ be a conformal mapping. Then φ has an extension to a quasiconformal mapping from $\mathbb{C}$ onto $\mathbb{C}$.*

Proof. Let f be a quasiconformal mapping with $\Omega = f(\mathbb{D})$ and $f(0) = \varphi(0)$. Set $F = f^{-1} \circ \varphi$. Then F is a homeomorphism of $\overline{\mathbb{D}}$ and F is quasiconformal on $\mathbb{D}$. Extend F to $\mathbb{C} \setminus \overline{\mathbb{D}}$ by reflection

$$\overline{F(z)} = \frac{1}{F(\frac{1}{\bar{z}})}.$$

Then F is a quasiconformal self map of $\mathbb{C}$ and $f \circ F$ is the desired quasiconformal extension of φ. ∎

A positive measure μ on a Jordan curve Γ is a **doubling measure** if there is a constant C such that

$$\mu(I) \leq C\mu(J)$$

whenever I and J are adjacent subarcs of Γ with $\mathrm{diam}(I) \leq \mathrm{diam}(J)$. Interchanging I and J, we see that μ is doubling if and only if

$$\mathrm{diam}(I) \leq C'\mathrm{diam}(J)$$

whenever I and J are adjacent arcs with $\mu(I) \leq \mu(J)$. If $f : \partial \mathbb{D} \to \partial \mathbb{D}$ is a homeomorphism, then f satisfies the Beurling–Ahlfors condition (3.5) if and only if $\mu(E) = |f(E)|$ is a doubling measure on $\partial \mathbb{D}$. Kahane's measure in Example 2.6 is a doubling measure on $\mathbb{R}$ singular to Lebesgue measure. From Jerison and Kenig [1982a] we have the following theorem.

Theorem 3.5. *Suppose Γ is a Jordan curve in the plane and suppose Ω_1 and Ω_2 are the two components of the complement $\mathbb{C}^* \setminus \Gamma$. Let $z_j \in \Omega_j$ and let $\omega_j(E) = \omega(z_j, E, \Omega_j)$. Then Γ is a quasicircle if and only if both ω_1 and ω_2 are doubling measures on Γ.*

Proof. Assume ω_1 and ω_2 are doubling measures and assume $z_2 = \infty \in \Omega_2$. Let $\varphi_1 : \mathbb{D} \to \Omega_1$ and $\varphi_2 : \mathbb{C}^* \setminus \overline{\mathbb{D}} \to \Omega_2$ be conformal mappings such that

$\varphi_2(\infty) = \infty$ and $\varphi_1(0) = z_1$, and define the **welding map** $h : \partial \mathbb{D} \to \partial \mathbb{D}$ by

$$h = \varphi_2^{-1} \circ \varphi_1.$$

Let I and J be adjacent subarcs of $\partial \mathbb{D}$ such that $|I| = |J| \le \pi$. Because ω_1 is doubling,

$$\text{diam}(\varphi_1(J)) \le C_1 \text{diam}(\varphi_1(I)).$$

Then because ω_2 is doubling,

$$\frac{|\varphi_2^{-1}(\varphi_1(J))|}{2\pi} = \omega_2(\varphi_1(J)) \le C_2 C_1 \omega_2(\varphi_1(I)) = \frac{|\varphi_2^{-1}(\varphi_1(I))|}{2\pi}.$$

Consequently

$$\frac{|h(J)|}{|h(I)|} \le C_1 C_2,$$

and after interchanging I and J we see that h satisfies (3.5). Now let $H : \mathbb{C} \to \mathbb{C}$ denote the quasiconformal extension of h given by Corollary 3.4 and define

$$\Phi_1(z) = \begin{cases} \varphi_1(z), & \text{if } |z| \le 1, \\ \varphi_2(H(z)), & \text{if } |z| > 1. \end{cases}$$

Then Φ_1 is a quasiconformal map such that $\Phi_1(\partial \mathbb{D}) = \Gamma$, and Γ is a quasicircle.

Conversely, assume Γ is a quasicircle and let $\varphi : \mathbb{D} \to \Omega_1$. Let I_1 and I_2 be adjacent arcs of Γ with $\text{diam}(I_1) = \text{diam}(I_2) \le \text{diam}(\Gamma)/4$. Let I_3 be an arc adjacent to I_2 with $\text{diam}(I_3) = \text{diam}(I_2)$ and $I_3 \cap I_1 = \emptyset$, and let w_j, w_{j+1} be the endpoints of I_j, $j = 1, 2, 3$. Then by the Ahlfors condition (2.5),

$$\text{dist}(I_1, I_3) \le |w_2 - w_3| \le A\text{diam}(I_2) = A\text{diam}(I_1) = A\text{diam}(I_3),$$

so that by Exercise 6(b) of Chapter IV and a rescaling,

$$d_{\mathbb{C}}(I_1, I_3) \le A'.$$

Set $J_k = \varphi^{-1}(I_k)$. We may suppose that $\omega(0, \cup_k J_k, \mathbb{D}) \le 1/2$. Then by Corollary 3.4 and (3.4), $d_{\mathbb{C}}(J_1, J_3) \le KA'$. Hence $\text{diam}(J_2) \le C \, \text{diam}(J_1)$, for otherwise an annulus of large modulus separates J_1 and J_3. Thus ω_1 is doubling. The proof for ω_2 is of course the same. ∎

The proof of doubling in Jerison and Kenig [1982a] is a different argument that applies to higher dimensional generalizations of quasidiscs known as **nontangentially accessible domains**. See Exercise 9 for a Jordan curve on which ω_2 is doubling but ω_1 is not doubling. A one-sided version of the definition of quasicircle yields domains called John domains, which are discussed in Exercise 13.

4. Chord-Arc Curves and the A^∞ Condition

A compact set K is called **Ahlfors regular** if there is a constant A such that

$$\Lambda_1(K \cap B(z, r)) \le Ar$$

for all $z \in K$. A Jordan curve Γ is called a **chord-arc curve** or a **Lavrentiev curve** if Γ is an Ahlfors regular quasicircle. Thus Γ is a chord-arc curve if and only if Γ is rectifiable and there is a constant A' such that for all $w_1, w_2 \in \Gamma$,

$$\frac{\ell(w_1, w_2)}{|w_1 - w_2|} \le A', \tag{4.1}$$

where $\ell(w_1, w_2)$ is the length of the shorter arc of Γ joining w_1 to w_2.

Figure VII.7 Chord-arc curve.

Note how (4.1) is the rectifiable analogue of (2.5). Curves satisfying (4.1) were introduced by Lavrentiev in [1936]. A Jordan curve Γ is called a **Lipschitz curve** if

$$\Gamma = \{e^{F(\theta) + i\theta} : 0 \le \theta \le 2\pi\},$$

where $F(\theta + 2\pi) = F(\theta)$ and $|F(\theta + t) - F(\theta)| \le A|t|$ for all t. The smallest constant A is called the **Lipschitz constant** for Γ. It is clear that every Lipschitz curve is a chord-arc curve. The object of this section is to prove a result, parallel to Theorem 3.5, that characterizes chord-arc curves via harmonic measure.

Let μ and ν be continuous positive measures on a curve Γ. We say μ and ν are A^∞**-equivalent** if there exists $\varepsilon < 1$ and $\eta < 1$ such that whenever I is a subarc of Γ and $E \subset I$,

$$\frac{\mu(E)}{\mu(I)} \le \varepsilon \implies \frac{\nu(E)}{\nu(I)} \le \eta. \tag{4.2}$$

To see that (4.2) defines an equivalence relation, simply replace E by $I \setminus E$. It is not hard to show that $\nu \ll \mu \ll \nu$ if μ and ν are A^∞-equivalent. If Γ is a chord-arc curve and if μ is A^∞-equivalent to arc length on Γ, then μ is a doubling measure. However, not every doubling measure on the unit circle

is A^∞-equivalent to arc length and Example 2.6 exhibits a doubling measure
singular to Lebesgue measure. A stopping time argument given in Exercise
14 shows that A^∞ is equivalent to this stronger looking condition: There exist
$\alpha > 0$ and $C > 0$ such that for all arcs $I \subset \Gamma$ and all $E \subset I$,

$$C^{-1} \left(\frac{\nu(E)}{\nu(I)} \right)^{\frac{1}{\alpha}} \leq \frac{\mu(E)}{\mu(I)} \leq C \left(\frac{\nu(E)}{\nu(I)} \right)^{\alpha}.$$

When $\mu = k(\theta)d\theta$ and $\nu = d\theta$ are A^∞-equivalent, we say k is an A^∞-**weight**
and write $k(\theta) \in A^\infty$.

Now let Γ be a rectifiable curve and let $\mu = k(z)ds$ be a positive finite
measure on Γ. We say $k(z)$ satisfies a **reverse Hölder condition** if there exist
$\delta > 0$ and $C > 0$ such that whenever $I \subset \Gamma$ is a subarc having length $\ell(I)$,

$$\left(\frac{1}{\ell(I)} \int_I k(z)^{1+\delta} ds \right)^{\frac{1}{1+\delta}} \leq C \frac{1}{\ell(I)} \int_I k(z)ds.$$

When $p < \infty$ we say $k(z)$ is an A^p-**weight** if there is $C > 0$ such that for all
arcs $I \subset \Gamma$,

$$\frac{1}{\ell(I)} \int_I k(z)ds \left(\frac{1}{\ell(I)} \int_I \left(\frac{1}{k(z)} \right)^{\frac{1}{p-1}} ds \right)^{p-1} \leq C. \qquad (4.3)$$

The A^p-condition (4.3) is important because of the famous theorems of Muckenhoupt and of Hunt, Muckenhoupt, and Wheeden telling us that when μ is
a measure on the unit circle and $1 < p < \infty$, the Hardy–Littlewood maximal
operator $f \to Mf$ or the conjugation operator $f \to \tilde{f}$ is bounded on $L^p(\mu)$ if
and only if $\mu = kdx$ where k is an A^p-weight. See Muckenhoupt [1972], Hunt,
Muckenhoupt and Wheeden [1973], or Garnett [1981].

Proposition 4.1. *Let Γ be a rectifiable curve and let $\mu = k(z)ds$ be a finite
positive measure on Γ. Then the following conditions are equivalent:*
(a) *μ and arc length measure ds are A^∞-equivalent,*
(b) *$k(z)$ satisfies a reverse Hölder condition, and*
(c) *There is p, $1 < p < \infty$, such that $k(z)$ is an A^p-weight.*
*There exist $\delta(\varepsilon, \eta)$, $p(\varepsilon, \eta)$, and $C = C(\varepsilon, \eta)$ such that when (a) holds with
constants ε and η, (b) holds for C and all $\delta < \delta(\varepsilon, \eta)$ and (c) holds for C and
all $p \geq p(\varepsilon, \eta)$.*

Some parts of the proof of Proposition 4.1 are easy applications of Hölder's
inequality but at one point a stopping time is needed. See Exercise 14, Coifman
and Fefferman [1974], Garnett [1981], or Torchinsky [1986].

Theorem 4.3 will show that every chord-arc curve satisfies the conditions of
Proposition 4.1. The easier Theorem 4.2 shows that Lipschitz curves satisfy a

strong form of condition (c).

Theorem 4.2. *Let Ω be a domain bounded by a Lipschitz curve Γ (as above), let $\varphi : \mathbb{D} \to \Omega$ be a conformal mapping, and let $\omega(E) = \omega(0, E, \Omega)$. Then there is a $p < 2$, depending only on the Lipschitz constant of Γ, such that $|\varphi'|$ and $1/|\varphi'|$ are A^p-weights on $\partial\mathbb{D}$ and $k = \dfrac{d\omega}{ds}$ is an $A^{\frac{1}{2-p}}$-weight on Γ.*

Proof. Let $\varphi : \mathbb{D} \to \Omega$ with $\varphi(0) = 0$. Then almost everywhere the magnitude of the angle between the tangent vector to Γ and the radius vector is

$$\left| \arg\left(\frac{e^{i\theta} \varphi'(e^{i\theta})}{\varphi(e^{i\theta})} \right) \right| \leq |\arg(1 + iA)| = \alpha < \frac{\pi}{2},$$

and so by Zygmund's theorem, Theorem II.3.1,

$$\sup_{a \in \mathbb{D}} \left(\int \left| \frac{z\varphi'}{\varphi} \right|^\lambda P_a d\theta \right)^{\frac{1}{\lambda}} \left(\int \left| \frac{\varphi}{z\varphi'} \right| P_a d\theta \right)$$

$$= \sup_{T \in \mathcal{M}} \left(\int \left| \frac{z\varphi'}{\varphi} \circ T \right|^\lambda dt \right)^{\frac{1}{\lambda}} \left(\int \left| \frac{\varphi}{z\varphi'} \circ T \right| dt \right) < \infty,$$

for all $\lambda < \pi/(2\alpha)$. If I is an interval in $\partial\mathbb{D}$, take $a \in \mathbb{D}$ such that $a/|a|$ is the center of I and $1 - |a| = |I|/2$. Then $P_a \geq C/|I|$ on I and since $|\varphi(z)/z|$ is bounded above and below on $\partial\mathbb{D}$,

$$\sup_I \left(\frac{1}{|I|} \int_I |\varphi'|^\lambda dt \right)^{\frac{1}{\lambda}} \left(\frac{1}{|I|} \int_I \left| \frac{1}{\varphi'} \right| dt \right) < \infty.$$

This shows that $|1/\varphi'| \in A^p$ with $p = 1 + 1/\lambda > 1 + 2\alpha/\pi$ and a similar argument shows that $|\varphi'| \in A^p$. Because $k = 1/|\varphi' \circ \varphi^{-1}|$, we obtain for $J = \varphi(I)$

$$\left(\frac{1}{\ell(J)} \int_J k\, ds \right) \left(\frac{1}{\ell(J)} \int_J \left(\frac{1}{k} \right)^{\frac{1}{q-1}} ds \right)^{q-1}$$

$$= \frac{\left(\frac{1}{|I|} \int_I |\varphi'|^{\frac{q}{q-1}} d\theta \right)^{q-1}}{\left(\frac{1}{|I|} \int_I |\varphi'| d\theta \right)^q}$$

$$\leq \left[\left(\frac{1}{|I|} \int_I \frac{1}{|\varphi'|} d\theta \right) \left(\frac{1}{|I|} \int_I |\varphi'|^{\frac{q}{q-1}} d\theta \right)^{\frac{q-1}{q}} \right]^q < \infty,$$

and $k \in A^q$ for $q > \pi/(\pi - 2\alpha)$. ∎

Theorem 4.3 (Lavrentiev). *Let* Ω *be the bounded domain bounded by a chord-arc curve* Γ, *let* $z_0 \in \Omega$, *and let* $\omega(E) = \omega(z_0, E, \Omega)$.

(a) *Then* ω *and arc length on* Γ *are* A^∞-*equivalent, with constants depending only on the chord-arc constant of* Γ *and* $\mathrm{dist}(z_0, \partial\Omega)/\mathrm{diam}(\Omega)$.

(b) *Moreover, if* $\varphi : \mathbb{D} \to \Omega$ *is a conformal mapping with* $\varphi(0) = z_0$, *then arc length and harmonic measure for* z_0 *on the curves* $\varphi(\{|z| = r\}), 0 < r < 1$ *are* A^∞-*equivalent with constants independent of* r.

Proof. We follow Jerison and Kenig [1982b]. To prove (a), let I be a small arc on $\partial\mathbb{D}$, form the box

$$Q = \{z = t\zeta : 1 - |I| < t < 1; \; \zeta \in I\},$$

and let z_Q be its center. Given that φ has a quasiconformal extension to $\mathbb{C}$, Proposition 3.1 tells us there is $r > 0$ such that

$$B(\varphi(z_Q), r) \subset \varphi(Q) \subset B(\varphi(z_Q), Cr), \tag{4.4}$$

where C depends only on the quasiconformality of the extension of φ. Since Γ is a chord-arc curve,

$$\ell(\varphi(I)) \leq A\,\mathrm{diam}(\varphi(I)) \leq A'r.$$

Let $J = (1 - |I|)I$ be the top edge of Q. By the Koebe estimates (I.4.13),

$$\ell(\varphi(J)) = \int_J |\varphi'(z)|ds \leq 4\mathrm{dist}(\varphi(J), \Gamma) \leq A'r.$$

Now let $L = (\zeta, \zeta_0)$, $\zeta_0 = (1 - |I|)\zeta$, be one of the edges in $\partial Q \setminus (I \cup J)$ of Q. Let $c = e^{8K}$, where K is the dilatation of the quasiconformal extension of φ from $\mathbb{C}$ to $\mathbb{C}$, and define $\zeta_k \in L$ to be the point of smallest modulus such that

$$|\varphi(\zeta_k) - \varphi(\zeta)| = c^{-k}|\varphi(\zeta_0) - \varphi(\zeta)|.$$

If $2^N \geq c$, then

$$|\varphi(\zeta_k) - \varphi(\zeta)| \leq 2^N |\varphi(\zeta_{k+1}) - \varphi(\zeta)|,$$

and because φ^{-1} also has a K-quasiconformal extension, Proposition 3.1 and induction yield

$$|\zeta_k - \zeta| \leq (2c)^N |\zeta_{k+1} - \zeta|. \tag{4.5}$$

Write $L_k = [\zeta_{k+1}, \zeta_k]$. Then by Proposition 3.1 and the definition of ζ_{k+1},

$$\mathrm{dist}(\varphi(L_k), \Gamma) \leq c|\varphi(\zeta_k) - \varphi(\zeta)| = c^{-k+1}|\varphi(\zeta_0) - \varphi(\zeta)|,$$

so that by the Koebe estimates

$$\ell(\varphi(L_k)) \le 4c^{-k+1}|\varphi(\zeta_0) - \varphi(\zeta)| \log\left(\frac{|\zeta_k - \zeta|}{|\zeta_{k+1} - \zeta|}\right).$$

With (4.5) we then obtain

$$\ell(\varphi(L)) \le |\varphi(\zeta_0) - \varphi(\zeta)| \sum_k 4c^{-k+1} N \log(2c) \le Cr.$$

Thus we have established

$$\ell(\partial\varphi(Q)) \le C'r. \tag{4.6}$$

With (4.6) and (4.4) we can apply Lavrentiev's estimate (VI.5.1) to the domain $\varphi(Q)$ and the point $\varphi(z_Q)$, and the bounds will not depend on Q. Let $E \subset \varphi(I)$. Given $\varepsilon > 0$ there exists $\delta > 0$ such that if $\frac{\ell(E)}{\ell(\varphi(I))} < \delta$ then by (VI.5.1)

$$\frac{\omega(z_Q, \varphi^{-1}(E), Q)}{\omega(z_Q, I, Q)} = \frac{\omega(\varphi(z_Q), E, \varphi(Q))}{\omega(\varphi(z_Q), \varphi(I), \varphi(Q))} < \varepsilon. \tag{4.7}$$

But if (4.7) holds with suitably small ε, then

$$\frac{\omega(z_0, E, \Omega)}{\omega(z_0, \varphi(I), \Omega)} = \frac{|\varphi^{-1}(E)|}{|I|} < 1/2,$$

and we conclude that ω and ds are A^∞-equivalent on $\partial\Omega$.

(b) By Lemma 4.4 below, the function

$$I(z) = \log \int_\alpha^\beta |\varphi'(ze^{i\theta})|d\theta + \log \frac{|z|/A'}{|\varphi(ze^{i\beta}) - \varphi(ze^{i\alpha})|}$$

is subharmonic on $\mathbb{D}$ for $A' > 0$. Because Γ is chord-arc, we can choose A' so that $I \le 0$ on $\partial\mathbb{D}$. Then by the maximum principle, $I(r) \le 0$, and therefore $\varphi(\{|z| = r\})$ is a chord-arc curve with a constant independent of r. ∎

Lemma 4.4. *If $f(z)$ is analytic and nowhere zero on $\mathbb{D}$ and if $0 \le \alpha < \beta < 2\pi$ then*

$$J(z) = \log \int_\alpha^\beta |f(ze^{i\theta})|d\theta$$

is subharmonic on $\mathbb{D}$.

Proof. Write $f = g^2$. Uniformly on compact sets, the function $J(z)$ is a limit of functions of the form

$$K(z) = \log \sum_{j=1}^n |g_j|^2,$$

and a calculation with Cauchy–Schwarz shows that

$$\frac{\partial^2}{\partial \bar{z} \partial z} K(z) \geq 0.$$

∎

Lemma 4.4 is also true without the assumption that f is nowhere zero, but the calculation is a bit more tedious.

Corollary 4.5. *Let Ω be a bounded domain bounded by a chord-arc curve and let $\varphi : \mathbb{D} \to \Omega$ be a conformal map. Then*
(a) *Ω is a Smirnov domain, and*
(b) *There is $q > 0$ such that $\frac{1}{\varphi'} \in H^q$.*
The index q in (b) depends only on the chord-arc constant of $\partial\Omega$ and on the dilatation of the quasiconformal extension of φ.

Proof. We first prove (b). By Theorem 4.3, the measures $d\theta$ and $|\varphi'(re^{i\theta})|d\theta$ are A^∞-equivalent, with constants independent of r. Thus by part (c) of Proposition 4.1, there is $q = \frac{1}{p-1}$, depending only on the constants of $\partial\Omega$, such that

$$\sup_r \int \left| \frac{1}{\varphi'(re^{i\theta})} \right|^q \frac{d\theta}{2\pi} < \infty$$

and (b) holds.

Recall that Ω is a Smirnov domain if and only if

$$\log |\varphi'(0)| = \int_{\partial\mathbb{D}} \log |\varphi'(e^{i\theta})| \frac{d\theta}{2\pi}.$$

Because $\varphi' \in H^1$, we have

$$\log |\varphi'(0)| \leq \int_{\partial\mathbb{D}} \log |\varphi'(e^{i\theta})| \frac{d\theta}{2\pi}.$$

But $(\varphi')^{-q} \in H^1$ by (b), so that we also have

$$\log |\varphi'(0)| \geq \int_{\partial\mathbb{D}} \log |\varphi'(e^{i\theta})| \frac{d\theta}{2\pi}.$$

Therefore Ω is a Smirnov domain and (a) holds. ∎

Theorem 4.6. *Let Γ be a Jordan curve and let Ω be the bounded component of $\mathbb{C} \setminus \Gamma$. Then Γ is a chord-arc curve if and only if the following three conditions all hold:*
(i) *Γ is a quasicircle,*

(ii) *Arc length on* Γ *is* A^∞*-equivalent to harmonic measure for any* $z \in \Omega$, *and*

(iii) Ω *is a Smirnov domain.*

Examples showing that Theorem 4.6 is sharp are given in Exercise 16.

Proof. We have already seen that (i), (ii), and (iii) hold when Γ is a chord-arc curve. Now assume (i), (ii), and (iii) hold. Let $\varphi : \mathbb{D} \to \Omega$, let $I \subset \partial\mathbb{D}$ be an arc such that $\mathrm{diam}(\varphi(I)) \leq \mathrm{diam}(\Gamma)/2$, let

$$Q = \{z = t\zeta : 1 - |I| < t < 1; \zeta \in I\},$$

and let z_Q be its center. Because ∂Q is a quasicircle and φ has a quasiconformal extension, there exists $r > 0$ and $C > 0$ such that

$$B(\varphi(z_Q), r) \subset \varphi(Q) \subset B(\varphi(z_Q), Cr)$$

and the endpoints w_1 and w_2 of $\varphi(I)$ satisfy

$$|w_1 - w_2| \geq \frac{r}{C}.$$

By (ii) and Proposition 4.1, $|\varphi'|$ satisfies the A^p-condition

$$\frac{1}{|I|}\int_I |\varphi'|d\theta \leq C\Big(\frac{1}{|I|}\int_I \Big|\frac{1}{\varphi'}\Big|^{\frac{1}{p-1}}d\theta\Big)^{1-p}$$

for some $p > 1$. But by Jensen's inequality,

$$\Big(\frac{1}{|I|}\int_I \Big|\frac{1}{\varphi'}\Big|^{\frac{1}{p-1}}d\theta\Big)^{1-p} \leq \exp\Big(\frac{1}{|I|}\int_I \log|\varphi'|d\theta\Big)$$

so that

$$\frac{1}{|I|}\int_I |\varphi'|d\theta \leq C\exp\Big(\frac{1}{|I|}\int_I \log|\varphi'|d\theta\Big)$$
$$\leq C\exp\Big(\int_{\partial\mathbb{D}} \log|\varphi'(e^{i\theta})|P_{z_Q}(\theta)d\theta\Big). \tag{4.8}$$

But by (iii) and (2.12),

$$\log|\varphi'(z_Q)| = \int_{\partial\mathbb{D}} \log|\varphi'(e^{i\theta})|P_{z_Q}(\theta)d\theta,$$

so that

$$\frac{1}{|I|}\int_I |\varphi'|d\theta \leq C|\varphi'(z_Q)| \leq \frac{2Cr}{|I|},$$

by Koebe's estimate. Therefore

$$\ell(\varphi(I)) = \int_I |\varphi'|d\theta \leq 2C^2|w_1 - w_2|$$

and Γ is a chord-arc curve. ∎

The left inequality in (4.8) is a "reverse Jensen's inequality" for A^p-weights. Pommerenke [1991] gives a more geometric proof of Theorem 4.6.

Theorem 4.7. *Let Γ be a quasicircle, let Ω be the bounded component of $\mathbb{C} \setminus \Gamma$, and let $\varphi : \mathbb{D} \to \Omega$ be a conformal mapping. Then Γ is a chord-arc curve if and only if arc length and harmonic measure for $\varphi(0)$ are A^∞-equivalent on $\varphi(\{|z| = r\}), 0 < r < 1$, with constants that are independent of r.*

Proof. If Γ is a chord-arc curve then by Theorem 4.3 arc length and harmonic measure on $\varphi(\{|z| = r\})$ are A^∞-equivalent uniformly in r. Conversely, the A^∞-equivalence implies conditions (i), (ii), and (iii) of Theorem 4.6 hold on $\varphi(\{|z| = r\})$, uniformly in r. Consequently the curves $\varphi(\{|z| = r\})$ are chord-arc uniformly in $r < 1$, and Γ is chord-arc. ∎

5. BMO Domains

Let $f \in L^1(\partial\mathbb{D})$. Recall from Appendix F that f has bounded mean oscillation, $f \in \text{BMO}$, if

$$\sup_I \frac{1}{|I|} \int_I |f - f_I| d\theta = \|f\|_{\text{BMO}} < \infty, \qquad (5.1)$$

where I is a subarc of $\partial\mathbb{D}$ and

$$f_I = \frac{1}{|I|} \int_I f d\theta$$

is the average of f over I. Condition (5.1) is very strong. By the John–Nirenberg theorem (see Exercise F.2), it implies that $f \in L^p(\partial\mathbb{D})$ for all $p < \infty$. Condition (5.1) also implies that if the Poisson integral $f(z)$ is analytic in $\mathbb{D}$, then

$$|f'(z)|(1 - |z|^2) \le C\|f\|_{\text{BMO}}.$$

See Exercise F.3. In other words,

$$\|f\|_{\mathcal{B}} \le C\|f\|_{\text{BMO}}$$

if $f(z)$ is analytic and $f(e^{i\theta}) \in \text{BMO}$. In this section we describe the simply connected domains Ω for which the Bloch function $g = \log(\varphi'(z)) \in \text{BMO}$. A second characterization will be given in Chapter X.

Let $k(\theta) \ge 0$ be a function on the unit circle $\partial\mathbb{D}$ and write $g = \log k$.

Lemma 5.1. *Suppose $k(\theta)d\theta$ is A^∞-equivalent to $d\theta$. Then $g = \log k \in$ BMO. Conversely, if $g \in$ BMO then there is a constant $b > 0$ such that $e^{bg}d\theta$ and $d\theta$ are A^∞-equivalent.*

Proof. Suppose $k(\theta)d\theta$ and $d\theta$ are A^∞-equivalent. Let $I \subset \partial\mathbb{D}$ be a subarc. By Proposition 4.1, k satisfies the A^p-condition (4.3),

$$\Big(\frac{1}{|I|}\int_I \exp(g - g_I)d\theta\Big)\Big(\frac{1}{|I|}\int_I \exp\big(\frac{g_I - g}{p-1}\big)d\theta\Big)^{p-1} \leq C \qquad (5.2)$$

By Jensen's inequality

$$\frac{1}{|I|}\int_I \exp(g - g_I)d\theta \geq 1 \quad \text{and} \quad \Big(\frac{1}{|I|}\int_I \exp\big(\frac{g_I - g}{p-1}\big)d\theta\Big)^{p-1} \geq 1.$$

Thus (5.2) holds if and only if each factor is bounded and we have

$$\exp\Big(\frac{1}{|I|}\int_I (g - g_I)d\theta\Big) \leq \frac{1}{|I|}\int_I \exp(g - g_I)d\theta \leq C'$$

and

$$\exp\Big(\frac{1}{|I|}\int_I \frac{g_I - g}{p-1}d\theta\Big) \leq \Big(\frac{1}{|I|}\int_I \exp\big(\frac{g - g_I}{p-1}\big)d\theta\Big)^{p-1} \leq C'.$$

Therefore

$$\sup_I \frac{1}{|I|}\int_I |g - g_I|d\theta < \infty$$

and $g \in$ BMO.

Conversely if $g \in$ BMO then by the John–Nirenberg theorem (Exercise F.2) there is $b > 0$ such that

$$\sup_I \frac{1}{|I|}\int_I e^{b|g - g_I|}d\theta < \infty.$$

Then $e^{bg} \in A^2$ and $e^{bg}d\theta$ and $d\theta$ are A^∞-equivalent by Proposition 4.1. ∎

Now let $\varphi : \mathbb{D} \to \Omega$ be a conformal mapping and write $g = \log(\varphi')$. We say Ω is a **BMO domain** if $g = \log(\varphi') \in$ BMO. We say Ω is a **chord-arc domain** if Ω is bounded by a chord-arc curve. If Ω is a chord-arc domain, then by Lemma 5.1 and Theorem 4.3, $Reg(e^{i\theta}) = \log|\varphi'(e^{i\theta})| \in$ BMO, so that $g \in$ BMO by Exercise F.3. More generally, Zinsmeister [1984] proved that every domain bounded by an Ahlfors regular Jordan curve is a BMO domain. However, not every Jordan BMO domain has an Ahlfors regular boundary. See Exercise 17.

Theorem 5.2. *Every chord-arc domain is a BMO domain. Conversely, there is $\beta > 0$ such that if $g \in$ BMO and $\|g\|_{\text{BMO}} < \beta$, then Ω is bounded by a chord-arc curve.*

Proof. We have just seen that $g = \log(\varphi') \in$ BMO when Ω is a chord-arc domain. Conversely, since $\|g\|_B \leq C\|g\|_{\text{BMO}}$, Theorem 2.3 gives us a β_0 such that $\partial\Omega$ is a quasicircle if $\|g\|_{\text{BMO}} < \beta_0$. By Lemma 5.1 there is β_1 such that if $\|g\|_{\text{BMO}} < \beta_1$, then $|\varphi'(re^{i\theta})| \in A^\infty$, uniformly in $r < 1$. Thus by Theorem 4.7, Ω is a chord-arc domain if $\|g\|_{\text{BMO}} \leq \min(\beta_0, \beta_1)$. ∎

Theorem 5.3. *The following are equivalent.*
(a) Ω *is a BMO domain.*
(b) *There exist $\delta > 0$ and $C > 0$ such that if $z_0 \in \Omega$ there is a subdomain $\mathcal{U} \subset \Omega$ such that*
   (i) $z_0 \in \mathcal{U}$,
   (ii) $\partial\mathcal{U}$ *is rectifiable and $\ell(\partial\mathcal{U}) \leq C\operatorname{dist}(z_0, \partial\Omega)$, and*
   (iii) $\omega(z_0, \partial\Omega \cap \partial\mathcal{U}, \mathcal{U}) \geq \delta$.
(c) *There exist $\delta > 0$ and $C > 0$ such that if $z_0 \in \Omega$ there is a subdomain $\mathcal{U} \subset \Omega$ such that*
   (i) $z_0 \in \mathcal{U}$ *and $\operatorname{dist}(z_0, \partial\Omega) \leq C\operatorname{dist}(z_0, \partial\mathcal{U})$,*
   (ii) $\partial\mathcal{U}$ *is chord-arc with constant at most C and $\ell(\partial\mathcal{U}) \leq C\operatorname{dist}(z_0, \partial\Omega)$, and*
   (iii) $\ell(\partial\Omega \cap \partial\mathcal{U}) \geq \delta\operatorname{dist}(z_0, \partial\Omega)$.

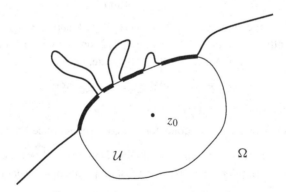

Figure VII.8 Chord-arc subdomain.

A map $G : \mathbb{C} \to \mathbb{C}$ is **bilipschitz** if there is a constant A such that

$$A^{-1}|z_1 - z_2| \leq |G(z_1) - G(z_2)| \leq A|z_1 - z_2|$$

whenever $z_1 \neq z_2$.

Corollary 5.4. *The family of* BMO *domains is invariant under bilipschitz homeomorphisms of the plane.*

Proof. Condition (c) is invariant under bilipschitz mappings.   ∎

Corollary 5.5. *Let* Γ *be a quasicircle, and let* Ω_1 *and* Ω_2 *be the components of* $C^* \setminus \Gamma$. *Then* Ω_1 *is a* BMO *domain if and only if* Ω_2 *is a* BMO *domain.*

Proof. Let $F : \mathbb{C} \to \mathbb{C}$ be a quasiconformal map such that $\Omega_1 = F(\mathbb{D})$. The **quasiconformal reflection** $G : \Omega_1 \to \Omega_2$ is defined by

$$G(z) = F\left(\frac{1}{\overline{(F^{-1}(z))}}\right),$$

and Ahlfors [1963] shows that G can be chosen to be bilipschitz.   ∎

Proof of Theorem 5.3. We show (a) $\Longrightarrow$ (b) $\Longrightarrow$ (c) $\Longrightarrow$ (a).

  (a) $\Longrightarrow$ (b): By (a) we may assume $\varphi^{-1}(z_0) = 0$. Indeed, if $T \in \mathcal{M}$ has $T(0) = \varphi^{-1}(z_0)$ then $\| \log T' \|_{\mathrm{BMO}} <$ Const, so that by the conformal invariance of BMO (see Exercise F.3),

$$g_T(z) = \log\big((\varphi \circ T)'(z)\big) = \log \varphi'(T(z)) + \log T'(z),$$

has $\|g_T\|_{\mathrm{BMO}} \le \|g\|_{\mathrm{BMO}} +$ Const.

  Fix $\alpha > 1$ and recall the nontangential maximal function

$$G(\zeta) = (g - g(0))_\alpha^*(\zeta) = \sup_{\Gamma_\alpha(\zeta)} |g(z) - g(0)|.$$

Let $\lambda > 0$ be large and form $E = \{\zeta : G(\zeta) \le \lambda\}$ and $\mathcal{V} = \bigcup_E \Gamma_\alpha(\zeta)$. If α is sufficiently large, then $\omega(z, E, \mathbb{D}) \le 1/2$ on $\partial \mathcal{V} \setminus E$. It follows that $\omega(0, E, \mathcal{V}) \ge \frac{|E|}{\pi} - 1 \ge \delta$ if λ is sufficiently large. On $\mathcal{V}$,

$$|\varphi'(z)| \le e^\lambda |\varphi'(0))| \le 4e^\lambda \mathrm{dist}(z_0, \partial\Omega),$$

and thus (b) holds for $\mathcal{U} = \varphi(\mathcal{V})$.

  (b) $\Longrightarrow$ (c): This is another cone domain construction. See Exercise 19.

  (c) $\Longrightarrow$ (a): Set $\varphi(0) = z_0$. Since the domain $\mathcal{U}$ in (c) has chord-arc boundary, φ has a non-zero angular derivative at almost all points of $\varphi^{-1}(\partial\mathcal{U} \cap \partial\Omega)$ and by Theorem 4.3,

$$\omega(z_0, \partial\mathcal{U} \cap \partial\Omega, \Omega) \ge \omega(z_0, \partial\mathcal{U} \cap \partial\Omega, \mathcal{U}) > \varepsilon = \varepsilon(C, \delta) > 0.$$

Consequently we can construct a cone domain

$$\mathcal{V} = \bigcup_E \Gamma_\alpha(\zeta) \subset \varphi^{-1}(\mathcal{U})$$

such that $|g(w) - g(0)| \leq M$ on $\mathcal{V}$ and $\omega(0, E, \mathbb{D}) \geq \varepsilon$, where $E \subset \partial \mathbb{D}$ is compact. In particular, $|g(\zeta) - g(0)| \leq M$ on E and $|E| \geq 2\pi\varepsilon$. Let I_1 be a component of $\partial \mathbb{D} \setminus E$ with center c_I and take $w_1 = (1 - \varepsilon|I|)c_I$ and $z_1 = \varphi(w_1)$. Then w_1 has bounded hyperbolic distance to $\mathcal{V}$, so that $|g(w_1) - g(0)| \leq CM$ since $\|g\|_B \leq 6$. Repeating this construction with z_1 in place of z_0, we obtain $E_1 \subset I_1$ with $\omega(w_1, E_1, \mathbb{D}) \geq \varepsilon$ and $|g(\zeta) - g(w_0)| \leq 2CM$ on E_1. Since

$$\omega(w_1, I_1 \cap E_1, \mathbb{D}) \geq \varepsilon - \omega(w_1, \partial \mathbb{D} \setminus I_1, \mathbb{D}) \geq \varepsilon - \frac{2}{\pi}\varepsilon \geq \frac{\varepsilon}{3},$$

we obtain $|I_1 \cap E_1| \geq c_2\varepsilon^2|I_1|$. We repeat this construction on all complementary intervals of E and of the newly constructed sets $E_j \cap I_j$. After n generations we arrive at a set F_n such that $|g(\zeta) - g(0)| \leq nCM$ on F_n and such that $|\partial \mathbb{D} \setminus F_n| \leq |\partial \mathbb{D} \setminus E| (1 - c_2\varepsilon^2)^n$. Thus

$$\omega(0, \{\zeta : |g(\zeta) - g(0)| > \lambda\}, \mathbb{D}) \leq c_3 e^{-c_4\lambda},$$

where c_3 and c_4 depend only on the constants in (c). Applying this argument to $\psi(z) = \varphi(\frac{z+w_0}{1+\overline{w}_0 z})$, we obtain

$$\omega(w_0, \{\zeta \in \partial \mathbb{D} : |g(\zeta) - g(w_0)| > \lambda \text{ and } |\zeta - w_0| \leq 2(1 - |w_0|)\}, \mathbb{D})$$
$$\leq c_5 e^{-c_4\lambda}.$$

That means that $g \in$ BMO. ∎

Notes

Section 1 follows Makarov [1990a]. Anderson, Clunie, and Pommerenke [1974] and Anderson and Pitt [1988] and [1989] have more on Bloch functions. The first Jordan domain not a Smirnov domain was constructed by Keldysh and Lavrentiev [1937]. We have only touched the beautiful theory of quasicircles. Gehring [1982] has much more, including seventeen different characterizations of quasicircles. Chord-arc curves originated in Lavrentiev [1936] and resurfaced in Pommerenke [1977], Coifman and Meyer [1979], and Jerison and Kenig [1982a] and [1982b]. It is an open problem to show that $\{\log \varphi' : \varphi : D \to \Omega \text{ is conformal and } \partial\Omega \text{ is chord-arc}\}$ is a connected subset of BMO. See Semmes [1986b], [1988a] and [1988b], Zinsmeister [1984], [1985], and [1989], Astala and Zinsmeister [1991], MacManus [1994], and Bishop and Jones [1994] for more about chord-arc curves and BMO domains. Theorem 5.3 is from Bishop and Jones [1994].

Exercises and Further Results

1. Let φ be a univalent function in $\mathbb{D}$ and suppose $0 < r < R < 1$.

(a) Use Theorem 2.1 and Exercise I.23(d) to show there is C independent of r and R such that

$$\int_{|\theta|<1-r} |\varphi'(Re^{i(\theta+t)})|d\theta \leq C\frac{(1-r)^3}{(1-R)^3}|\varphi'(re^{it})|.$$

(b) Use (a) to show

$$\int |\varphi'(Re^{it})|dt \leq C\frac{(1-r)^2}{(1-R)^2}\int |\varphi'(re^{it})|dt.$$

2. Let $g \in \mathcal{B}$. We say $g \in \mathcal{B}_0$ the **little Bloch space** if

$$\lim_{|z|\to 1} |g'(z)|(1-|z|^2) = 0.$$

A Jordan curve Γ is **asymptotically conformal** if

$$\sup_{w\in\Gamma(w_1,w_2)} \frac{|w - w_1| + |w - w_2|}{|w_1 - w_2|} \to 0 \text{ as } |w_1 - w_2| \to 0,$$

where $\Gamma(w_1, w_2)$ is the subarc of Γ having endpoints w_1 and w_2 and smaller diameter. Let $g = \log(\varphi')$ where φ is the conformal map from $\mathbb{D}$ to the domain bounded by a Jordan curve Γ. Prove Γ is asymptotically conformal if and only if $g \in \mathcal{B}_0$. (Pommerenke [1978]).

3. (a) The **spherical distance** between $z \in C^*$ and $w \in C^*$ is

$$S(z, w) = \inf_\gamma \int_\gamma \frac{2ds}{1 + |z|^2}$$

with the infimum taken over all rectifiable arcs joining z to w. If $|z| = 1$ then $S(0, z) = S(z, \infty) = \frac{\pi}{2}$.

(b) Let $|z_1| = |z_2| = 1$ and let Γ be the family of curves that separate $\{0, z_1\}$ from $\{z_2, \infty\}$. Using the metric $\rho(z) = \frac{2}{\pi}\frac{1}{1+|z|^2}$, prove

$$\lambda(\Gamma) \geq \frac{\pi}{4}.$$

(c) Use (b) to prove Proposition 3.1. Hint: We may assume $z = f(z) = 0$, $|z_1| = |z_2| = 1$, and $|f(z_1)| < |f(z_2)|$. Let Γ be the set of circles of radius $|f(z_1) < r < |f(z_2)|$. Then

$$\lambda(\Gamma) \leq \frac{1}{2\pi} \log \frac{|f(z_2)|}{|f(z_1)|},$$

while by (b)

$$\lambda(f^{-1}(\Gamma)) \geq \frac{\pi}{4}.$$

This proof is from Gehring [1982].

4. (a) Let $f : \mathbb{C} \to \mathbb{C}$ be quasiconformal and suppose $f(\partial\mathbb{D}) = \partial\mathbb{D}$. Use (3.4) and Exercise IV.5 to show that (3.5) holds for f.

(b) The converse, that f has a quasiconformal extension whenever (3.5) holds, is easier to prove on the line. Let $f : \mathbb{R} \to \mathbb{R}$ be an increasing homeomorphism such that for all $x \in \mathbb{R}$ and all $h > 0$,

$$\frac{1}{M} \leq \frac{f(x+h) - f(x)}{f(x) - f(x-h)} \leq M.$$

Extend F to the plane by defining

$$F(x+iy) = \frac{1}{2} \int_0^1 (f(x+ty) + f(x-ty))dt$$
$$+ \frac{i}{2} \int_0^1 (f(x+ty) - f(x-ty))dt$$

when $y \neq 0$ and $F(x) = f(x)$. Prove $F : \mathbb{C} \to \mathbb{C}$ is a homeomorphism.

(c) Prove F is ACL and satisfies (3.1) almost everywhere.

(d) Prove F is bilipschitz,

$$C^{-1}|z - w| \leq |F(z) - F(w)| \leq C|z - w|,$$

where $C = C(M)$.

See Beurling and Ahlfors [1956] or Ahlfors [1966].

5. Let Γ be a Jordan curve that satisfies the quasicircle condition (2.5) and let I and J be disjoint subarcs of Γ. Let Ω_1 and Ω_2 be the components of $C^* \setminus \Gamma$ with $\infty \in \Omega_2$.

(a) Prove

$$C^{-1}d_{\Omega_1}(I, J) \leq d_{\Omega_2}(I, J) \leq Cd_{\Omega_1}(I, J),$$

where the constant C depends only on the constant in (2.5). Elementary proofs can be found in Ahlfors [1963] and Gehring [1982].

(b) Let $\varphi_1 : \mathbb{D} \to \Omega_1$ and $\varphi_2 : \mathbb{C}^* \setminus \overline{\mathbb{D}} \to \Omega_2$ be conformal mappings such that $\varphi_2(\infty) = \infty$, and let $h : \partial\mathbb{D} \to \partial\mathbb{D}$ be the welding map $h = \varphi_2^{-1} \circ \varphi_1$. Using part (a) and Exercise IV.5 shows that h satisfies the Beurling–Ahlfors M-condition (3.5).

(c) Let $H : \mathbb{C} \to \mathbb{C}$ be the quasiconformal extension of h guaranteed by (b) and Theorem 3.2. Define

$$\Phi_1(z) = \begin{cases} \varphi_1(z), & \text{if } |z| \le 1; \\ \varphi_2(H(z)), & \text{if } |z| > 1. \end{cases}$$

Then Φ_1 is a quasiconformal map such that $\Phi_1(\partial\mathbb{D}) = \Gamma$, and Γ is a quasicircle.

(d) Let $1 < \alpha < 2$. Construct a quasicircle Γ_α such that

$$0 < \Lambda_\alpha(\Gamma_\alpha) < \infty.$$

See Gehring and Väisälä [1973].

(e) Exhibit a Jordan curve of positive area. It is known that every quasicircle has area zero.

6. (a) A domain Ω is an **M-extremal distance domain** (or **M-QED domain**) if whenever E and F are disjoint compact sets in $\overline{\Omega}$,

$$d_\Omega(E, F) \le M d_{\mathbb{C}}(E, F).$$

Note that the reverse inequality with $M = 1$ is trivial. Show the unit disc $\mathbb{D}$ is a 2-QED domain and show the constant 2 is sharp. It then follows by (3.3) that a K–quasidisc is an $M(K)$-QED domain. Now suppose Γ is a Jordan curve and define Ω_1 and Ω_2 as in Exercise 5. Assume that Ω_1 is a QED domain, and let I and J be disjoint arcs on Γ. Then

$$M^{-1} d_{\Omega_1}(I, J) \le d_{\Omega_2}(I, J) \le M d_{\Omega_1}(I, J),$$

in which the second inequality is obtained from taking conjugate extremal distances. From this it follows as in Exercise 5 that Γ is a quasicircle. See Gehring and Martio [1985]and Yang [1992] and [1994] for further results.

(b) Let φ be a conformal mapping from a quasidisc Ω to a simply connected domain W and let $U \subset W$ be a quasidisc. Then $\varphi^{-1}(U) \subset \Omega$ is also a quasidisc. This is easy from (a). See Fernández, Heinonen, and Martio [1989].

7. Let $\varphi : \mathbb{D} \to \Omega$ be conformal and let L be a line. Then there are $p > 1$ and $C > 0$, independent of Ω and φ, such that

$$\int_{L \cap \Omega} |(\varphi^{-1})'(w)|^p |dw| \le C.$$

This result is due to Fernández, Heinonen, and Martio [1989]. By Exercise III.16 we may assume $L = \partial U$ where $U \subset \Omega$ is a closed disc. Then by Exercise 6, $\varphi^{-1}(L)$ is a quasicircle. On the other hand, by (I.4.15) $\varphi^{-1}(L)$ is an Ahlfors regular curve. Therefore $\varphi^{-1}(L)$ is a chord-arc curve and $|(\varphi^{-1})'|$ is an A^∞-weight, which means that $(\varphi^{-1})' \in L^p(\partial U)$. Moreover, a check of the constants will show p and C are independent of Ω and φ.

8. Let Ω be a simply connected domain and let $\varphi : \mathbb{D} \to \Omega$ be conformal. For $z \in \mathbb{D}$ and $|z| > 1/2$, write $I(z) = \{\zeta \in \partial\mathbb{D} : |\zeta - z| < 2(1 - |z|)\}$.
   (a) If Ω is a quasidisc, then

$$C(K)(1 - |z|)|\varphi'(z)| \leq \operatorname{diam}\varphi(I(z)) \leq C(K)^{-1}(1 - |z|)|\varphi'(z)| \quad (\text{E.1})$$

and

$$\operatorname{dist}\big(\varphi(z), \varphi(I(z))\big) \leq C(K)(1 - |z|)|\varphi'(z)|,$$

where K is the smallest dilatation of a quasiconformal extension of φ to a quasiconformal $\Phi : \mathbb{C}^* \to C^*$ with $\Phi(\infty) = \infty$. Hint: Use (3.4) and estimate the modulus of $C^* \setminus \big(I(z) \cup B(z, (1 - |z|)/2)\big)$.
   (b) Conversely, if (E.1) holds whenever $|z| < 1/2$, then Ω is a quasidisc.
   (c) For $z \in \Omega$ write $d(z) = \operatorname{dist}(z, \partial\Omega)$ and for $\alpha > 1$ and $w \in \partial\Omega$ write

$$\Gamma_\alpha(w) = \{z \in \Omega : |z - w| < \alpha d(z)\}.$$

Then there are $\beta = \beta(\alpha, K) > 1$ and $\gamma(\alpha, K) > 1$ such that when $\zeta \in \partial\mathbb{D}$,

$$\varphi(\Gamma_\beta)(\zeta) \subset \Gamma_\alpha(\varphi(\zeta)) \subset \varphi(\Gamma_\gamma(\zeta)).$$

Hint: Use (a).
   (d) If $\partial\Omega$ is a chord-arc curve, then (E.1) can be improved to

$$C(\Omega, K)(1 - |z|)|\varphi'(z)| \leq \int_{I(z)} |\varphi'(\zeta)|ds \leq C(\Omega, K)^{-1}(1 - |z|)|\varphi'(z)|.$$

9. Let Γ be a Jordan curve in the plane, let Ω_1 and Ω_2 be the two components of $\mathbb{C}^* \setminus \Gamma$, so that $\infty \in \Omega_2$, let $\varphi_1 : \mathbb{D} \to \Omega_1$ and $\varphi_2 : \mathbb{C}^* \setminus \overline{\mathbb{D}} \to \Omega_2$ be conformal mappings, and let $h = \varphi_2^{-1} \circ \varphi_1$ be the welding map.
   (a) Let ω_j be harmonic measure for some point in Ω_j. Prove $h' = 0$ if and only if $\omega_1 \perp \omega_2$.
   (b) Prove there exists Γ such that ω_1 is doubling but ω_2 is not doubling. Hint: See Bishop's towers, Figure VI.10.

10. A Jordan curve in $\Gamma \subset C^*$ such that $\infty \in \Gamma$ is called a **quasicircle** if Γ is the quasiconformal image of $\mathbb{R}$, or equivalently, Γ is a Möbius image of a bounded quasicircle. Prove that a graph $\{y = f(x) : -\infty < x < \infty\}$ is a quasicircle if and only if f is a Zygmund function, $f \in Z^*$ (Jerison and Kenig [1982a]).

11. Let Γ be a quasicircle in C^* and assume $\infty \in \Gamma$ and write Ω_1 and Ω_2 for the complementary components of Γ. Ahlfors [1963] proved that in this case there exists a quasiconformal $G : \Omega_1 \to \Omega_2$ which is bilipschitz,

$$C^{-1}|z - w| \leq |G(z) - G(w)| \leq C|z - w|.$$

(a) If $\Gamma_1 \subset \overline{\Omega}_1$ is a rectifiable Jordan curve, then $\Gamma_2 = G(\Gamma_1) \subset \overline{\Omega}_2$ is rectifiable and

$$\Gamma_1 \cap \Gamma_2 = \Gamma_1 \cap \Gamma.$$

(b) Let ω_j be harmonic measure for some point $z_j \in \Omega_j$, and let K_j be the set of cone points for Ω_j. If $\omega_1(K_1) > 0$, then $\omega_1(K_1 \cap K_2) = \omega_1(K_1)$ and

$$\chi_{K_1 \cap K_2} \omega_1 \ll \chi_{K_1 \cap K_2} \Lambda_1 \ll \chi_{K_1 \cap K_2} \omega_2.$$

These results are from Bishop's thesis [1987].

12. Suppose the Jordan curve Γ is a finite union of linear images $w = Az + B$ of graphs

$$\Gamma_k = \{y = f_k(x)\}.$$

Let Ω_1 and Ω_2 be the components of $\mathbb{C}^* \setminus \Gamma$ and let ω_j be harmonic measure for some point in Ω_j. If $f_k \in Z^*$, then by Exercises 10 and 11 Γ is a quasicircle and $\omega_1 \ll \chi_{T_n} \Lambda_1 \ll \omega_2 \ll \omega_1$. See Exercise VI.9 and Bishop [1987].

13. Let Ω be a simply connected domain such that $\infty \notin \partial\Omega$. Then Ω is called a **John domain** if there is $z_0 \in \Omega$ and $c > 0$ such that for every $z_1 \in \Omega$ there is an arc $\sigma \in \Omega$ joining z_0 to z_1 such that

$$\text{dist}(z, \partial\Omega) \geq c|z - z_1|$$

for all $z \in \sigma$. Thus in Figure VII.9 a slit disc is a John domain but the teardrop is not a John domain. This exercise will survey some properties of John domains. See John [1961], Martio and Sarvas [1979] and especially Näkki and Väisälä [1991] for much more.

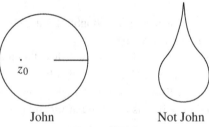

John   Not John

Figure VII.9

(a) Recall that a **crosscut** of Ω is a Jordan arc $\gamma \subset \Omega$ having both its endpoints in $\partial\Omega$. Every crosscut γ of Ω divides $\Omega \setminus \gamma$ into two components Ω_1 and Ω_2. Prove Ω is a John domain if and only if there is a constant C such that whenever γ is a crosscut of Ω,

$$\min_{j} \operatorname{diam}(\Omega_j) \leq C \operatorname{diam}(\gamma).$$

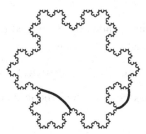

Figure VII.10 Both sides John.

(b) The boundary of a John domain Ω is locally connected. Consequently, if φ is a conformal map from $\mathbb{D}$ onto a John domain, then φ has a continuous extension to $\partial\mathbb{D}$. However, a John domain need not be a Jordan domain.

(c) A Jordan curve is a quasicircle if and only if both of its complementary components are John domains.

(d) Let Ω be a bounded simply connected domain and let $\varphi : \mathbb{D} \to \Omega$ be a conformal mapping. Recall $I(z) = \{\zeta \in \partial\mathbb{D} : |\zeta - z| < 2(1 - |z|)\}$ and write $B_z = \mathbb{D} \cap B(z, 2(1 - |z|))$. The next theorem is from Pommerenke [1982a]. See also Pommerenke [1964] and [1991].

Theorem. *The following are equivalent:*

(1) Ω *is a John domain.*

(2) *There is* C_1 *such that for whenever* $1/2 < |z| < 1$,

$$\operatorname{diam}\varphi(B_z) \leq C_1 \operatorname{dist}(\varphi(z), \partial\Omega).$$

(3) *There exist* $\alpha, 0 < \alpha \leq 1$, *and* C_2 *such that if* $0 < r < s < 1$ *and* $\zeta \in \partial\mathbb{D}$,

$$\frac{|\varphi'(s\zeta)|}{|\varphi'(r\zeta)|} \leq C_2 \left(\frac{1-s}{1-r}\right)^{\alpha-1}.$$

(4) *There exist* $\alpha, 0 < \alpha \leq 1$, *and* C_3 *such that if* $I \subset J$ *are subarcs of* $\partial\mathbb{D}$

$$\frac{\operatorname{diam}\varphi(I)}{\operatorname{diam}\varphi(J)} \leq C_3 \left(\frac{|I|}{|J|}\right)^{\alpha}.$$

To prove (1) $\implies$ (2), use Beurling's projection theorem to show that

$$\operatorname{dist}(\varphi(z), \varphi(J_j)) \leq K \operatorname{dist}(\varphi(z), \partial\Omega),$$

where J_j, $j = 1, 2$, are the two components of $2I(z) \setminus \frac{3}{2}I(z)$. Then by Exercise III.16(b) obtain two geodesic arcs γ_j with initial point z and terminal points in J_j, $j = 1, 2$, such that

$$\ell(\varphi(\gamma_j)) \leq C \mathrm{dist}(\varphi(z), \varphi(J_j)).$$

Write U_z for the domain bounded by γ_1, γ_2 and the arc of $\partial \mathbb{D}$ containing $I(z)$. When Ω is a John domain (a) shows

$$\mathrm{diam}(\varphi(U_z)) \leq C \mathrm{dist}(\varphi(z), \partial \Omega).$$

But because $B_z \setminus U_z$ has bounded hyperbolic diameter, Koebe's theorem gives

$$\mathrm{diam}\varphi(B_z \setminus U_z) \leq C \mathrm{dist}(\varphi(z), \partial \Omega)$$

even if Ω is not a John domain.

To show (2) $\Longrightarrow$ (3), we suppose $\zeta = 1$ and define

$$g(s) = \int_s^1 \int_{|y| \leq 1-x} |\varphi'(x + iy)|^2 dx dy.$$

By (2)

$$g(s) \leq \mathrm{Area}(\varphi(B_s)) \leq C_1^2 (1 - s)^2 |\varphi'(s)|^2.$$

By (I.4.14) there is an absolute constant $a > 0$ such that

$$g(s) \geq a(1 - s)^2 |\varphi'(s)|^2$$

and

$$g'(s) \leq -a(1 - s)|\varphi'(s)|^2,$$

since the sets $\{x + iy : |z| \leq 1 - x\}$ have bounded hyperbolic diameter. Therefore

$$\frac{d}{ds}(1 - s)^{-\alpha} g(s) \leq 0,$$

with $\alpha = a/2C_1^2$ and we obtain (3) with this α and $C_2 = a^{-1/2}C_1$. Note that when (2) holds, $\varphi \in C^\alpha$.

If (3) holds then by Koebe's estimate (1) holds with $z_0 = \varphi(0)$ and with $\sigma \varphi([0, \varphi^{-1}(z)])$. Thus (3) $\Longrightarrow$ (1) and a hyperbolic geodesic can be taken for the curve σ.

To show (3) $\Longrightarrow$ (4), let $I \subset J$ be subarcs of $\mathbb{D}$. We can assume $J = I(z)$ and $I = I(w)$. Then by (2)

$$\mathrm{diam}\varphi(I) \leq C_1(1 - |w|)|\varphi'(w)|.$$

Let $z' = \frac{|z|}{|w|} w$. Then by (3)

$$(1 - |w|)|\varphi'(w)| \le C_2 \left(\frac{1 - |w|}{1 - |z'|}\right)^\alpha |\varphi'(z')|.$$

But $|\varphi'(z')| \le a|\varphi'(z)|$ again via (I.4.14) and for arbitrary simply connected domains

$$\operatorname{diam}\varphi(I(z)) \ge a' \operatorname{diam}\varphi\left(B\left(z, \frac{1 - |z|}{2}\right)\right).$$

Therefore (4) holds with the exponent α from (3).

Now assume (4) and let $1/2 < |z| < 1$. Given $\varepsilon > 0$ there is $A > 0$ such that by Theorem V.3.5

$$|\varphi(\zeta) - \varphi(z)| \le A \operatorname{dist}(\varphi(z), \partial\Omega)$$

on $2I(z) \setminus \bigcup I_j$ where $|\bigcup I_j|/|2I(z)| \le \varepsilon$. Then if ε is sufficiently small (4) implies that $\max_j \operatorname{diam}(\varphi(I_j)) \le \frac{1}{2} \operatorname{diam}\varphi(2I(z))$ and hence

$$\operatorname{diam}\varphi(2I(z)) \le 2A \operatorname{dist}(\varphi(z), \partial\Omega) + \frac{1}{2} \operatorname{diam}\varphi(2I(z)).$$

It then follows from V.3.5 exactly as in the proof of (1) $\Longrightarrow$ (2) that $\operatorname{diam}\varphi(B_z) \le C \operatorname{dist}(\varphi(z), \partial\Omega)$. Therefore (2) holds.

(e) If Ω is a John domain then harmonic measure is doubling.

14. (a) If two measures μ and ν are A^∞-equivalent, then $\nu \ll \mu \ll \nu$.

(b) If μ and ν are A^∞-equivalent and if μ is a doubling measure, then ν is a doubling measure.

(c) There is a doubling measure on the unit interval singular to Lebesgue measure. Hint: Start with a quasicircle for which $\omega_2 \perp \omega_1$.

(d) Assume (4.2) holds for $\mu = dx$ and ν on $[0, 1]$. Let $N \ge 1$ and suppose $E \subset I \subset [0, 1]$ satisfy $\mu(E) < \left(\frac{\varepsilon}{2}\right)^N$. We may suppose I is a dyadic interval. Let $\{I_j\}$ be the maximal dyadic subintervals of I such that

$$\frac{\mu(E \cap I_j)}{\mu(I_j)} \ge \varepsilon.$$

By Lebesgue's theorem, $\mu\left(\bigcup I_j \setminus E\right) = 0$. Let I_j^* denote the dyadic interval containing I_j with $|I_j^*| = 2|I_j|$. Since I_j is maximal, $\nu(E \cap I_j^*) \le \eta\nu(I_j^*)$ by (4.2). Set $E_1 = \bigcup I_j^*$. Then $E \subset E_1$, $\nu(E) \le \nu(E_1)$, and

$$\frac{\mu(E_1)}{\mu(I)} \le 2 \sum \frac{\mu(I_j)}{\mu(I)} \le \frac{2}{\varepsilon}\mu(E) < \left(\frac{\varepsilon}{2}\right)^{N-1}.$$

Continuing by induction we obtain

$$\frac{\mu(E)}{\mu(I)} \le \left(\frac{\varepsilon}{2}\right)^N \implies \frac{\nu(E)}{\nu(I)} \le \eta^N,$$

which means

$$\frac{\nu(E)}{\nu(I)} \le \frac{1}{\eta}\left(\frac{\mu(E)}{\mu(I)}\right)^{\log\frac{1}{\eta}/\log\frac{2}{\varepsilon}}.$$

(e) Adapt the proof outlined in (d) to the case when μ and ν are continuous positive measures.

(f) If k is an A^∞-weight, then k satisfies the reverse Hölder condition. Fix I and λ large, and assume $\int_I k\,dx = 1$. Let $\{I_{j,1}\}$ be the maximal intervals in the dyadic decomposition of I such that

$$\lambda < f_{I_{j,1}} \le 2\lambda,$$

and set $E_1 = \bigcup I_{j,1}$. Then $k \le \lambda$ on $I \setminus E_1$ and

$$\frac{|E_1|}{|I|} \le \left(\frac{C}{\lambda}\right)\frac{1}{1-\lambda}.$$

Now let $\{I_{j,2}\}$ be the maximal intervals in the dyadic decomposition of I such that

$$\lambda^2 < f_{I_{j,1}} \le 2\lambda^2,$$

and set $E_2 = \bigcup I_{j,2} \subset E_1$. Then

$$\frac{|E_2|}{|E_1|} \le \left(\frac{2^{n-1}C^n}{\lambda^n}\right)^{\frac{1}{1-\lambda}}.$$

Continuing, we obtain the reverse Hölder condition

$$\left(\frac{1}{|I|}\int_I k^{1+\delta}dx\right)^{\frac{1}{1+\delta}} \le C\frac{1}{|I|}\int_I k\,dx$$

whenever $1 + \delta < \frac{1}{1-\alpha}$.

(g) Conversely, the reverse Hölder condition and Hölder's inequality yield A^∞ for $\alpha = \frac{\delta}{1+\delta}$.

(h) By Hölder's inequality, A^p implies A^∞ with $\alpha = \frac{p-1}{p}$. Conversely, A^∞ and its symmetry imply

$$\left(\frac{1}{\int_I k\,dx}\int_I \left(\frac{1}{k}\right)^\delta dx\right)^{\frac{1}{1+\delta}} \le \frac{C}{\int_I k\,dx}|I|,$$

from which A^p follows with $p = 1 + \frac{1}{\delta}$.

15. Let Ω be a Jordan domain, let $\varphi_1 : \mathbb{D} \to \Omega$ and $\varphi_2 : C^* \setminus \overline{\mathbb{D}} \to \mathbb{C}^* \setminus \overline{\Omega}$ be conformal mappings.

(a) Prove that the welding map $\varphi_2^{-1} \circ \varphi_1$ is in A^∞ if and only if Ω is a BMO domain and a quasidisc. (Astala and Zinsmeister [1991]).

(b) Prove $|\varphi'| \in A^\infty$ and Ω is a Smirnov domain if and only if there is a constant K such that

$$\int_{I_z} |\varphi'| ds \leq K \operatorname{dist}(\varphi(z), \partial\Omega)$$

where $I_z = \{\theta : |e^{i\theta} - \frac{z}{|z|}| < 1 - |z|\}$.

(c) Prove that Ω is an Ahlfors regular John domain if and only if Ω is a Smirnov domain and $|\varphi'|$ is an A^∞-weight. See Pommerenke [1982a] and Zinsmeister [1984].

16. (a) Example 2.6 is a quasicircle for which $|\varphi'|$ is an A^∞-weight, but it is not a Smirnov domain.

(b) Construct a Zygmund function $f(\theta)$ such that $f \geq 1$, $f' \in L^1$ but $f' \notin$ BMO. Then $\Omega = \{re^{i\theta} : 0 \leq r < f(\theta)\}$ is a quasidisc and a Smirnov domain, but $|\varphi'|$ is not in A^∞.

(c) Construct a Jordan Smirnov domain Ω such that $|\varphi'| \in A^\infty$ but Ω is not a quasidisc.

17. (a) Let φ map $\mathbb{D}$ to a domain bounded by an Ahlfors regular Jordan curve. Then for $p < 1/5$ there is C such that

$$\int_{I(z)} |\varphi'|^p ds \leq C(1 - |z|)|\varphi'(z)|^p.$$

Hint: By Hölder,

$$\int_{I(z)} |\varphi'|^p ds \leq \left(\int_{I(z)} \frac{|\varphi'(\zeta)|}{|\varphi(\zeta) - \varphi(z)|^2} ds \right)^p \left(\int_{I(z)} |\varphi(\zeta) - \varphi(z)|^{\frac{2p}{1-p}} ds \right)^{1-p}.$$

The first factor can be estimated directly, using the regularity of $\partial\mathbb{D}$, and the second factor can be handled with Prawitz's theorem.

(b) Under the assumptions in (a), $g = \log(\varphi') \in$ BMO. Hint: If p is small, then $(\varphi')^p = \psi'$ where ψ is a conformal map of $\mathbb{D}$ to a quasidisc. But by (a), $|\psi'| \in A^\infty$, so that $\log(\psi') \in$ BMO.

(c) Let φ be a conformal map from the half-plane onto the region above the curve $\Gamma = \{y = \sin(x^2), x \in \mathbb{R}\}$. Then $\log(\varphi') \in$ BMO, but Γ is not Ahlfors regular. See Zinsmeister [1984] and [1985].

18. The space VMO, for **vanishing mean oscillation**, consists of all $g \in$ BMO such that

$$\lim_{\delta \to 0} \sup_{|I| < \delta} \frac{1}{|I|} \int_I |g - g_I| ds = 0,$$

and VMOA = VMO ∩ BMOA. See Sarason [1975] or Garnett [1981]. A Jordan curve Γ is said to be **asymptotically smooth** if Γ is rectifiable and if

$$\lim_{\delta \to 0} \sup_{|w_1 - w_2| < \delta} \frac{\ell(w_1, w_2)}{|w_1 - w_2|} = 1,$$

whenever $w_1, w_2 \in \Gamma$ and $\ell(w_1, w_2)$ is the length of the shorter Γ subarc joining w_1 to w_2. Let $g = \log(\varphi')$ where φ maps $\mathbb{D}$ to a Jordan domain. Prove $g \in$ VMOA if and only if $\partial\Omega$ is asymptotically smooth. See Pommerenke [1978].

19. Let Ω be a Jordan domain such that $B(0, 1) \subset \Omega$. Assume $\ell(\partial\Omega) \leq M$ and let $E \subset \partial\Omega$ satisfy $\omega(0, E, \Omega) \geq \varepsilon$. Prove there is $\delta = \delta(\varepsilon, M)$ such that there is a chord-arc domain $\mathcal{U} \subset \Omega$ with $B(0, 1) \subset \mathcal{U}, \omega(0, E \cap \partial\mathcal{U}, \mathcal{U}) \geq \delta$ and $\ell(E \cap \partial\mathcal{U}) \geq \delta$. Hint: Take a union of fixed angle cones over a subset of E.

20. Let Γ be a Jordan curve bounding domains Ω_1 and Ω_2 and let $\varphi_j : \mathbb{D} \to \Omega_j$ be conformal. Suppose, as happens in Example 2.6, that $\log |\varphi_1'|$ is the Poisson integral of a negative singular measure on $\partial\mathbb{D}$, so that Ω_1 is badly not a Smirnov domain. Prove there is $c > 0$ such that $|\varphi_2'| \geq c$ on $\partial\mathbb{D}$, and that consequently Ω_2 is a Smirnov domain. Hint: Use the estimate (6.4). This result is due to Jones and Smirnov [1999].

VIII

Simply Connected Domains, Part Two

We begin the chapter with the famous theorems from Makarov [1985], [1987], and [1990a] on Hausdorff measures and harmonic measure in simply connected domains. Next we discuss the Brennan conjecture on the L^p integrability of derivatives of univalent functions and prove some related theorems from Carleson and Makarov [1994]. Then we give Baernstein's counterexample to the L^p version of the Hayman–Wu theorem.

1. The Law of the Iterated Logarithm for Bloch Functions

Theorem 1.1 (Makarov). *There is a constant $C > 0$ such that if $g(z)$ is a Bloch function on $\mathbb{D}$, then for a.e. $\zeta \in \partial\mathbb{D}$*

$$\limsup_{r \to 1} \frac{|g(r\zeta)|}{\sqrt{\log \frac{1}{1-r} \log\log\log \frac{1}{1-r}}} \le C\|g\|_{\mathcal{B}}. \tag{1.1}$$

Inequality (1.1) is called Makarov's law of the iterated logarithm because it was first proved by Makarov in [1985] and because it resembles the classical Khinchin law of the iterated logarithm in Feller [1968] or Chung [1974]. Consider for example the bad Bloch function

$$g(z) - \sum_{n=1}^{\infty} z^{2^n}. \tag{1.2}$$

The series (1.2) behaves much like a sum of independent identically distributed random variables, and Salem and Zygmund proved in [1950] that its partial sums $S_n(z) = \sum_{k=0}^{n} z^{2^k}$ satisfy

$$\limsup \frac{|S_n(\zeta)|}{\sqrt{n \log \log n}} = 1 \qquad (1.3)$$

for almost every $\zeta \in \partial\mathbb{D}$. But if $g(z)$ is defined by (1.2), then (1.3) is equivalent to

$$\limsup_{r \to 1} \frac{|g(r\zeta)|}{\sqrt{\log \frac{1}{1-r} \log \log \log \frac{1}{1-r}}} = 1,$$

which is an instance of (1.1).

Theorem 1.1 has a converse, proved by Jones in [1989].

Theorem 1.2. *If* $\|g\|_\mathcal{B} \le 1$ *and if there exist* $\beta > 0$ *and* $M < \infty$ *such that for all* $z_0 \in \mathbb{D}$,

$$\sup_{\{z : \rho(z, z_0) < M\}} \left(1 - |z|^2\right)|g'(z)| \ge \beta, \qquad (1.4)$$

then almost everywhere on $\partial\mathbb{D}$,

$$\limsup_{r \to 1} \frac{\operatorname{Re} g(r\zeta)}{\sqrt{\log \frac{1}{1-r} \log \log \log \frac{1}{1-r}}} \ge c(\beta, M) > 0. \qquad (1.5)$$

When φ is a conformal mapping from $\mathbb{D}$ onto a domain Ω, Theorem VII.2.4 gave a geometric condition on Ω that is necessary and sufficient for the Bloch function $g = \log(\varphi')$ to satisfy condition (1.4). For example, (1.4) holds when φ maps $\mathbb{D}$ to the domain inside the snowflake curve.

Proof of Theorem 1.1. This proof is due to Carleson (unpublished) and independently to Pommerenke [1986b]. We give Pommerenke's version, which yields the best known constant $C = 1$ in (1.1). A different proof, from Makarov [1990a], is given in Appendix J.

We may assume $g(0) = 0$ and $\|g\|_\mathcal{B} = 1$. Let $p \ge 0$ be an integer and consider the integral means

$$I_p(r) = \frac{1}{2\pi} \int_0^{2\pi} |g(re^{i\theta})|^{2p} d\theta.$$

The proof begins with the identity

$$\frac{d}{dr}(r I_p'(r)) = \frac{4p^2 r}{2\pi} \int_0^{2\pi} |g(re^{i\theta})|^{2p-2} |g'(re^{i\theta})|^2 d\theta, \qquad (1.6)$$

which is known as **Hardy's identity**. Hardy's identity can be proved by examining the Fourier series of $|g|^p$ or by differentiating $I_p(r)$ directly. A corollary of Hardy's identity is the inequality

$$I_p(r) \le p! \left(\log \frac{1}{1-r^2} \right)^p \le p! \left(\log \frac{1}{1-r} \right)^p, \tag{1.7}$$

valid for $||g||_B = 1$ and for integral $p \ge 0$. Indeed, (1.7) is trivial when $p = 0$, and to prove (1.7) when $p \ge 1$ we use (1.6) and induction to get

$$\frac{d}{dr}(rI_p'(r)) \le \frac{4p^2 r}{(1-r^2)^2} I_{p-1}(r)$$

$$\le \frac{4pp!r}{(1-r^2)^2} \left(\log \frac{1}{1-r^2} \right)^{p-1}$$

$$\le p! \frac{d}{dr} \left(r \frac{d}{dr} \left(\log \frac{1}{1-r^2} \right)^p \right),$$

and since $I_p(0) = 0$, an integration then yields (1.7).

Now let us apply the Hardy–Littlewood maximal theorem to the function $|g(re^{i\theta})|^p \in L^2$. By inequality (1.7), $g_r^*(e^{i\theta}) = \sup_{\rho < r} |g(\rho e^{i\theta})|$ satisfies

$$\frac{1}{2\pi} \int |g_r^*(e^{i\theta})|^{2p} d\theta \le Cp! \left(\log \frac{1}{1-r} \right)^p \tag{1.8}$$

with C independent of p. Next fix $\alpha > 1$ and set

$$A_p(r) = \frac{1}{1-r} \frac{1}{\left(\log \frac{1}{1-r} \right)^{p+1}} \frac{1}{\left(\log \log \frac{1}{1-r} \right)^\alpha}.$$

Then

$$\int_r^1 A_p(s) ds \ge \frac{C}{p} \frac{1}{\left(\log \frac{1}{1-r} \right)^p} \frac{1}{\left(\log \log \frac{1}{1-r} \right)^\alpha}, \tag{1.9}$$

while by (1.8)

$$\int_r^1 A_p(s) \int |g_s^*(e^{i\theta})|^{2p} d\theta ds$$

$$\le Cp! \int_r^1 \frac{1}{\left(\log \log \frac{1}{1-s} \right)^\alpha} \frac{1}{\left(\log \frac{1}{1-s} \right)} \frac{ds}{1-s} \le C_\alpha p!.$$

It follows, via Fubini's theorem and Chebychev's inequality, that the set

$$E_p = \left\{ \theta : \int_r^1 A_p(s) |g_s^*(e^{i\theta})|^{2p} ds > C_\alpha p^2 p! \right\}$$

satisfies $|E_p| \le \frac{1}{p^2}$. Therefore if $\theta \notin \bigcup_{p>p_0} E_p$, then by (1.9) and the definition of g_s^*,

$$\frac{|g(re^{i\theta})|}{\sqrt{\left(\log \frac{1}{1-r}\right) \log \log \log \frac{1}{1-r}}} \le \frac{C^{\frac{-1}{2p}} C_\alpha^{\frac{1}{2p}} p^{\frac{3}{2p}} (p!)^{\frac{1}{2p}} \left(\log \log \frac{1}{1-r}\right)^{\frac{\alpha}{2p}}}{\sqrt{\log \log \log \frac{1}{1-r}}}. \quad (1.10)$$

Finally setting $p = \log \log \log \frac{1}{1-r}$ in (1.10) and using Stirling's formula, we obtain (1.1) almost everywhere with constant $C = 1$. ∎

Jones' [1989] proof of Theorem 1.2 is elementary and similar to the classical proof of the law of the iterated logarithm for Bernoulli trials in Feller [1968]. We give his proof in martingale language in Appendix J.

2. Harmonic Measure and Hausdorff Measure

A real valued function $h(t)$ on an interval $(0, t_0)$ is called a **logarithmico-exponential function** or an **L-function** if $h(t)$ is defined by a finite algebraic combination of exponential functions and logarithm functions. The main theorem about L-functions is from Hardy [1954]: If h and g are L-functions, then the limit

$$\lim_{t \downarrow 0} \frac{h(t)}{g(t)}$$

exists in the extended interval $[0, \infty]$. Because the derivative of an L-function is again an L-function, it follows that every L-function is monotone on some interval $(0, t_1)$. Every measure function in this section will be assumed to be a positive, increasing and continuous L-function. When we are comparing harmonic measure ω to a Hausdorff measure Λ_h, we may also assume its measure function $h(t)$ satisfies

$$\lim_{t \to 0} \frac{h(t)}{t} = \infty. \quad (2.1)$$

Indeed if $\lim_{t \to 0} \frac{h(t)}{t} = 0$ then we have $\omega \perp \Lambda_h$ by Theorem VI.5.2, and if $0 < \lim_{t \to 0} \frac{h(t)}{t} < \infty$ then the situation is explained by Corollary VI.6.2. We write h^{-1} for the inverse function of h, so that $h^{-1}(h(t)) = t$.

Theorem 2.1 (Makarov). *Let h be a positive increasing L-function satisfying (2.1). Let φ be a conformal mapping from $\mathbb{D}$ onto a simply connected domain Ω and let ω denote harmonic measure for some point $w_0 \in \Omega$. Then*

(a) $\omega \ll \Lambda_h \iff \liminf\limits_{r \to 1} \dfrac{(1 - r^2)|\varphi'(r\zeta)|}{h^{-1}(1 - r)} > 0$ a.e. *on* $\partial \mathbb{D}$,

(b) $\omega \perp \Lambda_h \iff \liminf\limits_{r \to 1} \dfrac{(1 - r^2)|\varphi'(r\zeta)|}{h^{-1}(1 - r)} = 0$ a.e. *on* $\partial \mathbb{D}$, *and*

(c) *there is a set* $A \subset \partial \Omega$ *of* σ-*finite* Λ_h *measure such that* $\omega(A) = 1$ *if and only if*

$$\liminf_{r \to 1} \frac{(1 - r^2)|\varphi'(r\zeta)|}{h^{-1}(1 - r)} < \infty \text{ a.e. on } \partial \mathbb{D}.$$

Theorem 2.2 (Makarov). *There is an absolute constant* $C > 0$ *such that if*

$$h(t) = t e^{C\sqrt{\log \frac{1}{t} \log \log \log \frac{1}{t}}},$$

then for every simply connected domain Ω,

$$\omega \ll \Lambda_h. \tag{2.2}$$

Conversely, there is $c < C$ *such that if*

$$h(t) = t e^{c\sqrt{\log \frac{1}{t} \log \log \log \frac{1}{t}}}, \tag{2.3}$$

then there exists a Jordan domain Ω *for which*

$$\omega \perp \Lambda_h. \tag{2.4}$$

Note that Theorem 2.2 contains Theorem VI.5.1: If $\alpha < 1$, then $w \ll \Lambda_\alpha$. Let us derive Theorem 2.2 from Theorem 2.1 before we prove Theorem 2.1.

Proof of Theorem 2.2. Let Ω be any simply connected domain, let φ be a conformal map from $\mathbb{D}$ onto Ω and let $g = \log(\varphi')$. Then by Theorem VII.2.1, $\|g\|_{\mathcal{B}} \le 6$, and by Theorem 1.1 there is a constant $C > 0$ such that

$$\limsup_{r \to 1} \frac{|\mathrm{Re} g(r\zeta)|}{\sqrt{\log\left(\frac{1}{1-r}\right) \log \log \log\left(\frac{1}{1-r}\right)}} \le C \tag{2.5}$$

almost everywhere on $\partial \mathbb{D}$. Take

$$h_1^{-1}(t) - t e^{-C\sqrt{\log \frac{1}{t} \log \log \log \frac{1}{t}}}.$$

Then by (2.5),

$$\liminf_{r \to 1} \frac{(1 - r)|\varphi'(r\zeta)|}{h_1^{-1}(1 - r)} \ge 1$$

almost everywhere, and by Theorem 2.1(a), $\omega \ll \Lambda_{h_1}$. But then a little calculus shows that

$$h_1(t) = o\left(t e^{C\sqrt{\log \frac{1}{t} \log\log\log \frac{1}{t}}}\right)$$

as $t \to 0$, and (2.2) follows.

Conversely, let Ω be any Jordan domain that satisfies condition (c) of Theorem VII.2.4. The snowflake is one example. Let φ be the conformal mapping from $\mathbb{D}$ to Ω and let $g = \log(\varphi')$. Then by Theorem VII.2.4 and Theorem 1.2, there is $c_3 > 0$ such that

$$\limsup_{r \to 1} \frac{-\operatorname{Reg}(r\zeta)}{\sqrt{\log\left(\frac{1}{1-r}\right)\log\log\log\left(\frac{1}{1-r}\right)}} \geq c_3 > 0,$$

almost everywhere on $\partial\mathbb{D}$. If $c < c_3$, take c' with $c < c' < c_3$ and set

$$h_2^{-1}(t) = t e^{-c'\sqrt{\log \frac{1}{t} \log\log\log \frac{1}{t}}}.$$

Then

$$\liminf_{r \to 1} \frac{(1-r)|\varphi'(r\zeta)|}{h_2^{-1}(1-r)} = 0$$

almost everywhere, so that by (b) of Theorem 2.1,

$$\omega \perp \Lambda_{h_2}.$$

If (2.3) holds, then for small t

$$h(t) \leq h_2(t)$$

and thus (2.4) holds for $c < c_3$. ∎

Appendix K gives more examples where (2.4) holds.

The extremal constant

$$C_1 = \inf\{C : (2.2) \text{ holds}\}$$

is not known. However the above proof of (2.2) shows $C_1 < 6C_2$ where

$$C_2 = \inf\{C : (1.1) \text{ holds}\}$$

and the proof of Theorem 1.1 yields $C_2 \leq 1$, but the value of C_2 is not known. See Bañuelos and Moore [1999]. Bañuelos [1986] and Lyons [1990] have other proofs that $C_2 \leq 1$. The extremal

$$\sup\{c : (2.4) \text{ holds for some domain}\}$$

is also not known.

Theorem 2.1 itself is a consequence of the next lemma.

Lemma 2.3. *Let φ be the conformal mapping from $\mathbb{D}$ to a simply connected domain Ω and let $E \subset \partial\mathbb{D}$ be a Borel set.*
(a) *If*

$$\liminf \frac{(1-r)|\varphi'(r\zeta)|}{h^{-1}(1-r)} \leq A$$

almost everywhere on E, then there exists $E_1 \subset E$ such that

$$|E \setminus E_1| = 0$$

and

$$\Lambda_h(\varphi(E_1)) \leq 4A|E|.$$

(b) *There are constants $c_1 > 0$ and $q > 0$, not depending on E, such that if*

$$\liminf \frac{(1-r)|\varphi'(r\zeta)|}{h^{-1}(1-r)} > B > 0 \tag{2.6}$$

almost everywhere on E, then

$$\Lambda_h(\varphi(E)) \geq c_1 \frac{B}{(1+B)^{\frac{1}{2}}} |E|^q > 0.$$

Each of the six implications of Theorem 2.1 follows immediately from either part (a) or part (b) of Lemma 2.3, and the remaining details of the proof of Theorem 2.1 are left to the reader.

Proof of Lemma 2.3(a). The proof of (a) is almost the same as the proof of Theorem VI.5.2. We may assume $A > 0$. Fix $\alpha = \max\{2, \frac{1}{2A}\}$ and define

$$I(z) = \{\zeta \in \partial\mathbb{D} : |z - \zeta| < \alpha(1 - |z|)\}.$$

Let $\{z_n\}$ be a sequence in $\mathbb{D}$ such that

$$(1 - |z_n|)|\varphi'(z_n)| \leq Ah^{-1}(1 - |z_n|),$$

and such that $\{z_n\}$ is nontangentially dense on E. Fix $\delta > 0$. By the Vitali covering lemma there is a subsequence $\{z_k\}$ of $\{z_n\}$ so that the intervals $I(z_k)$ are pairwise disjoint,

$$(1 - |z_k|)|\varphi'(z_k)| < \frac{\delta}{2\alpha},$$

and

$$|E \setminus \bigcup I(z_k)| = 0.$$

Take $w_k = \varphi(z_k)$, $r_k = \alpha \operatorname{dist}(w_k, \partial\Omega)$, $B_k = B(w_k, r_k)$, and

$$V_\delta = \partial\Omega \cap \left(\bigcup B_k\right).$$

Then $r_k \le 2\alpha(1 - |z_k|)|\varphi'(z_k)| < \delta$ and

$$h(r_k) \le h\left(2\alpha A h^{-1}(1 - |z_k|)\right).$$

If $A \le \frac{1}{4}$, then

$$h(r_k) \le (1 - |z_k|),$$

and if $A > \frac{1}{4}$, then since $\frac{h(t)}{t}$ is decreasing by (2.1),

$$h(r_k) \le 4A(1 - |z_k|).$$

In either case we obtain

$$\sum h(r_k) \le 4A \sum |I(z_k)| \le 4A|E|.$$

Therefore $V = \bigcap V_{1/m}$ satisfies $\Lambda_h(V) \le 4A|E|$. On the other hand, because $\alpha \ge 2$, Lemma VI.5.3 implies that $|E \setminus \varphi^{-1}(V)| = 0$, and thus (a) holds for $E_1 = E \cap \varphi^{-1}(V)$. ∎

For the proof of (b) we need three additional lemmas.

Lemma 2.4. *Let φ be the conformal mapping from $\mathbb{D}$ onto a simply connected domain Ω, and let γ be a crosscut of Ω with endpoints $w_1, w_2 \in \partial\Omega$. Set $\zeta_j = \varphi^{-1}(w_j)$, $j = 1, 2$. If σ is the geodesic in $\mathbb{D}$ connecting ζ_1 to ζ_2, let $z_\sigma \in \sigma$ satisfy $|z_\sigma| = \inf_\sigma |z|$. Then*

$$\operatorname{diam}(\gamma) \ge c(1 - |z_\sigma|)|\varphi'(z_\sigma)|, \qquad (2.7)$$

for some absolute constant $c > 0$.

Proof. Applying a Möbius transformation to the disc and linear map to Ω, we may suppose that $\sigma = (-1, 1)$, $z_\sigma = 0$, $\varphi(0) = 0$ and $\varphi'(0) = 1$. We need to prove $\operatorname{diam}(\gamma) \ge c > 0$. Let $B_r = \{|z| \le r\}$. By the Koebe one-quarter theorem, $B_{\frac{1}{4}} \subset \Omega$. If $B(w_1, \operatorname{diam}(\gamma)) \cap B_{\frac{1}{5}} \ne \emptyset$, then $\operatorname{diam}(\gamma) \ge \frac{1}{20}$. On the other hand, if $B(w_1, \operatorname{diam}(\gamma)) \cap B_{\frac{1}{5}} = \emptyset$, then $\varphi^{-1}(\gamma)$ separates $B_{\frac{1}{24}} \subset \varphi^{-1}(B_{\frac{1}{6}})$ from a semicircle, say $T^+ = \partial\mathbb{D} \cap \{\operatorname{Im} z > 0\}$. Thus

$$d_{\mathbb{D}}(B_{\frac{1}{24}}, T^+) \ge d_{\mathbb{C}}(B_{\frac{1}{6}}, B(\zeta_1, \operatorname{diam}(\gamma))) \ge \log\left(\frac{C}{\operatorname{diam}(\gamma)}\right),$$

and (2.7) holds. ∎

The second lemma is from Carleson's fundamental paper [1973]. It is the key to many of the deeper results in this chapter.

Lemma 2.5. *Let Ω be a simply connected domain. Fix $w_0 \in \Omega$ and let $\zeta_0 \in \partial\Omega$. For*

$$0 < r < \frac{|w_0 - \zeta_0|}{2},$$

set

$$D = B(\zeta_0, r), \quad and \quad \widetilde{D} = B(\zeta_0, 2r),$$

and for any $M > 0$ take

$$k_0 = k_0(M) = \left[1 + \frac{M}{\pi}\right].$$

Then there exist $r_0 = r_0(M) > 0$ such that if $r < r_0$, then $\Omega \cap \partial\widetilde{D}$ contains

$$N \leq \frac{2\pi}{\log 2} k_0 \log \frac{1}{r}$$

crosscuts

$$\gamma_1, \cdots, \gamma_N$$

of Ω, such that each γ_j separates w_0 from a continuum $\beta_j \subset \partial\Omega$ and

$$\omega\left(D \cap \partial\Omega \setminus \bigcup_{j=1}^{N} \beta_j\right) < r^M, \tag{2.8}$$

where $\omega(E) = \omega(w_0, E, \Omega)$.

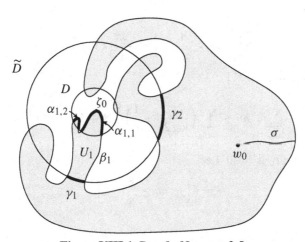

Figure VIII.1 Proof of Lemma 2.5.

Proof. Fix one curve $\sigma \subset \Omega \setminus \overline{D}$ connecting w_0 to $\partial\Omega$. Let $\{\gamma_j\}$ be the set of component arcs of $\Omega \cap \partial\tilde{D}$ having the property that there exists a curve $P \subset \Omega$ connecting w_0 to $w_j \in \Omega \cap D$ such that P first crosses $\Omega \cap \partial\tilde{D}$ through γ_j. By Exercise I.13, $\Omega \setminus \gamma_j$ consists of two simply connected components; take U_j to be that component with $w_0 \notin U_j$. Because Ω is simply connected and $\partial\Omega \setminus \overline{D} \neq \emptyset$, $U_j \cap U_k = \emptyset$ whenever $\gamma_j \neq \gamma_k$. Also by Exercise I.13, $\beta_j = (\partial U_j) \setminus \gamma_j \subset \partial\Omega$ is a continuum and γ_j separates β_j from w_0.

Now $U_j \cap \partial D$ is a union of arcs $\tau_{j,k}$, and by Exercise I.13, each $\tau_{j,k}$ separates w_0 from a continuum $\alpha_{j,k} \subset \beta_j$. Let $\alpha_j = \bigcup_k \alpha_{j,k}$ and let Γ_j be the family of paths in Ω joining σ to α_j. Then by Theorem IV 5.3 and Exercise VI.7,

$$\omega_j = \omega(\alpha_j) \leq \frac{8}{\pi} e^{-\pi\lambda(\Gamma_j)}, \tag{2.9}$$

even if Ω is not a Jordan domain or α_j is not connected. Let Γ_j' be the set of paths in U_j joining $\partial\tilde{D}$ to $\bigcup_k \tau_{j,k}$. Then every path in Γ_j contains a path from Γ_j', so that by the extension rule,

$$\lambda(\Gamma_j') \leq \lambda(\Gamma_j).$$

By the parallel rule,

$$\sum_j \frac{1}{\lambda(\Gamma_j')} \leq \frac{1}{d_{\tilde{D}\setminus\overline{D}}(\partial\tilde{D}, \partial D)} = \frac{2\pi}{\log 2}.$$

Hence for $k = 1, 2, \dots,$

$$\#\left\{j : \frac{1}{\lambda(\Gamma_j')} \geq \frac{1}{k \log\frac{1}{r}}\right\} \leq \frac{2\pi}{\log 2} k \log\frac{1}{r}$$

and

$$\#\left\{j : \omega_j \geq \frac{8}{\pi} r^{k\pi}\right\} \leq \frac{2\pi}{\log 2} k \log\frac{1}{r}. \tag{2.10}$$

Then since $k_0 = \left[1 + \frac{M}{\pi}\right]$,

$$\sum_{\omega_j \leq \frac{8}{\pi} r^{\pi k_0}} \omega_j \leq \sum_{k=k_0}^{\infty} \left(\frac{2\pi}{\log 2} \log\frac{1}{r}\right) \frac{8}{\pi} (k+1) r^{\pi k} \leq r^M$$

if $r < r_0$ and if r_0 is small. Let

$$N = \#\left\{j : \omega_j \geq \frac{8}{\pi} r^{\pi k_0}\right\}.$$

Then $N \le k_0 \frac{2\pi}{\log 2} \log \frac{1}{r}$ by (2.10). Select $\gamma_1, \dots, \gamma_N$ to have the largest ω_j. Then

$$\omega\left(D \cap \partial\Omega \setminus \bigcup_1^N \beta_j\right) \le \sum_{\omega_j \le \frac{8}{\pi} r^{\pi k_0}} \omega_j \le r^M. \qquad \blacksquare$$

Lemma 2.6. *Assume the L-function $h(t)$ satisfies (2.1) and*

$$Ct \le h(t) \le ct \exp\left(\log(1/t)\right)^{2/3}. \qquad (2.11)$$

Then there is $C_2 > 0$ such that if $0 < t < C_2$ then

$$\frac{B}{(1+B)^{\frac{1}{2}}} h(t/B) \le h(t), \qquad (2.12)$$

and there is $C_3 < \infty$ such that

$$\log(1/t) h\left(\frac{t}{\log(1/t)}\right) \le C_3 h(t). \qquad (2.13)$$

Proof. By (2.1), $h(t)/t$ is decreasing and (2.12) holds if $B \le 1$. Assume $B > 1$. Set $x = 1/t$ and $g(x) = \log\left(\frac{h(t)}{t}\right)$. Then by (2.11)

$$C'' \le g(x) \le (\log x)^{2/3}$$

for large x. Then since $g(x)$ and $(\log x)^{2/3}$ are both L-functions, Theorem 19 of Hardy [1954] gives

$$g'(x) \le c' \frac{2}{3x} (\log x)^{-1/3}$$

for x large and integration then yields (2.12) and (2.13). $\qquad \blacksquare$

Proof of Lemma 2.3(b). To prove (b) we may assume that $h(t)$ satisfies (2.11). Indeed, since (2.11) holds for every measure function of the form

$$h_1(t) = t e^{C\sqrt{\log \frac{1}{t} \log\log\log \frac{1}{t}}}$$

we will have established Theorem 2.2 if we prove (b) under the additional assumption (2.11). On the other hand, if (2.11) fails, then by Hardy's theorem on L-functions,

$$h_1(t) = o(h(t)).$$

But then we can use Theorem 2.2 (which would have been proved for the measure function h_1) to obtain $\Lambda_{h_1}(\varphi(E)) > 0$ and $\Lambda_h(\varphi(E)) = \infty$ whenever $E \subset \partial\mathbb{D}$ has $|E| > 0$. Therefore part (b) of Lemma 2.3 will also hold when (2.11) fails.

Now assume (2.6) holds for E with $|E| > 0$. Fix $\delta > 2$. By (VII.1.6) and Theorem VII.2.1 there is a constant $c > 0$ so that for $\zeta \in E$

$$\liminf_{z \in \Gamma_\delta(\zeta)} \frac{(1 - |z|)|\varphi'(z)|}{h^{-1}(1 - |z|)} > cB > 0.$$

Let $\Gamma_\delta^{\frac{1}{n}}(\zeta) = \Gamma_\delta(\zeta) \cap \{z : |z| > 1 - \frac{1}{n}\}$. For sufficiently large n, the closed set

$$E_n = \{\zeta \in E : \inf_{z \in \Gamma_\delta^{\frac{1}{n}}(\zeta)} \frac{(1 - |z|)|\varphi'(z)|}{h^{-1}(1 - |z|)} \geq (cB + \frac{1}{n})\} \subset E$$

satisfies $|E| < 2|E_n|$, and we may replace E by E_n and B by $B' = cB + \frac{1}{n}$. Cover $\varphi(E)$ by discs D_ν of radius $r_\nu < r_0$ such that

$$\sum h(r_\nu) \leq 2\Lambda_h(\varphi(E)).   \tag{2.14}$$

By choosing r_0 sufficiently small we can assume that $2\pi\omega(\partial \tilde{D}_\nu) < \frac{1}{n}$, $\text{dist}(\varphi(0), \varphi(E)) > 4r_0$, and $h(r_0)/r_0 > B$, since $h(t)/t$ is decreasing. Then by the $M = 1$ case of Lemma 2.5 there are subarcs $\gamma_j^{(\nu)} \subset \partial \tilde{D}_\nu$ for

$$1 \leq j \leq N(\nu) \leq \frac{2\pi}{\log 2} \log(\frac{1}{r_\nu}),$$

such that $\gamma_j^{(\nu)}$ is a crosscut of Ω separating w_0 from a continuum $\beta_j^{(\nu)} \subset \partial\Omega$ and

$$\omega(D_\nu \cap \partial\Omega \setminus \bigcup_1^{N(\nu)} \beta_j^{(\nu)}) \leq r_\nu \leq \frac{h(r_\nu)}{B}.   \tag{2.15}$$

On the other hand, if $\varphi(\zeta_1)$ and $\varphi(\zeta_2)$ are the endpoints of an arc $\beta_j^{(\nu)}$, then $|\zeta_1 - \zeta_2| < 2\pi\omega(\partial \tilde{D}_\nu) \leq \frac{1}{n}$. By discarding some $\beta_j^{(\nu)}$ if necessary, we will suppose that $\beta_j^{(\nu)} \cap E \neq \emptyset$. Thus $z_\sigma \in E$, since $\delta > 2$, where z_σ is the point closest to the origin on the geodesic connecting ζ_1 and ζ_2. By (2.6) and Lemma 2.4

$$h^{-1}(\omega(\beta_j^{(\nu)})) \leq h^{-1}(1 - |z_\sigma|) \leq \frac{c_2}{B} \text{diam}\gamma_j^{(\nu)},$$

so that

$$\sum_{j=1}^{N(\nu)} h^{-1}(\omega(\beta_j^{(\nu)})) \leq \frac{2c_2 r_\nu}{B}.$$

Fix v, set $t_j = h^{-1}(\omega(\beta_j^{(v)}))$, $t = \sum t_j \leq 2c_2 r_v/B$, and $s = t/\log 1/t$. Then

$$\sum_{j=1}^{N(v)} \omega(\beta_j^{(v)}) = \sum_{j=1}^{N(v)} h(t_j) = \sum_{t_j \leq s} h(t_j) + \sum_{t_j > s} h(t_j).$$

Because $N(v) \leq \frac{2\pi}{\log 2} \log(1/t)$, (2.13) gives

$$\sum_{t_j \leq s} h(t_j) \leq \frac{2\pi}{\log 2} \log(1/t) h\left(\frac{t}{\log(1/t)}\right) \leq C_4 h\left(\frac{2c_2 r_v}{B}\right).$$

Moreover, since $\frac{h(t)}{t}$ is decreasing, (2.13) also gives

$$\sum_{t_j \geq s} h(t_j) = \sum_{t_j \geq s} t_j \frac{h(t_j)}{t_j} \leq \sum_{t_j \geq s} t_j \frac{h(s)}{s} \leq C_3 h\left(\frac{2c_2 r_v}{B}\right).$$

Therefore by (2.12)

$$\sum_{j=1}^{N(v)} \omega(\beta_j^{(v)}) \leq C_4 \frac{(1+B)^{\frac{1}{2}}}{B} h(r_v). \tag{2.16}$$

Now (2.14), (2.15), and (2.16) yield

$$|E| = \omega(\varphi(E)) \leq C_1 \frac{(1+B)^{\frac{1}{2}}}{B} \Lambda_h(\varphi(E)),$$

which is assertion (b). $\blacksquare$

3. The Number of Bad Discs

Let Ω be a simply connected domain, normalized so that $\infty \in \Omega$ and $\operatorname{diam}(\partial\Omega) = 2$, and write $\omega(E) = \omega(\infty, E, \Omega)$. It follows from Beurling's projection theorem that

$$\omega(B(\zeta, \rho)) \leq \rho^{1/2}, \tag{3.1}$$

for all $\zeta \in \partial\Omega$ and all ρ. The power of ρ in inequality (3.1) is sharp if $\partial\Omega$ is a line segment with one end at ζ, because then $\omega(B(\zeta, \rho)) \geq c\rho^{1/2}$ for $\rho \leq 1$. In their [1994] paper, Carleson and Makarov estimated the number of times (3.1) can almost fail. For $\varepsilon > 0$ define $N(\Omega, \varepsilon, \rho)$ to be the maximum cardinality of sets of the form

$$\left\{\zeta_k \in \partial\Omega : B(\zeta_k, \rho) \text{ are pairwise disjoint and } \omega(B(\zeta_k, \rho)) \geq \rho^{\frac{1}{2}+\varepsilon}\right\}.$$

Theorem 3.1. *There are constants A and K such that for all Ω*

$$N(\Omega, \varepsilon, \rho) \le A\rho^{-K\varepsilon}. \tag{3.2}$$

The constant

$$K^* = \inf\{K : (3.2) \text{ holds for some constant } A\}$$

will play a leading role in the later sections of this chapter and determining K^* is an important open problem.

Proof. Fix δ small and consider the annuli

$$A_j(\zeta) = \{z : \delta^j < |z - \zeta| < \delta^{j-1}\},$$

having moduli $m = \frac{1}{2\pi}\log 1/\delta$. Let E_j denote any subarc of $\Omega \cap \{|z - \zeta| = \delta^j\}$ and write

$$m_j = \inf_{\{E_j, E_{j-1}\}} d_{\Omega \cap A_j}(E_j, E_{j-1})$$

and

$$X_j(\zeta) = m_j - m.$$

Then $X_j \ge 0$ and $X_j = 0$ if and only if $A_j \cap \partial\Omega$ lies on a line segment with endpoint ζ.

Lemma 3.2. *Given $\varepsilon > 0$, there is $n(\varepsilon) > 0$ such that if $n > n(\varepsilon)$, $\rho = \delta^n$ and*

$$\omega\big(B(\zeta, \rho)\big) \ge \rho^{\frac{1}{2}+\varepsilon},$$

then

$$\sum_{j=1}^{n-1} X_j(\zeta) \le \frac{3}{\pi}\varepsilon n \log\frac{1}{\delta}. \tag{3.3}$$

Proof. We use Lemma 2.5 with $w_0 = \infty$ and $M = \frac{1}{2} + 2\varepsilon$. Take n so large that

$$\rho^{-\varepsilon} \ge 1 + \frac{2\pi}{\log 2}\log\frac{1}{\rho}.$$

Then by Lemma 2.5 there is an arc $E \subset \{|z - \zeta| = 2\rho\} \cap \Omega$ that separates ∞ from a continuum $\beta \subset \partial\Omega$ for which

$$\omega(\beta) = \omega(\beta \cap B(\zeta, 2\rho)) \ge \frac{\rho^{\frac{1}{2}+\varepsilon} - \rho^{\frac{1}{2}+2\varepsilon}}{\frac{2\pi}{\log 2}\log\frac{1}{\rho}} \ge \rho^{\frac{1}{2}+2\varepsilon}.$$

Take $J = \{|z - \zeta| = 1\}$. Then J meets $\partial\Omega$ so that by Theorem IV.5.3,

$$\operatorname{dist}_\Omega(E, J) \leq 1 + \frac{1}{\pi}\left(\frac{1}{2} + 2\varepsilon\right) \log \frac{1}{\rho}.$$

Set $E_n = E$. Because Ω is simply connected we can by reverse induction (i.e., for $j = n - 1, n - 2, \ldots, 2, 1$) find arcs $E_j \subset \Omega \cap \{|z - \zeta| = \delta^j\}$ such that E_j separates E_{j+1} from J. Then by the serial rule,

$$\sum_{j=1}^{n-1} \operatorname{dist}_{\Omega \cap A_j}(E_j, E_{j-1}) \leq \operatorname{dist}_\Omega(E, J).$$

Hence

$$\frac{(n-1)}{2\pi} \log \frac{1}{\delta} + \sum_{j=1}^{n-1} X_j \leq 1 + \frac{1}{\pi}\left(\frac{1}{2} + 2\varepsilon\right) \log \frac{1}{\rho},$$

and (3.3) holds when $n \geq \frac{1}{2\varepsilon} + 1$. ∎

Lemma 3.3. *There are absolute constants $\sigma > 0$ and $\eta > 0$ such that if $\zeta \in \partial\Omega$, $k \geq 2$, and*

$$X_1(\zeta) \leq \delta^{k\sigma},$$

then for all $\zeta' \in \partial\Omega \cap A_1(\zeta)$ and all $2 \leq j \leq k$,

$$X_j(\zeta') \geq \eta \log \frac{1}{\delta}. \tag{3.4}$$

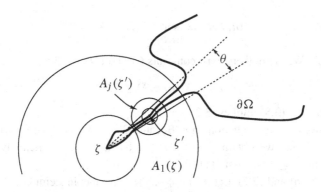

Figure VIII.2 Proof Lemma 3.3.

Proof. Since $\operatorname{diam}(\partial\Omega) = 2$ the continuum $\partial\Omega$ meets both boundary components of $A_1(\zeta)$. Let θ be the smallest angle of a sector of $A_1(\zeta)$ that contains

$A_1(\zeta) \cap \partial\Omega$. Then by Corollary V.4.3 or Exercise V.4,

$$X_1(\zeta) \geq c \frac{\theta^3}{(\log \frac{1}{\delta})^2},$$

where c is an absolute constant. Consequently

$$\theta < \frac{\delta^k}{2} \tag{3.5}$$

if $\sigma > 3$ and δ is sufficiently small. But now if $\zeta' \in A_1(\zeta) \cap \partial\Omega$ and $2 \leq j \leq k$, then the annulus $A_j(\zeta') \cap \partial\Omega$ contains two crosscuts of $A_j(\zeta')$ separated by angle $\frac{\pi}{3}$. Note this is even true for $|\zeta' - \zeta| > 1 - \delta$ because $\text{diam}(\partial\Omega) = 2$. That gives (3.4) with $\eta = \frac{1}{10\pi}$. ∎

To complete the proof of Theorem 3.1, we let D be any closed disc of radius 1 and let

$$N' = N'(\Omega, n, \lambda, \delta)$$

be the maximum number of points $\zeta_p \in D \cap \partial\Omega$ such that

$$|\zeta_p - \zeta_q| \geq \delta^n, \quad p \neq q \tag{3.6}$$

and

$$\sum_{j=1}^{n-1} X_j(\zeta_p) \leq \lambda. \tag{3.7}$$

Also let

$$N'(n, \lambda, \delta) = \sup_{\{\Omega, D\}} N'(\Omega, n, \lambda, \delta).$$

Fix δ small. We claim there exist constants C_1 and C_2 such that

$$N'(n, \lambda) \equiv N'(n, \lambda, \delta) \leq C_1 e^{C_2\lambda}. \tag{3.8}$$

By Lemma 3.2, (3.8) implies (3.2) with $K = \frac{3}{\pi}C_2$.

We prove (3.8) by induction on n. Because $\text{diam}(\partial\Omega) = 2$, we clearly have $N' \leq C\delta^{-2n}$ and we obtain (3.8) for $n \leq 2$ by taking C_1 sufficiently large. Now assume $n \geq 3$ and assume (3.8) holds for all $k < n$. Take $\zeta_1, \ldots \zeta_n \in \partial\Omega$ satisfying (3.6) and (3.7). Let σ and η be the constants in Lemma 3.3. There are three cases.

Case 1: $\min_{D \cap \partial\Omega} X_1(\zeta) > \delta^{2\sigma}$.

In this case

$$\sum_{j=2}^{n-1} X_j(\zeta_p) \le \lambda - \delta^{2\sigma}.$$

Cover $D \cap \partial\Omega$ by $c\left(\frac{1}{\delta}\right)^2$ discs of radius δ. Consider one such disc and suppose it contains points $\zeta_1, \cdots, \zeta_m$ that satisfy (3.6) and (3.7). Rescaling to a disc of radius $\frac{1}{2}$, we see by induction that

$$m \le N'(n-1, \lambda - \delta^{2\sigma}) \le C_1 e^{C_2(\lambda - \delta^{2\sigma})}.$$

Hence

$$N'(\Omega, n, \lambda) \le c\left(\frac{1}{\delta}\right)^2 e^{-C_2\delta^{2\sigma}} C_1 e^{C_2\lambda} \le C_1 e^{C_2\lambda},$$

if $C_2 \ge C_2(\delta)$.

Case 2: $\min_{D\cap\partial\Omega} X_1(\zeta) \le \delta^{n\sigma}$.

Fix $\zeta \in D \cap \Omega$ with $X_1(\zeta) \le \delta^{n\sigma}$. Then by Lemma 3.3, $X_j(z) \ge \eta \log \frac{1}{\delta}$ for $2 \le j \le n$ and all $z \in A_1(\zeta) \cap \partial\Omega$. Thus if $\lambda < (n-1)\eta \log \frac{1}{\delta}$, then all $\zeta_p \in D \cap \partial\Omega$ with (3.6) and (3.7) satisfy $\zeta_p \in B(\zeta, \delta)$, and by a rescaling

$$N'(\Omega, n, \lambda) \le N'(n-1, \lambda) \le C_1 e^{c_2\lambda}.$$

On the other hand, if $\lambda \ge (n-1)\eta \log \frac{1}{\delta}$, then by (3.6)

$$N'(\Omega, n, \lambda) \le c\left(\frac{1}{\delta}\right)^{2n} \le ce^{\frac{2n\lambda}{(n-1)\eta}} \le C_1 e^{C_2\lambda},$$

if C_1 and C_2 are sufficiently large.

Case 3: $\delta^{n\sigma} < \min_{D\cap\partial\Omega} X_1(\zeta) \le \delta^{2\sigma}$.

Let the minimum be attained at $\zeta \in D \cap \partial\Omega$, and choose $k \ge 2$ so that $X_1(\zeta) \in \left(\delta^{\sigma(k+1)}, \delta^{\sigma k}\right]$. Then by Lemma 3.3,

$$X_2, \ldots, X_k \ge \eta \log \frac{1}{\delta}$$

on $A_1(\zeta) \cap \partial\Omega$. By (3.5) we can cover $D \cap \partial\Omega \setminus B(\zeta, \delta)$ with $c\left(\frac{1}{\delta}\right)^k$ discs of radius δ^k, each centered in $A_1(\zeta)$. Then by rescaling and induction,

$$N'(\Omega, n, \lambda) \le N'(n-1, \lambda - \delta^{k\sigma}) + c\left(\frac{1}{\delta}\right)^k N(n-k, \lambda - (k-1)\eta \log \frac{1}{\delta})$$

$$\le C_1 e^{C_2(\lambda - \delta^{k\sigma})} + c\left(\frac{1}{\delta}\right)^k C_1 e^{C_2(\lambda - (k-1)\eta \log \frac{1}{\delta})}$$

$$\le C_1 e^{C_2\lambda}$$

provided $C_2 \ge C_2(\delta)$. ∎

4. Brennan's Conjecture and Integral Means Spectra

Let φ be a conformal mapping from the disc $\mathbb{D}$ onto a simply connected domain Ω. **Brennan's conjecture** is that

$$\iint_{\mathbb{D}} |\varphi'|^{2-p} dx dy = \iint_{\Omega} |(\varphi^{-1})'|^p dx dy < \infty \qquad (4.1)$$

whenever $\frac{4}{3} < p < 4$. When $p \notin (\frac{4}{3}, 4)$, the integral (4.1) diverges for the Koebe function $\varphi(z) = \frac{z}{(1-z)^2}$. When $p = 2$ the integral has value π. When $\frac{5}{3} < p < 3$, (4.1) is a consequence of the distortion estimate (I.4.17). See Metzger [1973]. When $\frac{4}{3} < p < 2$, (4.1) follows from the theorem of Prawitz, Exercise I.23(d), and a simple integration. Using the methods in Carleson [1973], Brennan [1978] established (4.1) in the range $3 < p < 3 + \eta$ for some unknown $\eta > 0$. In Appendix L we give Pommerenke's [1985a] elementary proof of (4.1) for $p < 3.399$. Brennan's conjecture is true for many special types of domains; see Exercise 8 and the paper [1998] of Barañski, Volberg, and Zdunik.

Brennan's conjecture is a point on a spectrum of questions about integral means of derivatives of univalent functions. Let φ be univalent on $\mathbb{D}$, let $t \in \mathbb{R}$, and define

$$\beta_\varphi(t) = \inf\Big\{\beta : \int |\varphi'(re^{i\theta})|^t d\theta = O\big(\frac{1}{1-r}\big)^\beta\Big\}. \qquad (4.2)$$

The function $\beta_\varphi(t)$ is the **integral means spectrum** of φ. Clearly $\beta_\varphi(t) \geq 0$. By Hölder's inequality, $\beta_\varphi(t)$ is a convex and continuous function. By the distortion theorem (I.4.17),

$$\beta_\varphi(t+s) \leq \begin{cases} \beta_\varphi(t) - s, & \text{if } s < 0, \\ \beta_\varphi(t) + 3s, & \text{if } s > 0. \end{cases} \qquad (4.3)$$

The Koebe function $k(z) = \frac{z}{(1-z)^2}$ has integral means spectrum

$$\beta_k(t) = \begin{cases} 3t - 1, & \text{if } t \geq \frac{1}{3}, \\ 0, & \text{if } -1 \leq t < \frac{1}{3}, \\ |t| - 1, & \text{if } t < -1. \end{cases} \qquad (4.4)$$

Define the (unbounded) **universal integral means spectrum** to be

$$B(t) = \sup\{\beta_\varphi(t) : \varphi \text{ is univalent}\}.$$

Thus $B(t)$ is the smallest number such that for all $\varepsilon > 0$ and all univalent φ there is constant $C(\varepsilon, \varphi)$ such that

$$\int |\varphi'(re^{i\theta})|^t d\theta \leq C(\varepsilon, \varphi)\big(\frac{1}{1-r}\big)^{B(t)+\varepsilon}.$$

Theorem I.4.5 yields the upper bound

$$B(t) \leq \begin{cases} 3t, & \text{if } t \geq 0, \\ |t|, & \text{if } t < 0, \end{cases} \tag{4.5}$$

and (4.4) shows that

$$B(t) \geq \begin{cases} 3t - 1, & \text{if } t \geq \frac{1}{3}, \\ 0, & \text{if } -1 \leq t < \frac{1}{3}, \\ |t| - 1, & \text{if } t < -1. \end{cases} \tag{4.6}$$

Feng and MacGregor [1976] proved $B(t) = 3t - 1$ for $t \geq 2/5$. See Exercise 7. Theorem 4.2 below shows that $B(t) = |t| - 1$ for $t < -K/2$, and Appendix L includes Pommerenke's [1985a] proof that

$$B(t) \leq -\frac{1}{2} + t + \left(\frac{1}{4} - t + 4t^2\right)^{1/2} \equiv \gamma(t) \leq 3t^2 + 7|t|^3 \tag{4.7}$$

for all t.

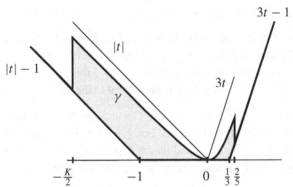

Figure VIII.3 Known bounds for $B(t)$.

Lemma 4.1. *Brennan's conjecture holds if and only if*

$$B(-2) = 1.$$

Proof. Assume $B(-2) = 1$ and let $2 < p < 4$. Then $B(2 - p) < \alpha < 1$ by convexity, so that for all φ

$$\int |\varphi'(re^{i\theta})|^{2-p} d\theta \leq C\left(\frac{1}{1-r}\right)^{\alpha}.$$

Then Fubini's theorem gives (4.1). Conversely, if (4.1) holds for $p < 4$ then because the means

$$\int |\varphi'(re^{i\theta})|^{2-p} d\theta$$

are increasing in r,

$$\int |\varphi'(re^{i\theta})|^{2-p}d\theta \le \frac{1}{1-r} \iint\limits_{\{r<|z|<1\}} |\varphi'(z)|^{2-p}dxdy$$

and $B(2-p) \le 1$. Therefore by continuity $B(-2) \le 1$, while (4.4) implies $B(-2) \ge 1$. ∎

The proof of Lemma 4.1 also shows

$$\sup\{p : (4.1) \text{ holds}\} = \inf\{p : B(2-p) \le 1\}.$$

Theorem 4.2. *Let K be a constant for which the conclusion (3.2) of Theorem 3.1 holds. Then for $t < -K/2$,*

$$B(t) = |t| - 1.$$

Define

$$K^* = \inf\{K : (3.2) \text{ holds for some constant } A\}. \tag{4.8}$$

By Theorem 4.2, Brennan's conjecture will be true if $K^* \le 4$. We will see conversely in Section 5 that Brennan's conjecture is false if $K^* > 4$. Section 5 also contains an example that shows $B(t) > |t| - 1$ for $t \in (-2, -1)$. Therefore $K^* \ge 4$ and Brennan's conjecture is equivalent to the conjecture that $K^* = 4$.

Proof. Fix $t < -K/2$ and assume $\infty \in \Omega$ and $\text{diam}(\partial\Omega) = 2$. Let $\varphi : \mathbb{D} \to \Omega$ be univalent and satisfy $\varphi(0) = \infty$. For $1 - r$ small we partition the circle $\{|z| = r\}$ into $\frac{C}{1-r}$ arcs I_j of equal length δ, with $\frac{1-r}{16} \le \delta < \frac{1-r}{8}$ and write z_j for the midpoint of I_j. Because $\| \log(\varphi')\|_B \le 6$, we have

$$\int_{I_j} |\varphi'|^t d\theta \sim \delta|\varphi'(z_j)|^t. \tag{4.9}$$

Let n and k be positive integers with n large and set

$$A(k) = \{j : \delta^{\frac{(k+1)}{n}} < |\varphi'(z_j)| \le \delta^{\frac{k}{n}}\}$$

and

$$N(k) = \text{card}\, A(k).$$

By the distortion theorem, $|\varphi'(z_j)| \ge \frac{1-r}{8} > \delta$, so $N(k) = 0$ for $k \ge n$. By Koebe's estimate if $j \in A(k)$ then

$$\varphi\left(B(z_j, \frac{\delta}{2})\right) \supset B\left(\varphi(z_j), \frac{\delta}{8}|\varphi'(z_j)|\right)$$

$$\supset B\left(\varphi(z_j), \frac{\delta^{1+\frac{k+1}{n}}}{8}\right) \supset \varphi\left(B(z_j, \frac{\delta^{1+\frac{1}{n}}}{32})\right).$$

Set $\rho = \frac{1}{10}\delta^{(1+\frac{k}{n})}$ and $\varepsilon = \frac{1}{2} - \frac{k}{n+k}$. Since $\delta^{\frac{1}{n}} \sim 1$ for large n, the discs $\{B(\varphi(z_j), \rho)\}_{j \in A(k)}$ are pairwise disjoint and

$$\omega\big(\infty, B(\varphi(z_j), \rho), \Omega\big) \geq \omega\big(0, B(z_j, \frac{\delta^{1+\frac{1}{n}}}{40}), \{|z| < 1 - r\}\big)$$

$$\geq C\delta = C\rho^{\frac{1}{2}+\varepsilon}.$$

By Theorem 3.1

$$\sum_{j \geq k} N(j) \leq A\delta^{\frac{-K}{2}(1-\frac{k}{n})}.$$

Therefore by (4.9) and a summation by parts,

$$\int |\varphi'(re^{i\theta})|^t d\theta \leq 2\pi + C \sum_{k=0}^{n-1} \delta^{\frac{k}{n}t+1} N(k)$$

$$\leq 2\pi + CA\delta^{1-\frac{K}{2}}\big(1 + \sum_{k=0}^{n-1} \delta^{\frac{k}{n}t}(1 - \delta^{-\frac{t}{n}})\delta^{\frac{K}{2}\frac{k}{n}}\big)$$

$$\leq C\big(\frac{t}{t+K/2}\big)\big(\frac{1}{1-r}\big)^{|t|-1},$$

since $t < -K/2$. ∎

5. *β* Numbers and Polygonal Trees

Let φ be the conformal map from $\mathbb{D}$ onto a simply connected domain Ω and let $\zeta_1 \neq \zeta_2$ be points of $\partial\mathbb{D}$ at which φ has nontangential limits $\varphi(\zeta_1)$ and $\varphi(\zeta_2)$. For $\varepsilon > 0$ let $B_\varepsilon(\zeta_j)$ be the component of $\Omega \cap B(\varphi(\zeta_j), \varepsilon)$ that contains some tail segment

$$\{\varphi(r\zeta_j) : r_0 < r < 1\}$$

of the image arc $\{\varphi(r\zeta_j) : 0 \leq r \leq 1\}$. (If $\varphi(\zeta_j) = \infty$, we let $B_\varepsilon(\zeta_j)$ be the corresponding component of $\Omega \cap \{|z| > 1/\varepsilon\}$.) Let $\Gamma_\varepsilon = \Gamma_\varepsilon(\Omega, \zeta_1, \zeta_2)$ be the family of paths in Ω joining $\overline{B_\varepsilon(\zeta_1)}$ to $\overline{B_\varepsilon(\zeta_2)}$, and let Γ_ε^* be the set of paths in $\mathbb{C}$ joining $\overline{B(\varphi(\zeta_1), \varepsilon)}$ to $\overline{B(\varphi(\zeta_2), \varepsilon)}$. If $\varepsilon < \delta$ let

$$\Gamma^j = \Gamma^j(\varepsilon, \delta) = \{\gamma \cap B_\delta(\zeta_j) : \gamma \in \Gamma_\varepsilon\},$$

$j = 1, 2$. Then by the basic rules for extremal lengths we have

$$\lambda(\Gamma_\varepsilon) \geq \lambda(\Gamma_\delta) + \lambda(\Gamma^1) + \lambda(\Gamma^2),$$

and

$$\lambda(\Gamma^j) \geq \frac{1}{2\pi} \log(\delta/\varepsilon), \quad j = 1, 2.$$

Then

$$
\begin{aligned}
0 \leq \ &\lambda(\Gamma_\delta) - \lambda(\Gamma_\delta^*) \\
&+ \lambda(\Gamma^1) - \frac{1}{2\pi} \log(\delta/\varepsilon) \\
&+ \lambda(\Gamma^2) - \frac{1}{2\pi} \log(\delta/\varepsilon) \\
\leq \ &\lambda(\Gamma_\varepsilon) - \lambda(\Gamma_\varepsilon^*) + O(\delta^2 + \varepsilon^2),
\end{aligned}
\tag{5.1}
$$

where we get $O(\delta^2 + \varepsilon^2)$ by comparing $\mathbb{C}^* \setminus \{B(\varphi(\zeta_1), \varepsilon) \cup B(\varphi(\zeta_2), \varepsilon)\}$ to an annulus bounded by concentric circles. Therefore the limit

$$\lim_{\varepsilon \to 0} \left(\lambda(\Gamma_\varepsilon) - \lambda(\Gamma_\varepsilon^*)\right)$$

exists in $[0, \infty]$. Define

$$\beta = \beta(\Omega, \zeta_1, \zeta_2) = \lim_{\varepsilon \to 0} e^{-2\pi(\lambda(\Gamma_\varepsilon) - \lambda(\Gamma_\varepsilon^*))}. \tag{5.2}$$

Then

$$0 \leq \beta(\Omega, \zeta_1, \zeta_2) \leq 1.$$

If $\beta > 0$ then for $\varepsilon < \delta$ small the two middle terms in (5.1) are small, and if $\frac{\delta}{\varepsilon}$ is fixed, then by Corollary V.4.3, $E_j \cap \left(B(\varphi(\zeta_j), \delta) \setminus B(\varphi(\zeta_j), \varepsilon)\right)$ is confined to a narrow sector with vertex $\varphi(\zeta_j)$, where E_j is the connected component of $\partial\Omega \cap B(\varphi(\zeta_j), \delta)$ which contains $\varphi(\zeta_j)$. Consequently $\beta(\Omega, \zeta_1, \zeta_2) > 0$ can hold for at most one point $\zeta_j \in \varphi^{-1}(\varphi(\zeta_j))$ and we can unambiguously define

$$\beta(\Omega, \varphi(\zeta_1), \varphi(\zeta_2)) = \beta(\Omega, \zeta_1, \zeta_2).$$

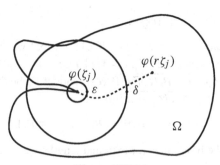

Figure VIII.4

It is clear that $\beta(\Omega, \zeta_1, \zeta_2)$ is invariant under Möbius transformations, and we will always take $\zeta_2 = \infty$.

Lemma 5.1. *Suppose $\psi : \mathbb{H} \to \Omega$ is a conformal mapping such that*

$$\frac{\psi(z)}{z^2} \to 1, \quad (z \to \infty). \tag{5.3}$$

Let $x \in \mathbb{R}$ and assume that $\psi(x) = a$, $\psi'(x) = 0$, and $\psi''(x)$ exists. Then

$$\beta(\Omega, a, \infty) = \frac{2}{|\psi''(x)|}. \tag{5.4}$$

Proof. If ε is small, $\psi^{-1}(B_\varepsilon(x))$ is approximately a half-disc of radius $\sqrt{\frac{2\varepsilon}{|\psi''(x)|}}$ while $\psi^{-1}(|z| = \frac{1}{\varepsilon})$ is approximately the semicircle of radius $\varepsilon^{-\frac{1}{2}}$. Then

$$\lambda(\Gamma_\varepsilon) \approx \frac{1}{\pi} \log\left(\frac{1}{\varepsilon}\left(\frac{|\psi''(x)|}{2}\right)^{\frac{1}{2}}\right)$$

while

$$\lambda(\Gamma_\varepsilon^*) \approx \frac{1}{2\pi} \log \frac{1}{\varepsilon^2},$$

and therefore (5.4) holds. ∎

For example, suppose $\partial\Omega$ is a finite connected union of line segments with extreme points $a_1, \ldots, a_n$ and ∞. In this case we say $\partial\Omega$ is a **polygonal tree**.

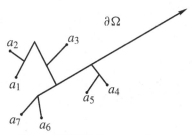

Figure VIII.5 Polygonal tree.

Let $\psi : \mathbb{H} \to \Omega$ satisfy (5.3) and let $a_j = \psi(x_j)$. Then

$$\beta(\Omega, a_j, \infty) = \frac{2}{|\psi''(x_j)|}.$$

On the other hand, if $a \notin \{a_j\}$,

$$\beta(\Omega, a, \infty) = 0.$$

When $\partial \Omega$ is a polygonal tree we write

$$\beta_j = \beta(\Omega, a_j, \infty).$$

Example 5.2. Let $0 < t < 1$ and set

$$\psi(z) = \sqrt{(z^2 - 1)(z^2 - t^2)}.$$

Then ψ is the conformal map from $\mathbb{H}$ to a domain Ω bounded by the half line $[-t, \infty]$ and the segment on the imaginary axis $[-ib, ib]$ where $b = (1 - t^2)/2$, and ψ satisfies (5.3).

$$ib$$

$$-t$$

$$-ib$$

Figure VIII.6

This example has three non-zero β numbers, and by (5.4) they are

$$\beta_1 = \beta(\Omega, -t, \infty) = \frac{2t}{1 + t^2}$$

and

$$\beta_2 = \beta_3 = \beta(\Omega, \pm ib, \infty) = \frac{1}{2} \frac{1 - t^2}{1 + t^2}.$$

Note that

$$\sum \beta_j^2 = \frac{1}{2} + \frac{2t^2}{(1 + t^2)^2} < 1$$

since $0 < t < 1$, but that for $p < 2$,

$$\sum \beta_j^p = \frac{2^p t^p + 2^{1-p}(1 - t^2)^p}{(1 + t^2)^p}$$

$$= 1 + 2^{1-p}(1 - t)^p + o((1 - t)^p) > 1$$

(5.5)

if $1 - t$ is small.

Theorem 5.3. *Let K^* be defined by (4.8) and let $p \geq 1$. Then the following are equivalent:*
(a) $\sum \beta_j^p \leq 1$ *for every polygonal tree.*
(b) $K^* \leq 2p$.
(c) $B(-p) = p - 1$.

From (5.5) and the theorem it follows that $K^* \geq 4$. Consequently we have the following corollary.

Corollary 5.4. *The following are equivalent:*
(a') $\sum \beta_j^2 \leq 1$ *for every polygonal tree.*
(b') $K^* = 4$.
(c') *Brennan's conjecture is true.*

Proof of Theorem 5.3. The implication (b) $\Longrightarrow$ (c) is Theorem 4.2, and we will prove (c) $\Longrightarrow$ (a) in Section 6.

(a) $\Longrightarrow$ (b): Assume (b) is false and let $p < q < K^*/2$. By the definition of K^* there exists, for any $A > 0$, $C > 0$, $\varepsilon > 0$, and $\rho > 0$, a simply connected domain Ω such that $\infty \in \Omega$ and $\operatorname{diam}(\partial\Omega) = 2$ and there exist

$$N \geq A\rho^{-2q\varepsilon} \left(\log \frac{1}{\rho}\right)^{2q+1} \tag{5.6}$$

points $\zeta_j \in \partial\Omega$ such that

$$|\zeta_j - \zeta_k| \geq 2\rho$$

and

$$\omega(\infty, B(\zeta_j, \rho) \cap \partial\Omega, \Omega) \geq \rho^{\frac{1}{2}+\varepsilon}.$$

By Lemma 2.5 there is an arc

$$\gamma_j \subset \Omega \cap \partial B(\zeta_j, 2\rho)$$

having endpoints in $\partial\Omega$ such that γ_j separates ∞ from a continuum $\alpha_j \subset \partial\Omega$ and

$$\omega(\infty, \alpha_j, \Omega) \geq c\frac{\rho^{\frac{1}{2}+\varepsilon}}{\log\frac{1}{\rho}}.$$

Here c denotes a constant having possibly different values at different occurrences.

Let $\varphi : \mathbb{D} \to \Omega$ be conformal with $\varphi(0) = \infty$. Then $\varphi^{-1}(\gamma_j)$ is an arc in $\mathbb{D}$ with endpoints on $\partial\mathbb{D}$. Let σ_j be the hyperbolic geodesic in $\mathbb{D}$ having the same endpoints as $\varphi^{-1}(\gamma_j)$ and let z_j be the midpoint of σ_j. Then

$$1 - |z_j| \sim \Lambda_1(\sigma_j) \geq \omega(\infty, \alpha_j, \Omega) \geq c\frac{\rho^{\frac{1}{2}+\varepsilon}}{\log\frac{1}{\rho}} \equiv \eta. \tag{5.7}$$

Set $r_j = (1 - |z_j|)/2$ and $B_j = B(z_j, r_j)$. By Koebe's estimate and the Gehring–Hayman Theorem, Exercise III.15,

$$\operatorname{diam}\varphi(B_j) \sim \operatorname{dist}(\varphi(z_j), \partial\Omega) \leq c\operatorname{diam}(\varphi(\sigma_j)) \leq c\operatorname{diam}(\gamma_j).$$

Therefore

$$\varphi(B_j) \subset B(\zeta_j, c\rho), \tag{5.8}$$

and hence no point lies in more than C_1 balls B_j. Also, by (I.4.17)

$$1 - |z_j| \le c|\varphi'(z_j)| \le \frac{c\rho}{1 - |z_j|}, \tag{5.9}$$

and by (5.7) and (5.9) we have

$$B_j \subset \{z : 1 - c\rho^{1/2} \le |z| \le 1 - \frac{\eta}{2}\}. \tag{5.10}$$

Moreover, (5.9) gives

$$\int_{B_j} |\varphi'|^{-q} dx dy \ge c r_j^2 |\varphi'(z_j)|^{-q} \ge c \frac{r_j^{2+q}}{\rho^q},$$

so that by (5.10), (5.8), and (5.6)

$$\int_{1-c\rho^{\frac{1}{2}} \le |z| \le 1-\frac{\eta}{2}} |\varphi'(z)|^{-q} dx dy \ge \frac{N}{C_1} \frac{\eta^{2+q}}{\rho^q} \ge cA\rho^{-2q\varepsilon} \left(\log \frac{1}{\rho}\right)^{2q+1} \frac{\eta^{2+q}}{\rho^q}.$$

Consequently there exists R,

$$1 - c\rho^{\frac{1}{2}} < R < 1 - \frac{\eta}{2} \tag{5.11}$$

such that

$$\int |\varphi'(Re^{i\theta})|^{-q} d\theta \ge c\rho^{-\frac{1}{2}} A\rho^{-2q\varepsilon} \left(\log \frac{1}{\rho}\right)^{2q+1} \frac{\eta^{2+q}}{\rho^q} \ge cA\left(\frac{1}{1-R}\right)^{q-1} \tag{5.12}$$

if ε is small.

From (5.12) it follows easily that (c) $\Longrightarrow$ (b); i.e., that $B(-p) > p - 1$ if $2p < K^*$; but let us continue and prove (a) $\Longrightarrow$ (b).

The left-hand side of (5.12) increases with R and $1 - \frac{1+R}{2}$ is comparable to $1 - R$. Thus we may suppose that $R = e^{-\frac{1}{n}} \sim 1 - \frac{1}{n}$ and that φ is analytic in a neighborhood of $\overline{\mathbb{D}}$ with $\varphi' \ne 0$ in $\overline{\mathbb{D}}$. Replacing φ with $-\varphi(-z)$ if necessary, we may suppose that

$$\left(\frac{1}{n}\right)^{q-1} \int_{\frac{\pi}{2}}^{\frac{3\pi}{2}} |\varphi'(e^{-\frac{1}{n}} e^{i\theta})|^{-q} \frac{d\theta}{2\pi} \ge cA. \tag{5.13}$$

Let φ_1 be the conformal map of $\mathbb{D}$ onto $\mathbb{D} \setminus [\frac{1}{e}, 1]$ such that $\varphi_1(0) = 0$ and

$\varphi(1) = \frac{1}{e}$. Then

$$\varphi_n(z) = \varphi_1(z^n)^{\frac{1}{n}} \equiv z \left(\frac{\varphi_1(z^n)}{z^n} \right)^{\frac{1}{n}}$$

is a conformal map of $\mathbb{D}$ onto $\mathbb{D} \setminus \cup \sigma_k$ where $\{\sigma_k\}$ are n equally spaced radial slits. Set $\zeta_k = e^{2\pi i k/n}$, $k = 1, \ldots, n$. Then $\varphi_n' = 0$ at $\{\zeta_k\}$ and $\varphi_n(\zeta_k) = R\zeta_k$. Moreover

$$|\varphi_n''(\zeta_k)| = n|\varphi_1''(1)||\varphi_1(1)|^{1-\frac{1}{n}} \sim cn.$$

Let φ_0 be the conformal map of $\mathbb{H}$ onto $\mathbb{D} \setminus [0, 1)$ given by

$$\varphi_0(z) = \frac{1}{(z + \sqrt{z^2 - 1})^2},$$

and set

$$\psi - C_0\, \varphi \circ \varphi_n \circ \varphi_0,$$

where

$$C_0 = \frac{4\varphi_1'(0)^{\frac{1}{n}}}{\lim_{z \to \infty} z\varphi(z)} \sim \frac{4e^{-\frac{1}{n}}}{\text{diam}(\partial\Omega)}.$$

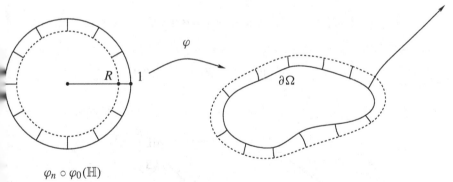

$\varphi_n \circ \varphi_0(\mathbb{H})$

Figure VIII.7

Then $\lim_{z \to \infty} \psi(z)/z^2 = 1$ and

$$\beta_k = \frac{2}{|\psi'(x_k)|} = \frac{2}{|\varphi'(R\zeta_k)||\varphi_n''(\zeta_k)||\varphi_0'(x_k)|},$$

where $x_k = \varphi_0^{-1}(\zeta_k)$. Since $|\varphi_0'(\varphi^{-1}(e^{i\theta}))| \le c_1 < \infty$ on $\{\frac{\pi}{2} \le \theta \le \frac{3\pi}{2}\}$,

$$\beta_k \geq \frac{c}{n|\varphi'(R\zeta_k)|}$$

provided $\frac{\pi}{2} \leq \arg \zeta_k \leq \frac{3\pi}{2}$.

By (5.13) and Koebe's estimate

$$\sum \beta_k^q \geq c \left(\frac{1}{n}\right)^{q-1} \sum \frac{1}{|\varphi'(R\zeta_k)|^q} \frac{1}{2n} \geq cA,$$

where the sums are taken over those k for which $\frac{\pi}{2} \leq \arg \zeta_k \leq \frac{3\pi}{2}$. Approximate $\psi(\mathbb{R})$ by a polygonal tree. Then (a) is violated because A is large, c is a constant and $p < q$. ∎

6. The Dandelion Construction and (c) $\Longrightarrow$ (a)

Let T be a polygonal tree with extreme points ∞ and $a_1, a_2, \ldots, a_N$. Let $a_0 \in T$ be such that the infinite segment $(a_0, \infty]$ is a relatively open subset of T in the direction $-e^{i\theta_0}$, and set $T_0 = T \setminus (a_0, \infty]$. The β numbers are dilation invariant, and we may suppose that $T_0 \subset B(a_0, 1)$. Choose $\varepsilon_0 > 0$ so that for each $j \geq 1$

$$T \cap B(a_j, \varepsilon_0) = [a_j, a_j - \varepsilon_0 e^{i\theta_j})$$

is a line segment and $B(a_j, \varepsilon_0) \cap B(a_k, \varepsilon_0) = \emptyset$ if $j \neq k$. Set $T_1 = T$ and for $\delta < \frac{1}{2}\varepsilon_0$ set

$$T_2 = T_1 \cup \bigcup (a_j + \delta e^{i(\theta_j + \theta_0)}(T_0 - a_0)).$$

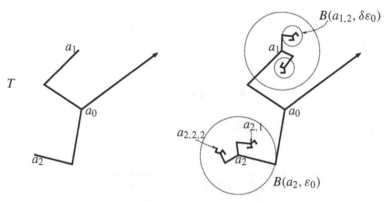

Figure VIII.8 Dandelion construction.

Then T_2 is a new polygonal tree with extreme points ∞ and the N^2 finite points

$$a_{j_1, j_2} = a_{j_1} + \delta e^{i(\theta_{j_1} + \theta_0)}(a_{j_2} - a_0).$$

For each pair j_1, j_2,

$$T_2 \cap B(a_{j_1, j_2}, \delta\varepsilon_0) = [a_{j_1, j_2}, a_{j_1, j_2} - \delta\varepsilon_0 e^{i\theta_{j_1, j_2}})$$

is again a line segment, and the balls $\{B(a_{j_1, j_2}, \delta\varepsilon_0)\}$ are pairwise disjoint. We continue, attaching a translation and rotation of $\delta^2 T_0$ at the N^2 finite extreme points of T_2 to obtain another polygonal tree $T_3 \supset T_2$ with N^3 finite extreme points, which we label a_{j_1, j_2, j_3}. This construction can be iterated indefinitely, yielding at stage n a tree $T_n \supset T_{n-1}$ having N^n finite extreme points $a_{j_1, j_2, \ldots, j_n}$, and T_{n+1} is constructed from T_n by attaching rotations and translations of T_0 dilated by the factor δ^n. Note that $B(a_{j_1, \ldots, j_n}, \delta^{n-1}\varepsilon_0) \subset B(a_{j_1, \ldots, j_{n-1}}, \delta^{n-2}\varepsilon_0)$ since $\delta^{n-1}\varepsilon_0 + \delta^{n-1} < \delta^{n-2}\varepsilon_0$ and so for each n the balls $\{B(a_{j_1, \ldots, j_n}, \delta^{n-1}\varepsilon_0)\}$ are pairwise disjoint.

Set

$$\Omega_n = \mathbb{C}^* \setminus T_n,$$

$$T_\infty = \overline{\bigcup_n T_n},$$

and

$$\Omega = \left(\bigcup \Omega_n\right)^\circ = \mathbb{C}^* \setminus T_\infty.$$

We call T_∞ the **dandelion** generated by T. Note, however, that T_∞ also depends on the choice of the dilation factor δ.

Write

$$J = (j_1, j_2, \ldots j_n)$$

for a multi-index of length $|J| = n$ in $\{1, 2, \ldots N\}^n$, and define

$$a_J = a_{j_1, j_2, \ldots, j_n},$$

$$\beta_J = \beta(\Omega_n, a_J, \infty),$$

$$\Omega_n^J = \Omega_n \cup \bigcup_{\substack{|I|=n \\ I \neq J}} B(a_I, 2\delta^n),$$

and

$$J' = (j_1, j_2, \ldots, j_{n-1}).$$

By the extension rule

$$\widetilde{\beta_J} \equiv \beta(\Omega_n^J, a_J, \infty) \le \beta_J,$$

and by the proof of (5.1)

$$\beta_J \le \beta_{J'}\beta_{j_n}(1 + O(\delta)). \tag{6.1}$$

The next lemma shows that the reverse inequality to (6.1) almost holds if the dilation factors are small enough.

Lemma 6.1. *Given* $\alpha > 0$ *there exists* $\delta = \delta(\alpha)$ *such that for every multi-index* J *of length n,*

$$\widetilde{\beta_J} \ge (1 - \frac{\alpha}{2})\widetilde{\beta_{J'}}\widetilde{\beta_{j_n}} \tag{6.2}$$

and

$$\beta_J \ge \widetilde{\beta_J} \ge (1 - \alpha)^n \beta_{j_1}\beta_{j_2}\cdots\beta_{j_n}. \tag{6.3}$$

Proof. We may choose δ small enough that

$$\widetilde{\beta_j} \equiv \beta(\Omega_1^j, a_j, \infty) \ge (1 - \frac{\alpha}{2})\beta_j,$$

for $j = 1, \ldots, N$. By the extension rule and induction, (6.2) implies (6.3) and we only prove (6.2). Because β numbers are translation invariant, we may assume $a_0 = 0$. Choose δ so small that

$$\delta \le \varepsilon_0 \left(\frac{1}{2\pi c}\log\frac{1}{1-\frac{\alpha}{2}}\right)^4,$$

where c is the constant given in Exercise IV.10. By construction,

$$\partial\Omega_n^J \cap \{z : \delta^{n-1} < |z - a_{J'}| < \delta^{n-2}\varepsilon_0\}$$

is a radial line segment and

$$\Omega_n^J \setminus B(a_{J'}, 2\delta^n) \supset \Omega_{n-1}^{J'} \setminus B(a_{J'}, 2\delta^{n-1}).$$

Set $\Sigma = \{z : |z - a_{J'}| = \delta^{n-\frac{3}{2}}\sqrt{\varepsilon_0}\}$. If ε is small then by Exercise IV.10 and the extension rule,

$$\lambda(\Gamma_\varepsilon) \equiv d_{\Omega_n^J}(\partial B(a_J, \varepsilon), \partial B(0, \tfrac{1}{\varepsilon}))$$

$$\leq d_{\Omega_n^J}(\partial B(a_J, \varepsilon), \Sigma) + d_{\Omega_n^J}(\Sigma, \partial B(0, \tfrac{1}{\varepsilon})) + c\left(\frac{\delta}{\varepsilon_0}\right)^{\frac{1}{4}}$$

$$\leq d_{\Omega_n^J}(\partial B(a_J, \varepsilon), \Sigma) + d_{\Omega_{n-1}^{J'}}(\Sigma, \partial B(0, \tfrac{1}{\varepsilon})) + \frac{1}{2\pi}\log\frac{1}{1 - \frac{\alpha}{2}}.$$

Removing a radial slit from an annulus does not affect the extremal distance between its bounding circles, so that as in the proof of (5.1)

$$d_{\Omega_{n-1}^{J'}}\left(\Sigma, \partial B(0, \tfrac{1}{\varepsilon})\right) - d_{\mathbb{C}}\left(\Sigma, \partial B(0, \tfrac{1}{\varepsilon})\right)$$

$$\leq d_{\Omega_{n-1}^{J'}}\left(\partial B(a_{J'}, \varepsilon), \partial B(0, \tfrac{1}{\varepsilon})\right) - d_{\mathbb{C}}\left(\partial B(a_{J'}, \varepsilon), \partial B(0, \tfrac{1}{\varepsilon})\right) + O(\varepsilon),$$

and

$$d_{\Omega_n^J}\left(\partial B(a_J, \varepsilon), \Sigma\right) - d_{\mathbb{C}}\left(\partial B(a_J, \varepsilon), \Sigma\right)$$

$$= d_{\Omega_1^{j_n}}\left(\partial B(a_{j_n}, \tfrac{\varepsilon}{\delta^{n-1}}), \partial B(a_0, \sqrt{\tfrac{\varepsilon_0}{\delta}})\right) - d_{\mathbb{C}}\left(\partial B(a_{j_n}, \tfrac{\varepsilon}{\delta^{n-1}}), \partial B(a_0, \sqrt{\tfrac{\varepsilon_0}{\delta}})\right)$$

$$\leq d_{\Omega_1^{j_n}}\left(\partial B(a_{j_n}, \tfrac{\varepsilon}{\delta^{n-1}}), \partial B(a_0, \tfrac{\delta^{n-1}}{\varepsilon})\right)$$

$$- d_{\mathbb{C}}\left(\partial B(a_{j_n}, \tfrac{\varepsilon}{\delta^{n-1}}), \partial B(a_0, \tfrac{\delta^{n-1}}{\varepsilon})\right) + O(\varepsilon)$$

and

$$d_{\mathbb{C}}\left(\partial B(a_J, \varepsilon), \Sigma\right) + d_{\mathbb{C}}\left(\Sigma, \partial B(0, \tfrac{1}{\varepsilon})\right) = d_{\mathbb{C}}\left(\partial B(a_J, \varepsilon), \partial B(0, \tfrac{1}{\varepsilon})\right) + O(\varepsilon).$$

Thus

$$\widetilde{\beta}_J = \lim_{\varepsilon \to 0} e^{-2\pi(\lambda(\Gamma_\varepsilon) - \lambda(\Gamma_\varepsilon^*))} \geq \widetilde{\beta}_{j_n}\widetilde{\beta}_{J'}(1 - \tfrac{\alpha}{2}),$$

for δ sufficiently small, which implies (6.2). ∎

Proof of (c) $\Longrightarrow$ (a): Fix $p > 0$ and assume (a) is false. Then there is a polygonal tree T with finite extreme points $a_1, a_2, \ldots, a_N$ such that

$$\sum_{j=1}^{N} \beta_j^p = A > 1.$$

We make the dandelion construction with dilation parameter δ so small, for $n \geq 2$, that Lemma 6.1 holds with α satisfying

$$(1 - \alpha)^P A > 1. \tag{6.4}$$

Let $\psi : \mathbb{H} \to \Omega$ be a conformal map satisfying (5.3) and set

$$\varphi(z) = \psi\left(i\left(\frac{1+z}{1-z}\right)\right) : \mathbb{D} \to \Omega.$$

For a multi-index $J = (j_1, \ldots, j_n)$, let

$$\varphi_J : \mathbb{D} \to \Omega_n^J,$$

with $\varphi_J(0) = \varphi(0)$ and $\varphi_J(1) = \infty$. If $f = \varphi^{-1}$, $f_J = \varphi_J^{-1}$, and $\zeta_J = f_J(a_J)$ then

$$|\varphi_J''(\zeta_J)| = \frac{8}{|1 - \zeta_J|^4} \frac{1}{\widetilde{\beta_J}} \leq \frac{C}{\widetilde{\beta_J}}. \tag{6.5}$$

By construction $B(a_J, \delta^{n-1}\varepsilon_0) \cap T_n$ is a line segment terminating at a_J, so that by the Schwarz reflection principle,

$$f_J(a_J + \delta^{n-1}\varepsilon_0 z^2) = \zeta_J + \left(\frac{2\delta^{n-1}\varepsilon_0}{\varphi_J''(\zeta_J)}\right)^{\frac{1}{2}} z + \ldots \tag{6.6}$$

is univalent on $\mathbb{D}$. By the growth theorem (I.4.16)

$$|f_J(a_J + \delta^{n-1}\varepsilon_0 z^2) - \zeta_J| = \left|\frac{2\delta^{n-1}\varepsilon_0}{\varphi_J''(\zeta_J)}\right|^{\frac{1}{2}} |z|(1 + O(|z|)), \tag{6.7}$$

uniformly in J. Set

$$U_J = \Omega_n^J \setminus B(a_J, \delta^n) \subset \Omega, \quad V_J = f_J(U_J) \subset \mathbb{D}.$$

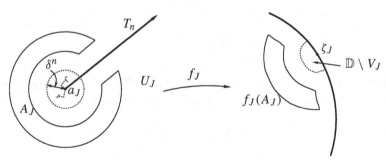

Figure VIII.9 The map f_J near a_J.

If $\tau_J : V_J \to \mathbb{D}$ is conformal with $\tau_J(0) = 0$ and $\tau_J(1) = 1$, then by Schwarz's lemma

$$1 - |f(z)| \geq 1 - |\tau_J \circ f_J(z)| \tag{6.8}$$

for $z \in U_J$. Set

$$A_J = \{z : 2\delta^n < |z - a_J| < 3\delta^n \text{ and } \mathrm{dist}(z, T) \geq \delta^n\}.$$

Then for $z \in A_J$,

$$\mathrm{dist}(z, \partial U_J) \geq c\, \mathrm{dist}(z, T),$$

so that by (6.8) and Koebe's estimate, Theorem I.4.3,

$$|f'(z)| \geq c|(\tau_J \circ f_J)'(z)|. \tag{6.9}$$

By (6.6) and the distortion theorem (I.4.17),

$$\left|\frac{d}{dz} f_J(a_J + \delta^{n-1}\varepsilon_0 z^2)\right| \geq \frac{1}{7}\left(\frac{2\delta^{n-1}\varepsilon_0}{|\varphi''(\zeta_J)|}\right)^{\frac{1}{2}}$$

for $|z| \leq 1/2$, and hence by (6.5)

$$|f_J'(z)| \geq C\left(\frac{\widetilde{\beta_J}}{\delta^n}\right)^{\frac{1}{2}}, \tag{6.10}$$

for $z \in A_J$. It follows from (6.7) that $\partial V_J \cap \mathbb{D}$ is approximately a semi-circle centered at ζ_J, so that by the distortion theorem there is a constant c_1 independent of J so that

$$|\tau_J'(w)| \geq c_1 \tag{6.11}$$

for $w \in f_J(A_J)$, and so by (6.9), (6.10), and (6.11)

$$|f'(z)| \geq C\left(\frac{\widetilde{\beta_J}}{\delta^n}\right)^{\frac{1}{2}}, \tag{6.12}$$

for $z \in A_J$.

Set

$$W = f(\cup_{|J|=n} A_J) \subset \mathbb{D}.$$

Then by the distortion theorem, (6.12), (6.3), and the binomial theorem,

$$\int_W (1 - |w|)^{p-2+2\varepsilon} |\varphi'(w)|^{-p} dxdy$$

$$= \sum_{|J|=n} \int_{A_J} (1 - |f(z)|)^{p-2+2\varepsilon} |f'(z)|^{p+2} dxdy$$

$$\geq C \sum_{|J|=n} \int_{A_J} (\delta^n)^{p-2+2\varepsilon} |f'(z)|^{2p+2\varepsilon} dxdy$$

$$\geq C\delta^{n\varepsilon} \sum_{|J|=n} \widetilde{\beta_J}^{p+\varepsilon}$$

$$\geq C \left[\delta^\varepsilon (1-\alpha)^{p+\varepsilon} \sum_{j=1}^N \beta_j^{p+\varepsilon} \right]^n.$$

By (6.4)

$$\delta^\varepsilon (1-\alpha)^{p+\varepsilon} \sum_{j=1}^N \beta_j^{p+\varepsilon} > 1$$

for ε sufficiently small, so that

$$\int_{\mathbb{D}} (1 - |w|)^{p-2+2\varepsilon} |\varphi'(w)|^{-p} dxdy = +\infty. \qquad (6.13)$$

Because $\int_0^1 (1-r)^{-1+\varepsilon} dr < \infty$ for $\varepsilon > 0$, (6.13) implies $B(-p) > p - 1$ and (c) $\Longrightarrow$ (a). ∎

7. Baernstein's Example on the Hayman–Wu Theorem

Exercise VII.5 described the improvement, by Fernández, Heinonen and Martio [1989], of the Hayman–Wu theorem: There is $p > 1$ and $C > 0$ such that whenever Ω is a simply connected domain, $f : \Omega \to \mathbb{D}$ is conformal, and L is a line,

$$\int_{L \cap \Omega} |f'|^p dx \leq C. \qquad (7.1)$$

The example $\Omega = \mathbb{D} \setminus [0, 1)$ shows that (7.1) fails when $p = 2$. A natural parallel exists between (7.1) and the Brennan problem; and in Havin, Hruščëv, and Nikol'skii [1984], Baernstein conjectured that (7.1) should hold for all $p < 2$. However, in [1989] Baernstein employed a dandelion construction to show that (7.1) fails for some $p < 2$.

Consider the tree

$$T = \{(1 + 2^{-\frac{1}{2}})e^{i\theta} : 0 \le \theta \le \pi\} \cup \{iy : y \ge 1 + 2^{-\frac{1}{2}}\}$$

and its truncation

$$T_0 = \{(1 + 2^{-\frac{1}{2}})e^{i\theta} : 0 \le \theta \le \pi\}$$

at the fork point $a_0 = i(1 + 2^{-\frac{1}{2}})$. Let $\delta > 0$ be a constant to be determined later. Let $T_1 = T$ and by induction assume T_k has 2^k finite extreme points a_J, all of which fall on the line

$$\left\{ \operatorname{Im} z = -(1 + 2^{-\frac{1}{2}}) \left(\frac{\delta - \delta^k}{1 - \delta} \right) \right\}.$$

Define

$$T_{k+1} = T_k \cup \bigcup (a_J + \delta^k (T_0 - a_0)),$$

and note that the induction hypothesis then holds for T_{k+1}. Set $T_\infty = \overline{\bigcup T_k}$, $\Omega = \mathbb{C}^* \setminus T_\infty$, $\Omega_k = \mathbb{C}^* \setminus T_k \subset \Omega$, and $L = \{\operatorname{Im} z = -(1 + 2^{-\frac{1}{2}})\frac{\delta}{1-\delta}\}$. Take conformal maps $f : \Omega \to \mathbb{D}$ and $f_k : \Omega_k \to \mathbb{D}$ such that $f(-i) = f_k(-i) = 0$. This construction differs from the dandelion construction in Section 6 because T is not polygonal, but it is not hard to see that all of the estimates from Section 6 still hold. Alternatively, approximate T_0 by a polygonal tree such that the final line segments ending at $z = \pm(1 + 2^{-\frac{1}{2}})$ are vertical and such that inequality (7.3) below holds.

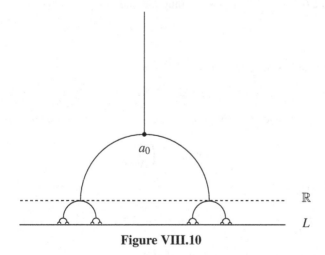

a_0

$\mathbb{R}$

L

Figure VIII.10

Theorem 7.1. *If δ is sufficiently small there exists p, $0 < p < 2$ such that*

$$\int_L |f'|^p dx = \infty. \tag{7.2}$$

Proof. The domain $\Omega_1 = \mathbb{C} \setminus T$ has two β numbers, $\beta_1 = \beta(\Omega_1, 1 + 2^{-\frac{1}{2}}, \infty)$ and $\beta_2 = \beta(\Omega_1, -(1 + 2^{-\frac{1}{2}}), \infty)$. Set

$$\psi_1 = z^2, \ \psi_2 = \frac{\sqrt{2}z - 1}{z + 1}, \ \psi_3(z) = \sqrt{z^2 - 1}, \ \psi_4(z) = i(1 + 2^{-\frac{1}{2}})\left(\frac{1+z}{1-z}\right),$$

and

$$\psi = \psi_4 \circ \psi_3 \circ \psi_2 \circ \psi_1.$$

Then ψ is a conformal map of $\mathbb{H}$ onto $\mathbb{C} \setminus T$ and

$$\lim_{z \to \infty} \frac{\psi(z)}{z^2} = 1$$

and $\psi(\pm 2^{-\frac{1}{4}}) = \pm(1 + 2^{-\frac{1}{2}})$. By explicit calculation,

$$|\psi''(2^{-\frac{1}{4}})| = |\psi''(-2^{-\frac{1}{4}})| = 8(\sqrt{2} - 1),$$

so that

$$A \equiv \beta_1 + \beta_2 = \frac{1 + \sqrt{2}}{2} > 1. \tag{7.3}$$

We will use the notation from Section 6. Choose $\alpha > 0$ so that $(1 - \alpha)A > 1$. By Lemma 6.1 with $p = 1$, and the binomial theorem,

$$\sum_{|J|=n} \widetilde{\beta_J} \geq [(1 - \alpha)A]^n, \tag{7.4}$$

for δ sufficiently small. Take such δ, fix n, and for $|J| = n$ set

$$I_J = L \cap A_J.$$

Then $|I_J| \geq C\delta^n$ and the intervals $\{I_J : |J| = n\}$ are pairwise disjoint. By (6.12)

$$\int_{I_J} |f'|^p dx \geq c\left(\frac{\widetilde{\beta_J}}{\delta^n}\right)^{\frac{p}{2}} \delta^n,$$

and by (7.4) if $p \leq 2$

$$\int_{\cup I_J} |f'|^p dx = \sum_{|J|=n} \int_{I_J} |f'|^p dx \geq C\left[\delta^{1-\frac{p}{2}}(1 - \alpha)A\right]^n.$$

But there exists $p_0 = p_0(\delta) < 2$ such that if $p > p_0$

$$\delta^{1-\frac{p}{2}}(1 - \alpha)A > 1$$

and therefore $f' \notin L^p(L)$. ∎

Notes

See Zygmund [1959] and M. Weiss [1959] for more about lacunary series similar to (1.2). Makarov's original [1985] proof of Theorem 1.1 used Fourier series instead of Hardy's inequality. For Theorem 2.1 see Makarov [1987] and [1990]. Theorem 2.2 is from Makarov [1985], but we have followed the proof in Makarov [1987]. A different proof by Rohde [1988] can also be found in Pommerenke [1991].

In [1987] Makarov extended his compression theorem VI.5.1 to other Hausdorff dimensions. If φ is univalent on $\mathbb{D}$ and if $E \subset \partial\mathbb{D}$, then for all $p > 0$,

$$\dim\varphi(E) \geq \frac{p\dim E}{\beta_\varphi(-p) + p + 1 - \dim E}.$$

See Exercise 5 for the proof.

Pommerenke's proof of (4.1) for $p < 3.399$ given Appendix L below has been improved by Bertilsson [1998] and [1999] to yield $p < 3.421$. Shimorin's recent paper [2003] establishes Brennan's conjecture up to $p < 3.7858$; Hedenmalm and Shimorin [2004] have improved Shimorin's estimate further.

Define the **bounded universal integral means spectrum** to be

$$B_b(t) = \sup\{\beta_\varphi(t) : \varphi \text{ is univalent and } \varphi(\mathbb{D}) \text{ is bounded}\}.$$

Thus $B_b(t) \leq B(t)$. Makarov [1998] proves

$$B(t) = \max\{B_b(t), 3t - 1\}.$$

See Exercise 9. Consequently there is a transition point t^* such that

$$B(t) = \begin{cases} 3t - 1, & \text{if } t \geq t^*, \\ B_b(t), & \text{if } t < t^*, \end{cases}$$

and by Exercise 7, $\frac{1}{3} < t^* < \frac{2}{5}$. By Exercise 9(a), $B_b(t) = t - 1$ for $t \geq 2$. Much deeper is the Jones and Makarov [1995] result that

$$B_b(t) = t - 1 + O(t - 2)^2, \text{ as } t \to 2,$$

which implies $B_b(2) = 1$ and $B_b'(2) = 1$. Near $t = 0$ the lower bound

$$B_b(t) = B(t) \geq 0.117t^2,$$

from Makarov [1986a] and Rohde [1989], complements the upper bound (4.7), and very recently the upper bound near $t = 0$ has been improved by Hedenmalm and Shimorin [2004] to

$$B_b(t) \le 0.437t^2.$$

Carleson and Jones [1992] proved that

$$B_b(1) = \sup\left\{\limsup_{n\to\infty} \frac{\log|na_n|}{\log n} : \varphi = \sum a_n z^n \text{ is bounded and univalent }\right\},$$

and conjectured that $B_b(1) = \frac{1}{4}$. Using the computer program "Zipper", see Marshall [1993], they obtained the estimate $B_b(1) \ge 0.21$. Computer experiments by Kraetzer [1996] for polynomial Julia sets also support the Carleson–Jones conjecture, but the best results known to date without computer assistance are

$$0.17 \le B_b(1) \le 0.4884$$

from Pommerenke [1975], Duren [1983] and Grinshpan and Pommerenke [1997], though Hedenmalm and Shimorin [2004] have improved the upper bound to 0.46 with the assistance of computer calculations. Brennan's conjecture and the Carleson–Jones conjecture would both follow from the beautiful and general conjecture in Kraetzer [1996]:

$$B_b(t) = \frac{t^2}{4}, \quad \text{for } |t| \le 2. \tag{BCJK}$$

Pommerenke [1999] calls (BCJK) the Brennan–Carleson–Jones–Kraetzer conjecture. Note that $g(t) = t^2/4$ is the only quadratic polynomial that satisfies both $g(0) = g'(0) = 0$ and $g(2) = g'(2) = 1$. It is not yet known if $B_b(t) = B_b(-t)$.

Binder [1997], [1998a] studied the counterpart of (BCJK) for complex parameters t: If φ is bounded and univalent, is

$$\int |\varphi'(re^{i\theta})^t|d\theta \le C(\varphi, \varepsilon)\left(\frac{1}{1-r}\right)^{\frac{|t|^2}{4}+\varepsilon}$$

for all $\varepsilon > 0$ and all $t \in \mathbb{C}$ with $|t| < 2$? Makarov [1998] characterized those functions $\beta(t)$ that have the form $\beta_\varphi(t)$ for some bounded univalent function φ. See Bertilsson [1999], Hedenmalm and Shimorin [2004], Makarov [1998], Pommerenke [1999], and the recent Jones [2005] for more complete discussions of Brennan's conjecture, integral means spectra, and related topics.

Exercises and Further Results

1. Let $g(z) = \sum_{k=0}^{\infty} z^{2^k}$, $z \in \mathbb{D}$ and $S_n(\zeta) = \sum_{k=0}^{n} \zeta^{2^k}$, $\zeta \in \partial\mathbb{D}$. Prove

$$\limsup \frac{|S_n(\zeta)|}{\sqrt{n \log\log n}} = 1 \qquad (1.3)$$

holds if and only if

$$\limsup_{r \to 1} \frac{|g(r\zeta)|}{\sqrt{\log\frac{1}{1-r} \log\log\log\frac{1}{1-r}}} = 1.$$

2. Prove Hardy's identity (1.6).

3. Pommerenke [1986b] shows that (1.1) fails with constant $C = 0.685$ by studying

$$g(z) = C_1 \sum_{n=1}^{\infty} z^{15^n}$$

with C_1 chosen so that $\|g\|_{\mathcal{B}} = 1$.

4. Let $\infty \in \Omega$ be a simply connected domain, let $\zeta_0 \in \partial\Omega$, let $K \subset \Omega$ be a continuum, and let $\alpha > 0$ and $\varepsilon > 0$. There is $\delta_0 > 0$ so that if $0 < \delta < \delta_0$ and $\omega(\infty, B(\zeta_0, \delta), \Omega) \geq \delta^\alpha$, then there is a crosscut $\beta \subset \partial B(\zeta_0, 2\delta)$ of Ω such that

$$\operatorname{dist}_\Omega(K, \beta) \leq \frac{\alpha + \varepsilon}{\pi} \log \frac{1}{\delta}.$$

Hint: Use the proof of Lemma 2.5.

5. Let φ be a conformal map from $\mathbb{D}$ to a simply connected domain Ω. The following results are in Makarov [1987]. See also Pommerenke [1991].

(a) Let $0 < t \leq 1$. Show that Lemma 2.5 is still true when the conclusion (2.8) is replaced with

$$M_{r^t}(\varphi^{-1}(D \cap \partial\Omega \setminus \bigcup_{j=1}^{N} \beta_j)) < C_t r^M.$$

Hint: Follow the proof of Lemma 2.5 but use Pfluger's theorem V.3.4. and Corollary D.3 from Appendix D to replace (2.9) with

$$M_{r^t}(\alpha_j) \leq C_t e^{-\pi\lambda(\Gamma_j)}.$$

(b) For all Borel $E \subset \partial\mathbb{D}$ and all $p > 0$,

$$\dim\varphi(E) \geq \frac{p \dim E}{\beta_\varphi(-p) + p + 1 - \dim E}.$$

We may assume Ω is a Jordan domain. Take $t < 1$, set

$$s = \frac{pt}{\beta_\varphi(-p) + p + 1 - t} > t$$

and

$$\tau = \frac{ps}{p+s} < s < 1$$

and assume $M_{r^s}(\varphi(E)) = 0$. Cover $\varphi(E)$ by discs $D_\nu = B(\zeta_\nu, r_\nu)$ so that $\sum r_\nu^s < \varepsilon$. For $M = s$ let γ_j^ν and β_j^ν, $1 \le j \le N(\nu) \le \frac{2\pi}{\log 2} \log(\frac{1}{r_\nu})$, be the crosscuts and continua given by the version of Lemma 2.5 proved in (a). Then by (a)

$$\sum_\nu M_{r^t}\left(\varphi^{-1}(D_\nu \cap \partial\Omega \setminus \bigcup_{j=1}^{N_\nu} \beta_j^\nu)\right) < C_t\varepsilon.$$

On the other hand, let $I_j^\nu = \varphi^{-1}(\beta_j^\nu)$ and $z_j^\nu = (1 - |I_j^\nu|)c_j^\nu$ where c_j^ν is the center of I_j^ν. Then by Hölder's inequality and Exercise IV.8,

$$\sum_{j=1}^{N_\nu}\left(|I_j^\nu|\,|\varphi'(z_j^\nu)|\right)^\tau \le C\left(\log(\frac{1}{r_\nu})\right)^{1-\tau}\left(\sum_{j=1}^{N_\nu} \operatorname{diam}\gamma_j^\nu\right)^\tau$$

$$\le C\left(\log(\frac{1}{r_\nu})\right)^{1-\tau} r_\nu^\tau$$

because the γ_j^ν are disjoint subarcs of $\partial B(\zeta_\nu, 2r_\nu)$. But taking $r = 1 - |I_j^\nu|$ we also have

$$|I_j^\nu|^{-p}|\varphi'(z_j^\nu)|^{-p} \le C|I_j^\nu|^{-p-1}\int|\varphi(re^{i\theta})|^{-p}d\theta \le C|I_j^\nu|^{(-\beta(-p)-p-1-\eta)}$$

for any $\eta > 0$. Together, these inequalities yield

$$\sum_\nu\sum_j |I_j^\nu|^t \le C\varepsilon,$$

which proves the assertion.

(c). Use (b) and (4.7) to prove $\dim\varphi(E) \ge 1$ whenever $\dim E = 1$ and show this result implies Theorem VI.5.1.

6. Prove (4.5) and (4.6).

7. Let φ be a univalent function on $\mathbb{D}$ and let $t > 2/5$. Then

$$\int|\varphi'(re^{i\theta})|^t d\theta \le C(t,\varphi)\frac{1}{(1-r)^{3t-1}}.$$

In other words, $B(t) = 3t - 1$. For $t > 1/2$, use Exercise I.23(d). For $2/5 < t \le 1/2$, set $p = 2/(2-t)$, $q = 2/t$, and fix an integer $k \ge 2$ such that $2pt(1 - 1/k) > 1$. Then write $g(z) = \left(\varphi(z^k)\right)^{1/k}$ and use Hölder's inequality, but estimate $\int |g(re^{i\theta})|^{pt(k-1)}d\theta$ and $\int |g'(re^{i\theta})|^2 d\theta$ with care. See Feng and MacGregor [1976] for the details.

8. (a) Let φ be a conformal mapping from $\mathbb{D}$ onto a simply connected domain Ω such that $\varphi(0) = 0$. The domain Ω and the map φ are said to be **starlike** if the segment $[0, w] \subset \Omega$ whenever $w \in \Omega$. This holds if and only if

$$\operatorname{Re}\left(\frac{z\varphi'(z)}{\varphi(z)}\right) > 0$$

on $\mathbb{D}$. More generally, φ and Ω are called **close-to-convex** if there exists starlike ψ such that

$$\operatorname{Re}\left(\frac{z\varphi'(z)}{\psi(z)}\right) > 0 \tag{E.1}$$

on $\mathbb{D}$, and φ is close-to-convex if and only if $\mathbb{C} \setminus \Omega$ is a union of closed half-lines meeting only at their endpoints. See Duren [1983] or Pommerenke [1991] for the proofs of these two geometric results.

(b) Let φ be starlike. Then for $0 < q < 2$,

$$\int_{\mathbb{D}} |\varphi'(z)|^{-q} dx dy < \infty.$$

Thus Brennan's conjecture is true for starlike univalent functions. For the proof, use (E.1) and the facts that if $F(z)$ is analytic on $\mathbb{D}$ and $\operatorname{Re} F(z) > 0$, then $F \in H^p$ for all $p < 1$ by Exercise A.4 and $|F(z)| \leq C/(1 - |z|)$ by the Schwarz lemma.

(c) If φ is close-to-convex and $\frac{4}{3} < q < 4$, then

$$\int_{\mathbb{D}} |\varphi'(z)|^{-q} dx dy < \infty.$$

Hint: Extend the proof of (b). Brennan [1978] attributes this result to B. Dahlberg and J. Lewis.

(d) If φ is close-to-convex and $E \subset \partial\mathbb{D}$ is a Borel set, then

$$\dim\varphi(E) \geq \frac{\dim E}{2 - \dim E}.$$

(e) If $0 < \alpha < 1$, there is a starlike domain Ω bounded by a rectifiable Jordan curve and $E \subset \partial\mathbb{D}$ such that $\Lambda_\alpha(E) > 0$ and

$$\dim\varphi(D) \leq \frac{\alpha}{2 - \alpha}.$$

See Makarov [1987] for (d) and (e).

9. The **bounded universal integral means spectrum** is

$$B_b(t) = \sup\Big\{\beta_\psi(t) : \psi \text{ is univalent and } \psi(\mathbb{D}) \text{ is bounded}\Big\}.$$

Thus $B_b(t) \le B(t)$.

(a) Suppose $\Omega = \psi(\mathbb{D})$ is bounded. Then

$$|\psi'(z)| \le \frac{4\operatorname{dist}\big(\psi(z), \partial\Omega\big)}{1 - |z|^2} = o\Big(\frac{1}{1 - |z|}\Big)$$

and

$$\frac{1}{|\psi'(z)|} = O\Big(\frac{1}{1 - |z|}\Big).$$

Consequently when ψ is bounded,

$$\beta_\psi(t \pm h) \le \beta_\psi(t) + s, \quad s > 0.$$

On the other hand, for $r \ge \frac{1}{2}$,

$$(1 - r)\int |\psi'(re^{i\theta})|^2 d\theta \le 2\int_r^1 \int_0^{2\pi} |\psi'(te^{i\theta})|^2 d\theta t dt \le 2\pi \operatorname{Area}(\Omega),$$

so that $\beta_\psi(2) \le 1$. It follows that

$$B_b(t) \le t - 1$$

for $t \ge 2$. Pommerenke [1999].

(b) Let $\Omega \subset \mathbb{C}^*$ be a simply connected domain such that $\partial\Omega$ is bounded and let $\varphi : \mathbb{D} \to \Omega$ be conformal. Then

$$\beta_\varphi(t) \le B_b(t),$$

for all real t. Hint: Consider the restrictions of φ to two subarcs of the circle.

(c) Let ψ be bounded and univalent and let $\delta = 1 - r$ be small. Partition $\{|z| = r\}$ into arcs I_k with $|I_k| = \frac{\delta}{2\pi r}$ and let z_k be the center of I_k. Then

$$\int |\psi'(re^{i\theta})| d\theta \sim \sum_j \Big(\frac{1}{\delta}\Big)^{a_j t - 1} N(a_j), \tag{E.2}$$

where

$$N(a) = \#\{I_k : |\psi'(z_k)| \sim \delta^{-a}\},$$

and

$$a_j = \frac{j}{\log\frac{1}{\delta}}.$$

Because the sum (E.2) has $O\left(\log \frac{1}{\delta}\right)$ terms, it compares to its largest term and we obtain

$$\beta_\psi(t) = \limsup_{\delta \to 0} \sup_j \left(a_j t - 1 + \frac{\log N(a_j)}{\log \frac{1}{\delta}}\right). \tag{E.3}$$

Moreover, let $\Omega = \psi(\mathbb{D})$, let $\alpha \geq \frac{1}{2}$, and define

$$f_\Omega(\alpha) = \lim_{\eta \downarrow 0} \limsup_{\rho \downarrow 0} \frac{\log N(\rho, \alpha, \eta)}{\log \frac{1}{\rho}},$$

where

$$N(\rho, \alpha, \eta) = \#\{\text{disjoint } B = B(\zeta, \rho), \zeta \in \partial\Omega, \rho^{\alpha+\eta} \leq \omega(B) \leq \rho^{\alpha-\eta}\}.$$

Then for $\alpha = \frac{1}{1-a}$,

$$\frac{f_\Omega(\alpha)}{\alpha} \sim \frac{\log N(a)}{\log \frac{1}{\delta}}. \tag{E.4}$$

Hence (E.3) gives

$$\beta_\psi(t) = \sup_\alpha \left(\frac{f_\Omega(\alpha) - t}{\alpha} - 1 + t\right)$$

and

$$B_b(t) = \sup_\alpha \left(\frac{F(\alpha) - t}{\alpha} - 1 + t\right) \tag{E.5}$$

where

$$F(\alpha) = \sup_{\Omega \text{ bounded}} f_\Omega(\alpha).$$

It is clear that $f_\Omega(\alpha)$ and $F(\alpha)$ are nondecreasing. See Makarov [1998] for (E.5) and several related results.

(d) Makarov [1998] also showed

$$B(t) = \max\{B_b(t), 3t - 1\}. \tag{E.6}$$

Therefore there is a transition point t^* such that

$$B(t) = \begin{cases} 3t - 1, & \text{if } t \geq t^*, \\ B_b(t), & \text{if } t < t^*, \end{cases}$$

and by Exercise 7, $\frac{1}{3} < t^* < \frac{2}{3}$.

To prove (E.6), we can assume by (b) that $\varphi = \frac{1}{\psi}$ where ψ is a bounded univalent function such that $\psi(0) = 1$. We consider the arcs I_k and their centers z_k from (c). For $0 < b \leq 2$ and $-1 < a < 1$ be of the form $\frac{j}{\log \frac{1}{\delta}}$ and define

$$N(a, b) = \#\{(a, b) : |\psi(z_k)| \sim \delta^b, |\psi'(z_k)| \sim \delta^{-a}\}.$$

Then

$$\int |\varphi'(re^{i\theta})|^t d\theta \sim \sum_{(a,b)} (\frac{1}{\delta})^{(a+2b)t-1} N(a, b).$$

In this sum the number of terms is $O\big((\log \frac{1}{\delta})^2\big)$, and therefore we have

$$\beta_\varphi(t) = \limsup_{\delta \to 0} \sup_{(a,b)} \big((a + 2b)t - 1 + \frac{\log N(a, b)}{\log \frac{1}{\delta}}\big). \qquad (E.7)$$

To estimate $N(a, b)$, fix a, b, and δ. By Lemma 2.5 we can assume

$$\{z : |z| = 1 - \delta, |\psi(z)| \sim \delta^b, |\psi'(z)| \sim \delta^{-a}\}$$

is an arc I of length $|I| = N(a, b)\delta = \delta^\tau$ and center c_I. Let $\rho = \delta^{1-a-b}$, let $z_I = (1 - |I|)c_I$, and define

$$T(z) = \frac{z + z_I}{1 + \overline{z_I} z}$$

and

$$\widetilde{\psi}(z) = \frac{\psi(T(z))}{\delta^b}.$$

Then on $T^{-1}(I)$, $|\widetilde{\psi}| \sim 1$ and $|\widetilde{\psi}'| \sim \delta^{\tau-a-b}$. Partition $T^{-1}(I)$ into $N(a, b)$ arcs of diameter (and harmonic measure) $\sim \delta^{1-\tau} = \rho^{\frac{1-\tau}{1-a-b}}$. Under $\widetilde{\psi}$ these arcs have images of diameter $\sim \delta^{1-a-b} = \rho$. It follows that

$$\frac{\log N(a, b)}{\log \frac{1}{\delta}} \leq (1 - a - b)F\big(\frac{1 - \tau}{1 - a - b}\big). \qquad (E.8)$$

By the Beurling projection theorem the worst case of (E.8) is given by $\tau = \frac{b}{2}$, and then (E.7) and (E.5) yield (E.6).

10. (a) Let T be a polygonal tree with finite extreme points $a_1, a_2, \dots, a_N$. Suppose that $f : \mathbb{D} \to \Omega = \mathbb{C} \setminus T$ is a conformal map such that $f(\zeta_k) = a_k$. Prove that

$$\lim_{r \to 1} (1 - r^2) \int_0^{2\pi} \frac{1}{|f'(re^{i\theta})|^2} \frac{d\theta}{2\pi} = \sum_{k=1}^{N} \frac{1}{|f''(\zeta_k)|^2}.$$

Prove an analogous statement with 2 replaced by p.

(b) Set $\beta_k = \beta(\Omega, a_k, \infty)$, where $\Omega = \mathbb{C} \setminus T$ and let $\psi : \mathbb{H} \to \Omega$ be a conformal map such that $\psi(z)/z^2 \to 1$ as $z \to \infty$. For fixed $t > 0$, define

$$f(w) = f_t(w) = \psi\left(it \cdot \frac{1+w}{1-w}\right) : \mathbb{D} \to \Omega.$$

Prove

$$\lim_{t\to\infty} \lim_{r\to1} (1-r^2) \int_0^{2\pi} \left|\frac{f'(0)}{f'(re^{i\theta})}\right|^2 \frac{d\theta}{2\pi} = 64 \sum_{k=1}^{N} \beta_k^2.$$

Prove

$$\lim_{r\to1} \lim_{t\to\infty} (1-r^2) \int_0^{2\pi} \left|\frac{f'(0)}{f'(re^{i\theta})}\right|^2 \frac{d\theta}{2\pi} = 64.$$

Note that $(f(w) - f(0))/f'(0)$ tends to the Koebe function uniformly on compact subsets of the disc as $t \to \infty$. Consequently, if the Koebe function is a local maximum for the mean of $1/|f'|^2$ on each circle of radius $r < 1$, then Brennan's conjecture is true. See Bertilsson [1999], p. 85, for the best result in this direction.

11. Let $\psi : \mathbb{H} \to \Omega$ be a conformal mapping.
    (a) For every $x \in \mathbb{R}$ and every $\alpha > 0$, the nontangential limit

$$J_\psi(x) = \lim_{\Gamma_\alpha(x)\ni z\to x} \left|\frac{z-x}{\psi'(z)}\right|$$

exists and is finite. Write $\Gamma_\alpha(\infty) = \{x + iy \in \mathbb{H} : |x| \le \alpha y\}$. Then also

$$J_\psi(\infty) = \lim_{\Gamma_\alpha(\infty)\ni z\to\infty} \left|\frac{\psi'(z)}{z}\right|$$

exists and is finite.
    (b) If $J_\psi(x) > 0$ and $J_\psi(\infty) > 0$, then $\lim_{\Gamma_\alpha(x)\ni z\to x} \psi(z) = a$ exists and is finite, and $\lim_{\Gamma_\alpha(\infty)\ni z\to\infty} \psi(z) = \infty$, and

$$\beta(\Omega, a, \infty) = J_\psi(x)J_\psi(\infty).$$

(c) Conversely, if $a \in \partial\Omega$ and $\beta(\Omega, a, \infty) > 0$, then there exists $\psi : \mathbb{H} \to \Omega$ such that $\lim_{\Gamma_\alpha(0)\ni z\to0} \psi(z) = a$, $J_\psi(0) > 0$, and $J_\psi(\infty) > 0$.
    See Bertilsson [1999] for (a), (b), and (c).

12. For $p > 0$, $B(-p) = p - 1$ if and only if there is a constant C such that for all normalized univalent functions $\varphi(z) = z + a_2 z^2 + \ldots$ and all $r < 1$,

$$\int |\varphi'(re^{i\theta})|^{-p} d\theta \le C\left(\frac{1}{1-r}\right)^{p-1}.$$

This observation is due to Bertilsson [1999]. Its proof is included in the argument following (5.12) together with (c) implies (a) of Theorem 5.3.

13. (a) It is a deep and difficult theorem of Astala [1994] that if $F : \mathbb{D} \to \mathbb{D}$ is K-quasiconformal, then for $p = \frac{2K}{K-1}$, the Jacobian J_F is weak L^p:

$$\text{Area}\{z \in \mathbb{D} : J_F(z) > \lambda\} \leq \frac{C}{\lambda^p}.$$

Eremenko and Hamilton [1995] have an alternate proof, also quite deep.
(b) Let Ω be a simply connected domain. Assume there exists $G : \Omega \to \mathbb{D}$ such that G is K-quasiconformal and locally Lipschitz,

$$\limsup_{w \to z} \frac{|G(z) - G(w)|}{|z - w|} \leq M < \infty,$$

for all $z \in \Omega$. Then whenever $\psi : \Omega \to \mathbb{D}$ is conformal map and $p < \frac{2K}{K-1}$,

$$\int_\Omega |\psi'(z)|^p dx dy < \infty.$$

Hint: Write $F = \psi \circ G^{-1}$, change variables and use (a). In particular, Brennan's conjecture is true if, for every $\varepsilon > 0$ and every simply connected domain Ω, there is a locally Lipschitz $2 + \varepsilon$ quasiconformal map of Ω onto $\mathbb{D}$. See Bishop [2002a] and [2002b].
(c) Suppose $\varphi : \mathbb{D} \to \Omega$ is a conformal map, and suppose there exists a K-quasiconformal $h : \mathbb{D} \to \mathbb{D}$ such that h is bilipschitz with respect to the hyperbolic metric on $\mathbb{D}$ and such that for some $C < \infty$,

$$1 - |h(z)| \leq C|\varphi'(z)|(1 - |z|)$$

for all $z \in \mathbb{D}$. Then $G = h \circ \varphi^{-1}$ satisfies the hypotheses in part (b). See Bishop [2002a] and Exercise 2, Appendix J.

IX

Infinitely Connected Domains

Let Ω be a domain such that $\infty \in \Omega$ and $\mathrm{Cap}(\partial\Omega) > 0$. In this chapter we study the relation between Hausdorff measures and harmonic measure on $\partial\Omega$. When Ω is finitely connected, the problem is well understood by Theorem VII.2.2 and by simple comparisons, and thus we assume that $\partial\Omega$ has infinitely many components.

In Section 1 we give the proof by Batakis of the Makarov–Volberg theorem that if $\Omega = \mathbb{C} \setminus K$ and K is a certain type of Cantor set then there exists $F \subset K$ such that $\omega(F, \Omega) = 1$ but $\dim_{\mathrm{Haus}}(F) < \dim_{\mathrm{Haus}}(K)$. In Section 2 we prove the result of Carleson [1985] and Jones and Wolff [1986] that if $K = \mathbb{C}^* \setminus \Omega$ is totally disconnected and satisfies some reasonable geometric hypotheses then there is $F \subset K$ such that $\omega(F, \Omega) = 1$ but $\dim_{\mathrm{Haus}}(F) < 1$. Then in Section 3 we discuss the [1988] Jones–Wolff theorem that for every Ω there exists $F \subset \partial\Omega$ with $\omega(F) = 1$ but $\dim_{\mathrm{Haus}}(F) \leq 1$.

1. Cantor Sets

Fix numbers $0 < a < b < 1/2$. For every sequence $\{a_n\}$ with

$$a \leq a_n \leq b \tag{1.1}$$

form the Cantor set $K = K(\{a_n\})$ as follows:

$$K = \bigcap K_n,$$

where $K_0 = [0, 1] \times [0, 1]$, $K_1 = \bigcup_{j=1}^{4} Q_j^1 \subset K_0$, and each Q_j^1 is a square of side $\sigma_1 = a_1$ containing a corner of $Q_0 = K_0$. At stage n

$$K_n = \bigcup Q_J^n,$$

315

where Q_J^n is a square of side $\sigma_n = a_1 a_2 \cdots a_n$ having its sides parallel to the axes, and $J = (j_1, j_2, \ldots, j_n)$ is a multi-index of length $|J| = n$ with $j_k \in \{1, 2, 3, 4\}$. Then $Q_J^n \supset \bigcup \{Q_{J,j}^{n+1} : j = 1, \ldots, 4\}$, and the $Q_{J,j}^{n+1}$ are components of K^{n+1} containing the four corners of Q_J^n. See Figure IX.1.

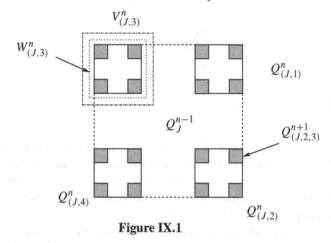

$$V_{(J,3)}^n$$

$W_{(J,3)}^n$

$Q_{(J,1)}^n$

Q_J^{n-1}

$Q_{(J,2,3)}^{n+1}$

$Q_{(J,4)}^n$

$Q_{(J,2)}^n$

Figure IX.1

It is clear from (1.1) and Chapter III that $\mathrm{Cap}(K) > 0$. It is also clear that $K(\{a_n\})$ has Hausdorff dimension

$$\alpha(\{a_n\}) = \sup\left\{\beta > 0 : \liminf_{n\to\infty} 4^n \sigma_n^\beta = \infty\right\}$$
$$= \inf\left\{\beta > 0 : \liminf_{n\to\infty} 4^n \sigma_n^\beta = 0\right\} \tag{1.2}$$

because the Hausdorff measure $\Lambda_\beta(K)$ can be estimated using coverings by the squares Q_J^n.

Set $\Omega = \mathbb{C}^* \setminus K$ and write $\omega(F) = \omega(\infty, F, \Omega)$.

Theorem 1.1. *There exists $F \subset K$ such that*

$$\omega(F) = 1 \tag{1.3}$$

but

$$\dim_{\mathrm{Haus}}(F) < \dim_{\mathrm{Haus}}(K). \tag{1.4}$$

The proof of Theorem 1.1 depends on four lemmas. Write $K_J^n = K \cap Q_J^n$ and take the squares

$$W_J^n = (1 + c_1) Q_J^n$$
$$V_J^n = (1 + c_2) Q_J^n$$

concentric with Q_J^n, where $0 < c_1 < c_2$ are constants that depend only on a and b and c_2 is so small that for some constant $c_3 > 0$

$$\text{dist}(V_{J_1}^n, V_{J_2}^n) \geq c_3 \sigma_n > 0$$

whenever $J_1 \neq J_2$.

Lemma 1.2. *There exist constants $\delta_1 > 0$ and $\delta_2 > 0$, depending only on a and b, such that for all $z \in \partial W_J^n$*

$$\omega(z, K_J^n, \Omega) \geq \omega(z, K_J^n, V_J^n \setminus K_J^n) \geq \delta_1, \tag{1.5}$$

and for all compact $E \subset K_J^n$

$$\omega(z, E, \Omega) \geq \omega(z, E, V_J^n \setminus K_J^n) \geq \delta_2 \omega(z, E, \Omega). \tag{1.6}$$

Proof. Let $g(z, w)$ be Green's function for V_J^n. Then by a change of scale there are constants C_1 and C_2 such that

$$g(z, \zeta) > C_1 \tag{1.7}$$

whenever $z \in \partial W_J^n$ and $\zeta \in K_J^n$, and

$$g(z, \zeta) \leq \log \frac{1}{|z - \zeta|} + \log \sigma_n + C_2 \tag{1.8}$$

whenever $z, \zeta \in K_J^n$. Let μ be the equilibrium distribution for K_J^n and consider the Green potential

$$U(z) = \int g(z, \zeta) d\mu(\zeta).$$

Then $U(z)$ is a positive harmonic function on $V_J^n \setminus K_J^n$. Since $\text{Cap}(K_J^n) \geq C_3 \sigma_n$ by Chapter III, (1.8) gives $0 < U(z) \leq C_4$ on K_J^n. On the other hand, $U(z) = 0$ on ∂V_J^n. Therefore on $V_J^n \setminus K_J^n$ we have

$$\omega(z, K_J^n, \Omega) \geq \omega(z, K_J^n, V_J^n \setminus K_J^n) \geq U(z)/C_4 \geq \frac{C_1}{C_4}$$

by (1.7), and (1.5) holds with $\delta_1 = \frac{C_1}{C_4}$. A very similar argument was used in Chapter III.

In (1.6) the leftmost inequality is obvious. To prove the rightmost inequality take $z_0 \in \partial W_J^n$ so that $\omega(z_0, E, \Omega) = \sup_{\partial W_J^n} \omega(z, E, \Omega)$. Then by the maximum principle and (1.5)

$$\omega(z_0, E, V_J^n \setminus K_J^n) = \omega(z_0, E, \Omega) - \int_{\partial V_J^n} \omega(\zeta, E, \Omega) d\omega(z, \zeta, V_J^n \setminus K_J^n)$$

$$\geq \omega(z_0, E, \Omega) - (1 - \delta_1) \omega(z_0, E, \Omega)$$

$$= \delta_1 \omega(z_0, E, \Omega)$$

and (1.6) follows by Harnack's inequality.   ∎

Choose the index j_{n+1} so that for every $J = (j_1, j_2, \ldots, j_n)$, $Q^{n+1}_{(J,1)}$ is the *upper right* corner of Q^n_J and $Q^{n+1}_{(J,4)}$ is the *lower left* corner.

Lemma 1.3. *There is $\eta > 0$ such that for all $n \geq 2$,*

$$\omega(\infty, K^n_{1,1,\ldots,1,1}, \Omega) \geq (1+\eta)\omega(\infty, K^n_{1,1,\ldots,1,4}, \Omega).$$

Proof. For $j = 1, 4$ write

$$u_j(z) = \omega(z, K^n_{1,1,\ldots,1,j}, \Omega)$$

and

$$U_j(z) = \omega(z, K^{n-1}_{1,1,\ldots,1,j}, \mathbb{C}^* \setminus K^n_{1,1,\ldots,1}).$$

Let L be the diagonal of $K^{n-1}_{1,1,\ldots,1}$ that separates $K^n_{1,1,\ldots,1,1}$ from $K^n_{1,1,\ldots,1,4}$ and let H be the half-plane bounded by L and containing $K^n_{1,1,\ldots,1,4}$. Then on $W = H \setminus K^{n-1}_{1,1,\ldots,1}$

$$U_4(z) - U_1(z) = \omega(z, K^n_{1,1,\ldots,1,4}, W).$$

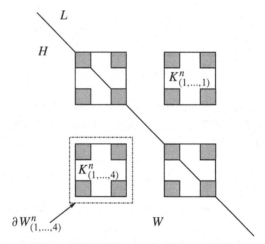

Figure IX.2 The proof of Lemma 1.3.

Therefore

$$U_4 - U_1 > 0 \qquad\qquad (1.9)$$

on W, while on $K^{n-1}_{1,1,\ldots,1,4}$

$$U_4 - U_1 > c_4 > 0 \qquad\qquad (1.10)$$

with constant c_4 independent of n, by Lemma 1.2 and Harnack's inequality. Now

$$u_1(z) - u_4(z) = U_1(z) - U_4(z) - \int_{K \setminus K^{n-1}_{1,1,\ldots,1}} (U_1(\zeta) - U_4(\zeta)) d\omega(z, \zeta, \Omega).$$

Hence by (1.9) and (1.10)

$$u_1(\infty) - u_4(\infty) \geq c_4 \omega(\infty, K^{n-1}_{1,1,\ldots,1,4}, \Omega).$$

On the other hand, by Lemma 1.2 and Harnack's inequality

$$\omega(z, K^{n-1}_{1,1,\ldots,1,4}, \Omega) \geq c_5$$

on $\partial W^n_{1,1,\ldots,1,4}$. Therefore $u_1(\infty) \geq (1 + c_4 c_5) u_4(\infty)$ by the maximum principle and the lemma holds with $\eta = c_4 c_5$. ∎

The next lemma is from Carleson [1985] and Makarov and Volberg [1986].

Lemma 1.4. *Let Ω be a domain such that $\infty \in \Omega$, and let*

$$A_1 \subset B_1 \subset A_2 \subset B_2 \subset \ldots A_n \subset B_n$$

be simply connected Jordan domains such that for each j

$$\Omega_j = B_j \setminus \overline{A_j} \subset \Omega$$

and

$$\mathrm{mod}(\Omega_j) \geq \alpha > 0. \tag{1.11}$$

Let u and v be positive harmonic functions on Ω such that

$$u = v = 0, \quad \text{on} \quad \partial \Omega \setminus A_1.$$

Then there exists $C = C(\alpha) > 0$ and $q = q(\alpha) < 1$ such that on $\Omega \setminus B_n$,

$$\left| \frac{u(z)/v(z)}{u(\infty)/v(\infty)} - 1 \right| \leq C q^n. \tag{1.12}$$

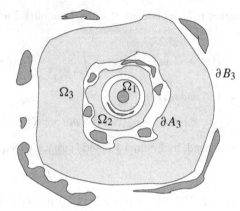

Figure IX.3

Proof. By (1.11) there is a conformal mapping

$$\varphi_j : \Omega_j \to \{1 < |z| < 1 + \beta_j\}$$

with $\varphi(\partial A_j) = \{|z| = 1\}$ and $\beta_j \geq \beta = e^{2\pi\alpha} - 1$. For $k = 1, 2$ set

$$\gamma_j^{(k)} = \varphi_j^{-1}\big(\{|z| = 1 + \frac{k\beta_j}{3}\}\big)$$

and

$$a_j^{(k)} = \frac{\max\{u(z)/v(z) : z \in \gamma_j^{(k)}\}}{\min\{u(z)/v(z) : z \in \gamma_j^{(k)}\}}.$$

By Harnack's inequality there is $C = C(\alpha)$ such that

$$1 \leq a_j^{(k)} \leq C.$$

We claim there is $\delta_3 > 0$ such that

$$a_j^{(2)} \leq a_j^{(1)}\big(1 - \delta_3(a_j^{(1)} - 1)\big). \tag{1.13}$$

Let us accept (1.13) for the moment and complete the proof of (1.12). By the maximum principle,

$$a_1^{(1)} \geq a_1^{(2)} \geq a_2^{(1)} \geq \ldots \geq a_n^{(2)}$$

because $u = v = 0$ on $\partial\Omega \setminus \overline{A_1}$. Write $a_j^{(2)} = 1 + b_j$. Then by (1.13)

$$1 + b_{j+1} \leq (1 + b_j)(1 - \delta b_j) \leq 1 + (1 - \delta)b_j$$

and $b_n \leq C(1 - \delta)^{n-1}$. Since on $\Omega \setminus B_n$

$$(1 + b_n)^{-1} \leq \left(\frac{u(z)}{v(z)}\right) \bigg/ \left(\frac{u(\infty)}{v(\infty)}\right) \leq (1 + b_n),$$

we therefore obtain (1.12).

To prove (1.13) we can assume $\min_{\gamma_j^{(1)}} \frac{u(z)}{v(z)} = 1$. Then on $\gamma_j^{(1)}$,

$$v(z) \leq u(z) \leq a_j^{(1)} v(z) = (1 + b)v(z)$$

where $b > 0$ by Harnack's inequality. Set

$$E_1 = \gamma_j^{(1)} \cap \{u \leq (1 + \frac{b}{2})v(z)\}$$

$$E_2 = \gamma_j^{(1)} \cap \{u > (1 + \frac{b}{2})v(z)\}$$

and let λ_z denote harmonic measure at z for the component of $\Omega \setminus \gamma_j^{(1)}$ containing $\gamma_j^{(2)}$. Since $u = v = 0$ on $\partial\Omega \setminus \overline{A_1}$ we have, for $z \in \gamma_j^{(2)}$,

$$u(z) \leq (1 + \frac{b}{2}) \int_{E_1} v(\zeta) d\lambda_z(\zeta) + (1 + b) \int_{E_2} v(\zeta) d\lambda_z(\zeta)$$

and

$$v(z) = \int_{E_1 \cup E_2} v(\zeta) d\lambda_z(\zeta).$$

Hence

$$u(z) \leq (1 + b)v(z) - \frac{b}{2}\lambda_z(E_1) \min_{E_1} v(\zeta)$$

and since by Harnack's inequality $v(z) \leq C_1 \min_{E_1} v(\zeta)$, with $C_1 = C_1(\alpha)$, we obtain

$$u(z) \leq v(z)a_j^{(1)}\left(1 - \frac{C_1 b \lambda_z(E_1)}{2(1 + b)}\right).$$

Similarly,

$$u(z) \geq v(z) + \frac{b}{2}\lambda_z(E_2) \min_{E_2} v(\zeta)$$

$$\geq v(z)\left(1 + C_2(\alpha)\lambda_z(E_2)\right),$$

and we conclude that

$$a_j^{(2)} \leq a_j^{(1)}\left(1 - C_3(\alpha)b\lambda_z(\gamma_j^{(1)})\right)$$

$$\leq a_j^{(1)}\left(1 - C_4(\alpha)b\right). \qquad \blacksquare$$

Lemma 1.5. *There is $N = N(a, b)$ such that for every n and J there exists $J' = J'(J) = (J, j_{n+1}, \ldots, j_{n+N})$ such that*

$$\omega(K_{J'}^{n+N}) \le \frac{1}{3} \frac{\omega(K_J^n)}{4^N}. \tag{1.14}$$

Proof. Fix J and N and for $j = 1, 4$ write

$$v_j(z) = \omega(z, K_{J,1,\ldots,1,j}^{n+N}, \Omega)$$

and

$$V_j(z) = \omega(z, K_{J,1,\ldots,1,j}^{n+N}, V_J^n \setminus K_J^n).$$

Let $z \in \partial W_J^n$. By Lemma 1.4,

$$\left| \frac{V_1(z)/V_4(z)}{v_1(\infty)/v_4(\infty)} - 1 \right| \le \left| \frac{V_1(z)/V_4(z)}{v_1(z)/v_4(z)} - 1 \right| + Cq^N.$$

But

$$\left| \frac{V_1(z)/V_4(z)}{v_1(z)/v_4(z)} - 1 \right| = \frac{v_4(z)}{V_4(z)} \left| \frac{V_1(z)}{v_1(z)} - \frac{V_4(z)}{v_4(z)} \right|$$

$$= \frac{v_4(z)}{V_4(z)} \int_{\partial V_J^n} \left| \frac{v_4(\zeta)}{v_4(z)} - \frac{v_1(\zeta)}{v_1(z)} \right| d\omega(z, \zeta, V_J^n \setminus K_J^n)$$

$$= \frac{v_4(z)}{V_4(z)} \int_{\partial V_J^n} \frac{v_4(\zeta)}{v_4(z)} \left| \frac{v_1(\zeta)/v_1(z)}{v_4(\zeta)/v_4(z)} - 1 \right| d\omega(z, \zeta, V_J^n \setminus K_J^n)$$

$$\le \frac{1}{\delta_2} (1 + Cq^N) Cq^N,$$

by Lemmas 1.2 and 1.4.

If we take $J = (1, 1, \ldots, 1)$ we get

$$\frac{v_1(\infty)}{v_4(\infty)} \ge 1 + \eta,$$

by Lemma 1.3 and if N is large,

$$\frac{V_1(z)}{V_4(z)} \ge 1 + \frac{\eta}{2}$$

by the two preceding estimates. However, $\frac{V_1(z)}{V_4(z)}$ does not depend on J. Therefore we have

$$\frac{v_1(\infty)}{v_4(\infty)} > 1 + \frac{\eta}{4}$$

for all J if N is large. Consequently

$$\inf_{J'} \frac{\omega(K_{J'}^{n+N})}{\omega(K_J^n)} \le \left(\frac{4^{N+1}}{4^{N+1}+\eta}\right)4^{-N}$$

and (1.14) holds if N is replaced by N^p, where $p = p(\eta, N)$. ∎

Conclusion of Proof of Theorem 1.1. On any K_J^n take two probabilities P_1 and P_2 such that

$$P_1(K_{J'(J)}^{n+N}) = \frac{1}{3}4^{-N} \quad \text{and} \quad P_2(K_{J'(J)}^{n+N}) = 4^{-N}.$$

Write $J = J_n(z)$ when $z \in K_J^n$. Let n_1 be large, take $n_{j+1} = n_j + j$ and set

$$E_j = \left\{z \in K : \frac{\#\{m \le j : z \in K_{J'(J(z))}^{(n_j+m)N+N}\}}{j} \ge \frac{2}{3}4^{-N}\right\}.$$

By the elementary theory of large deviations, Durrett [1996], pp. 70-76, there are constants C_1 and $C_2 > 0$ (depending on N) such that if n_1 is large

$$P_1(E_j) \le e^{-C_1 j} \tag{1.15}$$

and

$$P_2(E_j) \ge 1 - e^{-C_2 j}. \tag{1.16}$$

Set $F = \bigcup_{k=1}^{\infty} A_k$, where

$$A_k = \bigcap_{j=k}^{\infty} (K \setminus E_j). \tag{1.17}$$

Then

$$K \setminus F = \bigcap_{k=1}^{\infty} \bigcup_{j=k}^{\infty} E_j,$$

and by (1.14) and (1.15)

$$\omega(K \setminus F) \le \lim_{k \to \infty} \sum_{j=k}^{\infty} e^{-C_1 j} = 0.$$

On the other hand, since the E_j are independent it follows from (1.16) and (1.17) that for $j > k$, A_k can be covered by $e^{C_2(n_k-n_j)}4^{n_j N}$ squares of side $\sigma_{n_j N}$. Let $\alpha = \dim_{\text{Haus}}(K)$, let a be as in (1.1), and let $0 < \gamma < \alpha$ satisfy

$$e^{-C_2}\left(\frac{1}{a}\right)^{N(\beta-\gamma)} < 1$$

for some $\beta > \alpha$. Then by (1.1) and (1.2),

$$\Lambda_\gamma(A_k) \leq \lim_{j\to\infty} e^{C_2(n_k-n_j)} 4^{n_j N} \sigma_{n_j N}^\gamma = 0.$$

Therefore $\Lambda_\gamma(F) = 0$.   ∎

2. For Certain Ω, dim $\omega < 1$

Let $K \subset \mathbb{C}$ be compact and let $\Omega = \mathbb{C}^* \setminus K$. We assume there are constants α, C_1, and C_2 such that for all $z_0 \in K$ and all $0 < r < \text{diam}(K)$

$$\text{Cap}(B(z_0, r) \cap K) \geq C_1 r, \tag{2.1}$$

and assume there is a ring domain $A_r = A(z_0, r) \subset \Omega \cap \{r < |z - z_0| < \frac{r}{\alpha}\}$ bounded by rectifiable Jordan curves Γ_1 and Γ_2 such that

$$\ell(\Gamma_j) \leq C_2 r \tag{2.2}$$

and

$$\text{dist}(\Gamma_1, \Gamma_2) \geq \alpha r. \tag{2.3}$$

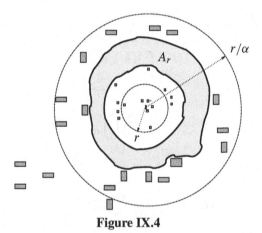

Figure IX.4

Theorem 2.1. *Assume K satisfies (2.1), (2.2), and (2.3). Then there exists $\delta = \delta(\alpha, C_1, C_2) > 0$ and there exists $F \subset K$ such that $\omega(\infty, F, \Omega) = 1$ but $\dim_{\text{Haus}}(F) \leq 1 - \delta$.*

The proof of Theorem 2.1 consists of three steps. We first construct a dyadic approximation of K by sets $Q_{n,j}$ called **cubes**. Next we employ a stopping time to select certain of the cubes $Q_{n,j}$. Then a lower bound for the harmonic measure of the stopped $Q_{n,j}$ is obtained from an analysis of the critical points of Green's function. Assume $\text{diam}(K) = 1$ and set $\omega(E) = \omega(\infty, E \cap K, \Omega)$. Throughout the proof c and C denote constants which may change with each occurrence but which depend only on α, C_1, and C_2.

Lemma 2.2. *For* $n = 1, 2, \ldots$ *there exist simply connected Jordan domains* $Q_{n,j}$ *such that*

$$K \subset \bigcup_j Q_{n,j}, \tag{2.4}$$

$$Q_{n,j} \subset \{z : \text{dist}(z, K) < 4^{-n}\}, \tag{2.5}$$

$$\text{dist}(\partial Q_{n,j}, K) \geq c4^{-n}, \tag{2.6}$$

$$\text{diam}(Q_{n,j}) \leq c\frac{4^{-n}}{\alpha}, \tag{2.7}$$

$$\text{dist}(Q_{n,j}, Q_{n,k}) \geq c4^{-n}, \tag{2.8}$$

and

$$\text{dist}(\partial Q_{n,j}, \partial Q_{n+1,k}) \geq c4^{-n}. \tag{2.9}$$

Moreover, $\partial Q_{n,j}$ *is a rectifiable Jordan curve of class* C^3 *with arc length parameterization* $z = z(s)$ *satisfying*

$$\left|\frac{d^2 z}{ds^2}\right| \leq C4^n \quad and \quad \left|\frac{d^3 z}{ds^3}\right| \leq C4^{2n}. \tag{2.10}$$

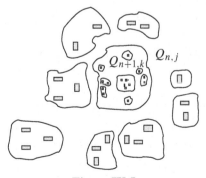

Figure IX.5

Proof. Let $\{S_{n,j}\}$ denote the components of $\{z : \text{dist}(z, K) < 4^{-n}\}$ and let $z_0 \in K \cap S_{n,j}$. Then by (2.3) $S_{n,j}$ is contained in the bounded component of $\mathbb{C}^* \setminus A(z_0, \frac{4^{-n}}{\alpha})$, and for $k \neq j$,

$$S_{n,k} \cap A(z_0, \frac{4^{-n}}{\alpha}) = \emptyset.$$

By (2.2), (2.3), and Exercise IV.5(d),

$$\text{mod}(A(z_0, \frac{4^{-n}}{\alpha})) = \frac{\log R_{n,j}}{2\pi} \geq C\alpha.$$

Let $f_{n,j}$ be a conformal mapping from $\{1 < |z| < R_{n,j}\}$ onto $A(z_0, \frac{4^{-n}}{\alpha})$ and let $\gamma_{n,j}$ be the core curve $f_{n,j}(\{z : |z| = \sqrt{R_{n,j}}\})$ of $A(z_0, \frac{4^{-n}}{\alpha})$. Then by Koebe's theorem there is $c = c(\alpha)$ such that

$$\text{dist}(\gamma_{n,j}, K) \geq c4^{-n} \tag{2.11}$$

and $\gamma_{n,j}$ satisfies the smoothness condition (2.10). Let $\{U_{n,j}\}$ be the bounded components of the complements of the components of $\bigcup \gamma_{n,j}$. Then each $U_{n,j}$ is simply connected. Let $\{V_{n,j}\}$ be the set of $U_{n,j}$ such that $K \cap U_{n,j} \neq \emptyset$ and $U_{n,j}$ is maximal with respect to set inclusion. Then by (2.11) and the smoothness of $\gamma_{n,j}$ there exists $Q_{n,j} \subset V_{n,j}$ satisfying (2.4)–(2.10) such that $K \cap V_{n,j} \subset Q_{n,j}$. ■

Like dyadic squares, the cubes $Q_{n,j}$ have the property

$$Q_{m,k} \subset Q_{n,j} \quad \text{or} \quad Q_{m,k} \cap Q_{n,j} = \emptyset \tag{2.12}$$

if $m > n$. Furthermore

$$K = \partial\Omega = \bigcap_n \bigcup_k Q_{n,k}.$$

Now fix integers M and m and set

$$\mathcal{B}_M = \left\{ Q_{n,k} : n \leq m, \, \omega(Q_{n,k}) \geq M4^{-n} \text{ and } Q_{n,k} \text{ is maximal} \right\},$$

$$\mathcal{G}_M = \left\{ Q_{m,j} : Q_{m,j} \not\subset Q_{n,k} \text{ for all } Q_{n,k} \in \mathcal{B}_M \right\}$$

and

$$\tilde{\Omega} = \mathbb{C}^* \setminus \left(\bigcup_{\mathcal{B}_M} (Q_{n,k}) \cup \bigcup_{\mathcal{G}_M} (Q_{m,j}) \right) \subset \Omega.$$

Write $\tilde{G}(z) = g_{\tilde{\Omega}}(z, \infty)$ and $\tilde{\omega}(E) = \omega(\infty, E, \tilde{\Omega})$.

Lemma 2.3. *There is a constant* $c = c(\alpha)$ *such that if* $Q_{n,j} \in \mathcal{B}_M \cup \mathcal{G}_M$, *then*

$$\widetilde{\omega}(Q_{n,j}) \leq c\omega(Q_{n,j}), \tag{2.13}$$

and on $\partial\widetilde{\Omega}$,

$$\frac{\partial\widetilde{G}}{\partial n} \leq cM. \tag{2.14}$$

Proof. By Lemma 2.2 there is open $\widetilde{Q}_{n,j} \supset Q_{n,j}$ with $K \cap \widetilde{Q}_{n,j} \subset Q_{n,j}$ and $\mathrm{dist}(\partial\widetilde{Q}_{n,j}, Q_{n,j}) \geq c4^{-n}$. Then $\omega(z, K \cap Q_{n,j}, \Omega) \geq c$ at all $z \in \partial\widetilde{Q}_{n,j}$ by (2.1) and the proof of Lemma 1.2. Then (2.13) follows by the maximum principle, since $\omega(z, Q_{n,j}, \Omega) \geq 0 = \omega(z, Q_{n,j}, \widetilde{\Omega})$ on $\partial\widetilde{\Omega} \setminus \widetilde{Q}_{n,j}$.

By construction $\omega(Q_{n,k}) \leq M4^{-n+1}$ for all $Q_{n,k} \in \mathcal{B}_M \cup \mathcal{G}_M$ and by (2.10)

$$\sup_{\partial Q_{n,k}} \frac{\partial\widetilde{G}}{\partial n} \leq c \inf_{\partial Q_{n,k}} \frac{\partial\widetilde{G}}{\partial n}.$$

Hence by (2.13) and Lemma 2.2,

$$\sup_{\partial Q_{n,k}} \frac{\partial\widetilde{G}}{\partial n} \leq \frac{2\pi c}{\ell(\partial Q_{n,k})} \widetilde{\omega}(\partial Q_{n,k}) \leq cM,$$

so that (2.14) holds. $\blacksquare$

The heart of the proof of Theorem 2.1 is the identity

$$\frac{1}{2\pi} \int_{\partial\widetilde{\Omega}} \frac{\partial\widetilde{G}}{\partial n} \log \frac{\partial\widetilde{G}}{\partial n} ds = \gamma + \sum_j \widetilde{G}(z_j), \tag{2.15}$$

in which $\{z_j\}$ is the set of critical points of $\widetilde{G}$ and $\gamma = -\log \mathrm{Cap}(\partial\widetilde{\Omega})$ is Robin's constant for $\partial\widetilde{\Omega}$. See Exercise III.2 for the proof. We assume $\mathrm{Cap}(K) < 1$ and take M so large that $\gamma > 0$ in (2.15). Then (2.14) and (2.15) yield

$$\sum_j \widetilde{G}(z_j) \leq \log(cM). \tag{2.16}$$

Now let $N = N(\alpha)$ be a positive integer to be determined later and define

$$\mathcal{U}_N = \{Q_{n,j} : Q_{p,k} \subset Q_{n,j} \text{ and } Q_{p,k} \in \mathcal{B}_M \cup \mathcal{G}_M \Rightarrow p > n + N\}.$$

If $N(\alpha)$ is sufficiently large, it follows from (2.1), (2.6), and (2.7) that whenever $Q_{n,j} \in \mathcal{U}_N$ there exist two disjoint cubes $Q_{n+N,k_1}, Q_{n+N,k_2} \notin \mathcal{B}_M \cup \mathcal{G}_M$ such that

$$Q_{n+N,k_1} \cup Q_{n+N,k_2} \subset Q_{n,j}. \tag{2.17}$$

Lemma 2.4. *There exist constants* c *and* N *such that*

(a) *If $Q_{n,j} \in \mathcal{U}_N$ then*

$$\inf_{\partial Q_{n,j}} \widetilde{G}(z) \geq c\widetilde{\omega}(Q_{n,j}). \qquad (2.18)$$

(b) *If $Q_{n+N,k} \subset Q_{n,j}$, then*

$$c \inf_{\partial Q_{n,j}} \widetilde{G} \leq \sup_{\partial Q_{n+N,k}} \widetilde{G} \leq \frac{1}{2} \inf_{\partial Q_{n,j}} \widetilde{G}. \qquad (2.19)$$

Proof. (a) There is $\widetilde{Q}_{n,j} \supset Q_{n,j}$ such that $\widetilde{Q}_{n,j} \setminus Q_{n,j} \subset \widetilde{\Omega}$ and

$$\operatorname{dist}(\partial \widetilde{Q}_{n,j}, \partial Q_{n,j}) \leq c \operatorname{dist}(\partial Q_{n,j}, \partial \widetilde{\Omega}).$$

By (2.1) and Lemma 2.2, $\widetilde{\omega}(z, Q_{n,j} \cap \partial \widetilde{\Omega}) \geq c$ on $\partial \widetilde{Q}_{n,j}$ and by Harnack's inequality and a comparison $\widetilde{G}(z, \zeta) > C$ for all $z \in Q_{n,j}$ and $\zeta \in \partial \widetilde{Q}_{n,j}$. Hence for $z \in Q_{n,j}$,

$$\widetilde{G}(z) \geq \int_{\partial \widetilde{Q}_{n,j}} G(z, \zeta) d\omega_{\widetilde{\Omega} \setminus \widetilde{Q}_{n,j}}(\infty, \zeta) \geq c\widetilde{\omega}(Q_{n,j}).$$

(b) Write

$$Q_{n,j} = Q_0 \supset Q_1 \supset Q_2 \supset \ldots \supset Q_N = Q_{n+N,k}$$

where $Q_p = Q_{n+p,k}$ for some k. By (2.1) and Lemma 2.2 there is $\eta > 0$ such that for all p and all $z \in \partial Q_p$,

$$\eta < \omega(z, \partial Q_{p-1}, Q_{p-1} \cap \widetilde{\Omega}) \leq 1 - \eta.$$

Hence by induction

$$\eta^N < \omega(z, \partial Q_0, Q_0 \cap \widetilde{\Omega}) \leq (1 - \eta)^N.$$

Also, by Lemma 2.2 and Harnack's inequality,

$$\sup_{\partial Q_0} \widetilde{G}(z) \leq C \inf_{\partial Q_0} \widetilde{G}(z),$$

so that since $\widetilde{G} = 0$ on $Q_0 \cap \partial \widetilde{\Omega}$,

$$c\eta^N \inf_{\partial Q_0} \widetilde{G} \leq \sup_{\partial Q_N} \widetilde{G} \leq C(1 - \eta)^N \inf_{\partial Q_0} \widetilde{G},$$

and if

$$C(1 - \eta)^N < \frac{1}{2}, \qquad (2.20)$$

then (2.19) holds.   ∎

Fix N so that (2.17) and (2.20) both hold, and note that the lower bound in (2.19) holds with $c = c(\alpha, N)$.

Lemma 2.5. *If $Q_{n,j} \in \mathcal{U}_{3N}$, there is a critical point $z_{n,j}$ of $\widetilde{G}$ such that*

$$z_{n,j} \in Q_{n,j} \setminus \bigcup Q_{n+3N,k}$$

and

$$\widetilde{G}(z_{n,j}) \geq c\widetilde{\omega}(Q_{n,j}).$$

Proof. There are 4 cubes $Q_{n,j} \supset Q_{n+N,j_1} \supset Q_{n+2N,j_2} \supset Q_{n+3N,j_3}$, which we relabel as $Q^1 \supset Q^2 \supset Q^3 \supset Q^4$, so that $Q^k \notin \mathcal{B}_M \cup \mathcal{G}_M$ and by Lemma 2.4

$$\sup_{\partial Q^{k+1}} \widetilde{G} < \frac{1}{2} \inf_{\partial Q^k} \widetilde{G}.$$

Take $a = \sup_{\partial Q^1} \widetilde{G}(z)$, $b = \inf_{\partial Q^4} \widetilde{G}(z)$, and $U = Q^1 \cap \{a < \widetilde{G} < b\} \subset \widetilde{\Omega}$. By (2.17) and (2.19) $Q^2 \setminus Q^3$ contains a component of $\partial U \cap \{\widetilde{G} = a\}$ and hence by Exercise II.17 there is a critical point $z_{n,j}$ of $\widetilde{G}$ in U and by (2.18) and (2.19)

$$\widetilde{G}(z_{n,j}) \geq c \inf_{\partial Q_{n,j}} \widetilde{G} \geq c\widetilde{\omega}(Q_{n,j}). \qquad \blacksquare$$

Proof of Theorem 2.1. By (2.8) and (2.12) each $Q_{p,j} \in \mathcal{B}_M \cup \mathcal{G}_M$ falls in $p - N$ cubes $Q_{n,k} \in \mathcal{U}_M$. Therefore by (2.16) and (2.18),

$$c \sum_{\mathcal{B}_M} n\widetilde{\omega}(Q_{n,k}) + c \sum_{\mathcal{G}_M} m\widetilde{\omega}(Q_{m,k}) \leq \log(cM),$$

and we can take m large and get

$$\sum_{\mathcal{G}_M} \widetilde{\omega}(Q_{m,k}) \leq \frac{1}{4}.$$

Set

$$\mathcal{B}_M^* = \left\{ Q_{n,k} \in \mathcal{B}_M : \omega(Q_{n,k}) \leq 4\widetilde{\omega}(Q_{n,k}) \right\}$$

so that

$$\sum_{\mathcal{B}_M \setminus \mathcal{B}_M^*} \widetilde{\omega}(Q_{n,k}) \leq \frac{1}{4}.$$

Then

$$\sum_{\mathcal{B}_M^*} \widetilde{\omega}(Q_{n,k}) \geq \frac{1}{2}$$

and on $\mathcal{B}_M^*$

$$\frac{1}{4} \leq \frac{\widetilde{\omega}(Q_{n,k})}{\omega(Q_{n,k})} \leq c$$

by (2.13). Therefore

$$\sum_{\mathcal{B}_M^*} \omega(Q_{n,k}) \geq \frac{1}{2c}.$$

Each $Q_{n,j} \in \mathcal{B}_M^*$ can be covered by a ball of radius

$$c4_{-n} \leq r_{n,j} \leq C4^{-n}$$

by Lemma 2.2 and

$$M4^{-n} \leq \omega(Q_{n,j}) \leq M4^{-n+1}$$

by the definition of $\mathcal{B}_M$. Therefore if M is large,

$$\sum r_{n,j} \log \frac{1}{r_{n,j}} \leq \frac{C \log M}{M} \tag{2.21}$$

and

$$\sum r_{n,j} \geq \sum \frac{\omega(Q_{n,j})}{M} \geq \frac{1}{2M}.$$

Choose β so that

$$\sum_{r_{n,j} \geq \beta} r_{n,j} \geq \frac{1}{4M}.$$

Then

$$\beta \geq M^{-c'}$$

because by (2.21)

$$\frac{\log \frac{1}{\beta}}{2M} \leq \sum_{r_{n,j} < \beta} r_{n,j} \log \frac{1}{r_{n,j}} \leq \frac{C \log M}{M}.$$

Consequently

$$\sum_{r_{n,j} \geq \beta} (r_{n,j})^{1-\delta} \leq \beta^{-\delta} \frac{c}{M} \sum \omega(Q_{n,j}) \leq C M^{1-c'\delta},$$

while

$$\sum_{r_{n,j} \geq \beta} \omega(Q_{n,j}) \geq c.$$

Taking $\delta > 0$ small, we see there exists, for $n = 1, 2, \ldots$ a sequence $B_{n,j}$ of balls of radii $r_{n,j}$ such that

$$\omega\left(\bigcup_j B_{n,j}\right) \geq c$$

and

$$\sum_j r_{n,j} \leq 2^{-n}.$$

Then

$$E = \bigcap_{m=1}^{\infty}\left(\bigcup_{n=m}^{\infty}\bigcup_j B_{n,j}\right)$$

satisfies

$$\omega(E) \geq c$$

and

$$\dim_{\text{Haus}}(E) \leq 1 - \delta.$$

By Harnack's inequality there is c' such that

$$\omega(z, E, \Omega) \geq c'$$

at all $z \in \cup_j \partial Q_{1,j}$. Now for every $Q_{n,j}$ given by Lemma 2.2 this construction yields a set $E_{n,j} \subset K \cap Q_{n,j}$ such that

$$\inf_{\partial Q_{n,j}} \omega(z, E_{n,j}, \Omega) \geq c$$

and

$$\dim_{\text{Haus}}(E_{n,j}) \leq 1 - \delta.$$

Now set

$$F = \bigcup_{n,j} E_{n,j}.$$

Then $\dim_{\text{Haus}}(E) \leq 1 - \delta$, and by the maximum principle $\omega(z, F, \Omega) \geq c$ for all $z \in \Omega$, which implies that $\omega(F) = 1$. ∎

3. For All Ω, dim $\omega \leq 1$

Let $K \subset \mathbb{C}$ be a compact set of positive capacity and set $\Omega = \mathbb{C}^* \setminus K$.

Theorem 3.1. *There exists $F \subset \partial\Omega$ such that $\omega(F) = 1$ and $\dim_{\mathrm{Haus}}(F) \le 1$.*

The proof consists of two lemmas and a domain modification construction.

Lemma 3.2. *Let $Q \subset \tilde{Q}$ be closed squares with the same center ζ_0, with sides parallel to the axes, and with sidelengths $\ell(Q)$ and $\ell(\tilde{Q}) = R\ell(Q)$. Assume*

$$\tilde{Q} \setminus Q \subset \Omega$$

and

$$E = Q \setminus \Omega$$

has $\mathrm{Cap}(E) > 0$. Fix $\varepsilon > 0$ and let B be a closed disc with center ζ_0 such that

$$e^{-\gamma_B} = \mathrm{Cap}(B) = (\ell(Q))^{-\varepsilon}\big(\mathrm{Cap}(E)\big)^{1+\varepsilon} = (\ell(Q))^{-\varepsilon}e^{-(1+\varepsilon)\gamma_E}. \qquad (3.1)$$

Set

$$\tilde{\Omega} = (\Omega \cup Q) \setminus B$$

and

$$\tilde{\omega} = \omega(\infty, \cdot, \tilde{\Omega}).$$

Then if $R = R(\varepsilon)$ is sufficiently large,

$$\tilde{\omega}(B) \ge C(\varepsilon)\omega(E), \qquad (3.2)$$

and for all $A \subset \partial\Omega \setminus \tilde{Q}$,

$$\tilde{\omega}(A) \ge \omega(A). \qquad (3.3)$$

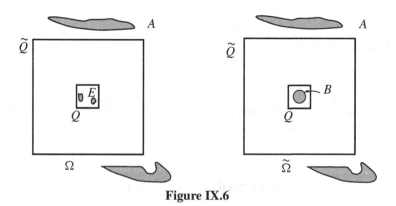

Figure IX.6

Proof. The sets $\Omega \cup Q$ and $\widetilde{\Omega}$ are open because $\widetilde{Q} \setminus Q \subset \Omega$. We may assume $\mathrm{Cap}(\partial\Omega \setminus \widetilde{Q}) > 0$, and after a change of scale we may assume $\ell(Q) = 1$. Let

$$g(z, \zeta) = \log \frac{1}{|z - \zeta|} + h(z, \zeta)$$

be Green's function for $\Omega \cup E$. Then

$$h(z, \zeta_0) = \int_{\partial\Omega \setminus \widetilde{Q}} \log |\zeta - \zeta_0| d\omega_{\Omega \setminus \widetilde{Q}}(z, \zeta) \geq \log\left(\frac{R-1}{2}\right), \tag{3.4}$$

and

$$|\nabla_\zeta h(z, \zeta)| = \mathrm{O}\left(\frac{1}{R}\right)$$

when $\zeta \in B \cup E$. Let μ_E and μ_B be the equilibrium distributions for E and B and fix $z_0 \in \partial Q$. Then on B,

$$u(z) = \int_B g(z, \zeta) d\mu_B(\zeta) = \gamma_B + h(z_0, \zeta_0) + \mathrm{O}\left(\frac{1}{R}\right),$$

and on E,

$$v(z) = \int_E g(z, \zeta) d\mu_E(\zeta) = \gamma_E + h(z_0, \zeta_0) + \mathrm{O}\left(\frac{1}{R}\right).$$

Furthermore $u = v = 0$ on $\partial\Omega \setminus \widetilde{Q}$ and $u(z) \geq v(z) + \mathrm{O}\left(\frac{1}{R}\right)$ on $\Omega \setminus \widetilde{Q}$ (because $\gamma_E > 0$). Hence on $\Omega \setminus \widetilde{Q}$,

$$
\begin{aligned}
\omega(z, B, \widetilde{\Omega}) &\geq \frac{u(z)}{\gamma_B + h(z_0, \zeta_0) + \mathrm{O}\left(\frac{1}{R}\right)} \\
&\geq \frac{v(z) + \mathrm{O}\left(\frac{1}{R}\right)}{\gamma_B + h(z_0, \zeta_0) + \mathrm{O}\left(\frac{1}{R}\right)} \\
&\geq \frac{\gamma_E + h(z_0, \zeta_0) + \mathrm{O}\left(\frac{1}{R}\right)}{(1 + \varepsilon)\gamma_E + h(z_0, \zeta_0) + \mathrm{O}\left(\frac{1}{R}\right)} \omega(z, E, \Omega),
\end{aligned}
$$

by (3.4), and with (3.4) that implies (3.2).

To prove (3.3) we assume $\zeta_0 = 0$. Write $U = \{|z| < \frac{R}{2}\}$, let

$$g(z, \zeta) = \log\left|\frac{1 - \frac{z}{R}\frac{\overline{\zeta}}{R}}{\frac{z}{R} - \frac{\zeta}{R}}\right|$$

be Green's function for U, let g_B be Green's function for $U \setminus B$, and let g_E be Green's function for $U \setminus E$. Then on $U \setminus B$

$$g_B(z, \zeta) = g(z, \zeta) - \int_{\partial B} g(w, \zeta) d\omega_{U \setminus B}(z, w)$$

$$\geq g(z, \zeta) - \left(\log 2 + \frac{C_2}{R}\right) \omega(z, \partial B, U \setminus B),$$

and on $U \setminus E$

$$g_E(z, \zeta) = g(z, \zeta) - \int_{\partial E} g(w, \zeta) d\omega_{U \setminus E}(z, w)$$

$$\leq g(z, \zeta) - \left(\log 2 + \frac{C_1}{R}\right) \omega(z, E, U \setminus E).$$

If $|z| = R/2$ and if R is large, then by hypothesis (3.1)

$$\omega(z, B, U \setminus B) \leq \omega(z, E, U \setminus E),$$

and hence

$$\frac{\partial g_B}{\partial n_\zeta} \geq \frac{\partial g_E}{\partial n_\zeta} \tag{3.5}$$

at $\zeta \in bU$.

Now suppose

$$\frac{\omega(z_0, A)}{\widetilde{\omega}(z_0, A)} = \sup_{|z| = R/2} \frac{\omega(z, A)}{\widetilde{\omega}(z, A)} = \lambda > 1.$$

Then by the maximum principle

$$\lambda \widetilde{\omega}(z, A) - \omega(z, A) > 0$$

on ∂U, because $\lambda \widetilde{\omega} > \omega$ on A. However, using (3.5) at z_0 then yields

$$0 = \lambda \widetilde{\omega}(z_0, A) - \omega(z_0, A)$$

$$= \frac{1}{2\pi} \int_{\partial U} \frac{\partial g_B(z_0, \zeta)}{\partial n} \lambda \widetilde{\omega}(\zeta, A) ds(\zeta) - \frac{1}{2\pi} \int_{\partial U} \frac{\partial g_E(z_0, \zeta)}{\partial n} \omega(\zeta, A) ds(\zeta)$$

$$> \frac{1}{2\pi} \int \frac{\partial g_B(z_0, \zeta)}{\partial n} (\lambda \widetilde{\omega}(\zeta, A) - \omega(\zeta, A)) ds(\zeta)$$

$$\geq 0,$$

which is a contradiction, and therefore (3.3) follows from the maximum principle. ∎

Lemma 3.3. *Write $g(z) = g_\Omega(z, \infty)$. Suppose $\Gamma \subset \{|z| < 1\}$ is a finite union of Jordan curves that separate $K = \mathbb{C} \setminus \Omega$ from ∞ and suppose there are con-*

stants $c_1, \ldots c_N$, possibly not distinct, such that $\Gamma = \bigcup_{j=1}^{N} \Gamma_j$ where

$$\Gamma_j \subset \{g(z) = c_j\}$$

is a Jordan curve. Then

$$I(\Gamma) = \frac{1}{2\pi} \int_\Gamma \frac{\partial g}{\partial n} \log |\nabla g| ds \geq -\log 2. \qquad (3.6)$$

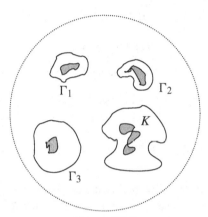

Figure IX.7

Proof. Replacing K by $\{g \leq \varepsilon\}$ for small ε, we can assume Ω is a finitely connected domain with smooth boundary. Let ζ_k be the critical points of g that lie outside Γ. Then by Green's theorem, almost exactly as in Exercise III.2,

$$I(\Gamma) = \sum_k g(\zeta_k) + \sum_j \frac{1}{2\pi} \int_{\Gamma_j} g \frac{\partial}{\partial n} \log |\nabla g| ds + \gamma$$

$$= \sum_k g(\zeta_k) + \sum_j \frac{c_j}{2\pi} \int_{\Gamma_j} \frac{\partial}{\partial n} \log |\nabla g| ds + \gamma,$$

where $\gamma = \gamma(\Gamma) \leq \gamma(K)$ is Robin's constant for Γ. Let α_j be the (finite) number of critical points of g inside Γ_j and let β_j be the (finite) number of components of K inside Γ_j. Then $\alpha_j = \beta_j - 1$, by Exercise II.17 and

$$\frac{1}{2\pi} \int_{\Gamma_j} \frac{\partial}{\partial n} \log |\nabla g| dz = \alpha_j - \beta_j = -1$$

by Green's theorem. Therefore

$$I(\Gamma) = \sum_k g(\zeta_k) - \sum_j c_j + \gamma.$$

Now write $n(c)$ for the number of components of $\{g = c\}$ outside Γ. Then

$$\sum g(\zeta_k) = \int_0^\infty (n(c) - 1)dc$$

and

$$\sum_j c_j = \int_0^\infty \sum_j \chi_{\{c \le c_j\}}(c)dc \le \int_0^{c_0} n(c)dc,$$

where $c_0 = \sup\{c : n(c) > 1\}$. Hence

$$I(\Gamma) \ge \gamma - c_0.$$

But since $\Gamma \subset \{|z| < 1\}$, all critical points fall into $\{|z| < 1\}$, and if $|z| < 1$ then

$$g(z) = \gamma + \int_K \log|z - \zeta|d\omega(\zeta) \le \gamma + \log 2.$$

Hence $c_0 \le \gamma + \log 2$ and (3.6) follows.   ∎

Proof of Theorem 3.1. We now modify the domain $\Omega = \mathbb{C}^* \setminus K$. We assume $K \subset \{|z| < 1/2\}$. Fix $\varepsilon > 0$ and fix an integer R so that $R > 2 + R(\varepsilon)$ where $R(\varepsilon)$ is given by Lemma 3.2. Choose a large constant M and a small constant $\rho > 0$ so that $M \le \log \frac{1}{\rho}$. Consider the grid $\mathcal{G}$ of dyadic squares having side ρ and lower left corners $(m + ni)\rho$. Partition $\mathcal{G}$ into R^2 subfamilies $\mathcal{G}_{p,q}$, where $1 \le p, q \le R$, by the rule

$$\mathcal{G} \ni Q \in \mathcal{G}_{p,q} \iff (m, n) \equiv (p, q) \bmod (R \times R).$$

Write

$$K_{p,q} = \bigcup_{\mathcal{G}_{p,q}} K \cap Q,$$

$$\Omega_{p,q} = \mathbb{C}^* \setminus K_{p,q},$$

and

$$\omega_{p,q}(E) = \omega(\infty, E, \Omega_{p,q}).$$

Then

$$\omega(E) \le \sum \omega_{i,j}(E \cap K_{p,q}),$$

and to prove Theorem 3.1 we may assume at the start that $\Omega = \Omega_{p,q}$ for some p, q. To modify $\Omega = \Omega_{p,q}$ we alternate two constructions:

disc construction: Let Q be a square of side ℓ. Replace $E = K \cap Q$ by a closed ball B with the same center as Q and with capacity

$$\text{Cap}(B) = \ell^{-\varepsilon}\left(\text{Cap} E\right)^{1+\varepsilon}.$$

Replacing E by B yields a new domain $\widetilde{\Omega}$ with a new harmonic measure $\widetilde{\omega}$ and (3.2) and (3.3) hold for $\widetilde{\omega}$ if $\partial\Omega \cap (RQ \setminus Q) = \emptyset$.

annulus construction: Given a square Q of side $\ell(Q)$ with sides parallel to the axes, replace Ω by $\widetilde{\Omega} = \Omega \cup (RQ \setminus Q)^{\circ}$ and K by $K \setminus (RQ \setminus Q)^{\circ}$. This yields a new harmonic measure $\widetilde{\omega} = \omega(\infty, \cdot, \widetilde{\Omega})$ which satisfies $\widetilde{\omega}(A) \geq \omega(A)$ whenever $A \cap (RQ \setminus Q) = \emptyset$.

To begin, take $\Omega = \Omega_{p,q}$ and perform the disc construction on every square $Q_j \in \mathcal{G}_{p,q}$. This yields a new domain Ω_1, whose complement is a union of discs B_j, and a new harmonic measure ω_1.

Now choose a maximal dyadic square Q^1 such that $\ell(Q^1) \geq \rho$ and

$$\omega_1(Q^1) \geq M\ell(Q^1).$$

If no such Q^1 exists, the domain modification stops. If Q^1 exists, perform the annulus construction on Q^1, obtaining a new Ω_1, and then perform the disc construction on Q^1 to Ω_1, replacing $K \cap Q^1$ by a disc B^1. Denote by Ω_2 and ω_2 the new domain and harmonic measure thusly obtained.

Next choose a maximal dyadic square Q^2 such that $\ell(Q^2) \geq \rho$,

$$Q^2 \not\subset Q^1,$$

and

$$\omega_2(Q^2) \geq M\ell(Q^2).$$

If no Q^2 exists, stop. If Q^2 exists, perform the annulus construction on Q^2. However, if $B^1 \cap \partial(RQ^2 \setminus Q^2) \neq \emptyset$, do not remove $B^1 \cap RQ^2 \setminus Q^2$ from Ω_2. Then perform the disc construction on Q^2 to Ω_2, replacing $K \cap Q^2$ by a disc B^2, and getting a new domain Ω_3 with harmonic measure ω_3.

By induction assume we have replaced squares $Q^1, \ldots, Q^{n-1}$ by discs $B^1, \ldots, B^{n-1}$ and formed new domains $\Omega_1, \ldots, \Omega_n$. Choose a maximal dyadic square Q^n such that $\ell(Q^n) \geq \rho$,

$$Q^n \not\subset Q^j, \quad \text{for all } j \leq n-1,$$

and

$$\omega_n(Q^n) \geq M\ell(Q^n).$$

If no Q^n exists, stop. If Q^n exists, perform the annulus construction on Q^n to Ω_n. However, if $j \leq n-1$ and $B^j \cap \partial(RQ^n \setminus Q^n) \neq \emptyset$, do not remove $B^j \cap RQ^n \setminus Q^n$ from Ω_n. Then perform the disc construction on Q^n to Ω_n, replacing $K \cap Q^n$ by a disc B^n, and getting a new domain Ω_{n+1} with harmonic measure ω_{n+1}.

Because there are only finitely many candidate squares and no square Q^j is repeated, the modification stops after a finite number of steps at a final domain $\Omega^* = \mathbb{C} \setminus K^*$ with harmonic measure ω^*. The complement K^* is the union of finitely many disjoint nonremoved discs

$$\{B^k : k \in S\} \cup \{B_j : j \in T\}.$$

If $k > j$ and $B^j \cap \partial(RQ^k \setminus Q^k) \neq \emptyset$, then $B^j \subset 2Q^j$ and $\ell(Q^k) \geq \ell(Q^j)$. Hence $B^j \cap R(\varepsilon)Q^k = \emptyset$, and (3.2) and (3.3) hold for each stage of the construction. Consequently

$$\omega^*(B^k) \geq C(\varepsilon)\omega(Q^k)$$

for all $k \in S$, and

$$\omega^*(Q_j) \geq C(\varepsilon)\omega(Q_j)$$

for all $j \in T$. By the construction

$$\omega^*(Q^k) \geq cM\ell(Q^k) \qquad (3.7)$$

for all $k \in S$, and

$$\omega^*(\{|z - z_0| < r\}) \leq CMr$$

if $z_0 \in Q = Q^j$ or Q_j and $r \geq \ell(Q)$.

Lemma 3.4.

$$K \subset \bigcup_S 2RQ^k \cup \bigcup_T Q_j. \qquad (3.8)$$

Proof. Let $Q \in \mathcal{G}_{p,q}$ and set $E = K \cap Q$. If $Q = Q_j, j \in T$ then (3.8) is obvious. If not there is a first index j_1 such that $Q^0 \subset RQ^{j_1} \setminus Q^{j_1}$ and Q^0 gets removed. If $j_1 \in S$, then (3.8) is true, and if $j_1 \notin S$ there is a first index $j_2 > j_1$ such that $Q^{j_1} \subset RQ^{j_2} \setminus Q^{j_2}$. However, in that case $\ell(Q^{j_2}) \geq 2\ell(Q^{j_1})$, because if $\ell(Q^k) = \ell(Q^j)$ then

$$Q^j \subset RQ^k \iff Q^k \subset RQ^j.$$

Since there exists an increasing chain $j_1 < j_2 < \ldots < j_n$ with $j_n \in S$ and $Q^{j_k} \subset RQ^{j_{k+1}} \setminus Q^{j_{k+1}}$, we conclude that $Q \subset 2RQ^{j_n}$. ∎

Now let $Q = Q^k$, $k \in S$ or $Q = Q_j$, $j \in T$, let B be the corresponding disc and assume $z \in 2Q \setminus B$. Then Green's function $g(z) = g_{\Omega^*}(z, \infty)$ has the form

$$g(z) = \int_B \log|z - \zeta|\,d\omega^*(\zeta) + \int_{K^* \setminus B} \log|z - \zeta|\,d\omega^* + \gamma(K^*)$$

$$= u(z) + v(z) + \gamma(K^*).$$

By (3.7) and a partial integration

$$|\nabla v(z)| \leq \int_{K^* \setminus B} \frac{d\omega^*(\zeta)}{|z - \zeta|} \leq CM \log \frac{1}{\ell(Q)}.$$

Let r denote the distance to the center z_0 of Q, and let $r(B)$ be the radius of B. Then simple calculations give

$$|\nabla u(z)| \leq \int_B \frac{d\omega^*(\zeta)}{|z - \zeta|}$$

and

$$-\frac{\partial u}{\partial r} = \int_B \text{Re}\left(\frac{(z - \zeta)\bar{z}}{|z - \zeta|^2 |z|}\right) d\omega^*(\zeta)$$

$$\geq \frac{\omega^*(B)}{|z - z_0|} - \frac{r(B)\omega^*(B)}{|z - z_0|^2}.$$

Set

$$\alpha = \alpha(B)\text{Max}\left(\frac{\omega^*(B)}{M^2 \log \frac{1}{\ell(Q)}}, 2r(B)\right).$$

If $\alpha > 2r(B)$, then $|\frac{\partial u}{\partial r}| > |\nabla v|$ on the circle $\sigma = \{|z - z_0| = \alpha\}$ and the level set $g = \inf_\sigma g(z)$ is a closed curve separating B from ∞. On this level curve

$$|\nabla g(z)| \leq \frac{\omega^*(B)}{\alpha} \leq CM^2 \log \frac{1}{\ell(Q)}. \tag{3.9}$$

If $\alpha = 2r(B)$ we take $\sigma = \partial B \subset \{g = 0\}$ for the level curve. Then because $R(\varepsilon)Q \setminus \bar{B} \subset \Omega^*$, $\sup_{\partial B}|\nabla g| \leq C \inf_{\partial B}|\nabla g|$ by reflection, and (3.9) also holds for σ.

Now $\Gamma = \bigcup_{S \cup T} \sigma$ separates K^* from ∞ and

$$\frac{1}{2\pi} \int_\Gamma \frac{\partial g}{\partial n} \log^+ |\nabla g| ds \leq C \sum_{S \cup T} \omega^*(B) \log^+\left(CM^2 \log \frac{1}{\ell(Q)}\right)$$

$$\leq C' \log \log \frac{1}{\rho}$$

because $M \leq \log \frac{1}{\rho}$. Therefore by Lemma 3.3,

$$\frac{1}{2\pi} \int_{\Gamma} \frac{\partial g}{\partial n} \big|\log |\nabla g|\big| ds \leq C \log \log \frac{1}{\rho}.$$

Consider the measure function $h(r) = r^{1+\varepsilon}$ and the Hausdorff content M_h defined in Appendix D. By (3.7)

$$M_h(K \cap \bigcup_S 2RQ^k) \leq (2R)^{1+\varepsilon} \sum_S (\ell(Q^k))^{1+\varepsilon}$$

$$\leq \frac{C^{1+\varepsilon}}{M} \sum_S \omega^*(Q^k) \leq \frac{CR^{1+\varepsilon}}{M}.$$

Let r_j be the radius of B_j, $j \in T$, and set

$$T_1 = \{j \in T : \omega^*(B_j) \geq \rho^{\varepsilon/2} r_j\}.$$

By (3.1) and (D.11) from Appendix D

$$M_h(K \cap Q_j) \leq r_j^{1+\varepsilon} \log\left(\frac{\mathrm{Cap}(K \cap Q_j)}{r_j}\right)$$

for any $j \in T$. Therefore

$$M_h\left(\bigcup_{T_1}(K \cap Q_j)\right) \leq C(1+\varepsilon)\rho^{\varepsilon/2} \sum_{T_1} \omega^*(B_j) \leq C\rho^{\varepsilon/2}.$$

Finally, we have

$$\omega\left(\bigcup_{T_2}(K \cap Q_j)\right) \leq \frac{1}{C(\varepsilon)} \sum_{T_2} \omega^*(B_j),$$

and by $(3.9)|\nabla g| \leq \rho^{\frac{\varepsilon}{2}}$ on ∂B_j. Consequently

$$\sum_{T_2} \omega^*(B_j) = \sum_{T_2} \frac{1}{2\pi} \int_{\partial B_j} \frac{\partial g}{\partial n} ds$$

$$\leq \frac{C}{\varepsilon \log \frac{1}{\rho}} \sum_{T_2} \int_{\partial B_j} \frac{\partial g}{\partial n} \big|\log |\nabla g|\big| ds$$

$$\leq \frac{C \log \log \frac{1}{\rho}}{\varepsilon \log \frac{1}{\rho}}.$$

Therefore, with ε fixed and $\eta > 0$ given, we can choose M and ρ so that there exists $A \subset K$ such that $M_h(A) < \eta$ while $\omega(K \setminus A) < \eta$. That means $\omega \perp \Lambda_h$, and the Theorem follows. ∎

Notes

Section 1 follows the elementary arguments in Batakis [1996]. Earlier Makarov and Volberg [1986] had given an explicit formula for

$$\dim \omega = \inf\{\alpha : \omega \perp \Lambda_\alpha\} \qquad (\text{N.1})$$

in terms of the critical points of Green's function. See also Volberg [1992], [1993] and Lyubich and Volberg [1995]. These papers rely on a beautiful connection between information theory and harmonic measure for self-similar sets originally due to Carleson [1985] and to Manning [1984] in the case of Julia sets that we regretfully must omit from this book. See Makarov [1990a] and [1998] for complete details and several further related results.

Theorem 2.1 is an unpublished result of Jones and Wolff [1986]. Carleson [1985] had obtained a similar theorem for certain Cantor sets using information theory and the identity (2.15) on Green's function critical points. Jones and Wolff later made a direct proof of the more general Theorem 2.1, still using (2.15). The Green's function identity (2.15) itself first appeared in Ahlfors' [1947] paper on analytic capacity. See Widom [1969], [1971a], and [1971b] and Jones and Marshall [1985] for other applications.

In the substantial paper [1993], Wolff extended Theorem 3.1 to show that for all Ω there is $F \subset \partial\Omega$ having $\sigma-$ finite linear measure and $\omega(F) = 1$.

Makarov's paper [1998] and the recent work [2003] by Binder, Makarov, and Smirnov explore the connection between other Hausdorff measures and harmonic measure in general domains. Let ω be harmonic measure for some domain Ω and define the **lower pointwise dimension** of ω at $z \in \partial\Omega$ as

$$\alpha_\omega(z) = \liminf_{\delta \to 0} \frac{\log \omega(B(z, \delta))}{\log \delta}.$$

Set

$$f_\omega(\alpha) = \dim\{z \in \partial\Omega : \alpha_\omega(z) \le \alpha\} \qquad (\text{N.2})$$

and define the **universal dimension spectra** to be

$$\Phi(\alpha) = \sup_{\Omega} f_\omega(\alpha)$$

and

$$\Phi_{\text{SC}}(\alpha) = \sup_{\Omega \text{ simply connected}} f_\omega(\alpha).$$

Makarov [1998] conjectures that for all $\alpha \ge 1$,

$$\Phi(\alpha) = \Phi_{\text{SC}}(\alpha). \qquad (\text{N.3})$$

Conjecture (N.3) is connected to the Brennan conjecture because Makarov [1998] proves that

$$\Phi_{SC}(\alpha) = \inf_{t \geq 0}\Big(\alpha(1 + B_b(t)) + (1 - \alpha)t\Big), \ (\alpha \geq 1),$$

where $B_b(t)$ is the bounded universal integral means spectrum defined in Exercise VIII.9.

To prove (N.3) it is enough to work with certain Cantor sets. Let D be a simply connected domain, let $\{\overline{D}_j\}$ be a finite collection of pairwise disjoint Jordan domains such that $\bigcup \overline{D}_j \subset D$, and suppose

$$F : \bigcup D_j \to D$$

is an analytic function such that each restriction $F|D_j$ is univalent. Write

$$F^{(n)} = F \circ F \circ \ldots \circ F$$

for the n-th iterate of F. The set

$$K = \bigcap_{n=1}^{\infty}(F^n)^{-1}\Big(\bigcup \overline{D}_j\Big)$$

is called a **conformal Cantor set**. Carleson and Jones [1992] and Makarov [1998] show that (N.3) holds if and only if

$$f_\omega(\alpha) \leq \Phi_{SC}(\alpha) \tag{N.4}$$

whenever $\mathbb{C}^* \setminus \Omega$ is a conformal Cantor set. In [2002] Binder, Makarov, and Smirnov established (N.4) for those conformal Cantor sets for which F is a polynomial.

Exercises and Further Results

1. (a) Let K be a Cantor set satisfying the hypotheses of Theorem 1.1. Then there exists $\sigma < \dim_{\text{Haus}}(K)$ such that there is $F \subset K$ with $\dim_{\text{Haus}}(F) = \sigma$ and $\omega(F) = 1$, but such that if $E \subset K$ has $\dim_{\text{Haus}}(E) < \sigma$ then $\omega(E) = 0$.
(b) However, if the definition of the Cantor set K at the beginning of Section 1 is generalized so that each square Q_J^n is replaced by four corner squares $Q_{J,j}^n$, $j = 1, 2, 3, 4$, having sidelengths

$$\ell(Q_{J,j}^n) = a_J^n \ell(Q_J^n)$$

that depend on J but satisfy

$$0 < a \leq a_J^n \leq b < 1/2,$$

then there is an example such that

$$\dim_{\text{Haus}}(\omega) = \dim_{\text{Haus}}(K).$$

This is due to Bishop and independently to Ancona and Batakis.
(c) If the Cantor set K is constructed with $a_n = \delta, 0 < \delta \leq 1/4$, harmonic measure on K had dimension greater than $1 - C\delta$ for some constant C. This result is due to Ancona.

See Batakis [1996] for the proofs of all these results.
2. Let Ω be a domain such that for some $\eta > 0$

$$\omega(w, B(w, 2\text{dist}(w, \partial\Omega)), \Omega) \geq \eta \qquad \text{(E.1)}$$

for all $w \in \Omega$. Then harmonic measure ω for Ω satisfies

$$\omega \perp \Lambda_h \qquad \text{(E.2)}$$

for every measure function $h(t)$ such that

$$\lim_{t \to 0} \frac{h(t)}{t} = 0.$$

The proof outlined below is a simple variation of the original Makarov argument given in Section VI.5.
(a) Let $\varphi : \mathbb{D} \to \Omega$ be a universal covering map (see Ahlfors [1973], Fisher [1983], or Gamelin [2001].) Let $\{z_n\} \subset \mathbb{D}$ be nontangentially dense on $\partial\mathbb{D}$ and let $w_n = \varphi(z_n)$. Then

$$\omega\left(\bigcup B(w_n, 2\text{dist}(w_n, \partial\Omega))\right) = 1.$$

Hint: Consider $u \circ \varphi$ when $1 - u$ is the harmonic measure above and use Exercises III.12 and VI.11.
(b) Let $\varphi : \mathbb{D} \to \Omega$ be a universal cover. If $w_0 = \varphi(z_0)$ then

$$\text{dist}(w_0, \partial\Omega) \leq |\varphi'(z_0)|(1 - |z_0|^2).$$

Hint: Since φ' has no zeros the monodromy theorem yields an inverse map $\psi : B(w, \text{dist}(w, \partial\Omega)) \to \mathbb{D}$ such that $\psi \circ \varphi(w) = w$ and $\psi(w_0)) = z_0$. Now apply the Schwarz lemma. Also see Lemma X.5.5.
(c) Use parts (a) and (b) and the proof of Theorem VI.5.2 to show that (E.1) implies (E.2). We do not know who first discovered this argument, which is mentioned in the introduction of Jones and Wolff [1988].
3. Condition (E.1) has many equivalents and a compact set $K \subset \mathbb{C}$ is called **uniformly perfect** if $\Omega = C^* \setminus K$ satisfies any one of them. Exercise III.11 gave a capacity condition equivalent to (E.1). Thirteen more equivalent conditions are listed below.

(i) There is a constant $c > 0$ such that if $\varphi : \mathbb{D} \to \Omega$ is a universal cover and if $w_0 = \varphi(z_0)$ then

$$\text{dist}(w_0, \partial\Omega) \geq c|\varphi'(z_0)|(1 - |z_0|^2).$$

In other words, the inequality in Exercise 2(b) can be reversed. See Pommerenke [1979].

(ii) Let $\varphi : \mathbb{D} \to \Omega$ be a universal cover map. The expression

$$d\rho(w) = \frac{|\varphi'(z)||dz|}{1 - |z|^2}, \quad \varphi(z) = w$$

which is independent of the choices of z or φ, is the infinitesimal form of the **hyperbolic metric** on Ω. In other word, the hyperbolic distance between z_1 and z_2 is

$$\rho(z_1, z_2) = \inf_\gamma \int_\gamma d\rho(w)$$

where the infimum is taken over all arcs γ in Ω that connect z_1 to z_2. Now K is uniformly perfect if and only if there is a constant $c > 0$ such that

$$\rho(z_1, z_2) \geq c \inf_\gamma \int_\gamma \frac{|dw|}{\text{dist}(w, \partial\Omega)}.$$

(iii) There is a constant $C > 0$ such that every closed curve in Ω not homotopic to a point in Ω has hyperbolic length a least C. See Pommerenke [1979].

(iv) There is a constant C such that if $R \subset \Omega$ is a ring and each component of $\mathbb{C}^* \setminus R$ meets K, then $\text{mod}(R) \leq C$. See Beardon and Pommerenke [1979].

(v) There is a constant C' such that if $A \subset \Omega$ is an annulus and each component of $\mathbb{C}^* \setminus A$ meets K, then $\text{mod}(A) \leq C$. Use (iii) and Exercise IV.5.

(vi) If $\varphi : \mathbb{D} \to \Omega$ is a universal cover, then $g = \log(\varphi')$ is a Bloch function. Pommerenke [1979].

(vii) There is a constant $c > 0$ such that if $\varphi : \mathbb{D} \to \Omega$ is a universal cover and if $\varphi(z_1) = \varphi(z_2)$ then $\rho(z_1, z_2) \geq c$. Pommerenke [1979].

(viii) There is a constant $\delta > 0$ such that if $\varphi : \mathbb{D} \to \Omega$ is a universal cover and if $\varphi^{-1}(w) = \{z_j\}$, then

$$\inf_k \prod_{j : j \neq k} \left| \frac{z_j - z_k}{1 - \overline{z}_k z_j} \right| \geq \delta.$$

Pommerenke [1984]. This the famous Carleson [1958] condition that $\{z_j\}$ be an interpolating sequence for H^∞. It also has many equivalents. See Garnett [1981], for example.

(ix) The domain Ω has a **strong barrier**; i.e., there is a nonconstant positive superharmonic function $U(z)$ on Ω and a constant $c > 0$ such that in the distribution sense

$$\Delta U(z) + \frac{cU(z)}{\text{dist}(z, \partial\Omega)^2} \leq 0.$$

See Ancona [1986].

(x) There is a constant C such that Hardy's inequality holds: If $u(z) \in C^1$ has compact support contained in Ω, then

$$\iint_\Omega \left(\frac{|u(z)|}{\text{dist}(z, \partial\Omega)}\right)^2 dxdy \leq C \iint_\Omega |\nabla u(z)|^2 dxdy.$$

Ancona [1986].

(xi) Let $\varphi : \mathbb{D} \to \Omega$ be a universal cover map. Then K is uniformly perfect if and only if there is a constant C such that whenever $D \subset \Omega$ is a disc, every component of $\varphi^{-1}(D)$ is a C-quasidisc. See Heinonen and Rohde [1995].

(xii) Let $g(z, \zeta)$ be Green's function for Ω. There is $\varepsilon > 0$ and $C < \infty$ such that for all $\zeta \in \Omega$

$$\sum_{\nabla_z g(z,\zeta)=0} g(z, \zeta)^{1+\varepsilon} < C.$$

See González [1992].

(xiii) Again let $\varphi : \mathbb{D} \to \Omega$ be a universal covering map. A **fundamental domain** for φ is a set $\mathcal{F} \subset \mathbb{D}$ such that $\varphi(\mathcal{F}) = \Omega$ and φ is one-to-one on $\mathcal{F}$. Then K is uniformly perfect if and only if there exists a fundamental domain $\mathcal{F}$ for φ such that $\partial\mathcal{F}$ is a quasicircle. González [1992].

The equivalence of (i), (ii), (v), (vi), and (viii) is independently due to Behrens [1976].

4. For $1 \leq j, k \leq N$ let $B_{j,k}$ be a closed disc with center $(\frac{j}{N}, \frac{k}{N})$ and radius $r_{j,k} < \frac{1}{2N}$, and set $K = \bigcup B_{j,k}$ and $\Omega = \mathbb{C}^* \setminus K$. Assume all discs $B_{j,k}$ have the same harmonic measure at ∞,

$$\omega(\infty, \partial B_{j,k}, \Omega) = \frac{1}{N^2}.$$

Prove there is C independent of N so that

$$\sum r_{j,k} \leq C.$$

Hint: Use Chapter III.

5. (a) Let ω be harmonic measure for some domain and define dimω by (N.1). Prove that if $\alpha = \dim\omega$, then $f_\omega(\alpha) = \alpha$, where f_ω is defined by (N.2).
(b) Using the estimate $B_b(t) \leq Ct^2$ from Chapter VIII, prove that Conjecture (N.3) implies

$$\Phi(\alpha) \leq \alpha - c(\alpha - 1)^2$$

for $1 \leq \alpha \leq 2$.

X

Rectifiability and Quadratic Expressions

First we introduce three topics, two old and one new.

(i) The first topic is the classical Lusin area function. The Lusin area function gives another description of H^p functions and another almost everywhere necessary and sufficient condition for the existence of nontangential limits. The area function is discussed in Section 1. In Appendix M we prove the Jerison–Kenig theorem that the area function determines the H^p class of an analytic function on a chord-arc domain.

(ii) The second topic is the characterizations of subsets of rectifiable curves in terms of certain square sums. These theorems, from Jones [1990], are proved in Sections 2 and 3.

(iii) The third topic is the Schwarzian derivative, which measures how much an analytic function deviates from a Möbius transformation. Section 4 is a brief introduction to the Schwarzian derivative.

Then we turn to the chapter's main goal, an exposition of the two papers [1990] and [1994] by Bishop and Jones. In Section 5 the Schwarzian derivative is estimated by the Jones square sums and by a second related quantity. In Section 6 rectifiable quasicircles are characterized by a quadratic integral akin to the Lusin function but featuring the Schwarzian derivative. In Sections 7, 8, and 9 the same quadratic integral gives new criteria for the existence of angular derivatives and further characterizations of BMO domains. Section 10 brings together most of the ideas from the chapter to prove a local version of the F. and M. Riesz theorem, and in Section 11 this F. and M. Riesz theorem leads us to the most general form of the Hayman–Wu theorem.

1. The Lusin Area Function

Let $F(z)$ be an analytic function on $\mathbb{D}$. The **area function** of F is

$$AF(\zeta) = A_\alpha F(\zeta) = \left(\iint_{\Gamma_\alpha(\zeta)} |F'(z)|^2 dxdy \right)^{\frac{1}{2}},$$

where $\alpha > 1$ and

$$\Gamma_\alpha(\zeta) = \{z \in \mathbb{D} : |z - \zeta| < \alpha(1 - |z|)\}.$$

$AF(\zeta)$ is called the area function because $(AF(\zeta))^2$ is the area of $F(\Gamma_\alpha(\zeta))$ when values are counted with their multiplicities. We will need a famous theorem of Calderón [1965].

Theorem 1.1. *Let $F(z)$ be analytic in $\mathbb{D}$ and let $0 < p < \infty$. Then $F \in H^p$ if and only if $AF \in L^p(\partial\mathbb{D})$, and there is a constant $c = c(p, \alpha)$ such that*

$$c\|AF\|_p^p \leq \|F - F(0)\|_{H^p}^p \leq c^{-1}\|AF\|_p^p. \tag{1.1}$$

A complete proof of Calderón's theorem will be given in Appendix M. See also Calderón [1965] or Stein [1970] and [1993]. In this section we prove Calderón's theorem for $p = 2$ on certain cone domains and we prove the local theorem that modulo sets of measure zero, $AF(\zeta) < \infty$ if and only if F has a nontangential limit at ζ.

On $\mathbb{D}$ the $p = 2$ case of Calderón's theorem has a simple Fourier series proof. By Fubini's theorem

$$\int_{\partial\mathbb{D}} (AF(\zeta))^2 ds = \iint_{\mathbb{D}} |F'(z)|^2 |\{\zeta : z \in \Gamma_\alpha(\zeta)\}| dxdy,$$

and by the definition of Γ_α

$$\frac{1}{c(\alpha)}(1 - |z|^2) \leq |\{\zeta : z \in \Gamma_\alpha(\zeta)\}| \leq c(\alpha)(1 - |z|^2).$$

Hence

$$\int_{\partial\mathbb{D}} (AF(\zeta))^2 ds \leq c(\alpha) \iint_{\mathbb{D}} |F'(z)|^2 (1 - |z|^2) dxdy$$

$$\leq c^2(\alpha) \int_{\partial\mathbb{D}} (AF(\zeta))^2 ds. \tag{1.2}$$

Now write $F(z) = \sum a_n z^n$. Then

$$\|F - F(0)\|_{H^2}^2 = 2\pi \sum_{n=1}^{\infty} |a_n|^2$$

and

$$\iint_{\mathbb{D}} |F'(z)|^2 (1 - |z|^2) dx dy = \pi \sum_{n=1}^{\infty} \frac{n}{n+1} |a_n|^2.$$

Therefore (1.1) holds at $p = 2$ and

$$\frac{1}{4} \|F - F(0)\|_{H^2}^2 \leq \iint_{\mathbb{D}} |F'(z)|^2 (1 - |z|^2) dx dy \leq \frac{1}{2} \|F - F(0)\|_{H^2}^2. \quad (1.3)$$

The case $p = 2$ also has a simple geometric proof using (1.2) and Green's theorem. See Exercise 8 of Chapter II.

Let U be a simply connected domain such that ∂U is a rectifiable curve, let $F(z)$ be analytic on U, and let φ be a conformal map from $\mathbb{D}$ onto U. By definition we say $F(z)$ is in $H^p(U)$, $0 < p < \infty$, if

$$F\big(\varphi(z)\big)(\varphi'(z))^{\frac{1}{p}} \in H^p(\mathbb{D}).$$

When $F \in H^p(U)$ we take

$$\|F\|_{H^p(U)}^p = \int_{\partial \mathbb{D}} |F\big(\varphi(e^{i\theta})\big)|^p |\varphi'(e^{i\theta})| d\theta.$$

By the F. and M. Riesz theorem, φ' has non-zero nontangential limit almost everywhere on $\partial \mathbb{D}$, and it follows that every $F(z) \in H^p(U)$ has nontangential limit $F(\zeta)$ arc length almost everywhere on ∂U and

$$\|F\|_{H^p(U)}^p = \int_{\partial \mathbb{D}} |F\big(\varphi(e^{i\theta})\big)|^p |\varphi'(e^{i\theta})| d\theta = \int_{\partial U} |F(\zeta)|^p ds.$$

Consequently the norm $\|F\|_{H^p(U)}^p$ does not depend on the choice of φ.

We define a **special cone domain** to be a cone domain having either the form

$$U = \bigcup_E \Gamma_\beta(\zeta), \quad (1.4)$$

where E is a compact subset of $\partial \mathbb{D}$, or the form

$$U = \bigcup_E \Gamma_\beta^h(\zeta), \quad (1.5)$$

where E is a compact subset of an arc $I \subset \partial\mathbb{D}$ such that $|I| \leq h < 1$ and where

$$\Gamma_\beta^h(\zeta) = \Gamma_\beta(\zeta) \cap \{z : 1 - |z| < h\}$$

with $\beta > 2$. See Figure X.1. When U is a special cone domain, we define the **central point** of U in case (1.4) by $c_U = 0$ and in case (1.5) by

$$c_U = (1 - \frac{3h}{4})c_I$$

where c_I is the center of the arc I. Every special cone domain U is connected and $U \ni c_U$ if $U \neq \emptyset$ because $\beta > 2$ and $h \geq |I|$.

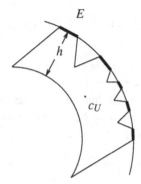

Figure X.1 The special cone domain $U = \bigcup_E \Gamma_\beta^h(\zeta)$.

When U is a domain and $z \in U$, we write $d(z) = \text{dist}(z, \partial U)$ and we define the **cone** (relative to U) at $\zeta \in \partial U$ to be

$$\Gamma_\alpha(\zeta, U) = \{z \in U : |z - \zeta| < \alpha d(z)\}.$$

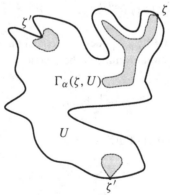

Figure X.2 $\Gamma_\alpha(\zeta, U)$ and $\Gamma_\alpha(\zeta', U)$.

If $F(z)$ is analytic on U we also define the area function

$$AF(\zeta) = A_\alpha F(\zeta, U) = \left(\iint\limits_{\Gamma_\alpha(\zeta, U)} |F'(z)|^2 dx dy \right)^{\frac{1}{2}}.$$

Theorem 1.2. *Let U be a special cone domain and let $F(z)$ be analytic on U. Then $F \in H^2(U)$ if and only if*

$$AF \in L^2(\partial U, ds)$$

and if and only if

$$\iint\limits_{U} |F'(z)|^2 d(z) dx dy < \infty. \tag{1.6}$$

There exist constants $C_1 = C_1(\beta) > 0$ and $C_2 > 0$ such that

$$C_1 \|F\|_{H^2(U)}^2 \leq |F(c_U)|^2 + \iint\limits_{U} |F'(z)|^2 d(z) dx dy \leq C_2 \|F\|_{H^2(U)}^2. \tag{1.7}$$

Furthermore, $F \in H^2(U)$ if and only if

$$\iint\limits_{U} |F''(z)|^2 d(z)^3 dx dy < \infty$$

and there is $C_3 = C_3(\beta) > 0$ such that

$$C_3^{-1} \iint\limits_{U} |F'(z)|^2 d(z) dx dy \leq |F'(c_U)|^2 + \iint\limits_{U} |F''(z)|^2 d(z)^3 dx dy$$
$$\leq C_3 \iint\limits_{U} |F'(z)|^2 d(z) dx dy. \tag{1.8}$$

Of course when $U = \mathbb{D}$ (1.6) is nothing but the case $p = 2$ of Calderón's theorem 1.1 above. Theorem 1.2 is a special case of a result from Kenig's thesis [1980], which generalized Calderón's theorem for $0 < p < \infty$ to the more general setting of Lipschitz domains. Later Jerison and Kenig [1982a] extended Calderón's theorem to domains bounded by chord-arc curves. See Appendix M.

The proof of Theorem 1.2 hinges on three geometric properties of special cone domains:

(i) For all $\alpha > 1$ there are constants $c_1 = c_1(\alpha, \beta, h)$ and $c_2 = c_2(\alpha, \beta, h)$ such that

$$c_1 d(z) \leq \Lambda_1(\{\zeta \in \partial U : z \in \Gamma_\alpha(\zeta, U)\}) \leq c_2 d(z). \tag{1.9}$$

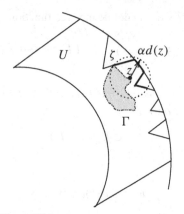

Figure X.3 $\Gamma = \Gamma_\alpha(\zeta, U)$ is shaded.

(ii) There exists $\delta = \delta(\beta, h)$ such that

$$B(c_U, \delta) \subset U. \tag{1.10}$$

(iii) Let $\varphi : \mathbb{D} \to U$ be a conformal mapping satisfying $\varphi(0) = c_U$. Then

$$\left| \arg\left(\frac{\varphi'(e^{i\theta})e^{i\theta}}{\varphi(e^{i\theta}) - c_U} \right) \right| = \left| \arg\frac{d\varphi}{d\theta} - \frac{\pi}{2} - \arg\left(\varphi(e^{i\theta}) - c_U\right) \right| \tag{1.11}$$

$$\leq \tau(\beta, h) < \frac{\pi}{2}.$$

Inequality (1.11) gives control almost everywhere on ∂U on the angle between the normal vector and the position vector emanating from c_U.

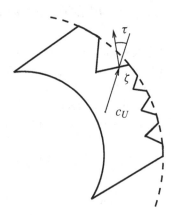

Figure X.4 The angle τ in (1.11).

Proof of Theorem 1.2. Again Fubini's theorem gives

$$\int_{\partial U} (AF(\zeta))^2 ds = \iint_U |F'(z)|^2 \Lambda_1(\{\zeta \in \partial U : z \in \Gamma_\alpha(\zeta, U)\}) dxdy,$$

and then (1.9) yields

$$\int_{\partial U} (AF(\zeta))^2 ds \le c(\alpha, U) \iint_U |F'(z)|^2 d(z) dxdy$$

(1.12)

$$\le c'(\alpha, U) \int_{\partial U} (AF(\zeta))^2 ds.$$

Therefore $AF \in L^2(\partial U)$ if and only if (1.6) holds. Notice (1.12) is the analogue of (1.2) for the special cone domain U.

To prove (1.7) define

$$A = \int_{\partial U} |F(\zeta)|^2 ds = \int_{\partial \mathbb{D}} |F \circ \varphi|^2 |\varphi'| d\theta,$$

$$B = \iint_{\mathbb{D}} |(F \circ \varphi)'(z)|^2 |\varphi'(z)|(1 - |z|^2) dxdy,$$

and

$$D = \iint_{\mathbb{D}} |F \circ \varphi|^2 \left| \frac{\varphi''(z)}{\sqrt{\varphi'(z)}} \right|^2 (1 - |z|^2) dxdy.$$

By the Koebe theorem,

$$B \le 8 \iint_U |F'(z)|^2 d(z) dxdy \le 8B.$$

We may assume $F(c_U) = 0$. By (1.3) we have

$$A \ge 2 \iint_{\mathbb{D}} \left| \frac{d}{dz} (F(\varphi(z)) \sqrt{\varphi'(z)}) \right|^2 (1 - |z|^2) dxdy$$

$$\ge \iint_{\mathbb{D}} |(F \circ \varphi)'(z)|^2 |\varphi'(z)|(1 - |z|^2) dxdy - \frac{1}{2}$$

$$\times \iint_{\mathbb{D}} |F \circ \varphi|^2 \left| \frac{\varphi''(z)}{\sqrt{\varphi'(z)}} \right|^2 (1 - |z|^2) dxdy$$

$$= B - \frac{D}{2}.$$

Write $g = \log(\varphi')$. By (1.11), $|\operatorname{Im} g| = |\arg(\varphi')| < 2\pi$, so that by Exercise 3 of Appendix F, $g \in \mathrm{BMO}$ and

$$\left|\frac{\varphi''}{\varphi'}\right|^2 (1 - |z|) dx dy = |g'(z)|^2 (1 - |z|) dx dy$$

is a Carleson measure with norm bounded by a fixed constant C. Therefore

$$D = \iint_{\mathbb{D}} |(F \circ \varphi)(z)|^2 |\varphi'(z)| |g'(z)|^2 (1 - |z|^2) dx dy \le CA$$

and

$$B \le (1 + 2C) A$$

with C independent of β and h.

The proof that $A \le C(\beta) B$ will exploit the inequality $\tau(\beta, h) < \frac{\pi}{2}$ in (1.11). Because $\tau(\beta, h) < \frac{\pi}{2}$ and $|\varphi(e^{i\theta})| \le 1$, we have

$$A \le \int_{\partial \mathbb{D}} |F \circ \varphi|^2 \left| \frac{\varphi'(e^{i\theta}) e^{i\theta}}{\varphi(e^{i\theta}) - c_U} \right| d\theta$$

$$\le \sec \tau \left| \int_{\partial \mathbb{D}} |F \circ \varphi|^2 \left(\frac{e^{i\theta} \varphi'(e^{i\theta})}{\varphi(e^{i\theta}) - c_U} \right) d\theta \right|.$$

Hence by Green's theorem (see Exercise 8 of Chapter II),

$$A \le \sec \tau \left| \iint_{\mathbb{D}} \Delta \left(|F \circ \varphi|^2 \frac{z \varphi'(z)}{\varphi(z) - c_U} \right) \log \frac{1}{|z|} dx dy \right|.$$

Write $h(z) = \frac{z \varphi'(z)}{\varphi(z) - c_U}$. Then

$$A \le 4 \sec \tau \iint_{\mathbb{D}} |(F \circ \varphi)'|^2 |h| \log \frac{1}{|z|} dx dy$$

$$+ 4 \sec \tau \iint_{\mathbb{D}} \left| (F \circ \varphi) \overline{(F \circ \varphi)'} h' \right| \log \frac{1}{|z|} dx dy,$$

and because $\left| \frac{z}{\varphi(z) - c_U} \right| \le c(\delta)$ by the maximum principle,

$$A \le 4 c(\delta)(\sec \tau) B + 4 \sec \tau \iint_{\mathbb{D}} \left| (F \circ \varphi) \overline{(F \circ \varphi)'} h' \right| \log \frac{1}{|z|} dx dy,$$

again by Exercise II.8. Now by (1.11), $h(z) = e^{V(z)}$ where $|\operatorname{Im} V| \le \tau < \frac{\pi}{2}$, and $h' = V'h$. Therefore by Cauchy–Schwarz

$$\iint_{\mathbb{D}} \left| (F \circ \varphi)\overline{(F \circ \varphi)'h'} \right| \log \frac{1}{|z|} dxdy$$

$$= \iint_{\mathbb{D}} \left| (F \circ \varphi)' \right| \left| (F \circ \varphi)V' \right| |h| \log \frac{1}{|z|} dxdy$$

$$\leq \left\{ \iint_{\mathbb{D}} \left| (F \circ \varphi)' \right|^2 |h| \log \frac{1}{|z|} dxdy \right\}^{\frac{1}{2}}$$

$$\times \left\{ \iint_{\mathbb{D}} \left| F \circ \varphi \right|^2 |V'|^2 |h| \log \frac{1}{|z|} dxdy \right\}^{\frac{1}{2}}$$

$$\leq \{c(\delta)B\}^{\frac{1}{2}} \left\{ \iint_{\mathbb{D}} \left| (F \circ \varphi)\sqrt{h} \right|^2 |V'|^2 \log \frac{1}{|z|} dxdy \right\}^{\frac{1}{2}}.$$

But since $|\text{Im} V| \leq \tau$

$$|V'|^2 \log \frac{1}{|z|} dxdy$$

is a Carleson measure with constant bounded by $c(\beta)$, and we conclude that

$$A \leq C(\beta)(B + B^{\frac{1}{2}} A^{\frac{1}{2}}),$$

so that (1.7) holds.

To prove (1.8) note that because $|F'(z)|^2$ is subharmonic we have

$$\left| F'(c_U) \right|^2 \leq c_1(\beta, h) \iint_U \left| F'(z) \right|^2 d(z) dxdy,$$

by property (ii) and

$$\left| F''(z) \right|^2 \leq \frac{2}{\pi \eta^4} \iint_{B(z,\eta)} \left| F'(w) \right|^2 dudv$$

for $z \in U$ and $\eta < d(z)$. Hence

$$\left| F''(z) \right|^2 d(z)^3 \leq \frac{32}{\pi d(z)} \iint_{B(z,\frac{d(z)}{2})} \left| F'(w) \right|^2 dudv.$$

Now if $|w - z| < \frac{d(z)}{2}$ then

$$\frac{d(z)}{2} \leq d(w) \leq \frac{3}{2} d(z)$$

so that by Fubini's theorem,

$$\iint\limits_{U} |F''(z)|^2 d(z)^3 dx dy \leq \frac{32}{\pi} \iint\limits_{U} |F'(w)|^2 \iint\limits_{B(w,d(w))} \frac{3}{2d(w)} dx dy du dv$$

$$\leq 48 \iint\limits_{U} |F'(w)|^2 d(w) du dv,$$

which establishes the right-hand inequality in (1.8).

To prove the other inequality in (1.8) we again use property (ii) to get

$$|F'(z) - F'(c_U)|^2 d(z) \leq C(\delta) \iint\limits_{U} |F''(z)|^2 d(z)^3 dx dy$$

for $z \in B(c_U, \frac{\delta}{2})$. Therefore

$$\iint\limits_{B(c_U, \frac{\delta}{2})} |F'(z) - F'(c_U)|^2 d(z) dx dy \leq c'(\delta) \iint\limits_{U} |F''(z)|^2 d(z)^3 dx dy.$$

When $z \in U \setminus B(c_U, \frac{\delta}{2})$ write $z_0 = c_U + \frac{\delta}{2} \frac{z - c_U}{|z - c_U|}$. Then

$$|F'(z) - F'(z_0)|^2 \leq \left(\int_{z_0}^{z} |F''(w)| ds \right)^2$$

$$\leq \int_{z_0}^{z} |F''(w)|^2 d(w)^{\frac{3}{2}} ds \int_{z_0}^{z} d(w)^{-\frac{3}{2}} ds$$

$$\leq C'(\beta) d(z)^{-\frac{1}{2}} \int_{z_0}^{z} |F''(w)|^2 d(w)^{\frac{3}{2}} ds$$

because by trigonometry $d(w) \geq d(z) + c_2(\beta)|z - w|$ if $w \in [z_0, z]$.

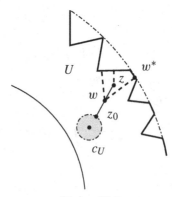

Figure X.5

Write w^* for the projection of $w \in U$ to ∂U along the radius $[0, \frac{w}{|w|}]$. Then again by trigonometry

$$d(z) \le c_3(\beta)|z - w^*|$$

when $z \in [w, w^*]$. Using polar coordinates we then obtain

$$\iint\limits_{U} |F'(z) - F'(c_U)|^2 d(z) dx dy$$

$$\le C(\delta) \iint\limits_{U} \left|F''(w)\right|^2 d(w)^3 dx dy$$

$$+ C''(\beta) \iint\limits_{U \cap \{|z - c_U| > \frac{\delta}{2}\}} |F''(w)|^2 d(w)^{\frac{3}{2}} \int_{w}^{w^*} |z - w^*|^{\frac{1}{2}} dr du dv$$

$$\le C''(\beta) \iint\limits_{U} |F''(w)|^2 d(w)^3 du dv,$$

which proves the other half of (1.8). ∎

We say two statements P and Q about points on $\partial \mathbb{D}$ are **almost everywhere equivalent** if the symmetric difference

$$\{P \text{ true}\} \Delta \{Q \text{ true}\}$$

has measure zero. For example, if $u(z)$ is a harmonic function on $\mathbb{D}$ and if $\alpha > 1$, then by Privalov's theorem, "$u_\alpha^*(\zeta) < \infty$" is almost everywhere equivalent to "u has nontangential limit at ζ". The next theorem is the parallel result for area functions.

Theorem 1.3. *Let $F(z)$ be an analytic function on $\mathbb{D}$ and let $\alpha > 1$. Then the statements*

$$AF(\zeta) < \infty$$

and

$$F \text{ has nontangential limit at } \zeta$$

are almost everywhere equivalent.

A simple corollary of Theorem 1.3 and Plessner's theorem is that Theorem 1.3 is also true when $F(z)$ is a harmonic function. The martingale analogue of Theorem 1.3 is the Lévy theorem J.3 in Appendix J. One half of Theorem 1.3 is due to Marcinkiewicz and Zygmund [1938] and the other half to Spencer [1943].

The proof of Theorem 1.3 depends on a fourth geometric property of the special cone domains having the form (1.4) and of their cones $\Gamma_\alpha(\zeta, U)$: If $\zeta \in E = \partial U \cap \partial \mathbb{D}$ and if $z \in U$, then $d(z) \le 1 - |z|$ and hence

$$\Gamma_\alpha(\zeta, U) \subset \Gamma_\alpha(\zeta).$$

Conversely, suppose $z \in \Gamma_\alpha(\zeta) \subset U$ and $z^* \in \partial \Gamma_\beta(\zeta)$ satisfy $|z - z^*| \le d(z)$, with $\alpha < \beta$. Then

$$1 - |z| \le d(z) + 1 - |z^*|$$

and

$$|z^* - \zeta| \le |z - \zeta| + d(z).$$

Hence

$$\frac{|z - \zeta|}{\alpha} \le d(z) + \frac{|z^* - \zeta|}{\beta} \le d(z) + \frac{|z - \zeta| + d(z)}{\beta},$$

so that

$$\Gamma_\alpha(\zeta) \subset \Gamma_\gamma(\zeta, U) \subset U \tag{1.13}$$

when

$$\gamma = \frac{1 + \frac{1}{\beta}}{\frac{1}{\alpha} - \frac{1}{\beta}}. \tag{1.14}$$

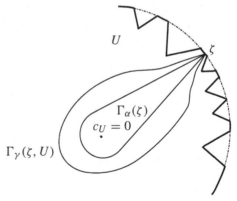

Figure X.6 Cones satisfying (1.13).

Proof of Theorem 1.3. First we prove the Marcinkiewicz–Zygmund half. Let $K \subset \partial \mathbb{D}$ be compact and assume F has nontangential limit at every $\zeta \in K$. Fix

$\alpha > 1$. Take $\beta > \alpha$ and define γ by (1.14). Taking K smaller, we can assume F is bounded on the cone domain

$$U = \bigcup_{\zeta \in K} \Gamma_\beta(\zeta).$$

Thus $F \in H^2(U)$ and by Theorem 1.2

$$A_\gamma \big(F(\zeta, U) \big) < \infty$$

almost everywhere on K. Therefore by (1.13)

$$A_\alpha F(\zeta) < \infty$$

almost everywhere on K.

To prove Spencer's half of the theorem, we can assume that for all $e^{i\theta} \in E$,

$$\iint_{\Gamma_\alpha(e^{i\theta})} |F'(z)|^2 dx dy \leq N$$

for some constant N. Let $K \subset E$ be a compact set such that for fixed δ,

$$\frac{|E \cap (\theta - h, \theta + h)|}{2h} \geq 3/4 \qquad (1.15)$$

whenever $0 < h < \delta$. By the theorem on points of density, it is enough to prove F has nontangential limit almost everywhere on K. Set

$$U = \bigcup_{e^{i\theta} \in K} \Gamma_\alpha(e^{i\theta}).$$

Then by (1.15) there is $c(\delta) > 0$ such that

$$2\pi N \geq \int_K \iint_{\Gamma_\alpha(e^{i\theta})} |F'(z)|^2 dx dy d\theta$$

$$= \iint_U |F'(z)|^2 \big| K \cap \{\theta : z \in \Gamma_\alpha(\theta)\} \big| dx dy$$

$$\geq c(\delta) \iint_U |F'(z)|^2 (1 - |z|^2) dx dy.$$

$$\geq c'(\delta) \iint_U |F'(z)|^2 \operatorname{dist}(z, \partial U) dx dy.$$

Therefore $F \in H^2(U)$ by Theorem 1.2, and F has nontangential limits almost everywhere on K by the F. and M. Riesz theorem. ∎

Theorem 1.4. *Let $F(z)$ be analytic in $\mathbb{D}$ and let $\alpha > 1$. Then*

$$\iint\limits_{\Gamma_\alpha(\zeta)} |F''(z)|^2 (1 - |z|^2)^2 dxdy < \infty \qquad (1.16)$$

is almost everywhere equivalent to

$$F \text{ has nontangential limit at } \zeta.$$

Proof. We outline two proofs. The first proof uses Theorem 1.2. Suppose

$$\iint\limits_{\Gamma_\alpha(\zeta)} |F''(z)|^2 (1 - |z|^2)^2 dxdy \le M$$

for all $\zeta \in E$. Given $\varepsilon > 0$ and $c < 1$, there exist a compact set $E_0 \subset E$ and a $\delta_0 > 0$ such that $|E \setminus E_0| < \varepsilon$ and

$$|B(\zeta, \delta) \cap E| \ge c\delta \qquad (1.17)$$

for all $\zeta \in E_0$ and all $\delta < \delta_0$. Set

$$U = \bigcup_{E_0} \Gamma_\alpha(\zeta).$$

By (1.17),

$$\left| E \cap \{|\zeta - z| < \alpha(1 - |z|)\} \right| \ge c'(1 - |z|^2)$$

when $z \in U$ and $1 - |z| \le \frac{\delta}{\alpha}$. Therefore

$$\iint\limits_{U} |F''(z)|^2 \text{dist}(z, \partial U)^3 dxdy \le \iint\limits_{U} |F''(z)|^2 (1 - |z|^2)^3 dxdy$$

$$\le C \int_{E} \iint\limits_{\Gamma_\alpha(\zeta)} |F''(z)|^2 (1 - |z|^2)^2 dxdy$$

$$\le CM.$$

Then by Theorem 1.2, $F \in H^2(U)$ and F has nontangential limits almost everywhere on E_0.

Conversely, assume F has nontangential limit at every $\zeta \in E$. Fix $\beta > \alpha$. Given $\varepsilon > 0$ there is a compact set $E_0 \subset E$ such that $|E \setminus E_0| < \varepsilon$ and F is bounded on

$$U = \bigcup_{E_0} \Gamma_\beta(\zeta).$$

Thus $F \in H^2(U)$ and by Theorem 1.2

$$\iint\limits_{U} |F''(z)|^2 d(z)^3 dxdy < \infty.$$

Let

$$U_0 = \bigcup_{E_0} \Gamma_\alpha(\zeta).$$

If $z \in U_0$, then $1 - |z|^2 \le c(\alpha, \beta)d(z)$. Consequently by Fubini's theorem

$$\int\limits_{E_0} \iint\limits_{\Gamma_\alpha(\zeta)} |F''(z)|^2 (1 - |z|^2)^2 dxdy$$

$$= \iint\limits_{U_0} |F''(z)|^2 (1 - |z|^2)^2 \big|\{\zeta \in E_0 : z \in \Gamma_\alpha(\zeta)\}\big| dxdy$$

$$\le C \iint\limits_{U} |F''(z)|^2 d(z)^3 dxdy,$$

and (1.16) holds almost everywhere on E_0. ∎

The second proof combines Theorem 1.3 with the pointwise result below.

Lemma 1.5. *Let $\alpha > 1$ and let $F(z)$ be analytic on $\Gamma_\alpha(\zeta)$.*
(a) If $1 < \beta < \alpha$ there is $C(\alpha, \beta)$ such that for all $\zeta \in \partial\mathbb{D}$,

$$\iint\limits_{\Gamma_\beta(\zeta)} |F''(z)|^2 (1 - |z|^2)^2 dxdy \le C(\alpha, \beta) \iint\limits_{\Gamma_\alpha(\zeta)} |F'(z)|^2 dxdy.$$

(b) There is C_α such that for all $\zeta \in \partial\mathbb{D}$,

$$\iint\limits_{\Gamma_\alpha(\zeta)} |F'(z)|^2 dxdy \le C_\alpha |F'(0)|^2 + C_\alpha \iint\limits_{\Gamma_\alpha(\zeta)} |F''(z)|^2 (1 - |z|^2)^2 dxdy.$$

The proof of Lemma 1.5, which is very like the proof of (1.8), is an exercise for the reader.

2. Squares Sums and Rectifiability

Let E be a bounded set. We seek geometric conditions that are necessary and sufficient for E to lie on a rectifiable curve. Clearly one necessary condition is $\Lambda_1(E) < \infty$, and if E is connected then $\Lambda_1(E) < \infty$ if and only if E is a subset of a rectifiable curve. In fact, a connected set E is the trace of a (self-intersecting) rectifiable closed curve of length ℓ if and only if $\Lambda_1(E) \leq \ell/4$; see Falconer [1985] or David and Semmes [1991]. However, there exist sets E such that $\Lambda_1(E) = 0$ but E is not contained in any rectifiable curve; see Exercise 3.

For $z \in E$ and $t > 0$, define

$$\beta_E(z, t) = \frac{1}{t} \inf \left\{ \sup_{E \cap B(z,t)} \operatorname{dist}(w, L) : L \text{ is a line through } z \right\}.$$

Thus $\beta_E(z, t)$ is a scale invariant measure of the least deviation $E \cap B(z, t)$ has from lines through the point z. Because $z \in E$, the width of the narrowest strip containing $E \cap B(z, t)$ is comparable to $t\beta_E(z, t)$.

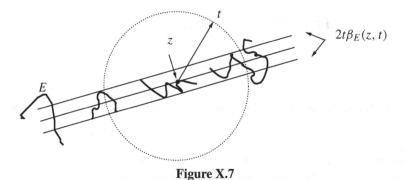

Figure X.7

Theorem 2.1. *There is a constant C such that if E is a bounded set and*

$$\int_0^{\operatorname{diam}(E)} \beta_E^2(z, t) \frac{dt}{t} \leq K \tag{2.1}$$

for all $z \in E$, then there is a rectifiable curve Γ such that $E \subset \Gamma$ and

$$\ell(\Gamma) \leq C e^{CK} \operatorname{diam}(E).$$

Theorem 2.5 will provide a converse of Theorem 2.1: If $E \subset \Gamma$ and Γ is a rectifiable curve, then (2.1) holds at Λ_1 almost every point of E. See Exercise 7 for a refinement of Theorem 2.1.

To simplify the proof of Theorem 2.1 we introduce a dyadic variant of $\beta_E(z, t)$. When Q is a dyadic square of side $\ell(Q)$, write aQ for the concentric square with side $\ell(aQ) = a\ell(Q)$ and define

$$\beta_E(aQ) = \frac{1}{\ell(aQ)} \inf \left\{ \sup_{E \cap aQ} \operatorname{dist}(z, L) : L \text{ is any line} \right\}. \tag{2.2}$$

Then $2\beta_E(aQ)\ell(aQ)$ is the width of the narrowest strip containing $E \cap aQ$.

Lemma 2.2. *There are constants c_1, c_2, and c_3 such that if $z \in E$, then for all $d > 0$,*

$$c_3 \int_0^d \beta_E^2(z,t) \frac{dt}{t} \leq \sum \{ \beta_E^2(3Q) : \ell(Q) \leq d, \ z \in Q \}$$

$$\leq c_1 + c_2 \int_0^d \beta_E^2(z,t) \frac{dt}{t}. \qquad (2.3)$$

Proof. By a change of scale we may assume $d = 1$. Note that if Q is a dyadic square containing z, then

$$3Q \subset B(z, 2\sqrt{2}\ell(Q))$$

and

$$\beta_E(3Q) \leq \frac{2\sqrt{2}}{3} \beta_E(z, 2\sqrt{2}\ell(Q)). \qquad (2.4)$$

Furthermore, if $t \leq s$ then

$$\beta_E(z,t) \leq \frac{s}{t} \beta_E(z,s). \qquad (2.5)$$

Write $Q_n = Q_n(z)$ for the unique dyadic square

$$Q_n = \left\{ j2^{-n} \leq x < (j+1)2^{-n}, \ k2^{-n} \leq y < (k+1)2^{-n} \right\}$$

such that $z \in Q_n$. Then by (2.4) and (2.5)

$$\sum_{n=0}^{\infty} \beta_E^2(3Q_n) \leq \sum_{n=0}^{\infty} \frac{8}{9} \beta_E^2(z, 2^{-n+\frac{3}{2}})$$

$$\leq \frac{32}{9 \log 2} \sum_{n=0}^{\infty} \int_{2^{-n+\frac{3}{2}}}^{2^{-n+\frac{5}{2}}} \beta_E^2(z,t) \frac{dt}{t}$$

$$\leq \frac{32}{9 \log 2} \left(\int_0^1 \beta_E^2(z,t) \frac{dt}{t} + \frac{5}{2} \log 2 \right).$$

To prove the lower estimate, note that when we reflect a strip S containing $E \cap 3Q$ about a line through $z \in E \cap Q$ parallel to the sides of S we obtain a strip having centerline through z and not more than twice as wide as S. Since $B(z, \ell(Q)) \subset 3Q$, we therefore have

$$\beta_E(3Q) \geq \frac{1}{6} \beta_E(z, \ell(Q)), \qquad (2.6)$$

and it follows from (2.5) and (2.6) that

$$\sum_{n=0}^{\infty} \beta_E^2(3Q_n) \geq \frac{1}{36} \sum_{n=0}^{\infty} \beta_E^2(z, 2^{-n})$$

$$\geq \frac{1}{144 \log 2} \sum_{n=0}^{\infty} \int_{2^{-n-1}}^{2^{-n}} \beta_E^2(z, t) \frac{dt}{t}$$

$$\geq \frac{1}{144 \log 2} \int_0^1 \beta_E^2(z, t) \frac{dt}{t}. \qquad \blacksquare$$

Proof of Theorem 2.1. In the proof we assume $\mathrm{diam}(E) \leq 1$. Then by (2.1) and Lemma 2.2

$$\sum_{\ell(Q)\leq 1, z\in Q} \beta_E^2(3Q) \leq c_1 + c_2 K \qquad (2.7)$$

for all $z \in E$. We first describe the basic construction underlying the proof. Let S be a closed rectangle with side lengths L and βL, where $\beta \leq 1$, and assume E meets each side of S. Partition each L side of S into three segments of length $L/3$, and define the middle third $M(S)$ as that subrectangle with each L side replaced by its middle third. There are two cases.

Case 1: There exists a point $p \in E \cap M(S)$. See Figure X.8. Cut S into two rectangles A_0 and A_1 by the line through p orthogonal to the L sides of S, and let $\alpha_j L$ be the length of the projection of A_j onto an L side of S. Then $1/3 \leq \alpha_j \leq 2/3$. Cover $A_j \cap E$ by a rectangle S_j such that $A_j \cap E$ meets each side of S_j and such that the shortest side of S_j is as small as possible. Then $S_0 \cap S_1 \cap E \neq \emptyset$ since $p \in S_0 \cap S_1$. Let L_j be the length of the longest side of S_j. By the Pythagorean theorem,

$$L_j \leq (\alpha_j^2 + \beta^2)^{1/2} L \leq (1 + 5\beta^2)\alpha_j L. \qquad (2.8)$$

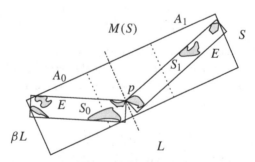

Figure X.8 Case 1.

Case 2: $E \cap M(S) = \emptyset$. Let A_0 and A_1 be the two components of $S \setminus M(S)$. See Figure X.9. Cover $A_j \cap E$ by a rectangle S_j such that each side of S_j meets $A_j \cap E$ and S_j has shortest side as small as possible. Join S_0 to S_1 by the shortest segment T having endpoints in $E \cap S_j$. Write $\alpha_0 = \alpha_1 = 1/2$, write L_j for the length of the longest side of S_j, and write $|T|$ for the length of T. Then (2.8) still holds and

$$|T| \leq (1 + \beta^2)L. \tag{2.9}$$

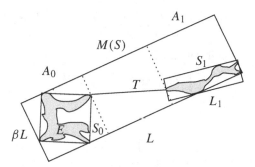

Figure X.9 Case 2.

In both cases

$$\alpha_0 + \alpha_1 = 1, \tag{2.10}$$

and there is a constant C such that S_j has shortest side of length $\beta_j L_j$ and

$$\beta_j \leq C\beta_E(3Q) \tag{2.11}$$

for any dyadic Q such that $Q \cap S_j \neq \emptyset$ and $\operatorname{diam}(S_j) \leq \ell(Q) < 2\operatorname{diam}(S_j)$, because in that case $S_j \subset 3Q$.

To start, assume $E \subset Q_0 = [0,1] \times [0,1]$, but assume no smaller dyadic Q contains E. Let S_0 be a rectangle such that $E \subset S_0$, E meets all four sides of S_0, and S_0 has smallest shortest side. Write its side lengths as L_0 and βL_0, where $\beta \leq 1$ and $\beta \leq C\beta(3Q_0)$.

Now perform the basic construction on S_0, getting two rectangles $S_{0,0}$ and $S_{0,1}$, and (in Case 2) a line segment T_0. Continue. After n steps we have rectangles S_I, $I = (i_1, i_2, \ldots, i_n)$, $i_j = 0, 1$, with longsides L_I, and we have weights $\alpha_I \in [1/3, 2/3]$ and factors $\beta_I \leq 1$. For these rectangles, conditions (2.8), (2.10), and (2.11) hold. If Case 2 is applied to S_I we also get a segment T_I of length $|T_I|$, for which (2.9) holds.

After $N = 25$ steps the diameter of a rectangle S_I drops by at least $1/2$. Indeed, if we begin with a rectangle that has side lengths 1 and $\beta \leq 1$ and diameter $d_0 = (1 + \beta^2)^{1/2}$, then in the worst case the construction yields a rectangle having sides α and $(\alpha^2 + \beta^2)^{1/2}$ and diameter $d_1 = (2\alpha^2 + \beta^2)^{1/2}$. Since $\alpha \leq 2/3$, we have $d_1 \leq (17/18)^{1/2} d_0$, and since $(17/18)^{25/2} \leq 1/2$, the claim holds. Consequently each dyadic square Q can occur at most N times during iterated applications of the basic construction.

For a multi-index $I = (i_1, i_2, \ldots, i_n)$ of size $|I| = n$ and for $m < n$, write $I_m = (i_1, i_2, \ldots, i_m)$. Let $\{I^{(1)}, \ldots, I^{(p)}\}$ be a set of multi-indices having size $|I^{(j)}| \leq n$ such that $I^{(j)} \not\subset I^{(k)}$ for $j \neq k$. Then by (2.8) and (2.10),

$$\sum L_{I^{(j)}} \leq L_0 \sum_{|I|=n} \Big(\prod_{1 \leq m \leq n} (1 + 5\beta_{I_m}^2) \Big) \Big(\prod_{1 \leq m \leq n} \alpha_{I_m} \Big).$$

By (2.11) and the hypothesis (2.7),

$$\prod_{1 \leq m \leq n} (1 + 5\beta_{I_m}^2) \leq C e^{5c_2 N K},$$

and by (2.10)

$$\sum_{|I|=n} \Big(\prod_{1 \leq m \leq n} \alpha_{I_m} \Big) = 1.$$

Therefore

$$\sum L_{I^{(j)}} \leq C e^{CK} \tag{2.12}$$

for some new constant C, and in particular

$$\sum_{|I|=n} L_I \leq C e^{CK}.$$

Next we estimate $\sum |T_I|$. Write

$$M_k = \sum \{|T_I| : \text{Case 2 has been used } k \text{ times on } I_1, I_2, \ldots, I_n = I\}.$$

To estimate $\sum M_k$ note that there is a constant $\beta_0 < 1$ such that if we use Case 2 and $L_0 + L_1 \geq 0.9L$, then $\beta \geq \beta_0$ and $L_0 + L_1 \leq CL$. Furthermore, by (2.7) the alternative $L_0 + L_1 \geq 0.9L$ can only hold for at most N' subindices I_m. Therefore by (2.9) and (2.12),

$$M_k \leq (1 + \beta_0^2) C^{N'} (0.9)^{k-N'} C e^{CK}$$

and

$$\sum_k M_k \leq C' e^{CK}. \tag{2.13}$$

At the nth stage of the construction we have a connected union Γ_n of rectangles and segments such that $E \subset \Gamma_n$. By (2.12) and (2.13) there is a continuous map $\sigma_n : [0, 1] \to \Gamma_n$ such that $\sigma_n([0, 1])$ has bounded length, such that $\sigma_n([0, 1])$ meets all four sides of each rectangle S_I, $|I| = n$, and such that σ_{n+1} is built by modifying σ_n on each $\sigma_n^{-1}(S_I)$, with $|I| = n$. Then $\lim \sigma_n$ defines a rectifiable curve containing E. ∎

An earlier characterization of subsets of rectifiable curves was also given by Jones in [1990]. For any set E define

$$\beta^2(E) = \mathrm{diam}(E) + \sum \{ \beta_E^2(3Q)\ell(Q) : Q \text{ dyadic}, \ \ell(Q) \leq \mathrm{diam}(E) \}.$$

Theorem 2.3. *Suppose E is a bounded set. Then E is contained in a rectifiable curve if and only if $\beta^2(E)$ is finite. Moreover, there are constants c_1, c_2 so that*

$$c_1\beta^2(E) \leq \inf_{\Gamma \supset E} \Lambda_1(\Gamma) \leq c_2\beta^2(E). \tag{2.14}$$

Theorem 2.3 is sometimes called the "traveling salesman theorem" because it bounds the total distance the salesman must travel to visit a finite set E of cities when each pair of cities in E is connected by a straight road. Exercise 6 describes the minimal spanning tree problem and the simple "greedy algorithm" which gives a shortest spanning tree. However the greedy algorithm gives no a priori bound like (2.14) for arbitrarily large sets. See Garey and Johnson [1979] and Lawler, Lenstra, Rinnooy Kan and Shmoys [1990] for more information about spanning trees and the classical traveling salesman problem.

We first prove the right-hand inequality of (2.14), that there exists a curve Γ with $\Lambda_1(\Gamma) \leq c_2\beta^2(E)$. The argument is like the proof of Theorem 2.1. Let R_n be the sum of the diameters of the rectangles at stage n. From (2.8) we have

$$L_0 + L_1 \leq L + C\beta_E^2(3Q)\ell(Q) \tag{2.15}$$

for any Q such that $Q \cap E \cap S \neq \emptyset$ and $\mathrm{diam}(S) \leq \ell(Q) \leq 2\,\mathrm{diam}(S)$. As noted before, each such Q can occur at most $N = 25$ times during iterations of the construction, so that

$$R_n \leq C\beta^2(E).$$

To estimate the lengths of the segments created by applications of Case 2, note again that if Case 2 is used and if $L_0 + L_1 \geq 0.9L$, then $\beta \geq \beta_0$ and $|T| \leq C\beta_E^2(3Q)\ell(Q)$. Thus the sum of the lengths of the middle segments created from applications of Case 2 with $\beta \geq \beta_0$ is at most $C\beta^2(E)$.

Now write

$$R_n = I_n + II_n$$

where I_n is the sum of the lengths of the rectangles at stage n to which Case 1 will be applied or for which $\beta \geq \beta_0$, and where II_n is the sum of the lengths of the rectangles to which Case 2 will be applied and for which $\beta < \beta_0$. Let T_{n+1} denote the sum of the lengths of the middle segments created in Case 2 with $\beta < \beta_0$. Then

$$R_{n+1} \leq I_n + C \sum \beta^2 (3Q) \ell(Q) + 0.9 I I_n,$$

where the sum is taken over all Q which intersect a rectangle R, created at stage n and such that $\mathrm{diam}(R) \leq \ell(Q) < 2\,\mathrm{diam}(R)$. Moreover by (2.9),

$$(0.1) T_{n+1} \leq (0.1) I I_n + C \sum \beta^2 (3Q) \ell(Q),$$

and so

$$R_{n+1} + (0.1) T_{n+1} \leq R_n + C \sum \beta^2 (3Q) \ell(Q).$$

Thus

$$(0.1) T_{n+1} \leq (R_n - R_{n+1}) + C \sum \beta^2 (3Q) \ell(Q),$$

and because R_n is bounded,

$$\sum_n T_n \leq C\,\mathrm{diam}(E) + \sum_{Q \text{ dyadic}} \beta^2 (3Q) \ell(Q).$$

The left-hand inequality in (2.14) will be proved in the next section. Here we only treat the special case of a Lipschitz graph.

Lemma 2.4. *Let Γ be the graph of a Lipschitz function. Then (2.14) holds for all $E \subset \Gamma$.*

Proof. We use an elegant argument from Jones [1990]. Assume that, after a translation and a dilation,

$$\Gamma = \{0 \leq x \leq 1, y = f(x)\},$$

where $f(0) = f(1)$ and

$$|f(x_2) - f(x_1)| \leq M|x_2 - x_1|.$$

It is enough to show the left side of (2.14) holds for $E = \Gamma$. Set

$$I_j^n = \left[\frac{j}{2^n}, \frac{(j+1)}{2^n} \right],$$

and

$$\Gamma_j^n = f(I_j^n),$$

and let J_j^n be the segment

$$J_j^n = \left[f\!\left(\frac{j}{2^n}\right), f\!\left(\frac{(j+1)}{2^n}\right) \right].$$

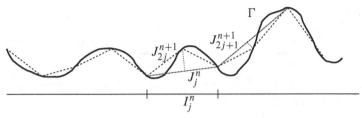

Figure X.10

Then J_{2j}^{n+1}, J_{2j+1}^{n+1}, and J_j^n are the three sides of a triangle and

$$\delta_{n,j} = 2^n \sup\left\{ \operatorname{dist}(z, J_j^n) : z \in J_{2j}^{n+1} \cup J_{2j+1}^{n+1} \right\}$$

satisfies

$$\ell(J_{2j}^{n+1}) + \ell(J_{2j+1}^{n+1}) - \ell(J_j^n) \geq c2^{-n}\delta_{n,j}^2,$$

with $c = c(M) > 0$, again by the Pythagorean theorem. Therefore

$$\sum_{m,k} c2^{-m}\delta_{m,k}^2 \leq 2\ell(\Gamma). \tag{2.16}$$

Now set

$$\beta_{n,j} = 2^n \sup\{\operatorname{dist}(z, J_j^n) : z \in \Gamma_j^n\} \leq \sum_{m=n}^{\infty} 2^{n-m} \sup\{\delta_{m,k} : I_k^m \subset I_j^n\}.$$

Write $p = m - n$ and

$$\sup\{\delta_{m,k} : I_k^m \subset I_j^n\} = \delta_{n+p,k}(n, j).$$

Then (2.16) and Minkowski's inequality give

$$
\left\{\sum_{n,j} 2^{-n}\beta_{n,j}^2\right\}^{\frac{1}{2}} \leq \left\{\sum_{n,j} 2^{-n} \left(\sum_{p=0}^{\infty} 2^{-p}\delta_{n+p,k}(n,j)\right)^2\right\}^{\frac{1}{2}}
$$

$$
\leq \sum_{p=0}^{\infty} \left\{\sum_{n,j} 2^{-n}2^{-2p}\delta_{n+p,k}^2(n,j)\right\}^{\frac{1}{2}} \tag{2.17}
$$

$$
\leq \sum_{p=0}^{\infty} 2^{-\frac{p}{2}} \left\{\sum_{m,k} 2^{-m}\delta_{m,k}^2\right\}^{\frac{1}{2}}
$$

$$
\leq 2c'\left(\ell(\Gamma)\right)^{\frac{1}{2}}.
$$

Extend f to $[-1, 2]$ by defining $f(x \pm 1) = f(x)$, and let $\Gamma(t)$ be the translate $\{y = f(x + t)\}$. We get numbers $\beta_{n,j}(t)$ that also satisfy (2.17). But then for fixed n,

$$\sum_{\ell(Q)=2^{-n-2}} \beta_\Gamma(3Q)^2 \leq C \int_{-1}^{1} \sum_{j} \beta_{n,j}(t)^2 dt,$$

because with probability $1/2$ there exists j such that $\{\text{Re} z - t : z \in Q\} \subset I_{n,j}$. That gives (2.14) for Γ. ∎

We make two observations about the construction. First, if E is a finite set, then the curve constructed consists of line segments joining the points of E, and these segments are the T_I from Case 2. Second, if E is connected, then by (2.14)

$$C_1 \beta^2(E) \leq \Lambda_1(E) \leq C_2 \beta^2(E).$$

If E is connected it is not necessary to construct the middle segments T_I in Case 2. Without the T_I, every stage of the construction yields a union of rectangles, each meeting E, and this union is connected because E is connected. Because the rectangles are shrinking, the constructed set is exactly E, although the parameterized curve may trace portions of E several times.

The next result, from Bishop and Jones [1994], is a converse to Theorem 2.1.

Theorem 2.5. *Let Γ be a Jordan curve. Then, except for sets of zero Λ_1 measure, z is a tangent point of Γ if and only if*

$$\int_0^1 \beta_\Gamma^2(z, t) \frac{dt}{t} < \infty. \tag{2.18}$$

Proof. Write $\text{Tn}(\Gamma)$ for the set of tangent points of Γ. In Chapter VI we proved that

$$\text{Tn}(\Gamma) = \bigcup_n K_n,$$

where, after an affine transformation,

$$K_n = \left\{ 0 \leq x \leq 1; \ y = f(x) = g(x) \right\},$$

where $f(x)$ and $g(x)$ are Lipschitz functions, and $f(x) \leq g(x)$, $0 < x < 1$. Moreover, there is a function $h(x)$ such that

$$f(x) \leq h(x) \leq g(x), \text{ on } 0 < x < 1, \tag{2.19}$$

and there is a neighborhood V_n of K_n such that

$$V_n \cap \Gamma = \left\{ y = h(x) : 0 < x < 1 \right\}.$$

By Lemma 2.4,

$$\sum \beta_f^2(3Q)\ell(Q) < \infty \text{ and } \sum \beta_g^2(3Q)\ell(Q) < \infty. \tag{2.20}$$

But whenever $K_n \cap Q \neq \emptyset$,

$$\beta_\Gamma^2(3Q) \leq C\left(\beta_f^2(3Q) + \beta_g^2(3Q)\right). \tag{2.21}$$

Indeed, we may assume that $\beta_g(3Q)$ is small and that $\beta_f(3Q) \leq \beta_g(3Q)$ by symmetry. Let S be a strip of width $\beta_g(3Q)\ell(Q)$ containing $3Q \cap \{y = g(x)\}$ and let T be a parallel strip having width $M\beta_g(3Q)\ell(Q)$ and having the same top edge as S. When M is large, $3Q \cap \{y = f(x)\} \subset T$ (for otherwise by (2.19) $\beta_f(3Q) > \beta_g(3Q)$) and consequently $3Q \cap \{y = h(x)\} \subset T$ and (2.21) holds with $C = C(M)$.

Now by (2.20) and (2.21)

$$\sum_{\ell(Q)<1} \beta_{\Gamma \cap V_n}^2(3Q)\ell(Q) < \infty.$$

Then (2.3) and integrating the left side of (2.18) against arc length on K_n yields inequality (2.18) at almost every point of K_n.

Conversely, let $E = \{z \in \Gamma : (2.18) \text{ holds}\}$. Then by Theorem 2.1 and the construction in Chapter VI,

$$E = E_0 \cup \bigcup_n E_n,$$

where $\Lambda_1(E_0) = 0$ and where, after an affine transformation,

$$E_n \subset \{(x, f_n(x)) : 0 < x < 1\},$$

and f_n is a Lipschitz function. We may suppose that $z = (0, f_n(0))$, that f_n is differentiable at $x = 0$, that $f_n'(0) = 0$, and that $(0, f_n(0))$ is a point of density of $E_n \cap \Gamma$. See Figure X.11.

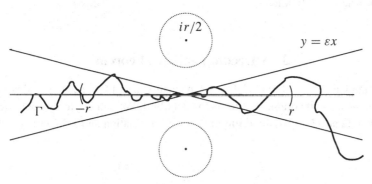

Figure X.11

For ε fixed and for r small,

$$\frac{\left|\{x \in (-r, r) : |f_n(x)| < \varepsilon|x|, (x, f_n(x)) \in \Gamma\}\right|}{2r} \geq (1 - \varepsilon)$$

and

$$\int_0^r \beta_\Gamma^2(0, t)\frac{dt}{t} < \varepsilon.$$

But then

$$\Gamma \cap \left(B\left(\frac{ir}{2}, \frac{r}{4}\right) \cup B\left(\frac{-ir}{2}, \frac{r}{4}\right)\right) = \emptyset$$

because otherwise $\beta_\Gamma(z, t) > \frac{1}{4}$ for all $t \in (\frac{r}{2}, r)$. Therefore z is a cone point for both components of $\mathbb{C} \setminus \Gamma$, and almost all such points lie in $\mathrm{Tn}(\Gamma)$. ∎

Theorem 2.5 is an improvement of Corollary VI.6.4. Let Ω_1 and Ω_2 be the components of $\mathbb{C} \setminus \Gamma$, let $t\theta_j(t)$ be the length of an arc of $\Omega_j \cap \{z : |z - w| = t\}$ that separates w from a point $z_j \in \Omega_j$ and let

$$\varepsilon(w, t) = \max\{|\pi - \theta_j(t)| : j = 1, 2\}.$$

Then clearly

$$\varepsilon(w, t) \leq C\beta(w, t).$$

Therefore Theorem 2.5 implies that

$$\int_0^1 \varepsilon^2(z, t)\frac{dt}{t} < \infty \tag{2.22}$$

at Λ_1 almost every tangent point of Γ, which is the assertion of Corollary VI.6.4. To our knowledge, it has not been determined if Γ has a tangent at almost every point where (2.22) holds.

3. A Decomposition Theorem

To complete the proof of the left-hand inequality in (2.14) we use the following decomposition theorem, also from Jones [1990]. We call a simply connected domain Ω an *M-Lipschitz domain* if, after a translation and a dilation, $0 \in \Omega$ and

$$\partial\Omega = \{r(\theta)e^{i\theta} : 0 \leq \theta \leq 2\pi\},$$

where

$$|r(\theta_1) - r(\theta_2)| \leq M|\theta_1 - \theta_2|$$

and

$$\frac{1}{M+1} \leq r(\theta) \leq 1.$$

Theorem 3.1. *There is a constant M such that whenever Ω is a simply connected domain with $\Lambda_1(\partial\Omega) < \infty$ there exists a rectifiable curve Γ such that*

$$\Omega \setminus \Gamma = \bigcup_{j=0}^{\infty} \Omega_j, \tag{3.1}$$

such that each Ω_j is an M-Lipschitz domain, and such that

$$\sum_j \Lambda_1(\partial\Omega_j) \leq M\Lambda_1(\partial\Omega). \tag{3.2}$$

Proof. The proof is in two steps. The first step is to use a corona construction as in Garnett [1981] or David and Semmes [1991] to obtain (3.1) and (3.2) with the Lipschitz domains Ω_j replaced by domains $\mathcal{U}_j$ bounded by uniformly chord-arc curves. The second step is to further decompose each chord-arc domain $\mathcal{U}_j$ into M-Lipschitz domains.

Let φ be a conformal map from $\mathbb{D}$ onto Ω. Write $F = \sqrt{\varphi'}$ and $g = \log(\varphi')$. The proof will use the familiar inequality

$$\|g\|_{\mathcal{B}} \leq 6 \tag{3.3}$$

many times. By the F. and M. Riesz theorem, $F \in H^2$ (see Exercise VI.1) and by (1.3) or Exercise II.8,

$$\iint_{\mathbb{D}} |\varphi'(z)||g'(z)|^2 \log\frac{1}{|z|} dxdy = 4 \iint_{\mathbb{D}} |F'(z)|^2 \log\frac{1}{|z|} dxdy$$
$$\leq 2\|F'\|_{H^2}^2 \tag{3.4}$$
$$\leq 2\Lambda_1(\partial\Omega).$$

Set $\mathcal{D}_0 = \{|z| < \frac{1}{2}\}$ and $\mathcal{U}_0 = \varphi(\mathcal{D}_0)$. Then by Theorem I.4.5, there is a constant M independent of Ω such that $\partial\mathcal{U}_0$ is an M-Lipschitz domain. Since $\varphi' \in H^1$, we also have

$$\Lambda_1(\partial\mathcal{U}_0) \leq \Lambda_1(\partial\Omega). \tag{3.5}$$

Next form the dyadic Carleson boxes

$$Q = \{re^{i\theta} : 1 - 2^{-n} \leq r < 1, \ \pi j2^{-(n+1)} \leq \theta < \pi(j+1)2^{-(n+1)}\},$$

$0 \leq j < 2^{n+1}$, of sidelength $\ell(Q) = 2^{-n}$ and their top halves

$$T(Q) = Q \cap \{1 - 2^{-n} \leq r < 1 - 2^{-(n+1)}\}$$
$$= Q \setminus \bigcup \{Q' : Q' \subset Q, Q' \neq Q\}.$$

Write z_Q for the center of $T(Q)$.

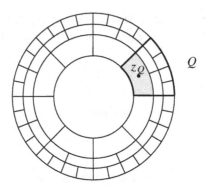

Figure X.12

Fix $\varepsilon > 0$ to be determined later and consider a Carleson box Q. If

$$\sup_{T(Q)} |g(z) - g(z_Q)| \geq \varepsilon, \tag{3.6}$$

we say Q is a **type 0** box and define $\mathcal{D}(Q) = T(Q)$. If Q is of type 0, then again by (3.3) $\partial \mathcal{U}_Q = \partial \varphi(\mathcal{D}_Q)$ is a chord-arc curve with chord-arc constant bounded by a fixed constant C_0.

If (3.6) fails for Q, define $G(Q)$ to be the set of maximal Carleson boxes $Q' \subset Q$ for which

$$\sup_{T(Q')} |g(z) - g(z_Q)| \geq \varepsilon, \tag{3.7}$$

and define

$$\mathcal{D}(Q) = Q \setminus \bigcup_{G(Q)} Q'.$$

Then $\partial \mathcal{D}(Q)$ is a chord-arc curve with chord-arc constant at most 4 and

$$\sup_{\mathcal{D}(Q)} |g(z) - g(z_Q)| \leq \varepsilon. \tag{3.8}$$

It follows that the domain $\mathcal{U}_Q = \varphi(\mathcal{D}(Q))$ is bounded by a chord-arc curve with constant at most 5 if ε is small.

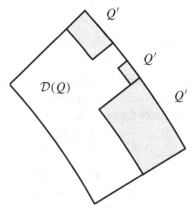

Figure X.13

Write $\{\mathcal{D}_j, j \geq 1\}$ for the set of maximal (with respect to set inclusion) regions $\mathcal{D}(Q)$. Then the family $\{\mathcal{D}_0\} \cup \{\mathcal{D}_j, j \geq 1\}$ is pairwise disjoint. Write $\mathcal{U}_j = \varphi(\mathcal{D}_j), j = 0, 1, \ldots$. To prove (3.2) we first show that

$$\sum_{j \geq 0} \Lambda_1(\partial \mathcal{U}_j) \leq M \Lambda_1(\partial \Omega). \tag{3.9}$$

If $\mathcal{D}_j = \mathcal{D}(Q)$ for Q of type 0, then by (3.3)

$$\begin{aligned}
\Lambda_1(\partial \mathcal{U}_j) &= \int_{\partial T(Q)} |\varphi'(z)| ds \\
&\leq c\ell(Q)|\varphi'(z_Q)| \\
&\leq c \iint_{T(Q)} |\varphi'(z)||g'(z)|^2 \log \frac{1}{|z|} dx dy.
\end{aligned}$$

Since Λ_1 almost every point lies in at most 2 sets $\partial \mathcal{U}_j$, (3.4) then gives

$$\sum_{\text{type 0}} \Lambda_1(\partial \mathcal{U}_j) \leq 2\Lambda_1(\partial \Omega). \tag{3.10}$$

If Q is not of type 0, we say Q has **type 1** if

$$\Lambda_1(\partial \mathbb{D} \cap \partial \mathcal{D}_Q) \geq \frac{\ell(Q)}{2}, \tag{3.11}$$

and we say Q has **type 2** if (3.11) fails.

When $\mathcal{D}_j = \mathcal{D}(Q)$ for Q of type 1, (3.8), (3.3), and (3.11) give

$$\Lambda_1(\partial \mathcal{U}_j) = \int_{\partial \mathcal{D}_j} |\varphi'(z)| ds \leq C \int_{b\mathbb{D} \cap \partial \mathcal{D}(Q)} |\varphi'(z)| ds.$$

Therefore

$$\sum_{\text{type 1}} \Lambda_1(\partial \mathcal{U}_j) \le C \int_{\partial \mathbb{D}} |\varphi'(z)| ds \le 2\Lambda_1(\partial \Omega). \tag{3.12}$$

Now assume $\mathcal{D}_j = \mathcal{D}(Q)$ where Q is a type 2 box. Then $\mathcal{D}_j$ is a chord-arc domain and $\partial \mathcal{D}_j$ is an Ahlfors regular curve, with constant independent of j. Let $\{J_k\}$ denote the top edges of the boxes in $G(Q)$. Then

$$\sum \Lambda_1(J_k) \ge \frac{\Lambda_1(\partial \mathcal{D}(Q))}{12}$$

because Q is of type 2. By (3.3), (3.7), and equicontinuity there are $\delta > 0$ and $E_k \subset J_k$ such that $|g(z) - g(z_Q)| \ge \delta$ on E_k and $\Lambda_1(E_k) \ge \delta \Lambda_1(J_k)$. Consequently

$$\int_{\partial \mathcal{D}(Q)} \frac{|F(z) - F(z_Q)|^2}{|F(z_Q)|^2} ds \ge C(\delta) \ell(Q),$$

and

$$\Lambda_1(\partial \mathcal{U}_j) = \int_{\partial \mathcal{D}(Q)} |F(z)|^2 ds \le C'(\delta) \int_{\partial \mathcal{D}(Q)} |F(z) - F(z_Q)|^2 ds.$$

We would like to estimate the latter integral via (1.7), but unfortunately $\mathcal{D}(Q)$ is not a Lipschitz domain. However $\mathcal{D}(Q)$ is a chord-arc domain, and hence Theorem M.1 of Appendix M gives[1]

$$\int_{\partial \mathcal{D}(Q)} |F(z) - F(z_Q)|^2 ds$$
$$\le C \iint_{\mathcal{D}(Q)} |F'(z)|^2 \Lambda_1\big(B\big(z, 2\text{dist}(z, \partial \mathcal{D}(Q))\big)\big) dx dy \tag{3.13}$$

Now because $\partial \mathcal{D}(Q)$ is Ahlfors regular, (3.13) gives

$$\int_{\partial \mathcal{D}(Q)} |F(z) - F(z_Q)|^2 ds \le C \iint_{\mathcal{D}(Q)} |F'(z)|^2 \text{dist}(z, \partial \mathcal{D}(Q)) dx dy$$
$$\le C \iint_{\mathcal{D}(Q)} |F'(z)|^2 \log \frac{1}{|z|} dx dy.$$

Therefore

$$\sum_{\text{type2}} \Lambda_1(\partial \mathcal{U}_j) \le C \iint_{\mathbb{D}} |F'(z)|^2 \log \frac{1}{|z|} dx dy \le C \Lambda_1(b\Omega). \tag{3.14}$$

---

[1] For a proof of (3.13) using Theorem 1.2. only, see Exercise 9.

Together (3.5), (3.10), (3.12), and (3.14) yield (3.9).

We have already showed that $\mathcal{U}_0 = \varphi(\mathcal{D}_0)$ is an M-Lipschitz domain. If $\mathcal{U}_j = \varphi(T(Q))$ for a type 0 box Q, partition $T(Q)$ into eight polar rectangles S_k by dividing the top edge of $T(Q)$ into four equal intervals and halving its radial side. Then by (3.3) $\varphi(S_k)$ is an M-Lipschitz domain and

$$\sum_j \Lambda_1(\partial\varphi(S_j)) \le C\Lambda_1(\partial\mathcal{U}_j).$$

Finally, suppose $\mathcal{U}_j = \varphi(\mathcal{D}(Q))$ for a box Q not of type 0. The construction is easier to explain in the upper half-plane $\mathbb{H}$ and we assume Q is a square in $\mathbb{H}$ with base and interval $I \subset \mathbb{R}$. See Figure X.14. Whenever $J = J_k$ is the top edge of a box $Q_k \in G(Q)$ we construct a tree $T = T(Q_k)$. Let S^1 be a segment of slope 1 joining the left endpoint of J to $\mathbb{R}$ and let S^2 be a segment of slope -1 joining the right endpoint of J to $\mathbb{R}$, and define

$$T^1 = J \cup S^1 \cup S^2.$$

Assume by induction that $T^n \supset T^{n-1}$ has been defined and that

$$T^n \cap \{\operatorname{Im} z = 4^{-n}\ell(J)\}$$

consists of 2^n points $z_{n,j}$ situated on segments $S^1_{n,j} \subset T_n$ of slope ± 1. Let $S^2_{n,j}$ be the segment of length $\sqrt{2}\, 4^{-n}\ell(J)$ orthogonal to $S^1_{n,j}$ that joins $z_{n,j}$ to $\mathbb{R}$ and set $T^{n+1} = T^n \cup \bigcup_j S^2_{n,j}$. Then $T = T(Q_k) = \bigcup_n T^n$ satisfies

$$\Lambda_1(T) = \left(1 + \sqrt{2}\Big(2 + \sum_{n=1}^{\infty} 2^n 4^{-n}\Big)\right)\ell(Q_k). \tag{3.15}$$

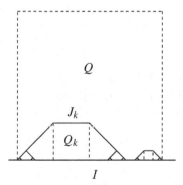

Figure X.14

Write

$$\mathcal{D}_j \setminus \bigcup_{G(Q)} T(Q_k) = \bigcup \mathcal{U}_{j,k}.$$

Then by (3.8) each $\varphi(\mathcal{U}_{j,k})$ is an M-Lipschitz domain, and by (3.15) and (3.8),

$$\sum_k \Lambda_1(\partial\varphi(\mathcal{U}_{j,k})) \le C|\varphi'(z_Q)|\ell(Q),$$

which establishes (3.1) and (3.2).   ∎

Corollary 3.2. *There exits a constant $M < \infty$ such that if Γ is a connected plane set with $\Lambda_1(\Gamma) < \infty$, then there exists a connected plane set $\tilde{\Gamma} \supset \Gamma$ such that $\Lambda_1(\tilde{\Gamma}) \le M\Lambda_1(\Gamma)$, the bounded components $\mathcal{D}_j$ of $\mathbb{C} \setminus \tilde{\Gamma}$ are M-Lipschitz domains with $\Gamma \subset \bigcup \partial\mathcal{D}_j$, and the boundary of the unbounded component of $\mathbb{C} \setminus \tilde{\Gamma}$ is a circle at least $3\sqrt{2}\Lambda_1(\Gamma)$ units from Γ.*

For the proof apply the theorem to each bounded component of the union of Γ, a line segment and a circle at least $3\sqrt{2}\Lambda_1(\Gamma)$ units from Γ.

Now let Γ be connected with $\Lambda_1(\Gamma) < \infty$, let $\{\mathcal{D}_j\}$ be the Lipschitz domains given by Corollary 3.2 and write $\Gamma_j = \partial\mathcal{D}_j$ and $\delta_j = \text{diam}\,\mathcal{D}_j$. Let Q be any dyadic square and define

$$\mathcal{F}(Q) = \{\Gamma_j : \Gamma_j \cap 3Q \ne \emptyset, \delta_j \ge \ell(Q)\},$$

and

$$\mathcal{G}(Q) = \{\Gamma_j : \Gamma_j \cap 3Q \ne \emptyset, \delta_j < \ell(Q)\}.$$

Then we have the following lemma.

Lemma 3.3. *There is a constant C_1 such that if $\ell(Q) \le \text{diam}(\Gamma)$,*

$$\beta_\Gamma^2(3Q) \le C_1 \sum_{\mathcal{F}(Q)} \beta_{\Gamma_j}^2(3Q) + C_1 \frac{1}{\ell^2(Q)} \sum_{\mathcal{G}(Q)} \text{Area}(\mathcal{D}_j). \qquad (3.16)$$

Proof. We may assume $\ell(Q) = 1$ and $\beta_\Gamma(3Q) > 0$, so that $3Q \cap \Gamma_j \ne \emptyset$ for some Γ_j and $3Q \subset \bigcup \overline{\mathcal{D}}_j$. If $\mathcal{F}(Q) = \emptyset$ then $\sum_{\mathcal{G}(Q)} \text{Area}(\mathcal{D}_j) \ge 9\ell^2(Q)$. Thus we can assume there exists $\Gamma_1 \in \mathcal{F}(Q)$. Let L be a line such that

$$d = \sup_{\Gamma_1 \cap 3Q} \text{dist}(z, L) \le \beta_{\Gamma_1}(3Q)\ell(3Q),$$

and let $z_0 \in \Gamma \cap 3Q$ have maximal distance $d_0 = \text{dist}(z_0, \Gamma_1)$. Let $z_1 \in \Gamma_1$ have minimal distance to z_0 and let $z_2 = \frac{z_0 + z_1}{2}$. There are three cases, as shown in Figure X.15.

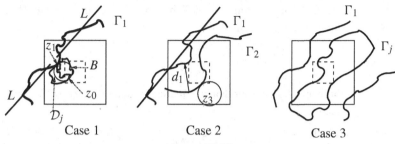

Case 1   Case 2   Case 3

Figure X.15

Case 1: $\mathcal{F}(Q) = \{\Gamma_1\}$.
Then

$$B = B(z_2, \frac{d_0}{2}) \subset \bigcup_{\mathcal{G}(Q)} \overline{\mathcal{D}}_j$$

and hence

$$\beta_\Gamma^2(Q) \le (d + d_0)^2 \le 8\beta_{\Gamma_1}^2(2Q) + 2d_0^2 \le 8\beta_{\Gamma_1}^2(2Q) + \frac{8}{\pi} \sum_{\mathcal{G}(Q)} \text{Area}(\mathcal{D}_j).$$

Case 2: $\mathcal{F}(Q) = \{\Gamma_1, \Gamma_2\}$ for disjoint $\mathcal{D}_1$ and $\mathcal{D}_2$.
In this case we may assume $\beta_{\Gamma_j}^2(3Q) < \varepsilon_0$, $j = 1, 2$, since otherwise (3.16) holds with $C_1 \sim \varepsilon_0^{-2}$. Let $d_1 = \sup_{\Gamma_2 \cap 3Q} \text{dist}(z, \Gamma_1)$. Then if ε_0 is sufficiently small

$$\beta_\Gamma(3Q) \le \beta_{\Gamma_1}(3Q) + \beta_{\Gamma_2}(3Q) + d_1$$

because for $j = 1, 2$, $\mathcal{D}_j$ is an M-Lipschitz domain with diameter $\delta_j \ge \ell(Q)$. Also because $\mathcal{D}_2$ is an M-Lipschitz domain there exists $z_3 \in (3Q) \setminus (\mathcal{D}_1 \cup \mathcal{D}_2)$ such that $\text{dist}(z_3, \Gamma_j) \ge cd_1$, $j = 1, 2$. Consequently there is constant C_2 such that

$$\beta_\Gamma^2(3Q) \le C_2(\beta_{\Gamma_1}^2(3Q) + \beta_{\Gamma_2}^2(3Q)) + C_2 d_1^2$$
$$\le C_2(\beta_{\Gamma_1}^2(3Q) + \beta_{\Gamma_2}^2(3Q)) + C_1 \sum_{\mathcal{G}(Q)} \text{Area}(\mathcal{D}_j).$$

Case 3: $\mathcal{F}(Q)$ contains at least three distinct Γ_j.
Then because each $\mathcal{D}_j$ is an M-Lipschitz domain there exists at least one $\Gamma_j \in \mathcal{F}(Q)$ such that $\beta_{\Gamma_j}(3Q) \ge C_1$, and (3.16) is clear in this case. ∎

To finish the proof of Theorem 2.3, let Γ be a rectifiable curve and let $\{\Gamma_j\}$ be as in Corollary 3.2. By Lemma 2.4

$$\sum_Q \beta_{\Gamma_j}^2(3Q)\ell(Q) \le c_2\ell(\Gamma_j).$$

If $\delta_j < 2^{-n}$ there are at most four dyadic squares Q such that $\ell(Q) = 2^{-n}$ and $\mathcal{D}_j \in \mathcal{G}(Q)$. Hence

$$\sum_Q \frac{1}{\ell(Q)} \sum_{\mathcal{G}(Q)} \text{Area}(\mathcal{D}_j) = \sum_j \text{Area}(\mathcal{D}_j) \sum_{\mathcal{D}_j \in \mathcal{G}(Q)} \frac{1}{\ell(Q)}$$

$$\le 4 \sum_j \text{Area}(\mathcal{D}_j) \sum_{m=0}^{\infty} (2^m \delta_j)^{-1}$$

$$\le 8 \sum_j \text{Area}(\mathcal{D}_j)(\delta_j)^{-1}$$

$$\le C \sum_j \ell(\Gamma_j) \le CM\ell(\Gamma).$$

Therefore by Lemma 3.3

$$\beta^2(\Gamma) \le C'\Lambda_1(\Gamma). \qquad \blacksquare$$

4. Schwarzian Derivatives

The **Schwarzian derivative** of a locally univalent analytic function is

$$\begin{aligned}
S\varphi(z) &= \left(\frac{\varphi''}{\varphi'}\right)'(z) - \frac{1}{2}\left(\frac{\varphi''(z)}{\varphi'(z)}\right)^2 \\
&= \left(\frac{\varphi'''(z)}{\varphi'(z)}\right) - \frac{3}{2}\left(\frac{\varphi''(z)}{\varphi'(z)}\right)^2.
\end{aligned} \qquad (4.1)$$

A calculation with the first expression in (4.1) shows that if

$$T(z) = \frac{az+b}{cz+d},$$

where $ad - bc \ne 0$, is a **Möbius transformation**, then

$$ST(z) = 0. \qquad (4.2)$$

Conversely, if $\mathcal{S}\varphi = 0$ on some disc, then for constants a and $A \neq 0$,

$$\frac{\varphi''(z)}{\varphi'(z)} = \frac{2}{a-z}, \quad \varphi'(z) = \frac{A}{(a-z)^2},$$

and $\varphi(z)$ is a Möbius transformation. A calculation with the second expression in (4.1) gives

$$\mathcal{S}(\varphi \circ \psi)(z) = \mathcal{S}\varphi(\psi(z))(\psi'(z))^2 + \mathcal{S}\psi(z), \qquad (4.3)$$

and therefore

$$\mathcal{S}(T \circ \varphi)(z) = \mathcal{S}\varphi(z) \qquad (4.4)$$

whenever T is a Möbius transformation. If φ is a locally univalent meromorphic function we define

$$\mathcal{S}\varphi = \mathcal{S}(1/\varphi)$$

in a neighborhood of a pole of φ. By (4.4), this definition is consistent with the original definition (4.1) of $\mathcal{S}\varphi$.

If φ is defined on the disc $\mathbb{D}$ and if $T \in \mathcal{M}$ is a self map of $\mathbb{D}$, then by (4.2) and (4.3)

$$(1-|z|^2)^2 \left| \mathcal{S}(\varphi \circ T)(z) \right| = (1-|z|^2)^2 \left| \mathcal{S}\varphi(T(z)) \right| \left| T'(z) \right|^2$$
$$= (1-|T(z)|^2)^2 \left| \mathcal{S}\varphi(T(z)) \right|.$$

Consequently the expression

$$(1-|z|^2)^2 \left| \mathcal{S}\varphi(z) \right| \qquad (4.5)$$

is a conformal invariant on $\mathbb{D}$.

Theorem 4.1. *If φ is univalent on $\mathbb{D}$, then*

$$(1-|z|^2)^2 \left| \mathcal{S}\varphi(z) \right| \leq 6. \qquad (4.6)$$

Conversely, if φ is analytic on $\mathbb{D}$ and if

$$\sup_{\mathbb{D}}(1-|z|^2)^2 \left| \mathcal{S}\varphi(z) \right| < 2, \qquad (4.7)$$

then φ is univalent and $\partial\varphi(\mathbb{D})$ is a quasicircle.

Proof. Suppose φ is univalent on $\mathbb{D}$. By (4.4) we may assume

$$\varphi(z) = z + a_2 z^2 + a_3 z^3 + \dots .$$

By the conformal invariance of (4.5) we only need to prove (4.6) at $z = 0$, where it becomes

$$\left| \varphi'''(0) - \frac{3}{2} (\varphi''(0))^2 \right| = |6a_3 - 6a_2^2| \leq 6.$$

But since the function $F(z) = \frac{1}{\varphi(z)} = \frac{1}{z} + \sum_{n=0}^{\infty} b_n z^n$ is univalent, the area theorem, Lemma I.4.2, gives

$$|a_3 - a_2^2| = |b_1| \leq 1,$$

which proves (4.6).

We omit the proof of the converse (4.7). Nehari first proved φ is univalent in [1949], and it is the famous theorem of Ahlfors and Weill [1962] that $\partial \varphi(\mathbb{D})$ is a quasicircle. See also Ahlfors [1963] and Lehto [1987]. The constants 6 and 2 in (4.6) and (4.7) are both sharp. However, if $\sup_{\mathbb{D}} (1 - |z|^2)^2 |\mathcal{S}\varphi(z)| \leq 2$, then φ is univalent (Nehari [1949]), and if $(1 - |z|^2)^2 |\mathcal{S}\varphi(z)| < 2$ for all $z \in \mathbb{D}$, then $\partial \varphi(\mathbb{D})$ is a Jordan curve (Gehring and Pommerenke [1984]). ∎

We will also need the local version of Theorem 4.1.

Theorem 4.2 (Ahlfors). *Let U be a K-quasidisc and let $\psi : U \to \mathbb{D}$ be conformal. Then there is $\varepsilon = \varepsilon(K) > 0$ such that if φ is meromorphic on U and if*

$$\sup_U \left(\frac{1 - |\psi(z)|^2}{|\psi'(z)|} \right)^2 |\mathcal{S}\varphi(z)| < \varepsilon,$$

then φ extends to a quasiconformal mapping of C^*. In particular, φ is one-to-one on U and $\partial \varphi(U)$ is a quasicircle.

We will not prove Theorem 4.2 either. See Ahlfors [1963] or Lehto [1987]. Lehto [1987] has a much fuller discussion of Theorems 4.1 and 4.2, which lead to the Bers identification of universal Teichmüller space with an open connected subset of the space of Schwarzians of univalent functions under the norm

$$\|\varphi\|_{\mathcal{S}} = \sup_{\mathbb{D}} (1 - |z|^2)^2 |\mathcal{S}\varphi(z)|.$$

Gehring [1977] proved a converse of Theorem 4.2: If U is a simply connected domain for which the conclusion of Theorem 4.2 holds for some $\varepsilon > 0$, then U is a quasidisc.

We always write

$$g(z) = \log(\varphi'(z)) \tag{4.8}$$

when φ is univalent on $\mathbb{D}$. Theorem 4.1 is reminiscent of Theorem VII.2.1, which said that for φ univalent

$$(1 - |z|^2)|g'(z)| = (1 - |z|^2)\left|\frac{\varphi''(z)}{\varphi'(z)}\right| \le 6, \tag{4.9}$$

and (4.7) resembles Becker's [1972] converse: If

$$\|g\|_{\mathcal{B}} = \sup_{\mathbb{D}}(1 - |z|^2)|g'(z)| < 1,$$

then φ is univalent. In fact, the interplay between the two conformal invariants

$$(1 - |z|^2)|g'(z)|$$

and

$$\left(1 - |z|^2\right)^2\left|\mathcal{S}\varphi(z)\right|$$

will be crucial throughout this chapter. Note that if $(1 - |z|^2)|g'(z)|$ is small on a hyperbolic ball B, then $\left(1 - |z|^2\right)^2\left|\mathcal{S}\varphi(z)\right|$ is also small on compact subsets of B. However, if $\varphi(z) = \frac{z-z_1}{z-z_2}$, then $\mathcal{S}\varphi = 0$, but

$$g'(z) = \frac{\varphi''(z)}{\varphi'(z)} = \frac{2}{z_2 - z} \tag{4.10}$$

is not small.

Lemma 4.3. *Let $\varepsilon > 0$, $a > 0$, $b > 0$, and $c > \max(a, b)$ be given. Then there is $\delta = \delta(\varepsilon, a, b, c)$ such that if φ is univalent on the hyperbolic ball $\{z : \rho(z, z_0) < c\}$, and if*

$$\left(1 - |z|^2\right)^2\left|\mathcal{S}\varphi(z)\right| < \delta \tag{4.11}$$

on the hyperbolic ball $\{z : \rho(z, z_0) < a\}$, then for some Möbius transformation T,

$$\left|T(\varphi(z)) - \frac{z - z_0}{1 - \bar{z}_0 z}\right| + \left|\varphi(z) - T^{-1}\left(\frac{z - z_0}{1 - \bar{z}_0 z}\right)\right| < \varepsilon \tag{4.12}$$

on the hyperbolic ball $\{z : \rho(z, z_0) < b\}$.

Proof. We may assume $z_0 = 0$. There is a unique Möbius transformation T so that

$$T \circ \varphi(0) = 0, \quad (T \circ \varphi)'(0) = 1 \text{ and } (T \circ \varphi)''(0) = 0, \tag{4.13}$$

and we replace φ by $T \circ \varphi$. By (4.8)

$$\mathcal{S}\varphi(z) = g''(z) - \frac{1}{2}(g'(z))^2, \tag{4.14}$$

and under the substitution

$$g' = -2\frac{h'}{h},$$

(4.14) becomes

$$h'' + \frac{S\varphi}{2}h = 0.$$

This differential equation has a unique solution satisfying

$$h(0) = 1 \text{ and } h'(0) = 0,$$

and the solution is obtained by power series: If

$$S\varphi(z) = \sum_{n=0}^{\infty} a_n z^n,$$

and

$$h(z) = 1 + \sum_{n=2}^{\infty} b_n z^n,$$

then for $n \geq 2$,

$$n(n-1)b_n = -\frac{a_{n-2}}{2} - \sum_{k=2}^{n-2} \frac{b_k a_{n-2-k}}{2}. \tag{4.15}$$

If δ is sufficiently small, then by (4.11) all a_n are small for small n. Thus for any $\tau > 0$ there is $\delta > 0$ so that (4.11) and (4.15) give

$$\sup_{\rho(z,0)<b} |h(z) - 1| < \tau. \tag{4.16}$$

Unwinding φ from h now gives (4.12) for ε, if (4.16) holds for $\tau = \tau(\varepsilon)$. ∎

5. Geometric Estimates of Schwarzian Derivatives

Let φ be the conformal map of $\mathbb{D}$ onto a bounded simply connected domain Ω. In this section we give two different geometric estimates for the invariant Schwarzian $\left(1 - |w|^2\right)^2 |S\varphi(w)|$. The two estimates are not equivalent, but both estimates measure how much Ω deviates from a half-plane or disc when Ω is observed from the vantage point $\varphi(w)$. The second estimate uses the numbers β_E from Section 1 while the first estimate uses a more function theoretic quantity $\eta_\Omega(z_\theta, t)$ defined immediately below.

Fix $e^{i\theta} \in \partial \mathbb{D}$ such that the nontangential limit $z_\theta = \varphi(e^{i\theta})$ exists. For $t > 0$, let $\mathcal{L}$ be the set of lines L such that

(a) $L \cap B(z_\theta, \frac{t}{2}) \neq \emptyset$,
   and such that one component $W_t(\theta)$ of $B(z_\theta, 2t) \setminus L$ satisfies both
(b) $W_t(\theta) \subset \Omega$, and
(c) when $\varphi\{(re^{i\theta} : 0 < r < 1\}$ is oriented by increasing r, $W_t(\theta)$ contains the first arc of $\varphi(\{re^{i\theta} : 0 < r < 1\}) \cap B(z_\theta, 2t) \setminus L$,

If $\mathcal{L} \neq \emptyset$, define

$$\eta_\Omega(z_\theta, t) = \frac{1}{2t} \inf_{L \in \mathcal{L}} \left\{ \sup \text{dist}(z, \partial\Omega) : z \in L \cap B(z_\theta, 2t) \right\},$$

and if $\mathcal{L} = \emptyset$, put $\eta_\Omega(z_\theta, t) = 1$.

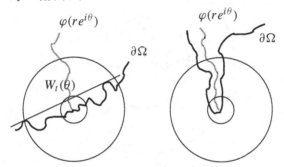

Figure X.16 On the left $\eta \sim \beta$; on the right $\eta = 1$ while β is small.

See also Exercise 12 for some remarks about the relation between $\eta_\Omega(z, t)$ and $\beta_\Omega(z, t)$.

Theorem 5.1. *Let $z_0 = \varphi(w_0) \in \Omega$ and let $z_\theta = \varphi(e^{i\theta}) \in \partial\Omega$ satisfy*

$$d = |z_\theta - z_0| \leq 2\text{dist}(z_0, \partial\Omega).$$

Then for s, $0 < s < 1$, there is $C = C(s)$, depending only on s, such that

$$\left(1 - |w_0|^2\right)^2 \left| S\varphi(w_0) \right| \leq C \sum_{k=0}^{\infty} \eta_\Omega(z_\theta, 2^k d) 2^{-ks}. \tag{5.1}$$

Proof. We may assume $z_0 = w_0 = 0$ and $d = 1$. Write $\eta_k = \eta_\Omega(z_\theta, 2^k d)$. Fix $\delta = \delta(s)$, $0 < \delta < \delta_0$ where δ_0 will be determined later. We can suppose $\eta_0 \leq \delta$, because otherwise (4.6) gives (5.1) with constant $C = \frac{6}{\delta}$. Since Ω is bounded, $\lim_{k \to \infty} \eta_k = 1$ and there exists a first integer N such that $\eta_{N+1} > \delta$.

Let L_0 be a line with $L_0 \cap B(z_\theta, \frac{1}{2}) \neq \emptyset$ such that L_0 gives the minimum η_0. The line L_0 exists because $\eta_0 \leq \delta$. Let H_0 be the component of $\mathbb{C} \setminus L_0$ such that $H_0 \cap B(z_\theta, 2) \subset \Omega$. Then $z_0 \in H_0$ by the first arc condition (c), because $z_0 = \varphi(0) \in B(z_\theta, 2)$. For $k \leq N$, let L_k be a line with $L_k \cap B(z_\theta, 2^{k-1}) \neq \emptyset$ such that L_k gives the minimum η_k. Again L_k exists because $\eta_k \leq \delta$. Let H_k be that component of $\mathbb{C} \setminus L_k$ with $H_k \cap B(z_\theta, 2^{k+1}) \subset \Omega$. Then by condition (c) $H_k \cap H_{k-1} \cap B(z_\theta, 2^{k-1}) \neq \emptyset$. Write $B_k = B(z_\theta, 2^k)$ and $A_k = B_{k+1} \setminus B_k$. Set $V_0 = H_0$ and for $1 \leq k \leq N$,

$$V_k = \text{int}\left(\overline{\left(H_0 \cap B(z_\theta, 2)\right) \cup \bigcup_{1 \leq j \leq k-1} (H_j \cap A_j) \cup \left((H_k \setminus B_k) \cap \Omega\right)}\right).$$

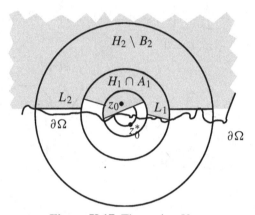

Figure X.17 The region V_2.

Then V_k is a simply connected domain, $V_k \subset \Omega$, and $z_0 \in V_k$ but $z_0^* \notin V_k$ if z_0^* is the reflection of z_0 through $L_0 = \partial V_0$.

By the distortion theorem, $|\varphi'(0)| \geq \text{dist}(z_0, \partial \Omega) \geq 1/2$, and if δ_0 is small then

$$\sigma = \varphi(\{|w| = 1/4\}) \subset \bigcap V_k.$$

Set $G(z) = g_\Omega(z, z_0)$ and $G_k(z) = g_{V_k}(z, z_0)$. Then the theorem is an immediate consequence of the next three lemmas:

Lemma 5.2. *There is a constant C, independent of φ, such that*

$$\left|\mathcal{S}\varphi(0)\right| \leq C \sup_\sigma |G(z) - G_0(z)|.$$

Lemma 5.3. *If $z \in \sigma$, then*

$$|G_N(z) - G_0(z)| \leq C(s) \sum_{k=0}^{\infty} \eta_k 2^{-ks}. \tag{5.2}$$

Lemma 5.4. *If $z \in \sigma$, then*

$$|G(z) - G_N(z)| \le C(s) \sum_{k=0}^{N-1} \eta_k 2^{-ks} + C(s) 2^{-Ns}. \qquad (5.3)$$

Proof of Lemma 5.2. By (4.6) we can assume

$$\varepsilon = \sup_{\sigma} |G(z) - G_0(z)|$$

is small, because otherwise the lemma holds with $C = \frac{6}{\varepsilon}$. Let $T(w) = \frac{z_0^* w}{w - 1}$ be the conformal map of $\mathbb{D}$ onto V_0. Then

$$\sup_{\sigma} \left| -\log |\varphi^{-1}(z)| + \log |T^{-1}(z)| \right| = \varepsilon,$$

so that on $\varphi^{-1}(\sigma) = \{w : |w| = 1/4\}$,

$$1 - \varepsilon \le \left| \frac{T^{-1}(\varphi(w))}{w} \right| \le 1 + \varepsilon.$$

Because $T^{-1}(0) = 0$, it follows via Taylor series that

$$\sup_{|w| \le 1/8} \left| T^{-1}(\varphi(w)) - \lambda w \right| \le C\varepsilon$$

for some constant λ with $|\lambda| = 1$ and therefore that

$$\left| \mathcal{S}\varphi(0) \right| = \left| \mathcal{S}(T^{-1} \circ \varphi)(0) \right| \le C\varepsilon,$$

which proves Lemma 5.2.   ∎

Proof of Lemma 5.3. First notice that by the definition of $\eta(z, t)$,

$$\operatorname{dist}\left(L_k \cap \partial B_k, L_{k-1} \cap \partial B_k \right) \le C 2^k (\eta_k + \eta_{k-1}). \qquad (5.4)$$

Let $r\Theta(r)$ be the length of $V_N \cap \{|z| = r\}$. For $k \ge 1$ and $2^k \le r \le 2^{k+1}$ we have

$$\{|z| = r\} \subset A_{k-1} \cup A_k \cup A_{k+1},$$

so that by (5.4)

$$\left| \Theta(r) - \pi \right| \le C(\eta_{k-1} + \eta_k + \eta_{k+1}) \le C\delta,$$

and

$$\Theta(r) \le \pi + C\delta.$$

Now choose δ so that

$$\frac{\pi}{\pi + C\delta} \ge s.$$

Then for $z \in \sigma$ and $k \geq 2$, Theorem IV.6.1 gives us

$$
\omega(z, A_k \cap \partial V_N, V_N) \leq C \exp\left(-\pi \int_1^{2^{k-1}} \frac{dr}{r\Theta(r)}\right)
$$

$$
\leq C \exp\left(\frac{-\pi}{\pi + C\delta} \int_1^{2^{k-1}} \frac{dr}{r}\right) \leq C2^{-ks}. \tag{5.5}
$$

In the same way, we also get

$$
\omega(z, (\partial V_N) \setminus B_N, V_N) \leq C2^{-Ns}.
$$

The function

$$
G_0(z) = \log\left|\frac{z - z_0^*}{z - z_0}\right|
$$

is harmonic on $\mathbb{C} \setminus \{z_0, z_0^*\}$ and when $k \geq 1$, $\sup_{A_k} |\nabla G_0(z)| \leq C2^{-k}$ since $|z_0 - z_\theta| \leq 2$. If $z \in A_k \cap \partial V_N$, then

$$
\operatorname{dist}(z, \partial V_0) \leq C2^k \sum_{n=0}^{\min(k,N)} \eta_n,
$$

and therefore

$$
G_0(z) \leq C \sum_{n=0}^{k} \eta_n.
$$

But then since $G_N - G_0$ is harmonic on V_N, (5.5) gives

$$
\left|G_N(z) - G_0(z)\right| = \left|\int_{\partial V_N} G_0(\zeta)d\omega(z, \zeta)\right|
$$

$$
\leq C \sum_{k=0}^{N} \left(\sum_{n=0}^{\min(k,N)} \eta_n\right) 2^{-ks}
$$

$$
\leq C \sum_k \eta_k \sum_{n \geq k} 2^{-ns}
$$

$$
\leq C(s) \sum_{k=0}^{\infty} \eta_k 2^{-ks},
$$

and that proves (5.2).   ∎

Proof of Lemma 5.4. Let Γ be a curve in V_N such that for $z \in \Gamma \cap A_k$

$$
c_1 \operatorname{dist}(z, L_k) \leq 2^k(\eta_{k-1} + \eta_k + \eta_{k+1}) \leq c_2 \operatorname{dist}(z, L_k),
$$

provided $\eta_{k-1} + \eta_k + \eta_{k+1} > 0$, and such that if $\zeta_k(s)$ is the arc length parameterization of $\Gamma \cap A_k$, then

$$|\zeta_k'| \le c, \text{ and } |\zeta_k''| \le c2^{-k}.$$

This curve Γ exists because of (5.4). If $\eta_{k-1} + \eta_k + \eta_{k+1} = 0$, then $\partial\Omega \cap A_{k+1}$
falls in a line segment and in this case let $\Gamma \cap A_k$ be a parallel segment such
that $\operatorname{dist}(\Gamma \cap A_k, \partial\Omega) \ge \delta 2^k$.

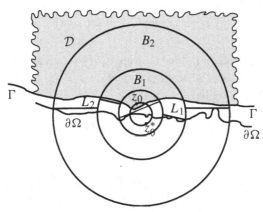

Figure X.18

Let $\mathcal{D} \subset V_N$ be the simply connected domain such that $z_0 \in \mathcal{D}$ and such that
$\partial\mathcal{D}$ consists of two arcs $\Gamma_1 = B_{N+1} \cap \Gamma$ and $\Gamma_2 \subset \partial B_{N+1}$. Partition $\partial\mathcal{D}$ into
arcs I_j such that

$$\operatorname{dist}(I_j, \partial\Omega) \sim \operatorname{diam}(I_j).$$

Then if $I_j \subset \Gamma_1 \cap A_k$,

$$\operatorname{diam}(I_j) \le c2^k(\eta_{k-1} + \eta_k + \eta_{k+1}).$$

Let $z \in \mathcal{D} \cap A_k$ satisfy $\operatorname{dist}(z, \Gamma) \sim 2^k$. Then for $I_j \subset A_k \cap \Gamma_1$, a simple
comparison using the smoothness of $\Gamma \cap A_k$ yields

$$\omega(z, I_j, \mathcal{D}) \le C \operatorname{diam}(I_j)2^{-k}.$$

By the proof of (5.5),

$$\omega(z_0, A_k \cap \Gamma_1, \mathcal{D}) \le C2^{-ks}.$$

Therefore,

$$\omega(z_0, I_j, \mathcal{D}) \le C \operatorname{diam}(I_j)2^{-k}2^{-ks} \le C2^{-ks}(\eta_{k-1} + \eta_k + \eta_{k+1}).$$

On the other hand, if $I_j \subset \Gamma_2$, then

$$\omega(z_0, I_j, \mathcal{D}) \le \omega(z_0, \Gamma_2, \mathcal{D}) \le C2^{-Ns},$$

also by the proof of (5.5). Altogether these inequalities give us

$$\omega(z_0, I_j, \mathcal{D}) \le 2^{-Ns} + \sum_{k=0}^{N} \eta_k 2^{-ks}. \tag{5.6}$$

Let $J_j = \varphi^{-1}(I_j)$. Then

$$\sup_{\partial D} G(z) = \sup_j \sup_{J_j} \log \frac{1}{|w|} \sim \sup_j \sup_{J_j}(1 - |w|). \tag{5.7}$$

Let $w^* \in \varphi^{-1}(\partial \mathcal{D})$ satisfy $1 - |w^*| = \sup_{\varphi(\partial \mathcal{D})}(1 - |w|)$ and say $w^* \in J_n$. Let $J_{n\pm1}$ denote the two arcs adjacent to J_n and set $J = J_{n-1} \cup J_n \cup J_{n+1}$. Then if $w \in \varphi^{-1}(\partial \mathcal{D}) \setminus J$, we have

$$|w| \ge |w^*| \text{ and } \rho(w, w^*) \ge C.$$

Set

$$\mathcal{U} = \varphi^{-1}(\mathcal{D}) \cap \left(\{|w| < |w^*|\} \cup \{\rho(w, w^*) < C\}\right).$$

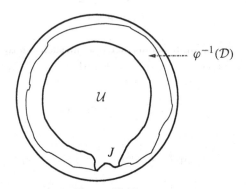

Figure X.19

Then by (5.6) and (5.7),

$$\sup_{\partial D} G(z) \le C_1(1 - |w^*|) \le C_2\omega(0, J, \mathcal{U})$$

$$\le C_2\omega(z_0, I_{n-1} \cup I_n \cup I_{n+1}, \mathcal{D})$$

$$\le C_2\left(2^{-Ns} + \sum_{k=0}^{N-1} \eta_k 2^{-ks}\right).$$

Therefore, since $\Omega \supset V_N \supset \mathcal{D} \supset \sigma$,

$$\sup_{\sigma} |G - G_N| = \sup_{\sigma} (G - G_N)$$

$$\leq \sup_{\partial \mathcal{D}} (G - G_N) \leq \sup_{\partial \mathcal{D}} G(z)$$

$$\leq C_2 \Big(2^{-Ns} + \sum_{k=0}^{N-1} \eta_k 2^{-ks} \Big).$$

That proves (5.3). ∎

The second estimate of $(1 - |w|^2)^2 |\mathcal{S}\varphi(w)|$ uses the β numbers from Section 2. It will also be valid in the more general situation in which φ is replaced by a covering map. Let E be a compact set having at least three points and let $\psi : \mathbb{D} \to \mathbb{C}^* \setminus E$ be a universal covering map, which by definition means that ψ is analytic, maps $\mathbb{D}$ onto $\mathbb{C}^* \setminus E$, and is locally univalent, i.e., $\psi'(z) \neq 0$. (It is a famous theorem of Koebe that such ψ exist when E has at least three points, see Ahlfors [1973], Fisher [1983], or Gamelin [2001].) Universal covering maps satisfy a weakened form of Theorem I.4.3.

Lemma 5.5. *If $\psi(w_0) = z_0$, and if ψ is univalent on $B(w_0, \eta(1 - |w_0|^2))$, then*

$$\text{dist}(z_0, E) \leq |\psi'(w_0)|(1 - |w_0|^2) \leq \frac{4}{\eta} \text{dist}(z_0, E).$$

Proof. We may take $z_0 = 0$ and $w_0 = 0$. Then the right-hand inequality is the Koebe estimate (I.2.11) applied to $\frac{\psi(\eta w)}{\eta \psi'(0)}$. By the monodromy theorem, ψ^{-1} has a single valued univalent branch on $B(0, \text{dist}(0, E))$ satisfying $\psi^{-1}(0) = 0$, and the left-hand inequality is the Schwarz lemma for this branch of ψ^{-1}. ∎

When $z_0 \notin E$, and let $Q = Q^0$ be the smallest dyadic square such that $z_0 \in Q$ and $\text{dist}(z_0, E) \leq \ell(Q)$. Let Q^n be that dyadic square of side $2^n \ell(Q)$ such that $Q \subset Q^n$ and set

$$\beta_n = \beta_E(10Q^n)$$

as defined in (2.2). Let S_n be a closed strip of width $10 \cdot 2^{n+1} \beta_n \ell(Q)$ such that

$$E \cap 10Q^n \subset S_n,$$

let L_n be the axis of S_n, and let E_n^* be the orthogonal projection of $E \cap 10Q_n$ onto L_n. Let m_n be the length of the largest interval in $(L_n \cap 10Q_n) \setminus E_n^*$ and set

$$\gamma_n = \gamma_E(Q_n) = \frac{m_n}{\ell(Q_n)}, \quad \delta_n = \beta_n + \gamma_n.$$

Clearly γ_n measures the largest gap in E_n^*.

Theorem 5.6. *Fix* $0 < \eta < 1$ *and* $0 < s < 1$. *Let* $\psi : \mathbb{D} \to \mathbb{C}^* \setminus E$ *be a universal covering map and* $w_0 \in \mathbb{D}$ *satisfy* $\psi(w_0) = z_0$. *Then there is* $C = C(\eta, s)$ *such that if* ψ *is univalent on* $B\big(w_0, \eta(1 - |w_0|)\big)$, *then*

$$\big(1 - |w_0|^2\big)^2 |S\psi(w_0)| \leq C \sum_{n=0}^{\infty} \delta_n 2^{-ns}. \tag{5.8}$$

Proof. The argument resembles the proof of Theorem 5.1. Fix $\delta = \delta(s)$. We can suppose $\delta \leq \delta_0$ where δ_0 is to be determined later, because otherwise (4.6) gives (5.8) with $C = 6/\delta$. If N denotes the first integer such that $\delta_{N+1} > \delta$ then $N > 1$ and N is finite, because eventually γ_n is large. We assume that $w_0 = 0$ and that $\ell(Q_0) = 1$.

Let S_0 be a closed strip of width $2\beta_0$ such that

$$E \cap 10Q^0 \subset S_0.$$

Since $\beta_0 < \delta$, $z_0 \notin S_0$. Let H_0 be the component of $\mathbb{C} \setminus S_0$ with $z_0 \in H_0$. For $n < N$, let S_n be a closed strip of width $2^{n+1}\beta_n$ such that

$$S_n \cap Q^n \neq \emptyset \text{ and } E \cap 10Q^n \subset S_n,$$

and let H_n be that component of $\mathbb{C} \setminus S_n$ such that

$$H_n \cap H_{n-1} \cap Q^{(n-1)} \neq \emptyset.$$

The component H_n exists because $2\beta_n + \beta_{n-1} < 1$, and H_n is unique because $\operatorname{diam}(E) \geq 2^n$.

This time write $A_n = 5Q^n \setminus 5Q^{n-1}$. Set $V_0 = H_0$ and for $1 \leq n \leq N$,

$$V_n = \operatorname{int}\Big(\overline{(H_0 \cap 5Q^0) \cup (H_1 \cap A_1) \cup \ldots \cup (H_n \cap A_n)}\Big).$$

Then V_n is a simply connected domain, $V_n \subset (\mathbb{C}^* \setminus E)$, and $z_0 \in V_n$ but if z_0^* is the reflection of z_0 through ∂V_0, then $z_0^* \notin V_n$. The analogue of (5.4) holds because $\beta_n + \beta_{n-1}$ is small.

We continue to assume $w_0 = z_0 = 0$. By the univalence hypothesis and the distortion theorem, $|\psi'(0)| \geq \operatorname{dist}(z_0, E)$ and

$$\psi(\{|w| \leq \eta/8\}) \subset \bigcap V_n.$$

Set

$$\sigma = \psi(\{|w| = \eta/8\}).$$

Define $G_n = g_{V_n}(z, z_0)$. If $T(w) = \frac{z_0^* w}{w-1}$ is the Möbius transformation from $\mathbb{D}$ to the half-plane V_0 with $T(0) = 0$, then

$$G_0(T(w)) = \log\left|\frac{1}{w}\right|.$$

Since $\psi'(w) \neq 0$, there exists, by the monodromy theorem, a unique branch of ψ^{-1} defined on V_N and satisfying $\psi^{-1}(0) = 0$. Using ψ^{-1}, we define the harmonic function $g(z)$, $z \in V_N \setminus \{0\}$ by

$$G(\psi(w)) = \log\left|\frac{1}{w}\right| \tag{5.9}$$

on $\psi^{-1}(V_N)$.

At this point the proofs of Lemmas 5.2, 5.3, and 5.4 can be repeated with φ replaced by ψ and with η_k replaced by δ_k to produce a proof of Theorem 5.6. Note that (5.7) holds for the new G because of the definition (5.9). We leave the details to the reader.   ∎

6. Schwarzian Derivatives and Rectifiable Quasicircles

Let φ be a conformal mapping from $\mathbb{D}$ onto a bounded Jordan domain Ω. We write $F = (\varphi')^{\frac{1}{2}}$ and, as always, $g = \log(\varphi')$.

Then

$$F' = \frac{1}{2} F g',$$

$$F'' = \frac{1}{2} F\left(g'' + \frac{1}{2}(g')^2\right) = \frac{1}{2} F(\mathcal{S}\varphi + (g')^2), \tag{6.1}$$

and

$$|\varphi'||\mathcal{S}\varphi|^2 \leq 2|\varphi'||\mathcal{S}\varphi + (g')^2|^2 + 2|\varphi'||g'|^4$$
$$\leq 8|F''|^2 + 8|F'|^2|g'|^2.$$

Because $\|g\|_\mathcal{B} \leq 6$, we therefore have

$$\iint_\mathbb{D} |\varphi'(z)||\mathcal{S}\varphi(z)|^2 (1 - |z|^2)^3 dx dy \leq 8 \iint |F''|^2 (1 - |z|^2)^3 dx dy$$

$$+ 8 \cdot 36 \iint |F'|^2 (1 - |z|^2) dx dy.$$

By Theorem 1.2 both terms on the right are bounded by $C\|F\|_{H^2}^2 = \|\varphi'\|_{H^1}$ and by the F. and M. Riesz theorem, $\|\varphi'\|_{H^1} = \ell(\Gamma)$. Therefore we have the following lemma.

Lemma 6.1. *There is a constant C such that if $\partial\Omega$ is rectifiable, then*

$$\iint_{\mathbb{D}} |\varphi'(z)||\mathcal{S}\varphi(z)|^2(1-|z|^2)^3 dxdy \leq C\ell(\partial\Omega).$$

When $\partial\Omega$ is a quasicircle, Bishop and Jones [1994] proved the converse of Lemma 6.1.

Theorem 6.2. *If Ω is a bounded quasidisc, then $\partial\Omega$ is rectifiable if and only if*

$$\iint_{\mathbb{D}} |\varphi'(z)||\mathcal{S}\varphi(z)|^2(1-|z|^2)^3 dxdy < \infty,$$

and there is a constant C_1 such that

$$C_1^{-1}\ell(\partial\Omega) \leq \operatorname{diam}(\Omega) + \iint_{\mathbb{D}} |\varphi'(z)||\mathcal{S}\varphi(z)|^2(1-|z|^2)^3 dxdy$$

$$\leq C_1\ell(\partial\Omega). \tag{6.2}$$

The constant C_1 depends only on the quasiconformality constant K of the quasicircle $\partial\Omega$. See Theorem VII.3.3.

Proof. By Lemma 6.1 and the trivial estimate $\operatorname{diam}(\Omega) \leq \ell(\partial\Omega)$, the right-hand inequality of (6.2) holds, even when $\partial\Omega$ is not a quasicircle.

To prove the left-hand inequality we must come to grips with the set where $|g'(z)|$ is large but $|\mathcal{S}\varphi(z)|$ is small. Let $\{S_j\}$ denote the Whitney squares for Ω and fix $\varepsilon > 0$ and $\delta > 0$. We write $S_j \in \mathcal{B}_{\varepsilon,\delta}$ and say S_j is a **bad square** if

$$\sup_{\varphi^{-1}(S_j)} |g'(z)|(1-|z|^2) \geq \varepsilon, \tag{6.3}$$

but

$$\sup_{\varphi^{-1}(S_j)} |\mathcal{S}\varphi(z)|(1-|z|^2)^2 \leq \delta. \tag{6.4}$$

Lemma 6.3. *There is $\varepsilon_0 > 0$ and $C > 0$, such that if $0 < \varepsilon < \varepsilon_0$ and $\delta > 0$, then*

$$\int_{\partial\mathbb{D}} |\varphi'| ds \leq C \operatorname{dist}(\varphi(0), \partial\Omega) + C\sum_{\mathcal{B}_{\varepsilon,\delta}} \ell(S_j)$$

$$+ C(1+\delta^{-2}) \iint_{\mathbb{D}} |\varphi'(z)||\mathcal{S}\varphi(z)|^2(1-|z|^2)^3 dxdy \tag{6.5}$$

Proof. Write $B = \bigcup\{\varphi^{-1}(S_j) : S_j \in \mathcal{B}_{\varepsilon,\delta}\}$. Using Koebe's theorem and using Theorem 1.2 twice with $F = (\varphi')^{\frac{1}{2}}$, we get

$$\int_{\partial \mathbb{D}} |\varphi'(z)|\,ds \sim \mathrm{dist}(\varphi(0), \partial\Omega) + \iint_{\mathbb{D}} |\varphi'(z)||g'(z)|^2(1 - |z|^2)\,dxdy, \quad (6.6)$$

and by (6.1)

$$\int_{\partial \mathbb{D}} |\varphi'(z)|\,ds \leq C_0\,\mathrm{dist}(\varphi(0), \partial\Omega)$$

$$+ C_0|F'(0)|^2 \qquad\qquad\qquad (6.7)$$

$$+ C_0 \iint_{\mathbb{D}} |\varphi'(z)|\left|S\varphi(z) + \frac{(g'(z))^2}{2}\right|^2(1 - |z|^2)^3\,dxdy,$$

for some constant C_0. Since $\|g\|_{\mathcal{B}} \leq 6$ we have

$$|F'(0)|^2 = \frac{|\varphi'(0)||g'(0)|^2}{4} \leq 36\,\mathrm{dist}(\varphi(0), \partial\Omega),$$

so that (6.7) gives

$$\int_{\partial \mathbb{D}} |\varphi'(z)|\,ds \leq C_1\mathrm{dist}(\varphi(0), \partial\Omega)$$

$$+ C_1 \iint_{\mathbb{D}} |\varphi'(z)||S\varphi(z)|^2(1 - |z|^2)^3\,dxdy$$

$$+ C_1 \iint_{B} |\varphi'(z)||g'(z)|^4(1 - |z|^2)^3\,dxdy$$

$$+ C_1 \iint_{\mathbb{D}\backslash B} |\varphi'(z)||g'(z)|^4(1 - |z|^2)^3\,dxdy,$$

for some constant C_1.

To estimate the integral over B we use $\|g\|_{\mathcal{B}} \leq 6$ and the Koebe theorem to get

$$\iint_{B} |\varphi'(z)||g'(z)|^4(1 - |z|^2)^3\,dxdy \leq 6^4 \sum_{\mathcal{B}_{\varepsilon,\delta}} \iint_{\varphi^{-1}(S_j)} \frac{|\varphi'(z)|}{1 - |z|^2}\,dxdy$$

$$\leq 6^4 C_1 \sum_{\mathcal{B}_{\varepsilon,\delta}} \ell(S_j).$$

For the remaining integral we note that on $\mathbb{D} \setminus B$ either (6.3) fails, or (6.4) fails but still $|g'(z)|(1 - |z|^2) \leq 6$. Furthermore, if (6.4) fails for S_j, then by

Theorem 4.1 and the Schwarz lemma, there are constants c_1 and c_2 such that

$$|S\varphi(z)|(1 - |z|^2)^2 \geq c_1\delta$$

on a disc $E_j \subset \varphi^{-1}(S_j)$ with $|E_j| \geq c_2|\varphi^{-1}(S_j)|$. Because

$$\sup_{\varphi^{-1}(S_j)} |\varphi'(z)| \leq c_3 \inf_{\varphi^{-1}(S_j)} |\varphi'(z)|$$

we consequently have

$$\iint_{\mathbb{D}\setminus B} |\varphi'(z)||g'(z)|^4(1 - |z|^2)^3 dxdy$$

$$\leq \varepsilon^2 \iint_{\mathbb{D}} |\varphi'(z)||g'(z)|^2(1 - |z|^2)dxdy$$

$$+ \frac{6^4 C}{\delta^2} \iint_{\mathbb{D}} |\varphi'(z)||S\varphi(z)|^2(1 - |z|^2)^3 dxdy.$$

Therefore

$$\int_{\partial\mathbb{D}} |\varphi'(z)|ds \leq C_1 \text{dist}(\varphi(0), \partial\Omega)$$

$$+ C_1\left(1 + \frac{6^4}{\delta^2}\right) \iint_{\mathbb{D}} |\varphi'(z)||S\varphi(z)|^2(1 - |z|^2)^3 dxdy \qquad (6.8)$$

$$+ 6^4 C_1 \sum_j \ell(S_j) + C_1\varepsilon^2 \iint_{\mathbb{D}} |\varphi'(z)||g'(z)|^2(1 - |z|^2)dxdy.$$

Then by (6.6) and (6.8),

$$\int_{\partial\mathbb{D}} |\varphi'(z)|ds \leq C_1 \text{dist}(\varphi(0), \partial\Omega) + C_2\varepsilon^2 \int_{\partial\mathbb{D}} |\varphi'(z)|ds$$

$$+ C_1\left(1 + \frac{6^4}{\delta^2}\right) \iint_{\mathbb{D}} |\varphi'(z)||S\varphi(z)|^2(1 - |z|^2)^3 dxdy$$

$$+ 6^4 C_1 \sum_{B_{\varepsilon,\delta}} \ell(S_j),$$

and whenever $C_2\varepsilon^2 < 1$, (6.5) holds with $C = \frac{6^4 C_1}{1 - C_2\varepsilon^2}$. ∎

Now assume Ω is a quasidisc with constant K. By a theorem of Ahlfors [1963], there is $A' = A'(K)$ such that whenever $\Gamma \subset \mathbb{D}$ is a circle, $\varphi(\Gamma)$ satisfies the Ahlfors three-point condition (VII.2.5) with constant A'.

When $T \in \mathcal{M}$, replacing φ by $\varphi \circ T$ does not change the integral in (6.2) and so we can assume

$$\operatorname{dist}(\varphi(0), \partial\Omega) \sim \operatorname{diam}(\Omega). \tag{6.9}$$

In that case we claim that if δ is sufficiently small, then every bad square is a bounded hyperbolic distance from $\varphi(0)$. By (6.9) and Lemma 6.3, that claim will prove the theorem.

Suppose S_j is a bad square such that $\varphi(z_0) \in S_j$ where $1 - |z_0|^2$ is small and where (6.3) holds at z_0. If δ is small, then by Lemma 4.3 there is a Möbius transformation

$$Tz = A\frac{z - z_1}{z - z_2}$$

such that

$$|T - \varphi| < \varepsilon \tag{6.10}$$

on the hyperbolic ball $B = \{z : \rho(z, z_0) \leq b\}$ and

$$(1 - |z_0|^2)\left|\frac{T''(z_0)}{T'(z_0)}\right| \geq \frac{\varepsilon}{2}.$$

Hence by (4.10), the pole z_2 of T satisfies

$$\operatorname{dist}(z_2, \partial B) \leq |z_2 - z_0| \leq \frac{4}{\varepsilon}(1 - |z_0|^2),$$

while for $b = b(\varepsilon)$ fixed so that $\sinh(2b) = \varepsilon^{-2}$, ∂B has euclidian radius

$$r \geq \varepsilon^{-2}(1 - |z_0|) + O\big((1 - |z_0|)^3\big).$$

Consequently, if $1 - |z_0|$ is small, there are adjacent arcs $I \subset \partial B$ and $J \subset \partial B$ with $\ell(I) = \ell(J) \geq \frac{2}{\sqrt{\varepsilon}}(1 - |z_0|)$ such that

$$\frac{\ell(T(I))}{\ell(T(J))} \geq \frac{C}{\varepsilon}.$$

But by (6.10), that contradicts (VII.2.5) for $\varphi(\partial B)$. $\blacksquare$

7. The Bishop–Jones $H^{\frac{1}{2}-\eta}$ Theorem

When Ω is not a quasidisc the condition

$$\iint_{\mathbb{D}} |\varphi'(z)||\mathcal{S}\varphi(z)|^2(1 - |z|^2)^3 dx dy < \infty \tag{7.1}$$

of Theorem 6.2 no longer implies that $\partial\Omega$ is rectifiable. For example, if Ω is a half-plane, then $\mathcal{S}\varphi = 0$ but $\partial\Omega$ is not rectifiable. However in [1994] Bishop and Jones obtained a sharp substitute theorem, and the proof of this theorem is the key to the deeper results in this chapter.

Theorem 7.1. *Let φ be the conformal mapping from $\mathbb{D}$ onto Ω. If*

$$B = |\varphi'(0)| + \iint_{\mathbb{D}} |\varphi'(z)||\mathcal{S}\varphi(z)|^2(1 - |z|^2)^3 dxdy < \infty,$$

then for any η, $0 < \eta < \frac{1}{2}$, $\varphi' \in H^{\frac{1}{2}-\eta}$, and

$$\|\varphi'\|_{H^{\frac{1}{2}-\eta}} \le C(\eta)B. \tag{7.2}$$

In particular, if the Bishop–Jones integral (7.1) is finite, then φ has non-zero angular derivative almost everywhere on $\partial\mathbb{D}$, the cone points of $\partial\Omega$ have full harmonic measure relative to Ω, and $\omega \ll \Lambda_1$ by Theorem VI.4.2. Theorem 7.1 is sharp; again the counterexample is the map φ from $\mathbb{D}$ to a half-plane.

For the applications to come, Theorem 7.1 will be less important than its local versions, Theorem 7.2 and Corollary 7.3. Recall we always write $g = \log(\varphi')$.

Theorem 7.2. *Let $E \subset \partial\mathbb{D}$ be compact. Let $U = \bigcup_{\zeta \in E} \Gamma_\beta(\zeta)$ be a cone domain, let $1 < \alpha < \beta$, and let $\varepsilon > 0$. Let φ be the conformal mapping from U onto a simply connected domain Ω and assume that at every $\zeta \in J \subset E$,*

$$\iint_{\Gamma_\alpha(\zeta)} |\varphi'(z)||\mathcal{S}\varphi(z)|^2(1 - |z|^2)^2 dxdy \le N < \infty. \tag{7.3}$$

Then there is $C(N, \varepsilon) < \infty$ and there exists $J_0 \subset J$ such that $|J_0| \ge (1 - \varepsilon)|J|$ and

$$\iint_{\Gamma_\alpha(\zeta)} |\varphi'(z)||g'(z)|^2 dxdy < C(N, \varepsilon) \tag{7.4}$$

at every $\zeta \in J_0$.

Corollary 7.3. *Let $E \subset \partial\mathbb{D}$ be compact. Let $U = \bigcup_{\zeta \in E} \Gamma_\beta(\zeta)$ be a cone domain, and let $1 < \alpha < \beta$. Let φ be the conformal mapping from U onto a simply connected domain Ω. Then at almost every $\zeta \in E$ for which*

$$\iint_{\Gamma_\alpha(\zeta)} |\varphi'(z)||\mathcal{S}\varphi(z)|^2(1 - |z|^2)^2 dxdy < \infty \tag{7.5}$$

we also have

$$\iint\limits_{\Gamma_\alpha(\zeta)} |\varphi'(z)||g'(z)|^2 dxdy < \infty. \tag{7.6}$$

The corollary follows by sending $\varepsilon \to 0$ and $N \to \infty$ in Theorem 7.2. Since $F(z) = (\varphi'(z))^{\frac{1}{2}}$ satisfies $|F'(z)|^2 = |\varphi'(z)||g'(z)|^2$, Theorem 1.3 and Corollary 7.3 imply φ has a non-zero angular derivative at almost every point where (7.5) holds.

The converse assertion, that (7.5) holds almost everywhere (7.6) holds, is very easy. Since $g = \log(\varphi')$ and $F = (\varphi')^{\frac{1}{2}}$ we have $|\varphi'||g'|^2 = 4|F'|^2$, so that (7.6) implies

$$\iint\limits_{\Gamma_\alpha(\zeta)} |F'(z)|^2 dxdy < \infty,$$

and by Lemma 1.5,

$$\iint\limits_{\Gamma_\delta(\zeta)} |F''(z)|^2(1 - |z|)^2 dxdy < \infty$$

for any δ, $1 < \delta < \alpha$. Also recall from Section 6,

$$|\varphi'||\mathcal{S}\varphi|^2 \le 8|F''|^2 + 8|g'|^2|F'|^2.$$

Since $\|g\|_\mathcal{B} \le 6$, we conclude that

$$\iint\limits_{\Gamma_\delta(\zeta)} |\varphi'(z)||\mathcal{S}\varphi(z)|^2(1 - |z|^2)^2 dxdy < \infty$$

if (7.6) holds at ζ, and a point of density argument then implies (7.5) holds almost everywhere that (7.6) holds.

The proof of Theorem 7.2 resembles the main argument in the proof of Theorem 7.1 and we will prove Theorem 7.1 first and then indicate the changes needed to get Theorem 7.2.

Proof of Theorem 7.1.. We take $B = 1$. Set $\Gamma(\zeta) = \{z : |z - \zeta| < 1 - |z|\}$. Given η, $0 < \eta < \frac{1}{2}$ and given $\lambda \ge \lambda_0 = \lambda_0(\eta) > 1$, we construct a region $\mathcal{R} \subset \mathbb{D}$ and a compact set $E \subset \partial\mathbb{D}$ such that

$$\bigcup_{\zeta \in E} \Gamma(\zeta) \subset \mathcal{R}, \tag{7.7}$$

$$|\partial\mathbb{D} \setminus E| \le C\lambda^{-1+2\eta}, \tag{7.8}$$

and

$$\iint_{\mathcal{R}} |\varphi'||g'|^2(1 - |z|^2)dxdy \le C\lambda, \tag{7.9}$$

where the constant C in (7.8) and (7.9) depends on η but not on λ.

Assume that we have built sets $\mathcal{R}$ and E satisfying (7.7), (7.8), and (7.9). Write $p = \frac{1}{2} - \eta$ and as usual take $F = (\varphi')^{\frac{1}{2}}$. Then $\varphi' \in H^p$ if and only if $F \in H^{2p}$ and

$$||\varphi'||_{H^p} = ||F||^2_{H^{2p}}.$$

Recall the area function

$$AF(\zeta) = \left(\iint_{\Gamma(\zeta)} |F'(z)|^2 dxdy\right)^{\frac{1}{2}}$$

$$= \left(\frac{1}{4}\iint_{\Gamma(\zeta)} |\varphi'(z)||g'(z)|^2 dxdy\right)^{\frac{1}{2}}.$$

By Theorem 1.1 $F \in H^{2p}$ if and only if $AF \in L^{2p}$ and

$$||F||_{H^{2p}} \le C_p||AF||_{2p}.$$

Since $\left|\{\zeta : z \in \Gamma(\zeta)\}\right| \le c(1 - |z|^2)$, we have by (7.7)

$$\int_E (AF)^2 d\theta \le C \iint_{\mathcal{R}} |\varphi'(z)||g'(z)|^2(1 - |z|^2)dxdy,$$

and thus by (7.8) and (7.9),

$$|\{\theta : AF(\theta) > t\}| \le |\partial\mathbb{D} \setminus E| + \frac{1}{t^2}\int_E (AF)^2 d\theta$$
$$\le C\lambda^{-1+2\eta} + C\frac{\lambda}{t^2}, \tag{7.10}$$

for all $\lambda \ge \lambda_0$. Take $\lambda = t^{\frac{2}{2p+1}}$ so that the two terms on the extreme right of (7.10) are equal. Then by (7.10) there is $t_0 = t_0(\lambda_0)$ such that

$$||AF||^{2p}_{2p} = 2p\int_0^\infty t^{2p-1}|\{\theta : AF(\theta) > t\}|dt$$
$$\le 2p\int_0^{t_0} 2\pi t^{2p-1}dt + C\int_{t_0}^\infty t^{\frac{4p^2-4p-1}{2p+1}}dt$$
$$\le C(\eta)$$

if $0 < \eta < 1$.

That proves (7.2), and the proof of Theorem 7.1 is reduced to constructing sets $\mathcal{R}$ and E that satisfy (7.7), (7.8), and (7.9).

We first construct $\mathcal{R}$. To start put $\{|z| < \frac{1}{2}\} \subset \mathcal{R}$ and notice that we have

$$\iint\limits_{|z|<\frac{1}{2}} |\varphi'(z)||g'(z)|^2(1 - |z|^2)dxdy \leq C|\varphi'(0)| \qquad (7.11)$$

because $\|g\|_\mathcal{B} \leq 6$.

As in Section 3 take the dyadic Carleson boxes

$$Q = \{re^{i\theta} : 1 - 2^{-n} \leq r < 1, \ \pi j 2^{-(n+1)} \leq \theta < \pi(j + 1)2^{-(n+1)}\},$$

$0 \leq j < 2^{n+1}$, of sidelength $\ell(Q) = 2^{-n}$ and their top halves

$$T(Q) = Q \cap \{1 - 2^{-n} \leq r < 1 - 2^{-(n+1)}\}$$
$$= Q \setminus \bigcup\{Q' : Q' \subset Q, Q' \neq Q\},$$

and write z_Q for the center of $T(Q)$. Fix δ and ε to be determined later. Say $Q \in \mathcal{L}$, for **large**, if

$$\sup_{T(Q)} (1 - |z|^2)^2 |S\varphi(z)| > \delta. \qquad (7.12)$$

When Q is large, define $\mathcal{D}(Q) = T(Q)$. Say $Q \in \mathcal{G}$, for **good**, if $Q \notin \mathcal{L}$ and

$$\sup_{T(Q)} (1 - |z|^2)|g'(z)| < \varepsilon.$$

Say $Q \in \mathcal{B}$, for **bad**, if $Q \notin \mathcal{L}$ and Q is not good, i.e.,

$$\sup_{T(Q)} (1 - |z|^2)|g'(z)| \geq \varepsilon.$$

If $Q \notin \mathcal{L}$, we call Q **maximal** if the next bigger $\tilde{Q} \supset Q$, $\ell(\tilde{Q}) = 2\ell(Q)$ satisfies $\tilde{Q} \in \mathcal{L}$ or if $\ell(Q) = \frac{1}{2}$. Write $\mathcal{M}$ for the set of maximal $Q \notin \mathcal{L}$. When $Q \in \mathcal{M}$ define

$$\mathcal{D}(Q) = Q \setminus \bigcup\{Q' \subset Q, Q' \in \mathcal{L}\}.$$

Then

$$\{z : \frac{1}{2} \leq |z| < 1\} = \bigcup_{\mathcal{L} \cup \mathcal{M}} \mathcal{D}(Q), \qquad (7.13)$$

and the sets under this union are disjoint.

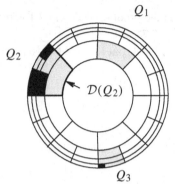

Figure X.20 $Q_1 \in \mathcal{L}$; $Q_2, Q_3, \in \mathcal{M}$.

To complete the construction of $\mathcal{R}$ we consider four cases and we estimate the contribution to (7.9) in each case.

Case I: $Q \in \mathcal{L}$.

Put $\mathcal{D}(Q) = T(Q) \subset \mathcal{R}$ and pass to the next level of boxes $Q' \subset Q$ with $\ell(Q') = \frac{1}{2}\ell(Q)$. Since $\|g\|_{\mathcal{B}} \le 6$, Theorem 4.1 and the Schwarz lemma yield

$$\iint\limits_{T(Q)} |\varphi'(z)||g'(z)|^2(1 - |z|^2)dxdy$$

$$\le C\left(\frac{6}{\delta}\right)^2 \iint\limits_{T(Q)} |\varphi'(z)||\mathcal{S}\varphi(z)|^2(1 - |z|^2)^3 dxdy \qquad (7.14)$$

for every Case I box.

Case II: $Q \in \mathcal{G} \cap \mathcal{M}$.

Put $\mathcal{D}(Q) \subset \mathcal{R}$ and pass to the maximal $Q' \subset Q$. Recall that $F = (\varphi')^{\frac{1}{2}}$ and $4|F'|^2 = |\varphi'|\,|g'|^2$. To estimate the contribution to (7.9) in Case II we need the inequality

$$\iint\limits_{\mathcal{D}(Q)} |\varphi'(z)|g'(z)|^2(1 - |z|^2)dxdy$$

$$= 4 \iint\limits_{\mathcal{D}(Q)} |F'(z)|^2(1 - |z|^2)dxdy \le C(\ell(Q))^3|F'(z_Q)|^2 \qquad (7.15)$$

$$+ C \iint\limits_{\mathcal{D}(Q)} |F''(z)|^2(1 - |z|^2)^3 dxdy.$$

Because inequality (7.15) very much resembles (1.7) of Theorem 1.2 its proof is left as an exercise. Remembering that $2|F''| = |F||\mathcal{S}\varphi + (g')^2|$ and that

$\|g\|_B \leq 6$, we obtain from (7.15)

$$\iint\limits_{\mathcal{D}(Q)} |\varphi'(z)||g'(z)|^2(1-|z|^2)dxdy$$

$$\leq C\ell(Q)|\varphi'(z_Q)| + C \iint\limits_{\mathcal{D}(Q)} |\varphi'(z)||\mathcal{S}\varphi(z)|^2(1-|z|^2)^3 dxdy \qquad (7.16)$$

$$+ C \iint\limits_{\mathcal{D}(Q)} |\varphi'(z)||g'(z)|^4(1-|z|^2)^3 dxdy$$

in Case II. We need the following lemma.

Lemma 7.4. *Assume* $\delta < \frac{\varepsilon^2}{2}$. *If* $Q \in \mathcal{G}$, *if* $Q' \subset Q$, *and if* $Q' \cap \mathcal{D}(Q) \neq \emptyset$, *then* $Q' \in \mathcal{G}$.

Accept Lemma 7.4 for a moment. Then we can bound the last term in (7.16) using

$$\iint\limits_{\mathcal{D}(Q)} |\varphi'(z)||g'(z)|^4(1-|z|^2)^3 dxdy \leq \varepsilon^2 \iint\limits_{\mathcal{D}(Q)} |\varphi'(z)||g'(z)|^2(1-|z|^2)dxdy.$$

Therefore if $C\varepsilon^2 < 1$ we have

$$\iint\limits_{\mathcal{D}(Q)} |\varphi'(z)||g'(z)|^2(1-|z|^2)dxdy$$

$$\leq C\ell(Q)|\varphi'(z_Q)| + C \iint\limits_{\mathcal{D}(Q)} |\varphi'(z)||\mathcal{S}\varphi(z)|^2(1-|z|^2)^3 dxdy. \qquad (7.17)$$

Consider the first term $C\ell(Q)|\varphi'(z_Q)|$ on the right side of (7.17). Because Q is maximal, either $\ell(Q) = 1/2$ or $Q \subset \tilde{Q}$, $\ell(\tilde{Q}) = 2\ell(Q)$ and (7.12) holds for $\tilde{Q}$. In the first case

$$\ell(Q)|\varphi'(z_Q)| \leq c|\varphi'(0)|,$$

and this first case can occur for at most four squares Q. In the second case Theorem 4.1 and the Schwarz lemma give

$$\iint\limits_{T(Q)} |\mathcal{S}\varphi(z)|^2(1-|z|^2)^3 dxdy \geq c\delta^2\ell(Q).$$

Because $\|g\|_B \leq 6$ it follows that the first term on the right-hand side of (7.17) is majorized by a constant multiple of the second term, and hence that

$$\iint\limits_{\mathcal{D}(Q)} |\varphi'(z)||g'(z)|^2(1-|z|^2)dxdy$$

$$\leq \frac{C}{\delta^2} \iint\limits_{\mathcal{D}(Q)} |\varphi'(z)||S\varphi(z)|^2(1-|z|^2)^3 dxdy. \tag{7.18}$$

Thus for Case II boxes (7.18) holds with at most four exceptions, when we have the additional term $c|\varphi'(0)|$, and for (7.9) this additional term is harmless if $\lambda_0 \geq 1$.

Proof of Lemma 7.4. Let $r_1 e^{i\theta} \in T(Q')$. There is $r_0 \geq \frac{1}{2}$ with $r_0 e^{i\theta} \in T(Q)$ and $[r_0 e^{i\theta}, r_1 e^{i\theta}] \subset \mathcal{D}(Q)$. Set

$$r = \sup\{s \leq r_1 : (1-t^2)|g'(te^{i\theta})| \leq \varepsilon \text{ on } [r_0, s]\}.$$

Then since

$$(1-t^2)^2|S\varphi(te^{i\theta})| \leq \delta$$

(4.14) gives

$$|g''(te^{i\theta})| \leq \frac{\delta + \frac{\varepsilon^2}{2}}{(1-t^2)^2} < \frac{d}{dt}\left(\frac{\varepsilon}{1-t^2}\right)$$

on $[r_0, r]$. Therefore

$$(1-r^2)|g'(re^{i\theta})| < \varepsilon$$

and $r = r_1$. ∎

The remaining two cases concern the bad boxes $Q \in \mathcal{B}$. We begin with a lemma that shows the bad boxes are sparsely distributed.

Lemma 7.5. *Given $\varepsilon > 0$ and an integer $n > 0$, there exist $C = C(\varepsilon)$ and $\delta = \delta(\varepsilon, n)$ such that if Q is a bad box for which*

$$\sup_{T(Q)} (1-|z|^2)^2|S\varphi(z)| \leq \delta, \tag{7.19}$$

and if

$$B(Q) = Q \cap \{1-|z| \geq 2^{-n}\ell(Q)\} \cap \{(1-|z|^2)|g'(z)| \geq \varepsilon\},$$

then there exists a hyperbolic geodesic σ such that

$$\sup_{B(Q)} \rho(z, \sigma) < C.$$

where ρ denotes hyperbolic distance in $\mathbb{D}$. Moreover, given $\eta > 0$ there is $\delta > 0$ such that if (7.19) holds for δ, then

$$\left(\frac{1-|z_0|}{1-|z_1|}\right)^{2-\eta} \le \left|\frac{\varphi'(z_1)}{\varphi'(z_0)}\right| \le \left(\frac{1-|z_0|}{1-|z_1|}\right)^{2+\eta} \tag{7.20}$$

if $z_0 \in \sigma \cap T(Q)$ and $z_1 \in \sigma \cap Q \cap \{2^{n/2}\ell(Q) \le 1 - |z_1| \le 2^{-n}\ell(Q)\}$.

Proof. If δ is small, then by Lemma 4.3 there is a Möbius transformation T such that $|\varphi - T|$ is small on $Q \cap \{1 - |z| \ge 2^{-n}\ell(Q)\}$, so small in fact that

$$(1-|z|^2)^2\left|\frac{T''(z)}{T'(z)} - g'(z)\right| \le \frac{\varepsilon}{2}$$

on $Q \cap \{1 - |z| \ge 2^{-n}\ell(Q)\}$. Let $z_0 \in T(Q) \cap \{|g'(z)|(1 - |z|^2) \ge \varepsilon\}$, and let z^* be the pole of T. Then by (4.10) the conclusions of Lemma 7.5 all hold when σ is the circular arc orthogonal to $\partial\mathbb{D}$ that passes through z_0 and $\frac{z^*}{|z^*|}$, because they all hold for T. ∎

Now fix $n \sim 10$ and $\beta = \beta(\varepsilon, \delta) \sim \delta^2$.
Case III: $Q \in \mathcal{B} \cap \mathcal{M}$ and

$$J = \iint\limits_{\mathcal{D}(Q)} |\varphi'(z)||S\varphi(z)|^2(1 - |z|^2)^3 dx dy \ge \beta\ell(Q)|\varphi'(z_Q)|.$$

Let $\sigma = \sigma_Q$ be the geodesic of Lemma 7.5 and choose points $w_j \in \sigma \cap B(Q)$, $j = 0, 1, \ldots, j_0$ such that $w_0 = z_0$ and $1 - |w_j| = 2^{-j}(1 - |z_0|)$. If possible, we let j^* be the least $j \le j_0$ such that

$$\sum_{j \le j^*}(1 - |w_j|)|\varphi'(w_j)| \sim \lambda J \ge \lambda\beta\ell(Q)|\varphi'(z_Q))|. \tag{7.21}$$

If j^* exists we take $Q^* = Q^*(Q) \subset Q$, $Q^* \in \mathcal{B}$, such that $w_{j^*} \in T(Q^*)$ and define $\widetilde{\mathcal{D}}(Q) = \mathcal{D}(Q) \setminus Q^*$. If j^* exists, then by (7.21) and the upper bound in (7.20),

$$\ell(Q^*(Q)) \le C\lambda^{-1+\eta}\ell(Q). \tag{7.22}$$

If no such j^* exists and if $w_{j_0} \in Q' \subset Q$ with $Q' \in \mathcal{B}$, we define $Q_1 = Q'$ and $\widetilde{\mathcal{D}}(Q) = \mathcal{D}(Q) \setminus Q_1$. If no such j^* exists and if no such Q' exists, we take $\widetilde{\mathcal{D}}(Q) = \mathcal{D}(Q)$. In each instance we put $\widetilde{\mathcal{D}}(Q) \subset \mathcal{R}$.

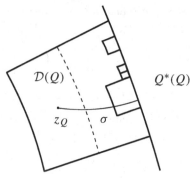

Figure X.21

By Lemmas 7.5 and 7.4 there is a cone Γ with vertex $\zeta \in \sigma \cap \partial\mathbb{D}$ such that if $Q' \in \mathcal{B}$, $Q' \subset Q$, and $Q' \cap \widetilde{\mathcal{D}}(Q) \neq \emptyset$, then there exists maximal dyadic $Q_j \in \mathcal{B}$, $Q_j \subset Q$, $Q_j \cap \widetilde{\mathcal{D}}(Q) \neq \emptyset$ such that $T(Q_j) \cap \Gamma \neq \emptyset$. Write

$$\iint_{\widetilde{\mathcal{D}}(Q)} |\varphi'(z)||g'(z)|^2(1-|z|^2)dxdy = \iint_{\widetilde{\mathcal{D}}(Q)\setminus\Gamma} + \iint_{\widetilde{\mathcal{D}}(Q)\cap\Gamma} .$$

By extending the radial edges of each maximal Q_j to Γ we partition $\widetilde{\mathcal{D}}(Q) \setminus \Gamma$ into chord-arc domains $\mathcal{D}_j \supset Q_j \cap \widetilde{\mathcal{D}}(Q) \setminus \Gamma$ with uniformly bounded chord-arc constants. Choose $z_j \in \Gamma \cap \partial\mathcal{D}_j$. Then by the proof for Case II and Harnack's inequality,

$$\iint_{\widetilde{\mathcal{D}}(Q)} |\varphi'(z)||g'(z)|^2(1-|z|^2)dxdy = \sum \iint_{\mathcal{D}_j} |\varphi'(z)||g'(z)|^2(1-|z|^2)dxdy$$

$$+ \iint_{\widetilde{\mathcal{D}}(Q)\cap\Gamma} |\varphi'(z)||g'(z)|^2(1-|z|^2)dxdy$$

$$\leq \sum \iint_{\mathcal{D}_j} |\varphi'(z)||\mathcal{S}\varphi(z)|^2(1-|z|^2)^3 dxdy$$

$$+ \sum (1-|z_j|^2)|\varphi'(z_j)|$$

$$+ \iint_{\widetilde{\mathcal{D}}(Q)\cap\Gamma} |\varphi'(z)||g'(z)|^2(1-|z|^2)dxdy.$$

The values $1 - |z_j|$ decrease geometrically and each z_j is a bounded hyperbolic distance from the geodesic σ. Therefore by Harnack's inequality and (7.21),

$$\sum (1 - |z_j|^2)|\varphi'(z_j)| \leq C \sum_{j \leq j^*} (1 - |w_j|^2)|\varphi'(w_j)| \leq C\lambda J.$$

Harnack's inequality also gives

$$\iint\limits_{\tilde{\mathcal{D}}(Q) \cap \Gamma} |\varphi'(z)||g'(z)|(1 - |z|^2) dx dy \leq C \sum (1 - |w_j|^2)|\varphi'(w_j)| \leq C\lambda J.$$

Since $\lambda > 1$, we therefore have

$$\iint\limits_{\tilde{\mathcal{D}}(Q)} |\varphi'(z)||g'(z)|^2(1 - |z|^2) dx dy$$

$$\leq C\lambda \iint\limits_{\tilde{\mathcal{D}}(Q)} |\psi'(z)||S\varphi(z)|^2(1 - |z|^2)^3 dx dy \tag{7.23}$$

if Q is a Case III box and if j^* exists.

If no such j^* exists and if there is no $Q' \subset Q$ with $Q' \in \mathcal{B}$ and $w_{j_0} \in Q'$, we stop the sum at j_0 and we still obtain (7.23). Finally, if no j^* exists but if $w_{j_0} \in Q' \in \mathcal{B}$, $Q' \subset Q$, we repeat the construction with Q replaced by $Q_1 = Q'$ and with a possibly new geodesic σ_1 containing $w_{j_0} \in Q_1$, possibly constructing a new Q_1^* or a new $Q_1' = Q_2$, and we obtain (7.23) for $\tilde{\mathcal{D}}(Q_1)$. We repeat the construction until we reach a case where a square Q_m^* is defined or a case where neither Q_m^* nor Q_m' is defined. If we reach a square Q_m^* we define $Q^*(Q) = Q_m^*$. Then (7.22) holds for $Q^*(Q)$. We put $\tilde{\mathcal{D}}(Q) = \mathcal{D}(Q) \setminus Q^*(Q) = \bigcup \tilde{\mathcal{D}}(Q_j)$ into $\mathcal{R}$. Note that (7.23) holds for each set $\tilde{\mathcal{D}}(Q_j)$ and that these sets are disjoint.

Case IV: $Q \in \mathcal{B} \cap \mathcal{M}$ and

$$J = \iint\limits_{\mathcal{D}(Q)} |\varphi'(z)||S\varphi(z)|^2(1 - |z|^2)^3 dx dy < \beta |\varphi'(z_Q)| \ell(Q).$$

Since $\beta \leq c\delta^2$, this case can only occur if $\ell(Q) = 1/2$, and thus for at most four boxes Q. Define Q^* by $Q^* \cap \sigma \neq \emptyset$ and

$$\ell(Q^*)|\varphi'(z_{Q^*})| \sim \lambda \ell(Q)|\varphi'(z_Q)| \tag{7.24}$$

and take $\tilde{\mathcal{D}}(Q) = \mathcal{D}(Q) \setminus Q^*$. If no such Q^* exists take $\tilde{\mathcal{D}}(Q) = \mathcal{D}(Q)$. Put $\tilde{\mathcal{D}}(Q) \subset \mathcal{R}$ and do not consider any smaller $Q' \subset Q$. The argument used in

Case III yields

$$\iint\limits_{\tilde{\mathcal{D}}(Q)} |\varphi'(z)||g'(z)|^2(1-|z|^2)dxdy \le C\lambda|\varphi'(0)|,$$

which is good enough because there are at most four such Q.

Proof of (7.7) and (7.9). By (7.13) we have

$$\mathcal{R} = \mathbb{D} \setminus \bigcup \{Q^*(Q) : Q \in \mathcal{B} \cap \mathcal{M}\}.$$

Since $|\varphi'(0)| \le 6$, (7.11), (7.14), (7.18), (7.24), and the many cases of (7.23) give (7.9) for $\mathcal{R}$ provided $\lambda \ge B$. Let $I^*(Q) \subset \partial\mathbb{D}$ be the base of $Q^*(Q)$ and set

$$E = \partial\mathbb{D} \setminus \bigcup \{3I^*(Q) : Q \in \mathcal{B} \cap \mathcal{M}\}. \tag{7.25}$$

Then (7.7) holds on E.

Proof of (7.8). To prove (7.8) we need an additional lemma.

Lemma 7.6. *Given $\eta > 0$ there is $C = C(\eta)$ such that if $Q \in \mathcal{B} \cap \mathcal{M}$ and if $Q^*(Q)$ exists, then*

$$\ell(Q) \le C\lambda^\eta \ell(\partial\mathcal{D}(Q) \cap \partial\mathbb{D}). \tag{7.26}$$

Proof of Lemma 7.6. By hypothesis and by (7.20) there exists Q^{**}, $Q^*(Q) \subset Q^{**} \subset Q$, such that

$$|\varphi'(z_{Q^{**}})|\ell(Q^{**}) \sim \lambda^\eta J, \tag{7.27}$$

and by the lower bound in (7.20)

$$\ell(Q^{**}) \ge C\lambda^{-\eta}\ell(Q). \tag{7.28}$$

Consider the chord-arc domain $\Omega = Q^{**} \cap \mathcal{D}(Q)$. By (7.27) and Corollary I.4.4, $\operatorname{diam}(\varphi(\Omega)) \ge C\lambda^\eta J$. By Theorem M.1 applied to $F = (\varphi')^{\frac{1}{2}}$ and by the proof of (7.22),

$$\ell(\varphi(\partial\Omega)) \le C|\varphi'(z_{Q^{**}})|\ell(Q^{**}) + C\lambda J \le C'|\varphi'(z_{Q^{**}})|\ell(Q^{**}).$$

Let $\mathcal{E} = \{Q' \in \mathcal{L} : \partial T(Q') \cap \partial\Omega \ne \emptyset\}$ and $A = \partial\Omega \cap \bigcup_{\mathcal{E}} \partial T(Q')$. Then by (7.12) and the Schwarz lemma,

$$\ell(\varphi(A)) = \int_A |\varphi'(z)|ds$$
$$\le C \sum_{\mathcal{E}} \iint\limits_{T(Q')} |\varphi'(z)||S\varphi(z)|^2(1-|z|^2)^3 dxdy \le J. \tag{7.29}$$

Since $\partial\varphi(\Omega)$ has length and diameter comparable to $\lambda^\eta J$ it follows from (7.29) and the Lavrentiev estimate (5.1) of Chapter VI that if $\lambda \geq \lambda_0(\eta)$ then $\omega(z_{Q^{**}}, A, \Omega) = \omega(\varphi(z_{Q^{**}}), \varphi(A), \varphi(\Omega))$ is small. Then since Ω is a chord-arc domain with bounded constants, we conclude that $\frac{\ell(A)}{\ell(Q^{**})}$ is small and hence that

$$\ell(\partial\Omega \cap \partial\mathbb{D}) \geq c\ell(Q^{**}),$$

and with (7.28) this implies (7.26). ∎

Finally, note that because the sets $\{\partial\mathcal{D}(Q) \cap \partial\mathbb{D} : Q \in \mathcal{B} \cap \mathcal{M}\}$ are pairwise disjoint, (7.26) and (7.22) give the inequality (7.8). ∎

Proof of Theorem 7.2. A point of density argument shows that there exists $J_1 \subset J, |J_1| \geq (1 - \frac{\varepsilon}{3})|J|$ such that if $W = \bigcup_{J_1} \Gamma_\beta(\zeta)$, then

$$\iint\limits_W |\varphi'(z)||\mathcal{S}\varphi(z)|^2(1 - |z|^2)^3 dxdy$$

$$\leq C \int_{J_1} \iint\limits_{\Gamma_\alpha(\zeta)} |\varphi'(z)||\mathcal{S}\varphi(z)|^2(1 - |z|^2)^2 dxdyds(\zeta) \leq CN.$$

Set

$$\mathcal{D} = \bigcup_{\zeta \in J_1} \Gamma_\beta(\zeta)$$

and

$$\mathcal{V} = \bigcup\{T(Q) : T(Q) \subset \mathcal{D}\}.$$

Then $\mathcal{V} \subset \mathcal{D}$. Define $\mathcal{L}, \mathcal{M}, \mathcal{G}$, and $\mathcal{B}$ as in the proof of Theorem 7.1, but include only those $T(Q)$ such that $T(Q) \subset \mathcal{D}$. For such Q define

$$\mathcal{D}(Q) = \mathcal{V} \cap Q \setminus \bigcup\{Q' \subset Q \cap \mathcal{D} : Q' \in \mathcal{L}\}.$$

Then the proof of Theorem 7.1 yields a set $E \subset J_1$, defined by (7.25) and a region $\mathcal{R} \subset \mathcal{V}$ so that (7.7), (7.8), and (7.9) hold, and so that $|E| \geq (1 - \frac{\varepsilon}{3})|J|$ and another point of density argument gives $J_0 \subset E$ for which $|J_0| \geq (1 - \varepsilon)|J|$ and for which (7.4) holds. ∎

8. Schwarzians and BMO Domains

Recall that a simply connected domain Ω is called a BMO domain if the mapping function $\varphi : \mathbb{D} \to \Omega$ satisfies

$$g = \log(\varphi') \in \text{BMO}.$$

The results in this chapter yield two characterizations of BMO domains that complement Theorem VII.5.3.

Theorem 8.1. *The following are equivalent.*

(a) Ω *is a BMO domain.*

(b) *There exist $\delta > 0$ and $C > 0$ such that for all $z_0 \in \Omega$ there is a subdomain $\mathcal{U} \subset \Omega$ such that*

   (i) $z_0 \in \mathcal{U}$,

   (ii) $\partial\mathcal{U}$ *is rectifiable and* $\ell(\partial\mathcal{U}) \le C \text{dist}(z_0, \partial\Omega)$, *and*

   (iii) $\omega(z_0, \partial\Omega \cap \partial\mathcal{U}, \mathcal{U}) \ge \delta$.

(c) *There exist $\delta > 0$ and $C > 0$ such that for all $z_0 \in \Omega$ there is a subdomain $\mathcal{U} \subset \Omega$ such that*

   (i) $z_0 \in \mathcal{U}$ *and* $\text{dist}(z_0, \partial\Omega) \le C \text{dist}(z_0, \partial\mathcal{U})$,

   (ii) $\partial\mathcal{U}$ *is chord-arc with constant at most C and* $\ell(\partial\mathcal{U}) \le C \text{dist}(z_0, \partial\Omega)$, *and*

   (iii) $\ell(\partial\Omega \cap \partial\mathcal{U}) \ge \delta \text{dist}(z_0, \partial\Omega)$.

(d) $|\mathcal{S}\varphi(z)|^2 (1 - |z|^2)^3 dxdy$ *is a Carleson measure on* $\mathbb{D}$.

(e) *There exist $\delta > 0$ and $C > 0$ such that for every $z_0 \in \mathbb{D}$ there exists a Lipschitz domain $\mathcal{V} \subset \mathbb{D}$ such that*

   (i) $z_0 \in \mathcal{V}$,

   (ii) $\omega(z_0, \partial\mathcal{V} \cap \partial\mathbb{D}, \mathcal{V}) \ge \delta$, *and*

   (iii) $\displaystyle\iint_{\mathcal{V}} |\varphi'(z)||\mathcal{S}\varphi(z)|^2 (1 - |z|^2)^3 dxdy \le C|\varphi'(z_0)|(1 - |z_0|^2)$.

Proof. Theorem VII.5.3 had the implications (a) $\Longrightarrow$ (b) $\Longrightarrow$ (c) $\Longrightarrow$ (a). Here we use the arguments of the previous section to treat (a) $\Longleftrightarrow$ (d), (a) $\Longrightarrow$ (e), and (e) $\Longrightarrow$ (b).

(a) $\Longrightarrow$ (d): This was first observed by Zinsmeister [1984]. By (4.14) we have

$$|\mathcal{S}(\varphi)| \le \frac{|g'|^2}{2} + |g''|.$$

It follows from (a) that $|g'(z)|^2(1 - |z|^2)dxdy$ is a Carleson measure and hence $|g'(z)|^4(1 - |z|^2)^3 dxdy$ is also a Carleson measure, because $g \in \mathcal{B}$. For any analytic function we have

$$\iint\limits_{T(Q)} |g''(z)|^2(1-|z|^2)^3dxdy \le C \iint\limits_{\tilde{T}(Q)} |g'(z)|^2(1-|z|^2)dxdy,$$

where $\tilde{T}(Q) = \{z : \text{dist}(z, T(Q)) \le \ell(Q)/4\}$. Thus $|g''(z)|^2(1-|z|^2)^3dxdy$ is also a Carleson measure and (d) holds.

(d) $\Longrightarrow$ (a): This is due to Astala and Zinsmeister [1991]. Because (a) and (d) are invariant under conformal self maps of $\mathbb{D}$, to prove (a) it is enough to show

$$\int |g(e^{i\theta}) - g(0)|^2d\theta \le C. \tag{8.1}$$

See Garnett [1981]. Let

$$A = \iint\limits_{\mathbb{D}} |g'(z)|^2(1-|z|^2)dxdy$$

and

$$B = \iint\limits_{\mathbb{D}} |g''(z)|^2(1-|z|^2)^3dxdy.$$

Then since $|g'(0)| \le 6$ we have by Fourier series,

$$B \le 12A \le 3B + 6^3\pi \tag{8.2}$$

and by Theorem 1.2 it will be enough to show $B \le C'$.

By (4.14) we have

$$B \le 2\iint\limits_{\mathbb{D}} |S\varphi(z)|^2(1-|z|^2)^3dxdy + \frac{1}{2}\iint\limits_{\mathbb{D}} |g'(z)|^4(1-|z|^2)^3dxdy. \tag{8.3}$$

Set $U = \{z \in \mathbb{D} : |g'(z)|(1-|z|^2) \le 1/2\}$. Then by (8.2)

$$\iint\limits_{U} |g'(z)|^4(1-|z|^2)^3dxdy \le \frac{A}{4} \le \frac{B}{12} + \frac{9}{2}\pi. \tag{8.4}$$

Set $V = \bigcup\{T(Q) : \sup_{T(Q)} |S\varphi(z)|(1-|z|^2)^2 \ge \delta\}$ where $\delta = \delta(1/4, 10)$ is given by Lemma 7.5. Then because $\|g\|_B \le 6$,

$$\iint\limits_{V} |g'(z)|^4(1-|z|^2)^3dxdy \le \frac{C}{\delta^2}\iint\limits_{V} |S\varphi(z)|^2(1-|z|^2)^3dxdy. \tag{8.5}$$

If $T(Q) \setminus (U \cup V) \ne \emptyset$ then by Lemma 7.5, $T(Q') \subset (U \cup V)$ for more than half of the Carleson boxes $Q' \subset Q$ with $\ell(Q') = 2^{-10}\ell(Q)$. Consequently

$\mathbb{D} \setminus (U \cup V) \subset \bigcup T(Q_j)$, where $\{Q_j\}$ is a sequence of Carleson boxes with $\sum \ell(Q_j) \leq C''$, with C'' independent of φ. Hence

$$\iint_{\mathbb{D}\setminus(U\cup V)} |g'(z)|^4 (1 - |z|^2)^3 dxdy \leq \sum_j C \iint_{T(Q_j)} (1 - |z|^2)^{-1} dxdy \leq C''.$$

(8.6)

Together (8.3), (8.4), (8.5), and (8.6) give us

$$\frac{23}{24} B \leq \left(2 + \frac{C}{\delta^2}\right) \iint_{\mathbb{D}} |\mathcal{S}\varphi(z)|^2 (1 - |z|^2)^3 dxdy + \frac{9}{4}\pi + C'',$$

which establishes (8.1).

(d) and (a) $\Longrightarrow$ **(e):** Because (e) is invariant under Möbius self maps of $\mathbb{D}$, we can suppose $z_0 = 0$. Then let $\mathcal{V}$ be the Lipschitz region constructed in the proof of (a) $\Longrightarrow$ (b) from Theorem VII.5.3 and note that $|\varphi'|$ is bounded above and below on $\mathcal{V}$. Then use (d).

(e) $\Longrightarrow$ **(b):** We may assume $z_0 = \varphi(0)$. We repeat the proof of Theorem 7.1, with $\mathbb{D}$ replaced by the Lipschitz domain $\mathcal{V}$ given by (e), just as we did in the proof of Theorem 7.2. We obtain a Lipschitz domain $\mathcal{R} \subset \mathcal{V}$ such that $\int_{\partial \mathcal{R}} |\varphi'| ds < \infty$, with $0 \in \mathcal{R}$ and with $\omega(0, \partial \mathbb{D} \cap \partial \mathcal{R}, \mathcal{R}) \geq \frac{\delta}{2}$. Then (b) holds for $\mathcal{U} = \varphi(\mathcal{R})$. ∎

9. Angular Derivatives

Let φ be a conformal mapping from $\mathbb{D}$ onto a simply connected domain Ω and let

$$G = \{\zeta \in \partial\mathbb{D} : \varphi \text{ has an angular derivative at } \zeta \text{ and } |\varphi'(\zeta)| \neq 0\}.$$

In this section we give several almost everywhere characterizations of G. By Theorem VI.6.1 we know that $\zeta \in G$ is almost everywhere equivalent to

$$\varphi(\zeta) \text{ is a cone point of } \Omega. \tag{9.1}$$

Furthermore, for any $\alpha > 1$, $\zeta \in G$ is almost everywhere equivalent to

$$\iint_{\Gamma_\alpha(\zeta)} |\varphi''(z)|^2 dxdy < \infty, \tag{9.2}$$

by the theorem of Marcinkiewicz, Zygmund, and Spencer, Theorem 1.3 above. Bishop and Jones [1994] give several other almost everywhere characterizations of G.

Theorem 9.1. *Let φ be a conformal map from $\mathbb{D}$ onto a simply connected domain Ω and let $g = \log(\varphi')$. Then the following are almost everywhere equivalent on $\partial\mathbb{D}$:*

(a) φ *has non-zero angular derivative at ζ.*

(b) $\displaystyle\iint_{\Gamma_\alpha(\zeta)} |\varphi'(z)| |\mathcal{S}\varphi(z)|^2 (1 - |z|^2)^2 dx dy < \infty.$

(c) $\displaystyle\iint_{\Gamma_\alpha(\zeta)} |\varphi'(z)| |g'(z)|^2 dx dy < \infty.$

(d) $\displaystyle\int_0^1 \eta^2(\varphi(\zeta), t) \frac{dt}{t} < \infty.$

(e) $\displaystyle\iint_{\Gamma_\alpha(\zeta)} |\mathcal{S}\varphi(z)|^2 (1 - |z|^2)^2 dx dy < \infty.$

Proof. The road map is: (a) $\Longrightarrow$ (b) $\Longrightarrow$ (c) $\Longrightarrow$ (a); then (d) $\Longrightarrow$ (a) $\Longrightarrow$ (d); and finally (b) $\Longrightarrow$ (e) $\Longrightarrow$ (a).

(a) $\Longrightarrow$ (b): Suppose $K \subset \partial\mathbb{D}$ is a compact set such that

$$\sup_{\Gamma_\alpha(\zeta)} |\varphi'(z)| \leq M$$

for fixed $M < \infty$ for all $e^{i\theta} \in K$. Form the cone domain

$$U = \bigcup_K \Gamma_\alpha(e^{i\theta}).$$

Then

$$\left| \{\zeta : z \in \Gamma_\alpha(\zeta)\} \right| \leq c_\alpha (1 - |z|),$$

and Fubini's theorem again gives

$$\int_K \iint_{\Gamma_\alpha(e^{i\theta})} |\varphi'(z)| |\mathcal{S}\varphi(z)|^2 (1 - |z|^2)^2 dx dy d\theta \tag{9.3}$$

$$\leq \iint_U |\varphi'(z)| |\mathcal{S}\varphi(z)|^2 (1 - |z|^2)^3 dx dy.$$

But the integral on the right in (9.3) is finite by Lemma 6.1 because $\partial\varphi(U)$ is rectifiable. Therefore (b) holds almost everywhere on K.

(b) $\Longrightarrow$ (c): This is Corollary 7.3.

(c) $\Longrightarrow$ (a): Because $|\varphi'| |g'|^2 = 4 |F'|^2$, where $F = (\varphi')^{\frac{1}{2}}$, Theorem 1.3 and Theorem VI.2.3 show φ' has non-zero nontangential limit almost everywhere that (c) holds.

(d) $\Longrightarrow$ **(a):** Fix $\alpha > 1$ large. If (d) holds on a set E of positive measure, then for any $\varepsilon > 0$ there exists $\delta > 0$ and compact $K \subset E$ such that

$$\int_0^\delta \eta^2(\varphi(\zeta), t) \frac{dt}{t} < \varepsilon$$

and $|E \setminus K| < \varepsilon$. Then by Theorem 4.1 and the Cauchy–Schwarz inequality,

$$|S\varphi(z)|(1 - |z|^2)^2 \le C\varepsilon^{\frac{1}{2}}, \tag{9.4}$$

on

$$V = \bigcup_K \Gamma_\alpha(\zeta) \cap \{\text{dist}(\varphi(z), \partial\Omega) \le \delta^{\frac{1}{2}}\varepsilon^{\frac{1}{4}}\}.$$

Let $U = V \cap B(\zeta, \delta^{\frac{1}{2}}\varepsilon^{\frac{1}{4}})$ where ζ is a point of density of K. If α is sufficiently large, then U is a quasidisc and by Theorem 4.2, $\varphi(U)$ is a quasidisc if ε is sufficiently small. But then by Exercise 12, $\beta(\varphi(\zeta), t) \le C\eta(\varphi(\zeta), t)$ on K. Therefore by (d), by Theorem 2.5, and by the F. and M. Riesz theorem, $\varphi(\partial U)$ has a tangent at $\varphi(\zeta)$ for almost every $\zeta \in K$. Hence (9.1) and (a) hold almost everywhere on K.

(a) $\Longrightarrow$ **(d):** Now suppose the angular derivative $\varphi'(\zeta)$ exists and is continuous on a compact set $K \subset \partial\mathbb{D}$. It is enough to show (d) holds almost everywhere on K. Let $\delta > 0$ be small and cover K by arcs $I_j = \{|\zeta - \zeta_j| \le \delta\}$ and form the cone domains

$$U_j = \{|z| \ge 1 - \delta\} \cap \bigcup_{\zeta \in K \cap I_j} \Gamma_2(\zeta).$$

If δ is small then U_j is connected and φ' is continuous and almost constant on $\overline{U}_j$, so that $\gamma_j = \varphi(\partial U_j)$ is a chord-arc curve. By Theorem 2.5

$$\int_0^1 \beta_{\gamma_j}^2(z_\theta, t) \frac{dt}{t} < \infty \tag{9.5}$$

almost everywhere on $K \cap I_j$. Let (9.5) hold at $e^{i\theta} \in K \cap I_j$ and let t be so small that $\beta_{\gamma_j}(z_\theta, 2t) < 1/16$. Let S be a strip of width $4t\beta_{\gamma_j}(z_\theta, 2t)$ containing $\gamma_j \cap B(z_\theta, 2t)$. Then one side L of S satisfies (a), (b), and (c) in the definitions of $\eta_{\varphi(U_j)}(z_\theta, t)$ and of $\eta_\Omega(z_\theta, t)$ because $\varphi(U_j) \subset \Omega$. Let $z \in L \cap B(z_\theta, 2t)$ and let z^* be the orthogonal (to L) projection of z onto γ_j. Then

$$\text{dist}(z, \partial\Omega) \le |z - z^*| + \text{dist}(z^*, \gamma_j \cap \varphi(K)),$$

and by the definition of U_j and the continuity of φ' on $\overline{U}_j$,

$$\text{dist}(z^*, \gamma_j \cap \varphi(K)) \le Ct\beta_{\gamma_j}(z_\theta, t).$$

Therefore $\eta_\Omega(z_\theta, t) \le C\beta_{\gamma_j}(z_\theta, t)$ and by (9.5), (d) holds at $e^{i\theta}$.

(b) $\Longrightarrow$ (e): This is clear because (b) $\Longrightarrow$ (a).

(e) $\Longrightarrow$ (a): We follow the proof of Theorem 8.1, (d) $\Longrightarrow$ (a). Assume (e) holds on a set E. Then there exists a compact set $K \subset E$ such that when $U = \bigcup_K \Gamma_\alpha(\zeta)$,

$$\iint\limits_{U} |\mathcal{S}\varphi(z)|^2 (1 - |z|^2)^3 dxdy < \infty. \tag{9.6}$$

Recall that $g = \log(\varphi')$. We will show (9.6) implies

$$B = \iint\limits_{U} |g''(z)|^2 d(z)^3 dxdy < \infty \tag{9.7}$$

where $d(z) = \text{dist}(z, \partial U)$. Since $|g'(0)| \le 6$, it will then follow from Theorem 1.2 that g, and also φ', have a finite non-zero nontangential limit almost everywhere on K. By (4.14)

$$B \le 2 \iint\limits_{U} |\mathcal{S}\varphi(z)|^2 d(z)^3 dxdy + \frac{1}{2} \iint\limits_{U} |g'(z)|^4 d(z)^3 dxdy,$$

and by Theorem 1.2

$$A = \iint\limits_{U} |g'(z)|^2 d(z) dxdy \le C(\alpha)(B + |g'(0)|^2).$$

Set $U_1 = U \cap \{z : |g'(z)|d(z) \le C(\alpha)^{-1/2}\}$. Then

$$\frac{1}{2} \iint\limits_{U_1} |g'(z)|^4 d(z)^3 dxdy \le \frac{B}{2} + \frac{|g'(0)|^2}{2}.$$

For fixed $\delta > 0$, set $U_2 = U \cap \{z : |\mathcal{S}\varphi(z)|(1 - |z|^2)^2 \ge \delta\}$. Then because $\|g\|_{\mathcal{B}} \le 6$,

$$\iint\limits_{U_2} |g'(z)|^4 d(z)^3 dxdy \le \iint\limits_{U_2} |g'(z)|^4 (1 - |z|^2)^3 dxdy$$

$$\le \frac{C}{\delta^2} \iint\limits_{U_2} |\mathcal{S}\varphi(z)|^2 (1 - |z|^2)^3 dxdy.$$

Finally, set $U_3 = U \setminus (U_1 \cup U_2)$. By Lemma 7.5, there is a sequence $\{Q_j\}$ of Carleson boxes such that $\sum \ell(Q_j) < \infty$ and $U_3 \subset \bigcup T(Q_j)$. But then

$$\iint_{U_3} |g'(z)|^4 (1 - |z|^2)^3 dxdy \le 6^4 \sum \ell(Q_j) < \infty,$$

and we obtain

$$\frac{B}{2} \le \left(2 + \frac{C}{\delta^2}\right) \iint |S\varphi|^2 (1 - |z|^2)^3 dxdy + 6^4 \sum \ell(Q_j).$$

Therefore (9.7) holds. ∎

10. A Local F. and M. Riesz Theorem

Theorem 10.1. *Let Ω be a simply connected domain, let $z_0 \in \Omega$ satisfy* $\operatorname{dist}(z_0, \partial\Omega) \ge 1$, *and let Γ be a connected set with $\Lambda_1(\Gamma) \le M < \infty$. Given $\delta > 0$ there exists $\varepsilon = \varepsilon(M, \delta) > 0$ such that if $E \subset \Gamma \cap \partial\Omega$ and $\omega(z_0, E, \Omega) > \delta$, then $\Lambda_1(E) > \varepsilon$.*

The F. and M. Riesz theorem is a simple corollary of Theorem 10.1. In [1980] Øksendal conjectured Theorem 10.1 and proved it where Γ is a line. See also Exercise III.14 and Øksendal [1981]. Shortly thereafter Kaufman and Wu [1982] proved the theorem when Γ is a chord-arc curve. But the full result was not proved until Bishop and Jones [1990]. The difficulty in the general case is to make effective use of the hypothesis $\Lambda_1(\Gamma) \le M$, and for this we will use Theorem 2.3.

Proof. Let $\psi : \mathbb{D} \to \mathbb{C}^* \setminus E$ be the universal covering map, $\psi(0) = z_0$. By the monodromy theorem there is a simply connected $G \subset \mathbb{D}$ such that $0 \in G$, ψ is univalent on G, and $\psi(G) = \Omega$. By hypothesis

$$\omega(0, \partial G \cap \partial\mathbb{D}, G) > \delta. \tag{10.1}$$

Then by Beurling's projection theorem

$$\operatorname{dist}(0, \partial G) \ge c(\delta), \tag{10.2}$$

and hence by Lemma 5.5,

$$|\psi'(0)| \le c'(\delta). \tag{10.3}$$

Let $\mathcal{G}$ denote the Fuchsian group $\{\gamma \in \mathcal{M} : \psi \circ \gamma = \psi\}$ and let I be the identity of $\mathcal{G}$, $I(z) = z$. The **normal fundamental domain** of $\mathcal{G}$ is defined by

$$\mathcal{F} = \{z \in \mathbb{D} : |\gamma'(z)| < 1 \text{ for all } \gamma \in \mathcal{G} \setminus \{I\}\},$$

and ψ is one-to-one on $\mathcal{F}$. Write $\mathcal{G} \setminus \{I\} = \{\gamma_j\}$ and $\gamma_j(z) = \frac{\lambda_j z + a_j}{1 + \overline{a_j}\lambda_j z}$, where $|\lambda_j| = 1$ and $\gamma_j(0) = a_j$. Then

$$|\gamma_j'(z)| = \frac{1 - |a_j|^2}{|1 + \overline{a_j}\lambda_j z|^2}$$

and the fundamental domain $\mathcal{F}$ has the form $\mathcal{D} \setminus \bigcup B_j$ where each B_j is a disc orthogonal to $\partial \mathbb{D}$.

Lemma 10.2. *The normal fundamental domain $\mathcal{F}$ satisfies*

$$\operatorname{dist}(0, \partial\mathcal{F}) \geq \frac{c(\delta)}{2} \tag{10.4}$$

and

$$|\partial\mathbb{D} \cap \partial\mathcal{F}| \geq 2\pi\delta. \tag{10.5}$$

There exist $\beta(\delta') > 1$ and a compact set $K \subset \partial\mathbb{D} \cap \partial\mathcal{F}$ such that

$$|K| \geq \pi\delta \tag{10.6}$$

and

$$\bigcup_K \Gamma_\beta(\zeta) \subset \mathcal{F}. \tag{10.7}$$

Proof. By (10.2), $|a_j| \geq c(\delta)$ for all j, and since

$$\inf\{|z| : |\gamma_j'(z)| = 1\} = \frac{1}{|a_j|} - \sqrt{\frac{1}{|a_j|^2} - 1} \geq \frac{|a_j|}{2},$$

that gives (10.4).

By Lemma I.2 from Appendix I and Plessner's theorem, ω_G almost every point of $\partial\mathbb{D} \cap \partial G$ is a cone point for G. Write K_G for the set of cone points of ∂G. Then by (10.1) and a comparison, $|\partial\mathbb{D} \cap K_G| \geq 2\pi\delta$. Because $\partial\mathbb{D} \cap K_G \cap \gamma_j(K_G) = \emptyset$ for all γ_j, we have

$$|\partial\mathbb{D} \cap \partial\mathcal{F}| \geq |\partial\mathbb{D} \cap \partial\mathcal{F} \cap K_G| + \sum |\partial\mathbb{D} \cap \partial\mathcal{F} \cap \gamma_j^{-1}(K_G \setminus \partial\mathcal{F})|$$

$$\geq |\partial\mathbb{D} \cap K_G|,$$

because $|(\gamma_j^{-1})'| \geq 1$ on $\gamma_j(\partial\mathcal{F}) \cap (\partial G \setminus \partial\mathcal{F})$. That gives (10.5).

X. Rectifiability and Quadratic Expressions

Write $I_j = B_j \cap \partial\mathbb{D}$. By (10.4) we have $|I_j| \leq \pi - \delta'' < \pi$, and if $\varepsilon > 0$ is small there is $\beta > 1$ such that if J_j is the arc concentric with I_j having $|J_j| = (1 + \varepsilon)|I_j|$, then $\Gamma_\beta(\zeta) \cap B_j = \emptyset$ for all $\zeta \in \partial\mathbb{D} \setminus J_j$. Choose $\varepsilon = \frac{\delta}{2(1-\delta)}$ and set

$$K = \partial\mathbb{D} \setminus \bigcup J_j.$$

Then (10.7) holds and (10.6) follows from (10.5). That proves Lemma 10.2. ∎

Fix $1 < \alpha < \beta$ and form the cone domain

$$U = \bigcup_K \Gamma_\alpha(\zeta).$$

If $\eta = \eta(\alpha, \beta)$ is sufficiently small, then $B(w, \eta(1 - |w|)) \subset \mathcal{F}$ for all $w \in U$. We claim that

$$\iint_U |\psi'(z)||\mathcal{S}\psi(z)|^2 (1 - |z|^2)^3 dxdy \leq CM. \tag{10.8}$$

If (10.8) holds then by Fubini's theorem

$$\int_K \iint_{\Gamma_\alpha(\zeta)} |\psi'(z)||\mathcal{S}\psi(z)|^2 (1 - |z|^2)^2 dxdyds(\zeta) < C'M,$$

and by Theorem 7.2 there exist $N < \infty$ and $K_0 \subset K$ such that $|K_0| > |K|/2$ and $F = \sqrt{\psi'}$ has area function $A(F) \leq N$ at all $\zeta \in K_0$. Take

$$U_0 = \bigcup_{K_0} \Gamma_\alpha(\zeta).$$

Then by Theorem 1.2, $\psi(U_0)$ is a Jordan domain having

$$\Lambda_1(\partial\psi(U_0)) \leq C(|\psi'(0)| + N) \leq C''(M, \delta),$$

$$\omega(z_0, E \cap \psi(U_0)) \geq \delta''(\alpha, \delta)$$

by (10.6), and $\text{dist}(z_0, \partial\psi(U_0)) \geq c''$.

Therefore by (10.3) and Lavrentiev's theorem, Exercise VI.2,

$$\Lambda_1(E) \geq c(\delta, \Lambda_1(\Gamma)),$$

and Theorem 10.1 is proved, provided (10.8) holds.

Proof of (10.8). Let $\{Q_j\}$ denote the Whitney squares in $\mathbb{C} \setminus E$. We may assume each Q_j is a dyadic square. Let Q_j^n denote the dyadic square $Q_j^n \supset Q_j$ such that $\ell(Q_j^n) = 2^n \ell(Q_j)$, and let $N(Q_j)$ be the first n such that

$$\delta_n(Q_j) = \beta_E(Q_j^n) + \gamma_E(Q_j^n) \geq 10^{-3}.$$

If $z \in U \cap \psi^{-1}(Q_j)$ then by Theorem 5.6,

$$(1 - |z|^2)^2 |S\psi(z)| \leq C_s \sum_{n=0}^{N(Q_j)} \delta_n(Q_j) 2^{-ns} \equiv \delta(Q_j),$$

for any $s < 1$. Therefore

$$\iint_U |\psi'(z)| |S\psi(z)|^2 (1 - |z|^2)^3 dxdy \leq \sum_{Q_j} \delta^2(Q_j) \iint_{U \cap \psi^{-1}(Q_j)} \frac{|\psi'(z)|}{(1 - |z|^2)} dxdy.$$

But by Lemma 5.5

$$\iint_{U \cap \psi^{-1}(Q_j)} \frac{|\psi'(z)|}{(1 - |z|^2)} dxdy \leq C\ell(Q_j),$$

and we obtain

$$\iint_U |\psi'(z)| |S\psi(z)|^2 (1 - |z|^2)^3 dxdy$$

$$\leq C \sum_{Q_j} \delta^2(Q_j) \ell(Q_j)$$

$$= 2C \sum_j \left(\sum_{n=0}^{N(Q_j)} \beta_E(Q_j^n) 2^{-ns} \right)^2 \ell(Q_j) \qquad (10.9)$$

$$+ 2C \sum_j \left(\sum_{n=0}^{N(Q_j)} \gamma_E(Q_j^n) 2^{-ns} \right)^2 \ell(Q_j).$$

The two sums on the right of (10.9) will be estimated using the next two lemmas. Write $\mathcal{Q}$ for the set of all dyadic squares.

Lemma 10.3. *For* $n \geq 0$,

$$\sum \{ \beta_E^2(Q_j^n) \ell(Q_j) : N(Q_j) \geq n \} \leq 2^n \sum_{\mathcal{Q}} \beta_E^2(Q) \ell(Q) \leq C 2^n \ell(\Gamma).$$

Proof. For each $Q \in \mathcal{Q}$ there are at most 4^n dyadic squares Q_j such that $Q_j^n = Q$ and for each of them $\ell(Q_j) = 2^{-n} \ell(Q)$. Therefore

$$\sum \{\beta_E^2(Q_j^n)\ell(Q_j) \,:\, N(Q_j) \geq n\}$$

$$= \sum_Q \beta_E^2(Q) \sum \{\ell(Q_j) \,:\, Q_j^n = Q, \, N(Q_j) \geq n\}$$

$$\leq \sum_Q \beta_E^2(Q) \sum \{\ell(Q_j) \,:\, Q_j^n = Q\}$$

$$\leq 2^n \sum_Q \beta_E{}^2(Q)\ell(Q) \leq C2^n \ell(\Gamma),$$

by Theorem 2.3.   ∎

We take $s > 1$. Then by Lemma 10.3 and the Cauchy–Schwarz inequality,

$$\sum_j \Big(\sum_{n=0}^{N(Q_j)} \beta_E \,(Q_j^n) 2^{-ns} \Big)^2 \ell(Q_j)$$

$$\leq \frac{1}{1-2^{-s}} \sum_n \sum_{\{j:N(Q_j)\geq n\}} \beta_E^2(Q_j^n) 2^{-ns} \ell(Q_j)$$

$$\leq \frac{1}{1-2^{-s}} \sum_n 2^{(1-s)n} \sum_Q \beta_E^2(Q)\ell(Q) \qquad (10.10)$$

$$\leq C\ell(\Gamma).$$

That bounds half of the right side of (10.9).

Lemma 10.4. *For $n \geq 0$,*

$$\sum \{\gamma^2(Q_j^n)\ell(Q_j) \,:\, N(Q_j) \geq n\} \leq C2^n \ell(\Gamma).$$

Proof. We can assume

$$\gamma(Q_j^n) \geq 10^5 \beta(Q_j^n) \qquad (10.11)$$

because Lemma 10.3 bounds the sum over the other Q_j^n. Using the Whitney squares of $\mathbb{C} \setminus E$, we partition $\Gamma \setminus E$ into disjoint subsets I_k such that

$$c_1 \operatorname{dist}(I_k, E) \leq \operatorname{diam}(I_k) \leq c_2 \operatorname{dist}(I_k, E)$$

for absolute constants c_1 and c_2. For each $Q = Q_j^n$ satisfying (10.11) there is $k = k(Q)$ such that $I_k \subset 10Q$ and

$$\gamma(Q) \leq c \frac{\operatorname{diam}(I_k)}{\ell(Q)}.$$

Then for each k and ℓ there is a bounded number of $Q \in \mathcal{Q}$ such that $k(Q) = k$ and $\ell(Q) = \ell$. Because $\ell(Q) \geq 10\,\mathrm{diam}(I_k)$, that gives

$$\sum_{k(Q)=k} \frac{1}{\ell(Q)} \leq \frac{C}{\mathrm{diam}(I_k)} \sum_{\nu=0}^{\infty} 2^{-\nu} \leq \frac{C}{\mathrm{diam}(I_k)}. \tag{10.12}$$

Again because there are at most 4^n squares Q_j with $Q_j^n = Q \in \mathcal{Q}$, we have by (10.11)

$$\sum \{\gamma^2(Q_j^n)\ell(Q_j) : N(Q_j) \geq n \text{ and } (10.11) \text{ holds}\}$$

$$= \sum_{(10.11) \text{ holds}} \gamma^2(Q) \sum \{\ell(Q_j) : Q_j^n = Q\}$$

$$\leq 2^n \sum_{(10.11) \text{ holds}} \gamma^2(Q)\ell(Q)$$

$$\leq C2^n \sum_{(10.11) \text{ holds}} \frac{\left(\mathrm{diam}(I_k(Q))\right)^2}{\ell(Q)}$$

$$= C2^n \sum_k \left(\mathrm{diam}(I_k)\right)^2 \sum_{\{Q:k(Q)=k\}} \frac{1}{\ell(Q)}$$

$$\leq C2^n \sum_k \mathrm{diam}(I_k) \leq C2^n \ell(\Gamma),$$

because the I_k are disjoint. ∎

To estimate the second sum in (10.9), we keep $s > 1$ and repeat the proof of (10.10) to get

$$\sum_j \left(\sum_{n=0}^{N(Q_j)} \gamma_E(Q_j^n)2^{-ns}\right)^2 \ell(Q_j) \leq C\ell(\Gamma).$$

That concludes the proof of (10.9) and therefore (10.8). ∎

11. Ahlfors Regular Sets and the Hayman–Wu Theorem

Recall that a compact set Γ is Ahlfors regular if there exists a constant $A > 0$ such that for every disc $B(z, r)$,

$$\Lambda_1(\Gamma \cap B(z, r)) \leq Ar.$$

Theorem 11.1. *Let Γ be a compact connected set. Then Γ is Ahlfors regular if* *and only if there is a constant $C(\Gamma)$ such that*

$$\Lambda_1(\varphi^{-1}(\Gamma \cap \Omega)) \leq C(\Gamma) \tag{11.1}$$

for every simply connected domain Ω and every conformal mapping φ from $\mathbb{D}$ onto Ω.

It is easy to see that (11.1) implies Γ is Ahlfors regular; take φ to be a linear map from $\mathbb{D}$ to $B(z, r)$. The proof of the converse uses Theorem 10.1 and Lemma 11.2 below.

Lemma 11.2. *Let Γ be an Ahlfors regular compact connected set and let Ω be a simply connected domain. Let $S = \{z_j\}$ be a sequence in $\Gamma \cap \Omega$ such that*

$$\sup_{\Gamma} \inf_{j} \rho(z, z_j) \leq M. \tag{11.2}$$

Set

$$K_j = \left\{ z : |z - z_j| \leq \frac{\mathrm{dist}(z_j, \partial\Omega)}{2} \right\},$$

and assume S has a finite partition $S = S_1 \cup \ldots \cup S_N$, so that if $z_j \in S_n$ and

$$\Omega_j = \Omega \setminus \left(\bigcup_{\{z_k \in S_n, k \neq j\}} K_k \right),$$

then

$$\inf_{j} \omega(z_j, \partial\Omega \cap \partial\Omega_j, \Omega_j) \geq \delta > 0. \tag{11.3}$$

Then for any conformal $\varphi : \mathbb{D} \to \Omega$,

$$\Lambda_1(\varphi^{-1}(\Gamma \cap \Omega)) \leq C(A, N, M, \delta).$$

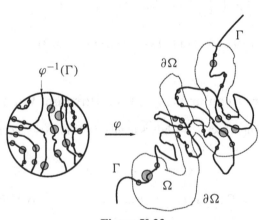

Figure X.22

Proof. Write $w_j = \varphi^{-1}(z_j)$ and $D_j = \{w \in \mathbb{D} : \rho(w, w_j) \le M\}$. Then by (11.2),

$$\Lambda_1(\varphi^{-1}(\Gamma \cap \Omega)) \le \sum \Lambda_1(D_j \cap \varphi^{-1}(\Gamma \cap \Omega)),$$

while by the Koebe estimates in Theorem I.4.3 and the Ahlfors regularity of Γ,

$$\Lambda_1(D_j \cap \varphi^{-1}(\Gamma \cap \Omega)) \le C(A, M)(1 - |w_j|).$$

Thus the lemma will be proved if we show

$$\sum_{S_n} 1 - |w_j| \le C. \tag{11.4}$$

We can assume $\rho(w_j, w_k) = \rho(z_j, z_k) \ge 1$ whenever $z_j \ne z_k$ by taking $M \ge 2$ in (11.2) and by removing any unnecessary z_j. Then the discs

$$\Delta_j^* = \{w : \rho(w, w_j) \le 1/3\}$$

are disjoint. Take $0 < \eta < \frac{1}{8}$ to be determined and set

$$\Delta_j = \{w : \rho(w, w_j) \le \eta\}.$$

Since $\eta < \frac{1}{8}$ we have $\Delta_j \subset \varphi^{-1}(K_j)$ by the estimates in Section I.4.

When proving (11.4) we may assume that $\{\Delta_j : z_j \in S_n\}$ is finite and that $0 \notin \cup \Delta_j$. Consider the harmonic function

$$u(z) = u_n(z) = \omega(z, \partial\mathbb{D}, \mathbb{D} \setminus \bigcup\{\Delta_j : z_j \in S_n\}).$$

By Green's theorem,

$$u(0) = \frac{1}{2\pi} \int_{\partial\mathbb{D}} u(e^{i\theta})d\theta - \sum_{S_n} \frac{1}{2\pi} \int_{\partial\Delta_j} \frac{\partial u}{\partial n} \log \frac{1}{|z|} ds,$$

where the normal derivatives are exterior to Δ_j. Hence

$$\sum_{S_n} \left(\inf_{\Delta_j} \log \frac{1}{|z|} \right) \left(\frac{1}{2\pi} \int_{\partial\Delta_j} \frac{\partial u}{\partial n} ds \right) \le \frac{1}{2\pi} \int_{\partial\mathbb{D}} u(e^{i\theta})d\theta - u(0) \le 1. \tag{11.5}$$

But

$$\inf_{\Delta_j} \log \frac{1}{|z|} \ge c_4(1 - |w_j|)$$

and in a moment we will prove the inequality

$$\frac{1}{2\pi} \int_{\partial\Delta_j} \frac{\partial u}{\partial n} ds \ge c_5. \tag{11.6}$$

Therefore (11.4) follows from (11.5) and (11.6).

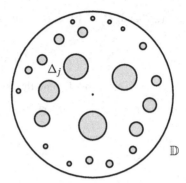

Figure X.23 The proof of (11.6).

To prove (11.6) we use the functions

$$U_j(z) = \omega(z, \partial\mathbb{D}, \mathbb{D} \setminus \bigcup\{\Delta_k, z_k \in S_n \setminus \{z_j\}\})$$

and

$$V_j(z) = \omega(z, \partial\Delta_j, \mathbb{D} \setminus \Delta_j).$$

By the maximum principle

$$u(z) \geq U_j(z) - V_j(z)$$

if $z \in \mathbb{D} \setminus \cup\{\Delta_k : z_k \in S_n\}$. All the annuli $\mathbb{D} \setminus \Delta_j$ have the same modulus, so that when $z \in \partial\Delta_j^*$,

$$V_j(z) = \varepsilon(\eta),$$

where $\lim_{\eta \to 0} \varepsilon(\eta) = 0$. By our hypothesis (11.3) and by Harnack's inequality

$$U_j(z) \geq c_6 = c_6(\delta) \quad \text{on } \partial\Delta_j^*.$$

Thus we can choose $\eta = \eta(\delta)$ so small that

$$u(z) \geq c_7 = c_6/2 \quad \text{on } \bigcup_{S_n} \Delta_j^*.$$

Let A_j be the annulus $(\Delta_j^*)^\circ \setminus \Delta_j$. Then

$$u(z) \geq c_7\omega(z, \partial\Delta_j^*, A_j) \quad \text{on } A_j,$$

so that

$$\frac{\partial u}{\partial n} \geq c_7\frac{\partial\omega(z, \Delta_j^*, A_j)}{\partial n} \quad \text{on } \partial\Delta_j.$$

But since all annuli A_j have the same modulus, we have

$$\frac{1}{2\pi} \int_{\partial \Delta_j} \frac{\partial \omega(z, \Delta_j^*, A_j)}{\partial n} ds = C(\eta) > 0.$$

Therefore (11.6) holds with $c_5 = C(\eta)c_7$. ∎

The moment he saw this proof of Lemma 11.2, Burgess Davis came up with a simple Brownian motion proof. See Exercise 15.

Proof of Theorem 11.1. Let $\{Q_j\}$ be the set of Whitney squares for Ω such that $Q_j \cap \Gamma \neq \emptyset$ and fix $z_j \in Q_j \cap \Gamma$. Given T large, we partition $\{z_j\}$ into $N = N(T)$ subsets $S_1, \ldots, S_N$ such that $\rho(z_j, z_k) \geq T$ whenever $z_j, z_k \in S_n$ and $j \neq k$. To prove Theorem 11.1 we verify condition (11.3) of Lemma 11.2 for $S_1 \cup S_2 \cup \ldots \cup S_N$.

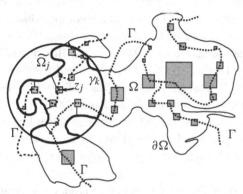

Figure X.24 The proof of Theorem 11.1.

Let $\widetilde{\Omega}_j$ be the simply connected component of $\Omega_j \cap B(z_j, 10 \operatorname{diam} Q_j)$ that contains z_j. By the Beurling projection theorem

$$\omega(z_j, \partial\widetilde{\Omega} \setminus B(z_j, 10 \operatorname{diam} Q_j), \widetilde{\Omega}_j) \leq 1/4.$$

Therefore (11.3) will hold if

$$\omega(z_j, (\partial\Omega_j \setminus \partial\Omega) \cap B(z_j, 10 \operatorname{diam} Q_j), \widetilde{\Omega}_j) \leq 1/4, \tag{11.7}$$

because (11.7) implies

$$\omega(z_j, \partial\Omega \cap \partial\Omega_j, \Omega_j) \geq \omega(z_j, \partial\Omega \cap B(z_j, 10 \operatorname{diam} Q_j), \widetilde{\Omega}_j) \geq 1/2.$$

To establish (11.7) we let γ_k be a line segment joining $Q_k \cap \Gamma$ to $\partial\Omega$ such that $\ell(\gamma_k) \leq 4 \operatorname{diam}(Q_k)$. Now since Γ is Ahlfors regular we can, given $\varepsilon > 0$, take T so large that

$$\Lambda_1(B(z_j, 10\,\mathrm{diam}\,Q_j) \cap (\partial\Omega_j \cup \bigcup \gamma_k)) \le \varepsilon\,\mathrm{diam}\,Q_j.$$

On the other hand,

$$\Gamma_j = (\Gamma \cap \widetilde{\Omega}_j) \cup \bigcup \{\gamma_k : z_k \in \partial\widetilde{\Omega}_j\}$$

is a connected set and $\Lambda(\Gamma_j) \le C\,\mathrm{diam}\,Q_j$. Therefore Theorem 10.1 implies inequality (11.7).   ∎

Notes

The books of Zygmund [1959] and Stein [1970], [1993] give more complete discussions of the Lusin area function. Our proof of Theorem 1.2 is from Coifman, Jones, and Semmes [1989]. We have derived Theorem 1.3 from Theorem 1.2, but Chapter XIV of Zygmund [1959] has a proof more like the arguments in our Chapter VI. Sections 2 and 3 are based on Jones [1990]. For further results and applications see Jones [1989], [1991]; Bishop and Jones [1997a], [1997b], [1997c]; David [1991], [1998]; David and Semmes [1991], [1993], [1997]; Fang [1994]; Garnett, Jones and Marshall [1992]; Mattila, Melnikov and Verdera [1996]; Okikiolu [1992]; and Pajot [2002].

We have barely touched the Schwarzian derivative. Lehto [1987] has an excellent introduction. See Thurston [1986] for a completely geometrical explanation and see Chuaqui and Osgood [1998] and Osgood [1997] for more recent results.

The recent work of Choi [2004] includes a proof of Theorem 10.1 that uses cone domain constructions instead of β numbers. Walden [1994] has the L^p version of Theorem 1.1: For some $p > 1$,

$$\int_{\Gamma \cap \Omega} |(\varphi^{-1})'(w)|^p |dw| < \infty.$$

Walden's theorem generalizes the Fernández, Heinonen and Martio theorem from Exercise VII.7. Lemma 11.2 is from Garnett, Gehring and Jones [1983]. See also Garnett [1986] and Exercise 15.

Exercises and Further Results

1. (a) If $f \in H^2(\mathbb{D})$ and $g(z) \in H^\infty(\mathbb{D})$ then by Green's theorem and the product rule,

$$\iint_{\mathbb{D}} |f(z)|^2 |g'(z)|^2 (1 - |z|) dx dy \leq \iint_{\mathbb{D}} |f(z)|^2 |g'(z)|^2 \log \frac{1}{|z|} dx dy$$

$$\leq C \|g\|_\infty^2 \|f\|_{H^2}^2.$$

(b) Use (a) to prove (1.7) *without* using Carleson measures. See Gamelin [1980] and Coifman, Jones and Semmes [1989].

2. (a) Prove Lemma 1.5.

(b) More generally, prove that for $n \geq 2$ and $\alpha > 1$,

$$\iint_{\Gamma_\alpha(\zeta)} |F'(z)|^2 dx dy \leq C(n, \alpha) \sum_1^{n-1} |F^{(j)}(0)|^2$$

$$+ C(n, \alpha) \iint_{\Gamma_\alpha(\zeta)} |F^{(n)}(z)|^2 (1 - |z|)^{2n-2} dx dy.$$

(c) If $n \geq 2$ and $1 < \beta < \alpha$, there is $C(n, \alpha, \beta)$ such that

$$\iint_{\Gamma_\beta(\zeta)} |F^{(n)}|^2 (1 - |z|)^{2n-2} dx dy \leq C(n, \alpha, \beta) \iint_{\Gamma_\alpha(\zeta)} |F'(z)|^2 dx dy.$$

3. Let K be the Cantor set obtained from $[0, 1]$ by repeatedly removing middle halves, and let $E = K \times K$. Then $\Lambda_1(E) < \infty$ but E is not contained in a rectifiable curve. Also construct an example with $\Lambda_1(E) = 0$. Hint: Alter the construction so that at stage n there are 4^n squares of side $\frac{4^{-n}}{n}$.

4. (a) Show that the integral in (2.1) is infinite at $z = 0$ when E is the union of two radii of the unit circle that meet at an angle less than π.

(b) Prove that if

$$\int_0^1 \beta_E^2(z, t) \frac{dt}{t} < \infty,$$

then $\lim_{t \to 0} \beta_E(z, t) = 0$.

(c) Suppose L_n are lines through z with

$$\beta_n = \beta_E(z, 2^{-n}) = \frac{1}{2^{-n}} \sup_{E \cap B(z, 2^{-n})} \text{dist}(w, L_n).$$

If L_n converges to a line L and $\beta_n \to 0$, then E has a tangent at z.

(d) Prove that if $\sum \beta_n < \infty$ (equivalently $\int_0^1 \beta_E(z,t)/t\,dt < \infty$) then E has a tangent at z.

5. (a) Construct a Jordan curve Γ such that $0 \in \Gamma$ and $\varepsilon(0,t) = 0$, $\quad t < 1$ but $\beta_\Gamma(0,t) \geq \beta_0 > 0$ $\quad t < 1$.

(b) Construct a Jordan curve Γ such that $0 \in \Gamma$,

$$\int_0^1 \beta_\Gamma{}^2(0,t)\frac{dt}{t} < \infty$$

but such that Γ fails to have a tangent at $z = 0$.

6. Let E be a finite set of points in $\mathbb{C}$. Define a **path** from $z_1 \in E$ to $z_n \in E$ to be a collection $\{[z_i, z_{i+1}] : 1 \leq i \leq n-1\}$ of line segments with endpoints $\{z_i\}_1^n \subset E$, and define a **circuit** to be a closed path (i.e. $z_1 = z_n$) such that $z_i \neq z_j$, for $1 \leq i < j \leq n-1$ and with $n \geq 4$. A **spanning tree** G for E is a collection of line segments with endpoints in E such that G contains a path connecting every pair of points $z_1, z_2 \in E$, and such that G contains no circuit in E. A **tour** of E is a closed path containing every point of E.

The **greedy algorithm** is the following construction of a spanning tree for E : Choose any $z_1 \in E$. Let $z_2 \in E$ satisfy $|z_1 - z_2| = \min_{z \in E \setminus \{z_1\}} |z - z_1|$, and put the line segment $e_1 = [z_1, z_2]$ into G. Having chosen distinct points $z_1, \ldots, z_{m-1}$ in E, choose $z_m \in E \setminus \{z_1, \ldots, z_{m-1}\}$ so that for some $k < m$,

$$|z_m - z_k| = \min\{|z - z_j| : z \in E \setminus \{z_1, \ldots, z_{m-1}\} \text{ and } 1 \leq j \leq m-1\},$$

and put the line segment $e_m = [z_k, z_m]$ into G. Continue this process until $G = E$.

(a) Prove that the greedy algorithm constructs a spanning tree of minimal possible length. Hint: Let M be a minimal spanning tree and let G be the spanning tree constructed by the greedy algorithm. Let k be the largest integer so that $e_1, \ldots, e_k \in G \cap M$. If we add e_{k+1} to M, then M contains a circuit all of whose segments have length no more that $|e_{k+1}|$. But by the construction of G, this circuit must contain and edge e with exactly one vertex in the collection of endpoints of $e_1, \ldots, e_k$ and length at least $|e_{k+1}|$. Replacing e with e_{k+1}, we obtain another minimal spanning tree M' with $e_1, \ldots, e_k, e_{k+1} \in G \cap M'$.

(b) Construct a tour T of E with $\ell(T) \leq 2\Lambda_1(G)$. Hint: Do induction on the number of edges.

(c) Prove that every tour of minimal length contains a spanning tree.

(d) If K is a connected set containing E with $\Lambda_1(K)$ minimal, then K contains a closed rectifiable curve Γ with $\ell(\Gamma) \leq 2\Lambda_1(K)$.

(e) Use the above to show that

$$2 \min\{\Lambda_1(K) : K \text{ connected } \supset E\}$$

$$\geq \min\{\ell(\Gamma) : \Gamma \text{ closed curve } \supset E\}$$

$$= \min\{\ell(T) : T \text{ tour of } E\} \tag{E.1}$$

$$> \Lambda_1(G)$$

$$\geq \min\{\Lambda_1(K) : K \text{ connected } \supset E\},$$

where G is the spanning tree constructed by the greedy algorithm. A connected set of minimal length containing E does not necessarily consist of line segments with vertices in E, but by (E.1) the greedy algorithm constructs a connected set $G \supset E$ of length no more that twice the minimal possible length. The greedy algorithm together with the construction in (b) constructs a tour of length no more than twice the length of a shortest tour.

(f) Show that the method of Section 2 constructs a spanning tree for E. Each segment comes from applications of Case 2.

(g) If E is a compact set in $\mathbb{C}$ contained in a rectifiable curve Γ, then replacing each arc in $\Gamma \setminus E$ with a straight line segment with endpoints in E, we obtain a rectifiable curve which is not longer than Γ. Given $\varepsilon > 0$, choose $z_1, \ldots, z_m \in E$ so that if $z \in E$ then $|z - z_j| < \varepsilon$ for some j. Applying the greedy algorithm to $\{z_1, \ldots, z_m\}$ we then obtain a tree G_ε such that $\Lambda_1(G_\varepsilon) \leq \ell(\Gamma)$, and $\operatorname{dist}(z, G_\varepsilon) < \varepsilon$ for every $z \in E$. Construct a tour T_ε of G_ε with $\ell(T_\varepsilon) \leq \Lambda_1(G_\varepsilon)$. Show that there is a limiting tour $T = \lim_{\varepsilon_j \to 0} T_{\varepsilon_j}$ with $\ell(T) \leq 2\ell(\Gamma)$.

Constructing a tour of minimal length containing a finite set E is called the traveling salesman problem. The greedy algorithm together with the construction in (b) gives a tour of length no more than twice the minimal possible length.

7. Recall from Chapter VII that a curve Γ is Ahlfors regular if there is a constant $A > 0$ such that for every disc $B(z, r)$, $\Lambda_1(\Gamma \cap B(z, r)) \leq Ar$. Let $E \subset \mathbb{C}$ be compact. Prove E is contained in an Ahlfors regular curve Γ if and only if there is a constant A' such that for any $z \in E$ and any $0 < R < \operatorname{diam} E$,

$$\int_{E \cap B(z,R)} \int_0^R \beta_E^2(w, t) \frac{dt}{t} d\Lambda_1(w) \leq A'R.$$

See Jones [1990]. Related results can be found in David and Semmes [1991] and Pajot [2002].

8. This exercise outlines the proof of the following result from Bishop and Jones [1997a].

Theorem. *Let E be a compact connected set such that*

$$\beta_E(3Q) \geq \beta_0 > 0 \tag{E.2}$$

for all dyadic squares Q such that $Q \cap E \neq \emptyset$ and $\ell(Q) \leq \frac{\text{diam}(E)}{3}$. Then

$$\dim(E) \geq 1 + c\beta_0^2, \tag{E.3}$$

where c is a constant independent of β_0.
(a) When E is a compact connected set define

$$\beta_k^2(E) = \text{diam}(E) + \sum \{\beta_E^2(3Q)\ell(Q) : Q \text{ is dyadic and } \ell(Q) \geq 2^{-k}\},$$

and

$$\ell_k(E) = \inf\{\ell(\Gamma) : \Gamma \text{ is rectifiable and } \inf_E \text{dist}(z, \Gamma) \leq 2^{-k-\frac{1}{2}}\}.$$

Then

$$C_1\beta_k^2(E) \leq \ell_k(E) \leq C_2\beta_k^2(E),$$

where the constants C_1 and C_2 do not depend on β_0. Hint: Apply Theorem 2.3 to

$$\Gamma_k(E) = \partial\left(\bigcup\{Q : Q \text{ dyadic}, \ell(Q) = 2^{-k}, \text{ and } Q \cap E \neq \emptyset\}\right).$$

(b) Let E be a compact connected set, let Q_N be a dyadic square of side $2^{-N} \leq \frac{\text{diam}(E)}{3}$ such that $Q \cap E \neq \emptyset$. Then for $n > N$, $3Q$ contains at least 2^{n-N} dyadic squares Q_n of side 2^{-n} such that $Q_n \cap E \neq \emptyset$.
(c) Now let E be a compact connected set such that (E.2) holds for every dyadic Q such that $Q \cap E \neq \emptyset$ and $\ell(Q) \geq \frac{\text{diam}(E)}{3}$. Fix a dyadic square Q_N of side $2^{-N} \leq \frac{\text{diam}(E)}{3}$ such that $Q_N \cap E \neq \emptyset$. Then by (b),

$$\beta_{N+k}^2(Q_N \cap E) \geq k\beta_0^2 2^{-N}.$$

Apply (a) to $E_Q = (E \cap 3Q_N) \cup \partial Q$ to show $3Q_N$ contains at least

$$N_k = ck\beta_0^2 2^k \tag{E.4}$$

dyadic squares $\{Q^j\}$ such that

$$\ell(Q^j) = 2^{-N-k},$$

$$3Q^j \cap 3Q^{j'} = \emptyset, \quad j \neq j',$$

and

$$Q^j \cap E \neq \emptyset,$$

where c is an absolute constant. Hint: Consider $\Gamma_{N+k}(E_Q)$.

(d) We continue to assume the compact connected set E satisfies (E.2). Fix $k = [\frac{1}{\beta_0^2}]$ and note that by (E.4)

$$N_k \geq 2^{(1+c\beta_0^2)k} \equiv \delta,$$

for some constant c. Assume $\mathrm{diam}(E) \geq 3$. By (c) and induction, there are generations $\mathcal{G}_n = \{Q_j^n\}, n = 0, 1, \ldots$ of dyadic squares of side 2^{-nk} such that

(i) For $n \geq 1$ each $Q_j^n \in \mathcal{G}_n$ is a subset of some $Q_j^{n-1} \in \mathcal{G}_{n-1}$,

(ii) Each $Q_j^{n-1} \in \mathcal{G}_{n-1}$ contains exactly N_k distinct squares $Q_j^n \in \mathcal{G}_n$, and

(iii) $Q_j^n \cap E \neq \emptyset$ for all $Q_j^n \in \mathcal{G}_n$.

(e) If the compact set E satisfies (i)–(iii), then

$$\dim(E) \geq 1 + c\beta_0^2. \tag{E.5}$$

Hint: Define the measure μ by

$$\mu(Q_j^n) = N_k^{-n} \leq 2^{-(1+c\beta_0^2)nk}.$$

Then μ has support E and $\mu(E) = 1$. Moreover, there is a constant $C = C(k)$ such that Frostman's Theorem D.1 can be applied to $C\mu$ to yield

$$\Lambda_{1+c\beta_0^2}(E) > 0.$$

(f) Except for the value of c in (E.5) the above theorem is sharp. If $\beta > 0$, this can be seen by a variation on the von Koch snowflake construction.

9. Let Q be a dyadic Carleson box in $\mathbb{D}$, let z_Q be the center of its top half $T(Q)$, let $G(Q)$ be a family of pairwise disjoint Carleson boxes $Q' \subset Q$, and let $\mathcal{D}(Q) = Q \setminus \bigcup_{G(Q)} Q'$. Let $F \in H^2$ have no zeros and $\|\log F\|_B \leq 6$. This exercise outlines a proof without using Appendix M of the inequality

$$\int_{\partial \mathcal{D}(Q)} |F(z) - F(z_Q)|^2 ds \leq C \iint_{\mathcal{D}(Q)} |F'(z)|^2 \mathrm{dist}(z, \partial \mathcal{D}(Q)) dx dy.$$

The construction is simpler in the upper half-plane and for this reason we assume the box Q and its subboxes Q' are dyadic squares in $\mathbb{H}$ that have one side in $\mathbb{R}$. Let $\{J_k\}$ be the set of maximal vertical line segments in $\partial \mathcal{D}(Q) \cap \bigcup_{G(Q)} \partial Q'$. For each, $J_k \subset \partial Q'(k)$ for a unique $Q'(k) \in G(Q)$. Let $W_k \subset Q'(k)$ be the triangle bounded by J_k, a segment of slope ± 2, and a segment I_k of length $\frac{\ell(J_k)}{2}$ on the top edge of $Q'(k)$. Define

$$\mathcal{U} = \mathcal{D}(Q) \cup \bigcup_k (W_k \cup J_k \cup I_k).$$

Then $\mathcal{U}$ is a Lipschitz domain and the proof of Theorem 1.2 can be adapted to yield (1.7) for $\mathcal{U}$. Now since $\log F$ is a Bloch function it is not hard to show

$$\int_{\partial \mathcal{D}(Q)} |F(z) - F(z_Q)|^2 ds \leq C \int_{\partial \mathcal{U}} |F(z) - F(z_Q)|^2 ds.$$

On the other hand,

$$\iint_{\mathcal{U}} |F'(z)|^2 \mathrm{dist}(z, \partial \mathcal{U}) dx dy \leq 2 \iint_{\mathcal{D}(Q)} |F'(z)|^2 \mathrm{dist}(z, \partial \mathcal{D}(Q)) dx dy.$$

Hint: The reflections of the W_k through the J_k are pairwise disjoint.

10. Let φ be univalent on $\mathbb{D}$ and let $g(z) = \log(\varphi'(z))$. If $\|g\|_B$ is small, then $\sup_{\mathbb{D}} \left(1 - |z|^2\right)^2 |\mathcal{S}\varphi|$ is small, but not conversely. On the other hand if $\sup_{\mathbb{D}} \left(1 - |z|^2\right)^2 |\mathcal{S}\varphi|$ is small, then there is a Möbius transformation T such that

$$\left(1 - |z|^2\right) \left| \frac{(T \circ \varphi)''}{(T \circ \varphi)'} \right|$$

is small.

11. Suppose $\mathcal{D} \subset \mathbb{D}$ is a special cone domain and let φ be univalent on $\mathcal{D}$. Then Lemma 4.3 holds for φ: Given $\varepsilon > 0$, $a > 0$, and $b > 0$, there exists $\delta = \delta(\varepsilon, a, b)$ such that if

$$\left(\mathrm{dist}(z, \partial \mathcal{D})\right)^2 |\mathcal{S}\varphi(z)| < \delta$$

on the hyperbolic ball $\{z : \rho_{\mathcal{D}}(z, z_0) < a\}$, then for some Möbius transformation T,

$$\left| T(\varphi(z)) - \frac{z - z_0}{1 - \bar{z}_0 z} \right| + \left| \varphi(z) - T^{-1} \left(\frac{z - z_0}{1 - \bar{z}_0 z} \right) \right| < \varepsilon$$

on the hyperbolic ball $\{z : \rho_{\mathcal{D}}(z, z_0) < b\}$.

12. Let Ω be a simply connected Jordan domain and let $z \in \partial\Omega$.
    (a) If Ω is a quasidisc, prove

$$\frac{1}{C} \eta_\Omega(z, t) \leq \beta_{\partial\Omega}(z, t) \leq C \eta_\Omega(z, t)$$

for small t, where the constant C depends only on the quasiconformality of $\partial\Omega$.

    (b) For any constant M make an example for which

$$\beta_{\partial\Omega}(z, t) \geq M \eta_\Omega(z, t).$$

(c) For any constant M make an example for which

$$\eta_\Omega(z, t) \geq M\beta_{\partial\Omega}(z, t).$$

13. (a) Let $\mathcal{D} \subset \mathbb{D}$ be a special cone domain and let φ be a conformal map from $\mathcal{D}$ onto a quasidisc Ω. Write $d(z) = \text{dist}(z, \partial\mathcal{D})$. Then

$$\frac{1}{C_1}\ell(\partial\Omega) \leq \text{diam}(\Omega) + \iint\limits_{\mathcal{D}} |\varphi'(z)||S\varphi(z)|^2 d(z)^3 dxdy.$$

(b) Let $\mathcal{D}$ be a chord-arc domain with $0 \in \mathcal{D}$ and $\text{dist}(0, \partial\mathcal{D}) \geq c\,\text{diam}(\mathcal{D})$. Write $d(z) = \text{dist}(z, \partial\mathcal{D})$. Let $F(z)$ be analytic on $\mathcal{D}$. Verify that $F \in H^2(\mathcal{D})$ if and only if

$$\iint\limits_{\mathcal{D}} |F'(z)|^2 d(z)dxdy < \infty,$$

and that

$$C||F||_{H^2(\mathcal{D})}^2 \leq |F(0)|^2 + \iint\limits_{\mathcal{D}} |F'(z)|^2 d(z)dxdy \leq \frac{1}{C}||F||_{H^2(\mathcal{D})}^2$$

and

$$C'||F||_{H^2(\mathcal{D})}^2 \leq |F(0)|^2 + |F'(0)|^2 + \iint\limits_{\mathcal{D}} |F''(z)|^2 d(z)^3 dxdy$$

$$\leq \frac{1}{C'}||F||_{H^2(\mathcal{D})}^2,$$

where C and C' depend only on the chord-arc constant of $\mathcal{D}$ and the constant c above. Hint: Show (1.9) holds for $\mathcal{D}$, and then use Theorem M.1 from Appendix M.

(c) Let $\mathcal{D}$ be as in part (b) and let φ be a conformal mapping from $\mathcal{D}$ to a domain Ω such that $\partial\Omega$ is rectifiable. Then

$$\iint\limits_{\mathcal{D}} |\varphi'(z)||S\varphi(z)|^2 d(z)^3 dxdy \leq C\ell(\partial\Omega).$$

14. Prove inequality (7.15) without using Theorem M.1.

15. (B. Davis) Let $\{w_j\} \subset \mathbb{D}$ satisfy $\rho(w_j, w_k) \geq 1$, let $\Delta_j = \{\rho(w, w_j) \leq \frac{1}{8}\}$, and assume

$$\inf_j \omega\left(w_j, \partial\mathbb{D}, \mathbb{D} \setminus \bigcup_{k \neq j} \Delta_k\right) \geq \delta > 0, \qquad \text{(E.6)}$$

just as in Lemma 11.2. Let P_z denote probability on the space of Brownian paths B_t starting at z and let E denote integration with respect to P_0. By

Kakutani's theorem the set

$$E_j = \{B_t \text{ hits } \Delta_j \text{ before hitting } \partial\mathbb{D}\}$$

has

$$P_0(E_j) \geq \frac{c(1 - |w_j|)}{2\pi}. \tag{E.7}$$

Set $N = \sum \chi_{E_j}$. The random variable N counts the number of Δ_j visited by B_t before B_t reaches $\partial\mathbb{D}$. By (E.7),

$$\sum (1 - |w_j|) \leq 2\pi E(N).$$

In terms of P_z our hypothesis (E.6) says that

$$\inf_{z \in \Delta_j} P_z\Big(B_t \text{ hits } \bigcup_{k \neq j} \Delta_j \text{ before hitting } \partial\mathbb{D}\Big) = \inf_{z \in \Delta_j} P_z\Big(\bigcup_{k \neq j} E_k\Big) \tag{E.8}$$

$$\leq 1 - \delta < 1.$$

Let $\tau_1 = \inf\{t : B_t \in \bigcup \Delta_j\}$ be the first time B_t hits $\bigcup \Delta_j$. Assuming that τ_n has been defined and that $B_{\tau_n} \in \Delta_{j(n)}$, let

$$\tau_{n+1} = \inf\{t > \tau_n : B_t \in \bigcup_{j \neq j(n)} \Delta_j\}$$

be the time of the first visit to a different disc Δ_j. Then the functions τ_n and $\tau_{\mathbb{D}} = \inf\{t : B_t \in \partial\mathbb{D}\}$ are measurable with respect to P_0.

Now let $M = \max\{n : \tau_n \leq \tau_{\mathbb{D}}\}$. Then $M - 1$ is the number of trips from one Δ_j to another taken by B_t before it hits $\partial\mathbb{D}$. Therefore $M \geq N$ and

$$E(N) \leq E(M) = \sum_{n=0}^{\infty} P_0(M \geq n).$$

Now by (E.8) and the independence of Brownian increments,

$$P_0(M \geq n + 1) = \sum_j P_0(M \geq n + 1; B_{\tau_n} \in \Delta_j)$$

$$\leq (1 - \delta) \sum_j P_0(M \geq n; B_{\tau_n} \in \Delta_j)$$

$$= (1 - \delta) P_0(M \geq n),$$

and by induction

$$\sum (1 - |w_j|) \leq 2\pi \sum_{n=1}^{\infty} (1 - \delta)^{n-1} = \frac{2\pi}{\delta}.$$

Appendices

A. Hardy Spaces

If $0 < p < \infty$, then by definition an analytic function $f(z)$ on the unit disc $\mathbb{D}$ is in the **Hardy space** H^p if

$$\sup_{0 < r < 1} \int |f(re^{i\theta})|^p d\theta = \|f\|_{H^p}^p < \infty. \tag{A.1}$$

For $p = \infty$ the Hardy space H^∞ is defined to be the space of bounded analytic functions on $\mathbb{D}$ with norm $\|f\|_\infty = \sup_{\mathbb{D}} |f(z)|$.

Write $f_r(z) = f(rz)$ and suppose $0 < r < s < 1$. Then $\|f_r\|_\infty \le \|f_s\|_\infty$. If $p < \infty$, then

$$|f_r(z)|^p \le \int_0^{2\pi} P_{\frac{r}{s}z} |f_s(e^{i\theta})|^p d\theta$$

because the function $|f(z)|^p$ is subharmonic. It then follows from Fubini's theorem that $\|f_r\|_p$ is an increasing function of r whenever $0 < p \le \infty$.

Theorem A.1. *Assume $1 \le p < \infty$ and let $f \in H^p$. Then*
(a) *The function $f(z)$ has a nontangential limit $f(e^{i\theta})$ almost everywhere $d\theta$ and*

$$\lim_{r \to 1} \int_0^{2\pi} |f(re^{i\theta}) - f(e^{i\theta})|^p d\theta = 0. \tag{A.2}$$

(b) *If $f \ne 0$, then*

$$\int \log|f(e^{i\theta})| d\theta > -\infty. \tag{A.3}$$

The case $p > 1$ of part (a) follows from Exercise I.11 of Chapter I, and the significance of Theorem A.1 lies in the case $p = 1$ and the inequality (A.3).

435

From part (a) it follows that

$$f(z) = \lim_{r \to 1} f(rz) = \int P_z(\theta) f(e^{i\theta}) d\theta,$$

and since $\|f_r\|_p$ is increasing in r, we obtain

$$\|f\|_{H^p}^p = \|f\|_{L^p(\partial\mathbb{D})}^p$$

for all $f \in H^p$. Theorem A.1 is also true for $f \in H^p$ for all $p < 1$. See Exercise A.1.

Part (a) of Theorem A.1 is due to F. and M. Riesz [1916], and part (b) is from F. Riesz [1923]. Part (b) has the following important corollary.

Corollary A.2. *If $f \in H^1$ and $f \not\equiv 0$, then $|f(e^{i\theta})| > 0$ almost everywhere.*

The proof of Theorem A.1 will use Theorem A.3 below, which is called **Hardy–Littlewood maximal theorem** for H^p functions:

Theorem A.3. *Let $\alpha > 1$ be fixed, let $0 < p \le \infty$, and let $f \in H^p$. Then the nontangential maximal function*

$$f_\alpha^*(\zeta) = \sup_{\Gamma_\alpha(\zeta)} |f(z)|$$

satisfies

$$\|f_\alpha^*\|_p \le A_\alpha \|f\|_{H^p}, \tag{A.4}$$

where the constant A_α depends only on α.

We first derive Theorem A.1 from Theorem A.3 and then return to the proof of Theorem A.3.

Proof of Theorem A.1(a). First suppose $p = 1$. By (A.1) and the Banach–Alaoglu theorem, there are $r_n \to 1$ so that $f(r_n e^{i\theta}) d\theta$ converges weak-star to a finite complex measure μ on $\partial\mathbb{D}$. Then

$$f(z) = \lim_{n \to \infty} f(r_n z) = \lim_{n \to \infty} \int P_z(\theta) f(r_n e^{i\theta}) d\theta = \int P_z(\theta) d\mu(\theta).$$

Write

$$d\mu = f d\theta + d\nu,$$

where $f(e^{i\theta}) \in L^1(d\theta)$ and ν is singular to Lebesgue measure. Then

$$f(z) = \int P_z(\theta) f(e^{i\theta}) d\theta + \int P_z(\theta) d\nu(\theta),$$

and by Fatou's theorem $\int P_z(\theta) f(e^{i\theta}) d\theta$ has nontangential limit $f(e^{i\theta})$ almost everywhere.

In a moment we will prove that

$$\lim_{\Gamma_\alpha(\zeta)\ni z\to\zeta} \int P_z(\theta) d\nu = 0 \tag{A.5}$$

almost everywhere. That will mean $f(z)$ has nontangential limit $f(e^{i\theta})$ almost everywhere, and (A.2) will then follow from (A.4) and dominated convergence.

To prove (A.5) we can assume $\nu > 0$. Fix $\alpha > 1$ and $\varepsilon > 0$. Take a compact set $K \subset \partial\mathbb{D}$ so that $|K| < \varepsilon$ and $\nu(\partial\mathbb{D} \setminus K) < \varepsilon$, and write $\nu_1 = \chi_K \nu$ and $\nu_2 = \nu - \nu_1$. Then $\lim_{z\to\zeta} \int P_z(\theta) d\nu_1(\theta) = 0$ for all $\zeta \in \partial\mathbb{D} \setminus K$. The proofs of Lemmas I.2.2 and I.2.4 show that

$$\left| \left\{ \zeta \in \partial\mathbb{D} : \sup_{\Gamma_\alpha(\zeta)} \int P_z(\theta) d\nu_2(\theta) > \lambda \right\} \right| \le \frac{3+6\alpha}{\lambda} \int d\nu_2 \le \frac{(3+6\alpha)\varepsilon}{\lambda}.$$

Taking $\lambda = \varepsilon^{1/2}$ then gives

$$\left| \left\{ \zeta \in \partial\mathbb{D} : \limsup_{\Gamma_\alpha(\zeta)\ni z\to\zeta} \int P_z(\theta) d\nu(\theta) \ge \varepsilon^{1/2} \right\} \right| \le |K| + (3+6\alpha)\varepsilon^{1/2}$$

$$\le \varepsilon + (3+6\alpha)\varepsilon^{1/2},$$

so that (A.5) holds almost everywhere.

If $1 < p < \infty$, part (a) is easier. There are $r_n \to 1$ so that $f_{r_n}(\theta) = f(r_n e^{i\theta})$ converges weakly in L^p to some function $f \in L^p$. Then

$$f(z) = \lim_{n\to\infty} f(r_n z) = \int P_z(\theta) f(e^{i\theta}) d\theta$$

because $P_z \in L^q, q = \frac{p}{p-1}$, and then Fatou's theorem, (A.4), and dominated convergence yield (A.2). $\blacksquare$

Proof of Theorem A.1(b). Part (b) follows from part (a). We may suppose $f(0) \ne 0$, by dividing f by z^n in case $f(0) = 0$. Then for $r < 1$ the familiar Jensen formula gives

$$\log|f(0)| \le \int \log|f(re^{i\theta})| \frac{d\theta}{2\pi}$$

$$= \int \log^+|f(re^{i\theta})| \frac{d\theta}{2\pi} - \int \log^-|f(re^{i\theta})| \frac{d\theta}{2\pi},$$

where $\log^- x = (\log(1/x))^+ = \max(-\log x, 0)$. By dominated convergence and part (a),

$$\lim_{r\to 1} \int \log^+|f(re^{i\theta})|d\theta = \int \log^+|f(e^{i\theta})|d\theta,$$

while by Fatou's lemma,

$$\int \log^-|f(e^{i\theta})|d\theta \le \liminf_{r\to 1} \int \log^-|f(re^{i\theta})|d\theta.$$

Therefore we have the important inequality

$$\log|f(0)| \le \int \log|f(e^{i\theta})|\frac{d\theta}{2\pi}, \qquad (A.6)$$

which implies (A.3). ∎

Proof of Theorem A.3. The proof will be divided into two cases, $p \ge 2$ and $0 < p < 2$. In the first case, $p \ge 2$, f is the Poisson integral of an L^p function $f(e^{i\theta})$ so that by Lemma I.2.2,

$$f_\alpha^*(e^{i\theta}) \le (1 + 2\alpha)Mf(e^{i\theta}), \qquad (A.7)$$

where Mf is the Hardy–Littlewood maximal function of $f(e^{i\theta})$. That means (A.4) is a consequence of the basic inequality

$$\|Mf\|_p \le C_p\|f\|_p, \qquad 1 < p \le \infty, \qquad (A.8)$$

which we will establish in a moment. In (A.8) the constants C_p depend on p, but $C_p \le 2\sqrt{6}$ for $p \ge 2$. Except for its range of p, (A.8) is just (A.4) with f_α^* replaced by Mf, and for this reason (A.8) is also known as the **Hardy–Littlewood maximal theorem**.

The second case, $0 < p < 2$, will be reduced to the first case by exploiting the subharmonicity of $\log|f(z)|$.

Case I: $p \ge 2$. We prove (A.8). Let $f \in L^p$. For $\lambda > 0$ set

$$m(\lambda) = |\{Mf > \lambda\}|.$$

The function $m(\lambda)$ is called the **distribution function** of Mf. By Fubini's theorem,

$$\int (Mf)^p d\theta = \int_0^{2\pi} \left\{ \int_0^{Mf(e^{i\theta})} p\lambda^{p-1}d\lambda \right\} d\theta = \int_0^\infty p\lambda^{p-1}m(\lambda)d\lambda. \qquad (A.9)$$

We also have

$$m(\lambda) \le \frac{3\|f\|_1}{\lambda}, \qquad (A.10)$$

from Lemma I.2.4. Now we use an important trick, due to Marcinkiewicz, to improve inequality (A.10). Set

$$f_1 = f \chi_{\{|f| > \frac{\lambda}{2}\}},$$

and

$$f_2 = f \chi_{\{|f| \le \frac{\lambda}{2}\}}.$$

Then $Mf \le M(f_1) + M(f_2)$, while $M(f_2) \le \|f_2\|_\infty \le \frac{\lambda}{2}$, so that by (A.10),

$$m(\lambda) \le \left| \left\{ M(f_1) > \frac{\lambda}{2} \right\} \right| \le \frac{6}{\lambda} \int_{\{|f| > \frac{\lambda}{2}\}} |f| d\theta.$$

Then from (A.9) and Fubini's theorem we get

$$\|Mf\|_p^p \le 6 \int_0^\infty p\lambda^{p-2} \left(\int_{\{|f| > \frac{\lambda}{2}\}} |f| d\theta \right) d\lambda$$

$$= 6 \int |f(e^{i\theta})| \int_0^{2|f(e^{i\theta})|} p\lambda^{p-2} d\lambda d\theta$$

$$= \frac{3p2^p}{p-1} \int |f|^p d\theta.$$

That proves (A.8), and because $\left(\frac{3p}{p-1}\right)^{\frac{1}{p}}$ is decreasing we also have the bound $C_p \le 2\sqrt{6}$ for $p \ge 2$. Together (A.8) and (A.7) prove (A.4) when $p \ge 2$ with $A_\alpha \le 2\sqrt{6}(1 + 2\alpha)$.

The Marcinkiewicz argument is also valid when $1 < p < 2$, but it unfortunately gives constants $2\left(\frac{3p}{p-1}\right)^{\frac{1}{p}}$ which are unbounded as $p \to 1$.

Case II. $0 < p \le 2$. We can assume $f \not\equiv 0$. Take $0 < r < 1$ so that f has no zeros on $\{|z| = r\}$ and set

$$U_r(z) = \frac{1}{2\pi} \int P_z(\theta) \log |f(re^{i\theta})| d\theta.$$

Then $\log |f(rz)| \le U_r(z)$ on $\mathbb{D}$. Since U_r is real and harmonic on $\mathbb{D}$, there is a unique harmonic **conjugate function** $\widetilde{U}_r$ such that $U_r + i\widetilde{U}_r$ is analytic and $\widetilde{U}_r(0) = 0$. Write

$$g_r(z) = e^{\frac{p}{2}(U_r + i\widetilde{U}_r)}.$$

Then $|f(rz)|^p \le |g_r(z)|^2$ and

$$\int |g_r|^2 d\theta = \int |f(re^{i\theta})|^p d\theta \le ||f||_p^p. \tag{A.11}$$

Let $r \to 1$, and let g be a weak limit of g_r in L^2. Then $g \in H^2$ because Poisson kernels are in L^2 and $||g||_{H^2}^{2/p} \le ||f||_{H^p}$ by (A.11). Moreover,

$$|f(z)| = \lim_{r \to 1} |f(rz)| \le |g(z)|^{2/p}.$$

Therefore $(f_\alpha^*)^p \le (g_\alpha^*)^2 \in L^1$, and

$$||f_\alpha^*||_p \le ||g_\alpha^*||_2^{2/p}.$$

Thus (A.4) holds for $0 < p < 2$ with the same bound on A_α as for the case $p = 2$.   ∎

Exercise A.1. (a) Suppose $0 < p < 1$, and suppose $f \in H^p$. Prove f has non-tangential limit $f(e^{i\theta})$ almost everywhere and if $f(0) \ne 0$,

$$\log|f(0)| \le \int_0^{2\pi} \log|f(e^{i\theta})| \frac{d\theta}{2\pi}.$$

(b) On the other hand, if $E \subset \partial \mathbb{D}$ is compact, if $|E| = 0$, and if $p < \infty$, then there is $f(z) \in H^p$ such that $\lim_{r \to 1} |f(re^{i\theta})| = \infty$ everywhere on E.

Exercise A.2. (a) For $p > 1$ and $\alpha > 1$, there exists $C(p, \alpha) > 0$ such that if $f \in L^p(\partial \mathbb{D})$ and $u = u_f$ then

$$||u_\alpha^*||_p \le C(p, \alpha)||f||_p.$$

(b) However there is $f \in L^1(\partial \mathbb{D})$ such that $\sup_{0 < r < 1} |u(re^{i\theta})| \notin L^1(\partial \mathbb{D})$.

Exercise A.3. Let μ be a Carleson measure on the unit disc. By Exercise I.19(a) it follows that

$$\int |u|^p d\mu \le C \int |f|^p d\theta$$

if $p > 1$ and if u is the Poisson integral of $f \in L^p$. It also follows from Exercise I.19(a) that

$$\int |f(z)|^p d\mu \le C \int |f(e^{i\theta})|^p d\theta$$

if $p > 0$ and if $f \in H^p$. These results are from Carleson [1958] and [1962].

Exercise A.4. If $F(z)$ is analytic on $\mathbb{D}$ and if $\mathrm{Re}\, F(z) > 0$ for all $z \in \mathbb{D}$, then $F \in H^p$ for all $p < 1$. Hint: Use Zygmund's theorem on conjugate functions from Chapter II. Somewhat deeper is the weak-type estimate

$$|\{\theta : |F_\alpha^*| > \lambda\}| \leq \frac{C}{\lambda}.$$

See for example Garnett [1981].

B. Mixed Boundary Value Problems

This appendix shows how to solve a mixed boundary value problem on a domain Ω by reducing it to the Dirichlet problem on a Riemann surface double of Ω.

Let Ω be a bounded finitely connected Jordan domain with $\partial\Omega = \bigcup_{j=1}^{m} \Gamma_j$, where each Γ_j is of class $C^{1+\alpha}$ and $\Gamma_j \cap \Gamma_k = \emptyset$ if $j \neq k$. Let v represent the steady-state temperature of a thin homogeneous plate in the shape of Ω. Then v is harmonic in Ω and the loss of heat through an arc J on $\partial\Omega$ is given by

$$\int_J \frac{\partial v}{\partial n} ds,$$

where n is the unit outer normal to $\partial\Omega$. Given a function f on $\partial\Omega$, the **Neumann problem** is to find $v \in C^1(\overline{\Omega})$ such that v is harmonic on Ω and $\partial v/\partial n = f$ on $\partial\Omega$. In other words, find a steady-state temperature on the plate that produces a prescribed heat flow f across the boundary. By Green's theorem f must satisfy the necessary condition

$$\int_{\partial\Omega} f ds = \int_{\partial\Omega} \frac{\partial v}{\partial n} ds = 0. \tag{B.1}$$

Except for issues of smoothness, (B.1) is the only condition necessary for the Neumann problem.

Theorem B.1. *Suppose Ω is a finitely connected Jordan domain such that $\partial\Omega \in C^{1+\alpha}$ for some $\alpha > 0$. Let $\beta > 0$, and suppose $f \in C^\beta(\partial\Omega)$ satisfies $\int_{\partial\Omega} f ds = 0$. Then there is a harmonic function $v \in C^1(\overline{\Omega})$ such that*

$$\frac{\partial v}{\partial n} = f \tag{B.2}$$

on $\partial\Omega$, and v is unique up to an additive constant.

If f satisfies (B.1) and is continuous but not necessarily Hölder continuous, there is an elementary argument that yields a weaker version of the conclusion $\partial v/\partial n = f$. See Exercise B.1.

Proof of Theorem B.1. The uniqueness is easy: we may assume $f = 0$ and, via a conformal mapping, we may assume $\partial\Omega$ consists of analytic curves. Each

$\zeta \in \partial \Omega$ has a neighborhood V and conformal map $\varphi : V \to \mathbb{D}$ so that

$$\varphi(V \cap \partial\Omega) = \mathbb{R} \cap \mathbb{D}$$

and

$$\varphi(V \cap \Omega) = \mathbb{D} \cap \{\text{Im} z > 0\}.$$

Then $v \circ \varphi^{-1} = \text{Im} H$ for some function H analytic on $\mathbb{D} \cap \{\text{Im} z > 0\}$. We can take $\text{Re} H = 0$ on $\mathbb{R} \cap \mathbb{D}$, since $(\text{Re} H)_x = (v \circ \varphi^{-1})_y = 0$ on $\mathbb{R} \cap \mathbb{D}$, by Corollary II.4.5 and the Cauchy–Riemann equations. By the Schwarz reflection principle, H extends to be analytic on $\mathbb{D}$, with

$$\text{Im} H(z) = \text{Im} H(\overline{z}). \tag{B.3}$$

Hence v extends to be harmonic on a neighborhood W of $\overline{\Omega}$. But then by (B.3), $\sup_W v(z)$ is attained on $\overline{\Omega}$ and v is constant.

To prove there exists harmonic $v \in C^1(\overline{\Omega})$ satisfying (B.2), first suppose that Ω is simply connected and let $\gamma(s)$ denote arc length parameterization of $\partial\Omega$ with positive orientation. Define

$$F(\gamma(s)) = \int_0^s f(\gamma(t))dt,$$

and let u be the solution to the Dirichlet problem with boundary data F. Then $\frac{\partial u}{\partial s} = f$ on $\partial\Omega$, by Corollary II.4.6, and $u = \text{Re} H$ for some analytic function H, so that by Corollary II.4.5, $v = -\text{Im} H \in C^{1+\varepsilon}(\overline{\Omega})$, for some $\varepsilon > 0$. Then by continuity and the Cauchy–Riemann equations,

$$\frac{\partial v}{\partial n} = \frac{\partial u}{\partial s} = \frac{\partial F}{\partial s} = f$$

on $\partial\Omega$, and (B.2) holds.

When Ω is not simply connected, this definition may not give a single valued function F, and not every harmonic function has a single valued conjugate function. To get around those difficulties, we need two lemmas.

Lemma B.2. *Let Ω be a bounded, finitely connected Jordan domain. Write $\partial\Omega = \bigcup_{j=1}^m \Gamma_j$, where $\Gamma_j \cap \Gamma_k = \emptyset$ for $j \neq k$ and each Γ_j is a Jordan curve. Let $\omega_j(z) = \omega(z, \Gamma_j, \Omega)$. If u is a harmonic function on Ω, then there exist unique $a_1, \ldots, a_{m-1}$ such that*

$$u(z) + \sum_{j=1}^{m-1} a_j \omega_j(z)$$

has a single valued conjugate function on Ω.

Proof. By Lemma II.2.2 we may assume each Γ_j is an analytic curve. If γ is a closed C^1 Jordan curve in Ω, then $\partial\tilde{u}/\partial s$ is a single valued function on γ, where $\tilde{u}$ is a locally defined harmonic conjugate of u, and there is a single valued conjugate function $\tilde{u}$ on the full domain Ω if and only if

$$\int_\gamma \frac{\partial\tilde{u}}{\partial s}ds = 0$$

for all such Jordan curves. If γ_1 and γ_2 are C^1 Jordan curves homologous in Ω, then by Cauchy's theorem,

$$\int_{\gamma_1-\gamma_2} \frac{\partial\tilde{u}}{\partial s}ds = \int_{\gamma_1-\gamma_2} \nabla u \cdot \mathrm{n}\, ds = \mathrm{Im}\int_{\gamma_1-\gamma_2}(u_x - iu_y)dz = 0,$$

since $u_x - iu_y$ is analytic on Ω. Let $\gamma_j \subset \Omega$ be homologous to Γ_j and define the **period** of $\tilde{u}$ over Γ_j by

$$P_j(\tilde{u}) - \int_{\gamma_j} \frac{\partial u}{\partial\mathrm{n}}ds.$$

We have just seen that the $P_j(\tilde{u})$ does not depend on the choice of γ_j. By Green's theorem,

$$\sum_{j=1}^m P_j(\tilde{u}) = 0,$$

so that u has a single valued conjugate if and only if

$$P_j(\tilde{u}) = 0 \quad \text{for } j = 1, \ldots, m-1.$$

By Schwarz reflection each ω_j extends to be harmonic on a neighborhood of $\overline{\Omega}$. Set

$$\alpha_{k,j} = \int_{\Gamma_k} \frac{\partial\omega_j}{\partial\mathrm{n}}ds. \tag{B.4}$$

Then Lemma B.2 is proved if we can find $a_1, \ldots, a_{m-1}$ such that

$$\sum_{j=1}^{m-1} \alpha_{k,j}a_j = -P_k(\tilde{u}), \quad 1 \le k \le m-1,$$

and the next lemma ensures that such $a_1, \ldots, a_{m-1}$ exist. ∎

Lemma B.3. *The matrix $\{\alpha_{k,j}\}_{j,k=1}^{m-1}$ is nonsingular.*

Proof. If $\sum \alpha_{k,j} b_j = 0$ then, as we have seen, there is an analytic function F on Ω such that

$$\operatorname{Re} F = \sum_{j=1}^{m-1} b_j \omega_j.$$

But then $\partial \operatorname{Im} F / \partial n = 0$ on $\partial \Omega$, and by the (already proven) uniqueness part of Theorem B.1, $\operatorname{Im} F$ is constant on Ω. Hence F is constant, and since $\operatorname{Re} F = 0$ on Γ_m, we have $b_j = \operatorname{Re} F|_{\Gamma_j} = 0$, and thus $\{\alpha_{k,j}\}$ is nonsingular. ∎

We return to the proof of the existence of v in Theorem B.1. By Lemma B.3, there exist $b_1, \ldots, b_{m-1}$ so that

$$f_1 = f - \sum_{j=1}^{m-1} b_j \frac{\partial \omega_j}{\partial n}$$

satisfies

$$\int_{\Gamma_k} f_1 ds = 0, \tag{B.5}$$

for $k = 1, \ldots, m - 1$. Since $\int_{\partial \Omega} f ds = 0$ and $\int_{\partial \Omega} \partial \omega_j / \partial n = 0$, the equality (B.5) also holds when $k = m$. On each Γ_j define

$$F(\zeta_j(s)) = \int_0^s f_1(\zeta_j(t)) dt,$$

where $\zeta_j(s)$ is the arc length parameterization of Γ_j with positive orientation. Then by (B.5), F is well defined on $\partial \Omega$ and $F \in C^{1+\varepsilon}(\partial \Omega)$ for some $\varepsilon > 0$. By Corollary II.4.6 there exists $u_1 \in C^{1+\varepsilon}(\overline{\Omega})$ with $\partial u_1 / \partial s = f_1$. Find $a_1, \ldots, a_{m-1}$, by Lemma B.2, so that

$$u_2 = u_1 + \sum_{j=1}^{m-1} a_j \omega_j = \operatorname{Re} H$$

for some function H analytic in Ω. Let $v_2 = -\operatorname{Im} H$. If $\zeta \in \partial \Omega$, there is a simply connected domain $N \subset \Omega$ bounded by a $C^{1+\alpha}$ curve, so that $\partial N \cap \partial \Omega$ is a neighborhood of ζ in $\partial \Omega$. By Corollary II.4.5, $v_2 \in C^{1+\varepsilon}(\overline{\Omega})$ for some $\varepsilon > 0$. Again continuity and the Cauchy–Riemann equations yield

$$\frac{\partial v_2}{\partial n} = \frac{\partial u_2}{\partial s} = \frac{\partial u_1}{\partial s} = \frac{\partial F}{\partial s} = f_1$$

Now let

$$v = v_2 + \sum_{j=1}^{m-1} b_j \omega_j.$$

Then

$$\frac{\partial v}{\partial n} = \frac{\partial v_2}{\partial n} + \sum_{j=1}^{m-1} b_j \frac{\partial \omega_j}{\partial n} = f_1 + \sum_{j=1}^{m-1} b_j \frac{\partial \omega_j}{\partial n} = f,$$

and Theorem B.1 is proved. ∎

The solutions to the Neumann and Dirichlet problems combine to solve mixed boundary value problems.

Theorem B.4. *Let Ω be a finitely connected Jordan domain with $\partial\Omega \in C^{1+\alpha}$ for some $\alpha > 0$. Let $f_1 \in C(\partial\Omega)$, let $\beta > 0$, and let $f_2 \in C^\beta(\partial\Omega)$ be such that $\int_{\partial\Omega} f_2 ds = 0$. Let J be a finite union of open arcs in $\partial\Omega$ with pairwise disjoint closures. Then there is a harmonic function $u \in C^1(\Omega \cup J) \cap C(\overline{\Omega})$ such that*

$$u = f_1 \quad \text{on } \partial\Omega \setminus J, \tag{B.6}$$

and

$$\frac{\partial u}{\partial n} = f_2 \quad \text{on } J. \tag{B.7}$$

If $\partial\Omega \setminus J \neq \emptyset$, the function u is unique.

Proof.

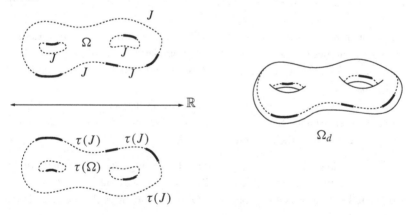

Figure B.1. The doubled Riemann surface Ω_d.

Form the **doubled Riemann surface** Ω_d by attaching two copies of Ω along the set J. In other words, assume $\Omega \subset \{\text{Im} z > 0\}$, let $\tau(z) = \overline{z}$ map Ω to the reflected domain $\tau(\Omega)$ and set $\Omega_d = \left(\Omega \cup J \cup \tau(\Omega \cup J)\right) / \sim$, where the union denotes a disjoint union and where $z \sim w$ if and only if $w = \tau(z) = \overline{z}$ and $z \in J$. The sets Ω and $\tau(\Omega)$ are declared open sets in Ω_d. To define a basic neighborhood of $p \in J$, let W be a simply connected neighborhood of

p in $\overline{\Omega}$ such that $W \cap \partial\Omega \subset J$ and let $\varphi : W \cap \Omega \to \mathbb{D}_+ = \mathbb{D} \cap \{\operatorname{Im} z > 0\}$ be a conformal map so that the continuous extension of φ to $W \cap \partial\Omega$ satisfies $\varphi(W \cap \partial\Omega) = (-1, 1)$. The companion neighborhood $\tau(W \cap \Omega)$ in $\tau(\Omega)$ is identified with the reflection $\mathbb{D}_-$ of $\mathbb{D}_+$ through $(-1, 1)$ so that $\tau(z)$ corresponds to $\overline{\varphi(z)}$. The full disc $V = \left(W \cup \tau(W)\right)/\sim$ then defines a neighborhood of p in Ω_d, with local coordinate

$$\Phi(z) = \begin{cases} \varphi(z), & z \in W, \\ \overline{\varphi(\overline{z})}, & z \in \tau(W). \end{cases}$$

Thus, f is analytic on V if and only if $f \circ \Phi^{-1}$ is analytic on $\mathbb{D}$. For example, if Ω is the unit disc $\mathbb{D}$, then Ω_d is conformally equivalent to $\mathbb{C}^* \setminus (\partial\mathbb{D} \setminus J)$. Notice that if v is harmonic on V and if $v(\tau(z)) = v(z)$, then $\partial v \circ \varphi^{-1}/\partial y = 0$ on $(-1, 1)$, and since $\partial\Omega \in C^{1+\alpha}$, it then follows from Theorem II.4.3 that $\partial v/\partial n = 0$ on $J \cap \partial W$, where the derivative is now taken with respect to the original coordinates on W.

We may assume $\partial\Omega$ consists of analytic curves. Choose a domain $\Omega_1 \subset \Omega$ with $\partial\Omega \setminus \partial\Omega_1 = J$, such that $\Omega_d \setminus \tau(\Omega_1)$ is (conformally equivalent to) a domain $\Omega_2 \supset \Omega$ with $\Omega_2 \cap \partial\Omega = J$. On Ω_d we can apply the Schwarz alternating method from the proof of Theorem II.1.1, with σ and σ' augmented by $\partial\Omega_1 \cap \Omega$ and $\partial\Omega_2 \setminus \partial\Omega$, respectively, and therefore the Dirichlet problem can be solved on Ω_d.

By Theorem B.1, there is a $u_1 \in C^1(\overline{\Omega})$ with $\partial u_1/\partial n = f_2$ on $\partial\Omega$. Let u_2 solve the Dirichlet problem on Ω_d for the boundary values $f_1 - u_1$ on $\partial\Omega_d$. Then, by the remarks above, $\partial u_2/\partial n = 0$ on J, and therefore $u_1 + u_2$ satisfies conditions (B.6) and (B.7).

To prove uniqueness when $\partial\Omega \setminus J \neq \emptyset$, we can suppose $f_1 = f_2 = 0$. Then if u satisfies (B.6) and (B.7), u extends to be harmonic on Ω_d with $u = 0$ on $\partial\Omega_d$. Hence $u = 0$ by the maximum principle. ∎

We do not know the origin of the idea of reducing mixed boundary value problems on Ω to Dirichlet problems on Riemann surface doubles of Ω. Ohtsuka [1970] credited conversations between Ohtsuka and Strebel. See also Tsuji [1959], p.31.

Exercise B.1. (a) Let f be real valued and continuous on $\partial\mathbb{D}$, and assume $\int f d\theta = 0$. Then there is $v(z) \in C^1(\overline{\mathbb{D}})$, harmonic on $\mathbb{D}$, with $\partial v/\partial r = f$ on $\partial\mathbb{D}$ if and only if $\widetilde{f}$ is continuous on $\partial\mathbb{D}$.

(b) If $f \in C(\partial\mathbb{D})$ is real valued and if $\int_0^{2\pi} f(e^{it}) dt = 0$, set

$$H(z) = \int_0^{2\pi} \frac{e^{i\theta} + z}{e^{i\theta} - z} \int_0^\theta f(e^{it}) dt \frac{d\theta}{2\pi}.$$

Then

$$H(z) = -\int_0^{2\pi} \left[\frac{2}{i} \log \left(e^{i\theta} - z \right) - \theta \right] f(e^{i\theta}) \frac{d\theta}{2\pi},$$

and

$$\frac{\partial H}{\partial r}(z) = -\frac{i}{r} \int_0^{2\pi} \left(\frac{e^{i\theta} + z}{e^{i\theta} - z} - 1 \right) f(e^{i\theta}) \frac{d\theta}{2\pi}.$$

Thus $v = -\operatorname{Im} H \in C(\overline{\mathbb{D}})$ and $\partial v / \partial r = f$ on $\partial \mathbb{D}$. This gives an alternate proof of Theorem B.1 when $\Omega = \mathbb{D}$ without the assumption that $f \in C^\beta$. However, it does not prove $v \in C^1$ and in fact v may not be C^1 in this situation.

Exercise B.2. Use Green's function to find the kernel $N(z, \zeta)$, $z \in \Omega$, $\zeta \in \partial \Omega$, such that under the conditions of Theorem B.1,

$$v(z) = \int_{\partial \Omega} N(z, \zeta) f(\zeta) ds.$$

Exercise B.3. Prove that the matrix $\{\alpha_{k,j}\}_{j,k=1}^{m-1}$ defined by (B.4) is symmetric, $\alpha_{k,j} = \alpha_{j,k}$.

C. The Dirichlet Principle

A real valued function on $\partial \mathbb{D}$ is **piecewise continuous** if f is continuous except on a finite set E but f has finite left and right limits at each point of E. A real valued function $u(z)$ is **piecewise smooth** on a domain Ω if $u(z)$ is continuous on Ω and if $u(z)$ is continuously differentiable on $\Omega \setminus J$, where $J = J(u)$ is a finite union of (closed or open) analytic arcs in Ω. The **Dirichlet integral** of u or the **squared Dirichlet norm** of u is

$$D(u) = D_\Omega(u) = \iint_\Omega |\nabla u|^2 dx dy = \iint_{\Omega \setminus J} |\nabla u|^2 dx dy.$$

If $u = \operatorname{Re} F$ and F is analytic on Ω, then

$$D(u) = \iint_\Omega |F'|^2 dx dy = \operatorname{Area} F(\Omega),$$

where each value $F(z)$ is counted according to its multiplicity. If $D(u) < \infty$ and $D(v) < \infty$, the **Dirichlet inner product** is

$$D(u, v) = \iint_\Omega \nabla u \cdot \nabla v \, dx dy = \iint_\Omega (u_x v_x + u_y v_y) dx dy.$$

It satisfies the Cauchy–Schwarz inequality,

$$D(u, v) \le \sqrt{D(u)}\sqrt{D(v)},$$

and

$$D(u \pm v) = D(u) + D(v) \pm 2D(u, v). \tag{C.1}$$

Given $f \in C(\partial\Omega)$, set

$$\mathcal{D}_f = \big\{v : v \text{ is piecewise smooth in } \Omega \text{ and } \lim_{z \to \zeta} v(z) = f(\zeta), \text{ for all } \zeta \in \partial\Omega\big\}.$$

The **Dirichlet principle** is the assertion that the extremal problem

$$D(u) = \inf\{D(v) : v \in \mathcal{D}_f\}, \tag{C.2}$$

has *unique* solution $u \in \mathcal{D}_f$, and that $u = u_f$ is the solution of the Dirichlet problem on Ω for the boundary value f. Unfortunately, this formulation of the Dirichlet principle is not true. Here is a counterexample: If $f \in C(\partial\mathbb{D})$ has Fourier series

$$f(e^{i\theta}) \sim \sum_{-\infty}^{+\infty} a_n e^{in\theta}, \tag{C.3}$$

then its Poisson integral is

$$u(re^{i\theta}) = \sum_{-\infty}^{+\infty} a_n r^{|n|} e^{in\theta}, \qquad 0 \le r < 1,$$

and

$$D(u) = 2\pi \sum_{-\infty}^{+\infty} |n||a_n|^2. \tag{C.4}$$

Hence there exist $f \in C(\partial\mathbb{D})$ such that $D(u) = \infty$. For example,

$$f(e^{i\theta}) = \sum_{k=1}^{\infty} \frac{1}{k^2} \cos(2^k \theta)$$

is in $C(\partial\mathbb{D})$, but

$$D(u) = \pi \sum_{k=1}^{\infty} \frac{2^k}{k^4} = \infty.$$

However this example is the only thing wrong with Dirichlet principle. We shall see that (C.2) holds whenever

$$\mathcal{D}_f \cap \{v : D(v) < \infty\} \ne \emptyset.$$

Lemma C.1. *Let f be piecewise continuous on $\partial\mathbb{D}$ and let u be the Poisson integral of f. Let $E \subset \partial\mathbb{D}$ be the finite set of discontinuities of f, and set*

$$\mathcal{D}_f = \{v : v \ bounded \ and \ piecewise \ smooth \ on \ \mathbb{D}, \lim_{z \to \zeta} v(z)$$

$$= f(\zeta), \ for \ all \ \zeta \in \partial\mathbb{D} \setminus E\}.$$

If $\mathcal{D}_f \cap \{v : D(v) < \infty\} \neq \emptyset$, then $D(u) < \infty$, and u is the unique element of $\mathcal{D}_f$ of minimal Dirichlet norm. That is,

$$D(u) = \inf\{D(v) : v \in \mathcal{D}_f\},$$

and if $v \in \mathcal{D}_f$ satisfies $D(v) = D(u)$, then $v = u$.

Proof. Let f have Fourier series (C.3) and set

$$u_N(re^{i\theta}) = \sum_{-N}^{N} a_n r^{|n|} e^{in\theta}.$$

Then $D(u_N) < \infty$. If $v \in \mathcal{D}_f$ and $D(v) < \infty$, then $D(v - u_N) < \infty$, and by Green's theorem and dominated convergence,

$$D(u_N, v - u_N) = \lim_{r \to 1} \int_0^{2\pi} (v - u_N)(re^{i\theta}) \frac{\partial u_N}{\partial r}(re^{i\theta}) d\theta$$

$$= \int_0^{2\pi} (v - u_N)(e^{i\theta}) \frac{\partial u_N}{\partial r}(e^{i\theta}) d\theta.$$

But on $\partial\mathbb{D}$, $v - u_N = f - u_N$ has Fourier series $\sum_{|n| \geq N+1} a_n e^{in\theta}$, so that by the Parseval identity,

$$D(u_N, v - u_N) = 0.$$

Consequently

$$D(v) = D(u_N) + D(v - u_N) \tag{C.5}$$

by (C.1), and

$$D(u_N) \leq D(v).$$

Since $|\nabla u_N| \to |\nabla u|$ on $\mathbb{D}$, Fatou's lemma gives

$$D(u) \leq \liminf D(u_N) \leq D(v)$$

and we conclude both that $D(u) < \infty$ and that u has minimal Dirichlet norm in $\mathcal{D}_f$.

To prove that u is the unique element of $\mathcal{D}_f$ of minimal Dirichlet norm, note that when $D(u) < \infty$, (C.4) implies that $D(u - u_N) \to 0$, and consequently that $D(v - u_N) \to D(v - u)$. Hence by (C.5) we have

$$D(v) = D(u) + D(v - u) \qquad (C.6)$$

whenever $v \in \mathcal{D}_f$ and $D(v) < \infty$. Therefore if $v \in \mathcal{D}_f$ and $D(v) = D(u)$, then $D(v - u) = 0$ and $v - u$ is constant on Ω. But then since $v(z) - u(z) \to 0$ when $z \to \zeta \in \partial\mathbb{D} \setminus E$, that means $v = u$. ∎

Because the Dirichlet integral is a conformal invariant, Lemma C.1 also holds with any simply connected Jordan domain in place of $\mathbb{D}$.

Lemma C.2. *Let I be an open arc on $\partial\mathbb{D}$, and let f be a bounded continuous function on I. Set*

$$\mathcal{C}_f = \{v : v \text{ is piecewise smooth on } \mathbb{D} \text{ and } \lim_{z \to \zeta} v(z) = f(\zeta), \text{ for all } \zeta \in I\}.$$

If

$$\mathcal{C}_f \cap \{v : D(v) < \infty\} \neq \emptyset,$$

then $\mathcal{C}_f$ contains a unique element $u(z)$ of minimal Dirichlet norm and $u(z)$ solves the mixed boundary value problem:

$$\begin{cases} \Delta u = 0, & \text{on } \Omega, \\ u = f, & \text{on } I, \\ \dfrac{\partial u}{\partial r} = 0, & \text{on } \partial\mathbb{D} \setminus \overline{I}. \end{cases} \qquad (C.7)$$

The set $\partial\mathbb{D} \setminus \overline{I}$ is called a **free boundary** because nothing is required of a function in $\mathcal{C}_f$ on $\partial\mathbb{D} \setminus \overline{I}$.

Proof. Conformally map $\mathbb{D}$ to $\mathbb{D}^+ = \mathbb{D} \cap \{\text{Im } z > 0\}$ so that $\partial\mathbb{D} \setminus I$ is mapped to $[-1, 1] \subset \partial\mathbb{D}^+$. Because Dirichlet integrals are conformally invariant and because the map is conformal across $\partial\mathbb{D} \setminus \overline{I}$, this does not change the extremal problem $\inf\{D(v) : v \in \mathcal{C}_f\}$ nor the condition (C.7) expected of its solution. Continue to write $f(\zeta)$ for the boundary data on $\partial\mathbb{D}^+ \setminus [-1, 1]$. Let $U(z)$ be the Poisson integral on $\mathbb{D}$ of

$$g(\zeta) = \begin{cases} f(\zeta), & \text{Im } \zeta \geq 0, \\ f(\overline{\zeta}), & \text{Im } \zeta < 0. \end{cases}$$

Then U satisfies (C.7) on $\mathbb{D}^+$ and $D(U) = \inf\{D(V) : V \in \mathcal{D}_g\}$ by Lemma C.1. If $v \in \mathcal{C}_f$ were continuous up to $(-1, 1)$, its reflection

$$V(z) = \begin{cases} v(z), & \text{Im } z \geq 0, \\ v(\overline{z}), & \text{Im } z < 0, \end{cases}$$

would be in $\mathcal{D}_g$ and the lemma would follow from Lemma C.1. However, because v may not be continuous up to $(-1, 1)$ we must first make some minor adjustments.

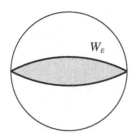

Figure C.1

For $\varepsilon > 0$, set $\gamma_\varepsilon = \mathbb{D}^+ \cap \{z : |z + \frac{i}{\varepsilon}| = \sqrt{1 + \frac{1}{\varepsilon^2}}\}$, and

$$W_\varepsilon = \left\{z : \left|z + \frac{i}{\varepsilon}\right| < \sqrt{1 + \frac{1}{\varepsilon^2}}\right\} \cap \left\{z : \left|z - \frac{i}{\varepsilon}\right| < \sqrt{1 + \frac{1}{\varepsilon^2}}\right\},$$

and for $z \in W_\varepsilon$, let $z^* \in \gamma_\varepsilon$ satisfy $\mathrm{Re}z^* = \mathrm{Re}z$. For small $\delta > 0$,

$$\int_0^\delta \int_{\gamma_\varepsilon} |\nabla v|^2 ds d\varepsilon \leq C \iint_{W_\delta \cap \mathbb{D}^+} |\nabla v|^2 dxdy,$$

and there is ε, $0 < \varepsilon < \delta$, so that $\varepsilon \int_{\gamma_\varepsilon} |\nabla v|^2 ds \leq \delta$. Set

$$V_\delta(z) = \begin{cases} v(z), & z \in \mathbb{D}^+ \setminus W_\varepsilon, \\ v(\bar{z}), & \bar{z} \in \mathbb{D}^+ \setminus W_\varepsilon; \\ v(z^*), & z \in W_\varepsilon. \end{cases}$$

Then $V_\delta \in \mathcal{D}_g$, so that

$$D(U) \leq D(V_\delta).$$

But

$$D_{\mathbb{D}^+}(v) = \frac{1}{2} \lim_{\delta \to 0} D_{\mathbb{D}}(V_\delta),$$

and hence $D_{\mathbb{D}^+}(U) \leq D_{\mathbb{D}^+}(v)$.

The proof of uniqueness is the same as for Lemma C.1. If $D(v) = D_{\mathbb{D}^+}(U)$, then $v - U$ is constant on $\mathbb{D}^+$, and because $\lim_{z \to \zeta}(v(z) - U(z)) = 0$ for all $\zeta \in \partial\mathbb{D} \cap \{\mathrm{Im}z > 0\}$, then $v = U$. ∎

Theorem C.3. *Let Ω be a finitely connected Jordan domain, let E be a non-empty finite union of open arcs on $\partial\Omega$ and let $f \in C(\overline{E})$. Set*

$$\mathcal{C}_f = \big\{ v : v \text{ is piecewise smooth on } \Omega, \ \lim_{z \to \zeta} v(z) = f(\zeta), \text{ for all } \zeta \in E \big\},$$

and assume

$$\mathcal{C}_f \cap \{ v : D(v) < \infty \} \neq \emptyset.$$

Then the extremal problem

$$D(u) = \inf\{ D(v) : v \in \mathcal{C}_f \} \tag{C.8}$$

has unique solution u, and u satisfies

$$\Delta u = 0 \text{ on } \Omega,$$

and

$$u = f \text{ on } \overline{E}.$$

Moreover, if for some $\alpha > 0$, $\partial\Omega \setminus E$ consists of $C^{1+\alpha}$ arcs, then

$$\frac{\partial u}{\partial n} = 0 \quad \text{on } \partial\Omega \setminus \overline{E}. \tag{C.9}$$

Proof. After a conformal mapping, we may assume $\partial\Omega$ consists of analytic Jordan curves. By Theorem II.4.3 this assumption does not change (C.9).

Let $\{u_n\} \subset \mathcal{C}_f$ be a minimizing sequence for the extremal problem (C.8), so that

$$D(u_n) \to \inf\{ D(v) : v \in \mathcal{C}_f \}.$$

Then by (C.1),

$$D\Big(\frac{u_n - u_m}{2}\Big) + D\Big(\frac{u_n + u_m}{2}\Big) = \frac{D(u_n) + D(u_m)}{2},$$

and $D(u_n - u_m) \to 0$ because $\mathcal{C}_f$ is a convex set. Hence there is a vector $(\alpha, \beta) \in L^2(\Omega)$ such that

$$\iint_\Omega \big\{ |(u_n)_x - \alpha|^2 + |(u_n)_y - \beta|^2 \big\} dxdy \to 0. \tag{C.10}$$

Moreover, (α, β) is independent of the minimizing sequence $\{u_n\}$. Fix an open disc D such that $\overline{D} \subset \Omega$, and let U_n be the solution of the Dirichlet problem on D with boundary value u_n, and replace $u_n \in \mathcal{C}_f$ by

$$U_n \chi_D + u_n \chi_{\Omega \setminus D} \in \mathcal{C}_f.$$

Then the new sequence $\{u_n\}$ is still a minimizing sequence for (C.8). The familiar estimate for harmonic functions

$$\sup_{\{|z-z_0|<r/2\}} |U(z)| \le \left(\frac{4}{\pi r^2}\right)^{\frac{1}{2}} \left(\iint_{\{|z-z_0|<r\}} |U(z)|^2 dxdy \right)^{\frac{1}{2}},$$

which follows from the Cauchy–Schwarz inequality and the mean value property, now shows that both sequences $\{(u_n)_x\}$ and $\{(u_n)_y\}$ converge to harmonic functions, uniformly on compact subset of D. Since D is any disc in Ω, we conclude that α and β are (after alterations on a set of measure zero) harmonic functions on Ω and

$$\frac{\partial \alpha}{\partial y} = \lim_n \frac{\partial^2 u_n}{\partial x \partial y} = \frac{\partial \beta}{\partial x}.$$

We claim there exists a harmonic function u such that

$$\nabla u = \left(\frac{\partial u}{\partial x}, \frac{\partial u}{\partial y}\right) - (\alpha, \beta). \qquad (C.11)$$

Now (C.11) holds if

$$\int_\gamma \alpha dx + \beta dy = 0$$

for every smooth simple closed curve $\gamma \subset \Omega$. Fix such a curve γ and take a simply connected domain W such that ∂W is piecewise analytic, such that $\overline{W} \subset \Omega$, and such that $\gamma \setminus W$ is a single point p. See Figure C.2.

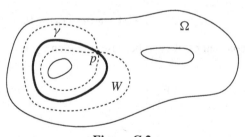

Figure C.2

As before, we can suppose the minimizing functions u_n are harmonic on a neighborhood of p. In W let U_n be the solution of the Dirichlet problem for the boundary value u_n. Then by Corollary II.4.6, ∇U_n is bounded on $\gamma \cap W$. Set $v_n = U_n \chi_W + u_n \chi_{\Omega \setminus W}$. Then v_n is also a minimizing sequence, and by dominated convergence

$$\int_\gamma \alpha dx + \beta dy = \lim_n \int_\gamma \frac{\partial v_n}{\partial x} dx + \frac{\partial v_n}{\partial y} dy = \lim_n \big(u_n(p) - u_n(p)\big) = 0.$$

Therefore there exists a harmonic function $u(z)$ satisfying (C.11). Because $E \neq \emptyset$ and (C.10) holds, an integration shows $\lim u_n(z)$ exists for all $z \in \Omega$. Then since $\nabla u_n \to \nabla u$ on Ω, we may assume $\lim u_n(z) = u(z)$ for all $z \in \Omega$.

We now claim that $u \in C_f$ and that u satisfies (C.9). Let $\zeta \in E$. There is $\delta > 0$ such that

$$W = \Omega \cap B(\zeta, \delta)$$

is a Jordan domain whose boundary consists of one arc $\gamma_1 = \overline{B(\zeta, \delta)} \cap \partial\Omega \subset E$ and a second arc $\gamma_2 = \Omega \cap \partial B(\zeta, \delta)$. On W let v_n be the solution of the Dirichlet problem with boundary value $u_n \in C(\partial W)$, and let v be the solution with boundary value $f\chi_{\gamma_1} + u\chi_{\gamma_2}$. Then $v_n \to v$ and $\nabla v_n \to \nabla v$ on W. By Lemma C.1, $D_W(v_n) \leq D_W(u_n)$, and in fact by (C.6)

$$D_W(u_n) - D_W(v_n) = D_W(u_n - v_n).$$

But

$$D_W(v_n) + D_\Omega(u_n) - D_W(u_n) \geq \beta = \lim_n D_\Omega(u_n),$$

because $v_n\chi_W + u_n\chi_{\Omega\setminus W} \in C_f$, and therefore $D_W(u_n - v_n) \to 0$. With (C.10) that gives

$$D_W(u - v_n) \to 0,$$

so that again by Fatou's lemma,

$$D_W(u - v) \leq \liminf_n D_W(u - v_n) = 0.$$

Hence $u - v$ is constant on W, and because $v = u$ on γ_2, $u = v$ in W. Consequently

$$\lim_{z \to \zeta} u(z) = f(\zeta),$$

and $u \in C_f$.

To verify (C.9) let γ_1 be a component of $\partial\Omega \setminus \overline{E}$ and let $\gamma_2 \subset \Omega$ be a Jordan arc connecting the endpoints of γ_1 so that $\gamma_1 \cup \gamma_2$ bounds a Jordan domain $W \subset \Omega$. Because $u \in C_f$ has minimal Dirichlet norm and functions in C_f are unrestricted on γ_1,

$$D_W(u) = \inf\{D_W(v) : v \text{ piecewise smooth on } W, v = u \text{ on } \gamma_2\}$$

and Lemma (C.2) now implies $\partial u/\partial n = 0$ on γ_1. ∎

Corollary C.4. *Let Ω be a finitely connected Jordan domain, and let E and F be finite unions of open arcs on $\partial\Omega$ such that $\overline{E} \cap \overline{F} = \emptyset$. Then the extremal problem*

$$\inf\Big\{ D_\Omega(v) : v \text{ piecewise smooth on } \Omega,$$
$$\limsup_{z\to\zeta\in E} v(z) \le 0, \ \liminf_{z\to\zeta\in F} v(z) \ge 1 \Big\} \tag{C.12}$$

has a unique solution, which satisfies

$$\Delta u(z) = 0 \quad \text{on } \Omega, \tag{C.13}$$

and

$$u = 0 \quad \text{on } E \quad \text{and} \quad u = 1 \quad \text{on } F. \tag{C.14}$$

If $\partial\Omega$ consists of $C^{1+\alpha}$ curves, then

$$\frac{\partial u}{\partial n} = 0 \quad \text{on } \partial\Omega \setminus \overline{(E \cup F)}. \tag{C.15}$$

Proof. This is a corollary of Theorem B.4 and of Theorem C.3. We will give both proofs; a third proof will be given in Appendix H. By conformal invariance, $\partial\Omega$ can be assumed to consist of analytic Jordan curves.

To derive the corollary from Theorem C.3, let $\varepsilon > 0$ and take $h \in C^1(\mathbb{R})$ so that $h(x) = x$ on $[0, 1]$, and so that $0 \le h' \le 1$ and $-\varepsilon < h < 1 + \varepsilon$ on $\mathbb{R}$. When v is in the class (C.12), so also is $h \circ v$ and $D(h \circ v) \le D(v)$. Therefore the extremum (C.12) is the same for the subclass in which $\lim_{z\to\zeta\in E} v(z) = 0$ and $\lim_{z\to\zeta\in F} v(z) = 1$, and by Theorem C.3 the unique extremal for the subclass is the function u satisfying (C.13) and (C.14).

To obtain the corollary from Theorem B.4 we follow Courant's book [1950]. Let $u(z)$ be the solution given by Theorem B.4 of the mixed boundary value problem (C.13), (C.14), and (C.15). Then $u(z)$ is a harmonic measure on a Riemann surface double of Ω, so that $\partial u/\partial n$ is real analytic on the open set $E \cup F \subset \partial\Omega$.

At an endpoint ζ of $E \cup F$, $|\nabla u| = O(|z - \zeta|^{-1/2})$. Indeed, assume that $\zeta \in \partial E$ and that there is a neighborhood W of ζ and a conformal map ψ of W onto D with $\psi(\zeta) = 0$, $\psi(E \cap W) = [0, 1)$, and $\psi(W \cap \Omega) = \mathbb{D}^+$ or $\mathbb{D}^-$. The function $U(w) = u \circ \psi^{-1}(w^2)$ can be reflected from a quarter disc first across the imaginary axis and then the real axis, to yield a function harmonic in $\mathbb{D} \setminus \{0\}$. Because u is bounded, the singularity at $w = 0$ is removable, and $U(w) = \operatorname{Im} F(w)$ with F analytic on $\mathbb{D}$ and $F(0) = 0$. Because $u > 0$ on Ω, F has only a simple zero and $F(w) = aw + O(|w|^2)$, with $a > 0$. Thus

$$u(z) = \operatorname{Im} F(\sqrt{\psi(z)}) = a(\operatorname{Im}\sqrt{\psi(z)}) + O(|z - \zeta|),$$

and $|\nabla u(z)| = O(|\psi(z)|^{-1/2})$.

Hence for any v in the class (C.12) and any sequence $\Omega_n \uparrow \Omega$ of domains with smooth boundaries such that $\overline{\Omega_n} \subset \Omega$, we have

$$D_\Omega(v - u, u) = \lim_{n \to \infty} D_{\Omega_n}(v - u, u) = \lim_{n \to \infty} \int_{\partial\Omega_n} (v - u)\frac{\partial u}{\partial n} ds$$

$$= \int_{E \cup F} (v - u)\frac{\partial u}{\partial n} ds.$$

Because u is a harmonic measure we have $0 \le u(z) \le 1$ by Exercise H.5(a), and hence $\partial u/\partial n \le 0$ on E and $\partial u/\partial n \ge 0$ on F. Therefore $D_\Omega(v - u, u) \ge 0$, and

$$D_\Omega(v) = D_\Omega(u) + D_\Omega(v - u) + 2D_\Omega(v - u, u) \ge D_\Omega(u),$$

with equality holding if and only if $v = u$. ∎

D. Hausdorff Measure

The **two-dimensional Hausdorff measure** of a set $E \subset \mathbb{C}$ is defined to be $\Lambda_2(E) = \frac{1}{\pi}\text{Area}(E)$, where

$$\text{Area}(E) = \inf\left\{\sum \pi r_j^2 : E \subset \bigcup B(z_j, r_j)\right\} \tag{D.1}$$

is the area of E. To define other Hausdorff measures we replace πr^2 in (D.1) by other functions of r. A **measure function** is a continuous and increasing function $h(r)$ on $[0, \infty)$. With respect to the measure function h, the **Hausdorff content** of a set E is

$$M_h(E) = \inf\left\{\sum h(r_j) : E \subset \bigcup B(z_j, r_j)\right\}. \tag{D.2}$$

The Hausdorff content M_h assigns finite mass to every bounded set, but the content M_h is usually not finitely additive. For example if $h(r) = 2r$, then $M_h([0, 1]) = M_h([0, 1] + i\varepsilon) = 1$, while

$$M_h\big([0, 1] \cup ([0, 1] + i\varepsilon)\big) = \sqrt{1 + \varepsilon^2}.$$

To obtain a measure, we change the definition (M.2) by requiring that the cover of E be comprised of small balls only. Define

$$\Lambda_h^\delta(E) = \inf\left\{\sum h(r_j) : E \subset \bigcup B(z_j, r_j) \text{ and } r_j \le \delta\right\}.$$

If $\delta_2 < \delta_1$, then $\Lambda_h^{\delta_2} \ge \Lambda_h^{\delta_1}$ because $\Lambda_h^{\delta_2}$ entails fewer covers. Consequently the limit

$$\Lambda_h(E) = \lim_{\delta \to 0} \Lambda_h^\delta(E)$$

exists in $[0, \infty]$, and $\Lambda_h(E)$ is the definition of the **h-Hausdorff measure** of E. When $h(r) = r^\alpha$, we call $\Lambda_h(E)$ the **α-dimensional Hausdorff measure** of E and write $\Lambda_h(E) = \Lambda_\alpha(E)$. For example, $\Lambda_2(E) = \frac{1}{\pi}\text{Area}(E)$ as was noted above, $\Lambda_0(E)$ is the number of points in E, and if E lies on a rectifiable curve, then $\Lambda_1(E)$ is half the arc length of E. For any measure function h, Λ_h is a metric outer measure in the sense of Carathéodory:

$$\Lambda_h(A \cup B) = \Lambda_h(A) + \Lambda_h(B), \quad \text{if } \text{dist}(A, B) > 0,$$

and therefore Λ_h is a countably additive measure for which all Borel sets are measurable. See Wheeden and Zygmund [1977] or Folland [1984].

Note that if $\alpha < \beta$, then

$$\delta^{\beta - \alpha} \Lambda_\alpha^\delta(E) \geq \Lambda_\beta^\delta(E),$$

so that

$$\Lambda_\alpha(E) < \infty \implies \Lambda_\beta(E) = 0. \tag{D.3}$$

The **Hausdorff dimension** of E is

$$\dim(E) = \inf\{\alpha : \Lambda_\alpha(E) = 0\}.$$

By (D.3) we have

$$\Lambda_\alpha(E) = \begin{cases} 0, & \text{if } \alpha > \dim(E), \\ \infty, & \text{if } \alpha < \dim(E). \end{cases}$$

Exercise D.1 shows that for $E \subset \mathbb{C}$, $\dim(E)$ can be any number in $[0, 2]$.

Hausdorff content can be used to find $\dim(E)$. Clearly $M_h(E) \leq \Lambda_h(E)$. On the other hand, if $M_h(E) = 0$, then for any $\varepsilon > 0$ there is a cover $\{B(z_j, r_j)\}$ of E such that $\sum h(r_j) < \varepsilon$, and $r_j \leq \inf\{t : h(t) \geq \varepsilon\} \to 0$ as $\varepsilon \to 0$ because h is increasing. It follows that $\Lambda_h(E) = 0$. Thus for any measure function h,

$$M_h(E) = 0 \quad \text{if and only if} \quad \Lambda_h(E) = 0, \tag{D.4}$$

and

$$\dim(E) = \inf\{\alpha : M_{r^\alpha}(E) = 0\}. \tag{D.5}$$

There is a dyadic version of M_h that is easier to work with. For integers j, k, and n, form the **dyadic square**

$$Q = \left[\frac{j}{2^n}, \frac{j+1}{2^n}\right) \times \left[\frac{k}{2^n}, \frac{k+1}{2^n}\right).$$

It has side length $\ell(Q) = 2^{-n}$ and lower left corner $(j2^{-n}, k2^{-n})$. For each n, the dyadic squares Q with $\ell(Q) = 2^{-n}$ are a disjoint paving of the plane. Every dyadic square Q is contained in a unique dyadic square $\widetilde{Q}$ with $\ell(\widetilde{Q}) = 2\ell(Q)$. Therefore the dyadic squares have this important property:
If Q_1 and Q_2 are two dyadic squares, then

$$\begin{cases} Q_1 \subset Q_2, \\ Q_2 \subset Q_1, \quad \text{or} \\ Q_1 \cap Q_2 = \emptyset. \end{cases}$$

Now define

$$m_h(E) = \inf\left\{\sum h(\ell(Q_p)) : E \subset \bigcup Q_p, \text{ and each } Q_p \text{ is a dyadic square}\right\}.$$

Every dyadic square of side r is contained in a ball of radius r while each ball of radius r is contained in 25 dyadic squares Q with $r/2 < \ell(Q) \le r$. Therefore

$$M_h(E) \le m_h(E) \le 25 M_h(E), \tag{D.6}$$

and (D.4) and (D.5) also hold with m_h in place of M_h.

Using m_h yields an easy proof that every $E \subset \mathbb{C}$ has Hausdorff dimension at most two. Partition the unit square Q into 4^n smaller squares Q_j such that $\ell(Q_j) = 2^{-n}$. Then if $\alpha > 2$,

$$\lim_n \sum_j (2^{-n})^\alpha = \lim_n 2^{n(2-\alpha)} = 0,$$

and $m_{r^\alpha}(Q) = 0$. Thus $\Lambda_\alpha(Q) = 0$ and since Λ_α is countably additive, $\Lambda_\alpha(\mathbb{C}) = 0$. This shows that $\dim(E) \le 2$ for all $E \subset \mathbb{C}$.

The following theorem is useful for computing dimensions and estimating capacities.

Theorem D.1 (Frostman). *Let h be a measure function. If μ is a positive measure such that*

$$\mu(B(z, r)) \le h(r) \tag{D.7}$$

for all z and r, then

$$\mu(E) \le \Lambda_h(E)$$

for every Borel set E. *Conversely, if E is a compact set, and if $M_h(E) > 0$, then there exists a positive measure μ, supported on E, such that (D.7) holds for all z and r and such that*

$$\mu(E) \geq \frac{M_h(E)}{225}. \tag{D.8}$$

Proof. Suppose $\mu(B(z,r)) \leq h(r)$ for all z and r. If $E \subset \bigcup B(z_j, r_j) = \bigcup B_j$ then

$$\mu(E) \leq \mu(\bigcup B_j) \leq \sum \mu(B_j) \leq \sum h(r_j),$$

and that proves the first assertion of the theorem.

Now suppose E is compact, and without loss of generality suppose E is contained in the unit square. For each $n \geq 0$, define the measure μ_n so that for every dyadic Q_n with $\ell(Q_n) = 2^{-n}$,

$$\mu_n(Q_n) = \begin{cases} h(2^{-n}), & \text{if } Q_n \cap E \neq \emptyset, \\ 0, & \text{if } Q_n \cap E = \emptyset, \end{cases}$$

and such that on Q_n the measure μ_n is absolutely continuous to area measure and has constant density. Each Q_{n-1} with $\ell(Q_{n-1}) = 2^{-n+1}$ is the union of 4 squares $Q_{n,j}$ with $\ell(Q_{n,j}) = 2^{-n}$. If

$$\mu_n(Q_{n-1}) = \sum_j \mu_n(Q_{n,j}) > h(2^{-n+1}),$$

then reduce the density on each $Q_{n,j}$ so that

$$\mu_n(Q_{n-1}) = \sum \mu_n(Q_{n,j}) = h(2^{-n+1}),$$

and call this new measure $\mu_n^{(n-1)}$. Now repeat this process for squares Q_k with $\ell(Q_k) = 2^{-k}$ and obtain measures $\{\mu_n^{(k)}\}$, for $k = n - 2, \ldots, 0$. We arrive at a measure $\mu_n^{(0)}$ satisfying $\mu_n^{(0)}(Q) \leq h(\ell(Q))$ for all dyadic squares Q with $\ell(Q) \geq 2^{-n}$.

The sequence $\{\mu_n^{(0)}\}$ has a subsequence $\{\mu_{n_j}^{(0)}\}$ which converges weak-star to a measure μ. Then because $\mu_n^{(0)}$ has support in $\{z : \text{dist}(z, E) \leq 2 \cdot 2^{-n}\}$, supp $\mu \subset E$. Because weak-star convergence implies $\mu(U) \leq \liminf \mu_n(U)$ for all open U, we have $\mu(Q) \leq 9h(\ell(Q))$ for every dyadic square Q. Therefore since each $B(z, r)$ is contained in 25 dyadic squares Q with $r/2 < \ell(Q) \leq r$, we obtain (D.6) for the new measure $\mu/225$.

By construction, the measure $\mu_n^{(0)}$ is supported on a union of dyadic squares Q with $\ell(Q) = 2^{-n}$ such that either

$$\mu_n^{(0)}(Q) = h(\ell(Q)),$$

or Q is contained in a larger dyadic square $\tilde{Q}$ with

$$\mu_n^{(0)}(\tilde{Q}) = h(\ell(\tilde{Q})).$$

Thus

$$\mu_n^{(0)}(E) \geq \inf \sum h(Q_j), \tag{D.9}$$

where the infimum is taken over all disjoint collections of dyadic squares $\{Q_j\}$ satisfying $\ell(Q_j) \geq 2^{-n}$ and $\bigcup Q_j \supset E$. Then by (D.9) and (D.6)

$$\mu(E) = \lim \mu_n^{(0)}(E) \geq m_h(E) \geq M_h(E),$$

and dividing μ by 225 yields (D.8). ∎

Corollary D.2. *Let h be a measure function such that*

$$\int_0^1 \frac{h(r)}{r} dr < \infty. \tag{D.10}$$

If E is a compact set such that if $\Lambda_h(E) > 0$, then

$$\mathrm{Cap}(E) \geq \mathrm{diam}(E) \exp\left(-\frac{225}{M_h(E)} \int_0^{\mathrm{diam}(E)} \frac{h(r)}{r} dr\right). \tag{D.11}$$

Proof. If (D.10) holds and if $\Lambda_h(E) > 0$ then by (D.4), $M_h(E) > 0$ and by Theorem D.1, there is a measure μ, supported on E, for which (D.7) and (D.8) hold. Fix $z \in E$ and write $m(r) = \mu(B(z, r))$. Then

$$U_\mu(z) = \int_E \log \frac{1}{|\zeta - z|} d\mu(\zeta)$$

$$= \lim_{\varepsilon \to 0} \int_\varepsilon^{\mathrm{diam}(E)} \log \frac{1}{r} dm(r),$$

and by a partial integration,

$$U_\mu(z) \leq \mu(E) \log \frac{1}{\mathrm{diam}(E)} + \lim_{\varepsilon \to 0} \int_\varepsilon^{\mathrm{diam}(E)} \frac{m(r)}{r} dr$$

$$\leq \mu(E) \log \frac{1}{\mathrm{diam}(E)} + \int_0^{\mathrm{diam}(E)} \frac{h(r)}{r} dr.$$

Hence by Theorem III.4.1,

$$\gamma(E) \leq I\left(\frac{\mu}{\mu(E)}\right) \leq \frac{1}{\mu(E)} \int_0^{\mathrm{diam}(E)} \frac{h(r)}{r} dr + \log \frac{1}{\mathrm{diam}(E)},$$

which yields (D.11). ∎

Corollary D.2 shows that Cap(E) is positive if dim(E) > 0. The next corollary is a quantitative version of that result.

Corollary D.3. *If* $0 < \alpha \leq 2$, *there is a constant* $C = C_\alpha > 0$ *such that if E is compact then*

$$\text{Cap}(E) \geq C\left(M_{r^\alpha}(E)\right)^{\frac{1}{\alpha}}. \qquad (D.12)$$

Proof. By a change of scale $z \to \lambda z$, $\lambda > 0$, we may assume $M_{r^\alpha}(E) = 1$. Let μ be a measure on E satisfying $\mu(B(z, r)) \leq r^\alpha$ for all balls $B(z, r)$ and $\mu(E) \geq \frac{1}{225}$. Then by a partial integration

$$\begin{aligned}
U_\mu(z) &\leq \int_{\{|\zeta - z| \leq 1\}} \log \frac{1}{|\zeta - z|} d\mu(\zeta) \\
&= \int_0^1 \frac{\mu((B(z, r))}{r} dr \\
&\leq \int_0^1 r^{\alpha - 1} dr = \frac{1}{\alpha},
\end{aligned}$$

and $\gamma(E) \leq 225\alpha$ by Theorem III.4.1. That gives (D.12) with $C_\alpha \geq e^{-225\alpha}$. ∎

Theorem D.4. *Let*

$$h(r) = \left(\log \frac{1}{r}\right)^{-1}. \qquad (D.13)$$

If Cap(E) > 0, *then* $\Lambda_h(E) = \infty$.

Theorem D.4 is due to Erdös and Gillis [1937]. It is almost a converse to Corollary D.2. For example, when $p > 1$ and $0 < r < 1/3$, the measure function $h(r) = (\log 1/r)^{-1} (\log \log 1/r)^{-p}$, satisfies the hypothesis (D.10) of Corollary D.2.

Proof. If Cap(E) > 0, then

$$\lim_{r \to 0} \int_{E \cap B(z, r)} \log \frac{1}{|\zeta - z|} d\mu_E(\zeta) = 0$$

almost everywhere μ_E. Then by Egoroff's theorem there exists compact $J \subset E$, such that $\mu_E(J) > 0$ and there exists a positive decreasing function $K(r)$ such that

$$\lim_{r \to 0} \frac{\log 1/r}{K(r)} = 0,$$

and

$$\sup_{z \in J} \int_E K(|\zeta - z|) d\mu_E(\zeta) \leq M \leq \infty.$$

Suppose $r_j \leq \delta$ and

$$J \subset \bigcup_1^N B(z_j, r_j).$$

Then

$$\mu_E(J) \leq \sum \mu_E(B(z_j, r_j)) \leq M \sum \frac{1}{K(r_j)} \leq M \sup_{r \leq \delta} \left(\frac{\log 1/r}{K(r)}\right) \sum h(r_j),$$

and sending $\delta \to 0$ shows $\Lambda_h(J) = \infty$.   ∎

Exercise D.1. (a) The Cantor set, defined in Example III.4.7, has Hausdorff dimension

$$\alpha = \frac{\log 2}{\log 3}.$$

Recall that $K = \bigcap K_n$ where $K_0 = [0, 1]$ and K_n is obtained from K_{n-1} by removing the middle one-third of each interval in K_{n-1}. If $\delta = 3^{-n}/2$, then we can cover K_n by 2^n balls of radius δ and

$$\Lambda_\alpha^\delta(K) \leq \Lambda_\alpha^\delta(K_n) \leq 2^n \left(\frac{3^{-n}}{2}\right)^\alpha = 2^{-\alpha}.$$

Letting $n \to \infty$, we obtain $\Lambda_\alpha(K) \leq 2^{-\alpha}$ and $\dim(K) \leq \alpha$. For the opposite inequality, let μ_n be a constant multiple of one-dimensional Lebesgue measure restricted to K_n. Choose the constant so that each of the 2^n intervals I in K_n has mass

$$\mu_n(I) = 2^{-n}.$$

Let μ be a weak-star limit of μ_n. Then

$$\mu(K) = \lim_m \mu(K_m) = \lim_m \lim_n \mu_n(K_m) = 1.$$

Let $B = B(z, r)$ and choose m so that

$$3^{-m} \leq 2r < 3^{-m+1}.$$

The distance between two intervals in K_{m-1} is at least 3^{-m+1}, so B intersects at most one interval in K_{m-1}. Then for $n > m$,

$$\mu_n(B) \leq 2^{-m+1} \leq 2^{\alpha+1} r^\alpha.$$

Letting $n \to \infty$, we obtain

$$\mu(B) \le 2^{\alpha+1} r^{\alpha},$$

and by Theorem D.1,

$$1 = \mu(K) \le \Lambda_{\alpha}(K) 2^{\alpha+1}.$$

Thus $2^{-\alpha-1} \le \Lambda_{\alpha}(K) \le 2^{-\alpha}$ and $\dim(K) \ge \alpha$. A more careful estimate shows that $\Lambda_{\alpha}(K) = 2^{-\alpha}$. An alternate proof can be found in Falconer [1985].

(b) If, when we construct K_n from K_{n-1}, we remove the middle intervals of relative length $1 - 2\delta$, $0 < \delta < 1/2$, instead of the middle one-third, then we obtain $\dim(E) = \log 2 / \log 1/\delta$.

(c) We describe a general construction of planar Cantor sets. Let K_0 be a square in $\mathbb{R}^2$ with edge length $l_0 = 1$ and let $l_{j+1} < l_j/2$, $j = 0, 1, \ldots$. The (closed) set K_j consists of 4^j squares of side length l_j, each sitting in a corner of a square in K_{j-1}. The sets K_1, K_2, and K_3 are shown in Figure D.1.

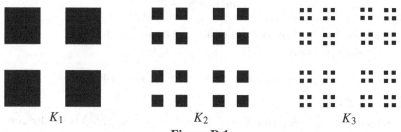

$$K_1 \qquad\qquad K_2 \qquad\qquad K_3$$

Figure D.1

Let $K = \bigcap K_j$ and let h be a measure function with

$$h(l_j) = 4^{-j}, \tag{D.14}$$

for $j = 0, 1, \ldots$ Then

$$\frac{1}{36} \le \Lambda_h(K) \le 1. \tag{D.15}$$

To prove (D.15), cover K_j by 4^j discs of radius l_j. Then for $\delta = l_j$,

$$\Lambda_h^{\delta}(K) \le \Lambda_h^{\delta}(K_j) \le 4^j h(l_j) = 1.$$

Letting $j \to \infty$, we obtain $\Lambda_h(K) \le 1$. To prove the lower bound on $\Lambda_h(K)$, let μ_j be a constant multiple of area measure restricted to K_j. Choose the constant so that each of the 4^j squares S in K_j has mass

$$\mu_j(S) = 4^{-j}.$$

Let μ be a weak-star limit of μ_j. Then

$$\mu(K) = \lim_m \mu(K_m) = \lim_m \lim_j \mu_j(K_m) = 1.$$

Let $B = B(z, r)$ and choose m so that

$$l_m \le r < l_{m-1}.$$

Then B intersects at most nine disjoint squares of side l_{m-1} and hence at most nine squares in K_{m-1}. Thus for $n > m$

$$\mu_n(B) \le 9 \cdot 4^{-m+1} \le 36h(r).$$

Letting $n \to \infty$, we obtain $\mu(B) \le 36h(r)$ and by Theorem D.1

$$1 = \mu(K) \le 36\Lambda_h(K),$$

and (D.15) holds.

(d) If $l_j = a^j$, $0 < a < 1/2$, then the Hausdorff dimension of $K = K_a$ is

$$\alpha = \dim(K_a) = \frac{\log 4}{\log 1/a}.$$

This shows that the Hausdorff dimension can be any number in $[0, 2]$.

(e) More generally, if

$$\alpha = \sup\{t : \liminf 4^j l_j^t = \infty\} = \inf\{t : \liminf 4^j l_j^t = 0\},$$

then $\dim(K) = \alpha$.

(f) Let $\Omega = \mathbb{C}^* \setminus K$, with K as in part (d). Prove $g_\Omega(z, \infty)$ is Hölder continuous, and find its best Hölder exponent.

Exercise D.2. (a) Assume the compact set E can be covered by $A(r)$ balls of radius at most r and

$$\int_0^1 \frac{1}{rA(r)} dr = \infty. \tag{D.16}$$

Then $\mathrm{Cap}(E) = 0$. Hint: Without loss of generality, we may assume that $A(r)$ is a decreasing function of r, and that $E \subset B(0, 1/2)$. Then $\log 1/|z - \zeta| \ge 0$ for all $z, \zeta \in E$. Assume $\mathrm{Cap}(E) > 0$, and let μ be the equilibrium distribution for E. Then

$$I(\mu) = \int_E \int_E \log \frac{1}{|z - \zeta|} d\mu(\zeta) d\mu(z)$$

$$= \int_E \int_0^1 \log \frac{1}{r} dm(z, r) d\mu(z),$$

where $m(z, r) = \mu(B(z, r))$. Show

$$\int_E \log \frac{1}{\varepsilon} m(z, \varepsilon) d\mu(z) \leq \int_E \int_0^\varepsilon \log \frac{1}{r} dm(z, r) d\mu(z) \to 0,$$

as $\varepsilon \to 0$, and

$$I(\mu) = \int_E \int_0^1 \frac{m(z, r)}{r} dr d\mu(z)$$

$$\geq \sum_{n=0}^\infty \int_{2^{-n-1}}^{2^{-n}} \int_E m(z, 2^{-n-1}) d\mu(z) \frac{dr}{r}. \tag{D.17}$$

Now cover E by $A_{n+2} = A(2^{-n-2})$ balls $B(z_k, 2^{-n-2})$. We may suppose that no point is in more than 16 balls, by modifying the collection of balls. Then by (D.17)

$$I(\mu) \geq \sum_{n=0}^\infty \int_{2^{-n-1}}^{2^{-n}} \frac{1}{16} \sum_{k=1}^{A_{n+2}} \int_{B(z_k, 2^{-n-2})} m(z, 2^{-n-1}) d\mu(z) \frac{dr}{r}$$

$$\geq \sum_{n=0}^\infty \int_{2^{-n-1}}^{2^{-n}} \frac{1}{16} \sum_{k=1}^{A_{n+2}} m(z_k, 2^{-n-2})^2 \frac{dr}{r}$$

$$= \frac{\log 2}{16} \sum_{n=0}^\infty \sum_{k=1}^{A_{n+2}} m(z_k, 2^{-n-2})^2.$$

But

$$1 = \mu(E)^2 \leq \left(\sum_{k=1}^{A_{n+2}} \mu(B(z_k, 2^{-n-2})) \right)^2 \leq A_{n+2} \sum_{k=1}^{A_{n+2}} m(z_k, 2^{-n-2})^2.$$

Hence

$$\infty > I(\mu) \geq \frac{\log 2}{16} \sum_{n=0}^\infty \frac{1}{A_{n+2}} \geq C \int_0^1 \frac{1}{r A(r)} dr,$$

which contradicts (D.16).

(b) Let K be the set constructed in Exercise D.1(c). Then $\mathrm{Cap}(K) > 0$ if and only if

$$\sum_{m=1}^\infty 4^{-m} \log \frac{1}{l_m} < \infty. \tag{D.18}$$

Hint: Suppose (D.17) holds. As in (D.14) let h be a measure function with

$h(l_m) = 4^{-m}$. Then

$$\int_0^1 \frac{h(r)}{r}dr \le \sum_{m=1}^{\infty} h(l_{m-1}) \log \frac{l_{m-1}}{l_m} = 3\sum_m 4^{-m} \log \frac{1}{l_m} < \infty.$$

Since $\Lambda_h(K) > 0$, Corollary D.2 implies $\text{Cap}(K) > 0$.

But if (D.9) fails we can cover K by 4^m balls of radius l_m, so let A be decreasing with $A(l_m) = 4^m$. Then

$$\int_0^1 \frac{1}{rA(r)}dr \ge \sum_{m=1}^{\infty} 4^{-m} \log \frac{l_{m-1}}{l_m} = \infty,$$

and by part (a), $\text{Cap}(K) = 0$.

(c) Cantor sets can also be used to show that (D.10) is necessary in Corollary D.2. Suppose h is a measure function such that $h^{-1}(t/4) < h^{-1}(t)/2$ and

$$\int_0^1 \frac{h(r)}{r}dr = \infty,$$

then there exists a compact set K with $\Lambda_h(K) > 0$ and $\text{Cap}(K) = 0$. Hint: Take $l_m = h^{-1}(4^{-m})$, construct the Cantor set as in D.1(c) and apply part (b).

Further examples show that no complete description of capacity can be given in terms of Hausdorff measures. See Carleson [1967a] for (a), (b), and (c).

E. Transfinite Diameter and Evans Functions

Transfinite diameter is a generalization of diameter that gives a geometric interpretation of capacity. Let E be a compact set. For $z_1, \ldots, z_n \in E$ set

$$D_n(z_1, \ldots, z_n) = \left(\prod_{1 \le i < j \le n} |z_i - z_j| \right)^{\frac{2}{n(n-1)}}. \tag{E.1}$$

The exponent $p_n = n(n-1)/2$ is chosen because the product D_n has p_n factors, so that D_n is homogeneous of degree 1:

$$D_n(\lambda z_1, \ldots, \lambda z_n) = \lambda D_n(z_1, \ldots, z_n), \qquad \lambda > 0.$$

The **diameter of order n** is defined by

$$d_n = d_n(E) = \max_{z_1, \ldots, z_n \in E} D_n(z_1, \ldots, z_n).$$

The usual diameter of E is $d_2(E)$. A second quantity related to d_n is

$$m_n(E) = \max_{z_1,\ldots,z_n \in E} \ \min_{z \in E} \frac{1}{n} \sum_j \log \frac{1}{|z - z_j|}. \tag{E.2}$$

Theorem E.1. *If E is compact then*

$$\mathrm{Cap}(E) = \lim_n d_n(E) = \lim_n e^{-m_n(E)}.$$

The limit $d_\infty(E) = \lim_n d_n(E)$ is called the **transfinite diameter** of E.

Proof. We first compare (E.1) and (E.2). Choose $z_1, \ldots, z_n \in E$ so that $D_n(z_1, \ldots, z_n) = d_n(E)$ and let

$$T_j(z) = \frac{1}{n-1} \sum_{i;i \neq j} \log \frac{1}{|z - z_i|}.$$

Write

$$\log \frac{1}{d_n(E)} = \frac{2}{n(n-1)} \sum_{i,j:i<j} \log \frac{1}{|z_j - z_i|}. \tag{E.3}$$

If we view $\{z_i : i \neq j\}$ as fixed, then the terms in (E.3) which entail z_j are

$$\frac{2}{n(n-1)} \sum_{i:i \neq j} \log \frac{1}{|z_j - z_i|}.$$

Thus z_j is chosen so as to minimize T_j on E, and hence $T_j(z_j) \leq m_{n-1}(E)$. This implies

$$\log \frac{1}{d_n(E)} = \frac{1}{n} \sum_{j=1}^n \frac{1}{n-1} \sum_{i:i \neq j} \log \frac{1}{|z_j - z_i|}$$

$$= \frac{1}{n} \sum_{j=1}^n T_j(z_j) \leq m_{n-1}(E). \tag{E.4}$$

Now let $\mu_n = \frac{1}{n} \sum \delta_{z_n}$. Then

$$I_N(\mu_n) \equiv \iint \min\left(\log \frac{1}{|z - \zeta|}, N\right) d\mu_n(z) d\mu_n(\varsigma)$$

$$\leq \frac{1}{n^2} \sum_{i,j:i \neq j} \log \frac{1}{|z_j - z_i|} + \frac{N}{n}$$

$$\leq \log \frac{1}{d_n(E)} + \frac{N}{n}.$$

Take a subsequence $\{\mu_{n_j}\}$ converging weak-star to a measure $\nu \in P(E)$. Then

$$I_N(\nu) \equiv \iint \min\left(\log\frac{1}{|z-\zeta|}, N\right) d\nu(z)d\nu(\zeta) \leq \liminf_{n\to\infty} \log\frac{1}{d_n(E)}.$$

Letting $N \to \infty$ we obtain by Theorem III.4.1

$$\gamma(E) \leq I(\nu) \leq \liminf_{n\to\infty} \log\frac{1}{d_n(E)}. \tag{E.5}$$

Suppose now that $\mathrm{Cap}(E) > 0$, and let μ be the equilibrium distribution for E. Then for any $z_1, \ldots, z_n \in E$

$$\inf_{z\in E} \frac{1}{n}\sum_{j=1}^{n}\log\frac{1}{|z-z_j|} \leq \frac{1}{n}\sum_{j=1}^{n}\int_E \log\frac{1}{|z-z_j|}d\mu(z)$$

$$= \frac{1}{n}\sum_{j=1}^{n}U_\mu(z_j) \leq \gamma(E),$$

and hence

$$m_n(E) \leq \gamma(E). \tag{E.6}$$

By (E.4), (E.5) and (E.6),

$$\gamma(E) = \lim_n \log\frac{1}{d_n(E)} = \lim_n m_n(E),$$

which proves the theorem when $\mathrm{Cap}(E) > 0$.

If $\mathrm{Cap}(E) = 0$, take E_j decreasing to E such that $\mathrm{Cap}(E_j) > 0$. Then by (E.4)

$$e^{-m_{n-1}(E)} \leq d_n(E),$$

and by (E.5)

$$\limsup_{n\to\infty} d_n(E) \leq \limsup_{n\to\infty} d_n(E_j) \leq e^{-\gamma(E_j)},$$

and the Theorem follows. ∎

The extremal (E.2) can be used to construct, for any compact set E of zero capacity, a potential that is infinite exactly on E.

Theorem E.2 (Evans). *Suppose E is compact. If $\mathrm{Cap}(E) = 0$ there exists a probability measure μ supported on E such that the potential $U = U_\mu$ of μ satisfies*

$$\lim_{z\to\zeta_0} U(z) = +\infty, \tag{E.7}$$

for all $\zeta_0 \in E$. Conversely, if there exist a function U harmonic on $\mathbb{C} \setminus E$ and a constant $K > 0$ such that (E.7) holds and

$$U(z) \geq K \log \frac{1}{|z|} \qquad \text{(E.8)}$$

for $|z|$ sufficiently large, then $\text{Cap}(E) = 0$.

See Evans [1933]. A potential U_μ satisfying (E.7) is called an **Evans function** for E.

Proof. Suppose $\text{Cap}(E) = 0$. Choose $z_1, \ldots, z_n \in E$ so that

$$\min_{z \in E} \frac{1}{n} \sum_{j=1}^{n} \log \frac{1}{|z - z_j|} = m_n(E),$$

and let $\mu_n = \frac{1}{n} \sum_{j=1}^{n} \delta_{z_j}$. Then

$$U_{\mu_n}(z) = \int_E \log \frac{1}{|z - \zeta|} d\mu(\zeta) \geq m_n(E) \to \infty.$$

Choose a subsequence $\{U_{n_j}\}$ of $\{U_n\}$ so that $U_{n_j}(z) \geq 2^j$ for all $z \in E$, and let

$$\mu = \sum_{j=1}^{\infty} 2^{-j} \mu_{n_j}.$$

Then

$$U_\mu(z) = \sum_{j=1}^{\infty} 2^{-j} U_{n_j}(z)$$

satisfies (E.7).

Conversely, suppose there is U harmonic in $\mathbb{C} \setminus E$ satisfying (E.7) and (E.8) and suppose that $\text{Cap}(E) > 0$. Let μ be the equilibrium distribution for E and let U_μ be its potential. Then by the maximum principle

$$U(z) - 2K U_\mu(z) \equiv +\infty,$$

for $z \in \mathbb{C} \setminus E$, which is impossible. Thus $\text{Cap}(E) = 0$. ∎

Exercise E.1. If f is analytic in a region Ω, then the **cluster set of f at $\zeta \in \partial\Omega$** is

$$Cl(f, \zeta) = \{w : \text{there exists } \{z_n\} \subset \Omega, \text{ with } \lim z_n = \zeta \text{ and } \lim f(z_n) = w\}.$$

If $K \subset \partial\Omega$ then the **cluster set of f at K** is

$$Cl(f, K) = \bigcup_{\zeta \in K} Cl(f, \zeta).$$

Suppose f is bounded and analytic on $\mathbb{D} \setminus K$ where K is compact and $\mathrm{Cap}(Cl(f, K)) = 0$. Prove f extends to be analytic and bounded on $\mathbb{D}$. Radó [1924] proved this result in the case $Cl(f, K) = \{0\}$. Hint: First show we may suppose f is bounded and analytic on $\mathbb{C} \setminus K$, by considering

$$f_1(z) = \int_{\partial B(0, 1-2\varepsilon)} \frac{f(\zeta)}{\zeta - z} \frac{d\zeta}{2\pi i}.$$

By Cauchy's integral formula $f - f_1$ extends to be bounded and analytic on $\mathbb{C} \setminus K$. When f is bounded and analytic in a neighborhood of ∞, f extends to be analytic at ∞, and we may suppose that $f(\infty) \notin Cl(f, K)$ by replacing f with $f \circ \varphi$ for some linear fractional transformation φ if needed. Now let U be an Evans function for $E = Cl(f, K)$ and consider $V = U \circ f$. Then V is constant by Theorem E.2 and Corollary III.8.4, and hence f is constant.

F. Martingales, Brownian Motion, and Kakutani's Theorem

Let $(X, \mathcal{M}, P)$ be a **probability space**, that is, a measure space with $P(X) = 1$, and let $\mathcal{N} \subset \mathcal{M}$ be a sub σ-algebra. For $f \in L^1(X, \mathcal{M}, P)$ there is a unique $\mathcal{N}$-measurable function F such that

$$\int \chi_S f \, dP = \int \chi_S F \, dP \tag{F.1}$$

for all $S \in \mathcal{N}$. The function F is called the **conditional expectation of f given** $\mathcal{N}$, and it is written as

$$F = E(f|\mathcal{N}).$$

To prove F exists (and is unique almost everywhere) we use the Radon–Nikodym theorem: The signed measure $\mathcal{N} \ni S \to \nu(S) = \int \chi_S f \, dP$ is absolutely continuous to P on the σ-algebra $\mathcal{N}$. Hence there is an $\mathcal{N}$-measurable F so that

$$\nu(S) = \int \chi_S F \, dP,$$

and F is unique almost everywhere. We will have $F \neq f$, unless f happens to be $\mathcal{N}$-measurable. If $f \in L^2(X, M, P)$, then F is the orthogonal projection of f onto the closed subspace $L^2(X, \mathcal{N}, P)$ of $\mathcal{N}$-measurable functions.

Example F.1. Let $(X, \mathcal{M}, P) = ([0, 2\pi), \mathcal{B}, d\theta/2\pi)$, where $\mathcal{B}$ is the algebra of Borel sets. Let $0 = \theta_0 < \theta_1 < \ldots < \theta_n = 2\pi$, set

$$I_j = [\theta_{j-1}, \theta_j),$$

and let $\mathcal{N}$ be the σ-algebra generated by $\{I_1, I_2, \ldots, I_N\}$. The $\mathcal{N}$-measurable functions are the step functions constant on the I_j, and

$$E(f|N) = \sum_{j=1}^{N} \alpha_j \chi_{I_j},$$

where

$$\alpha_j = \frac{1}{\theta_j - \theta_{j-1}} \int_{I_j} f \, d\theta$$

is the average of f over I_j.

Now let

$$\mathcal{M}_1 \subset \mathcal{M}_2 \subset \ldots \subset \mathcal{M}$$

be an increasing sequence of σ-algebras such that

$$\bigcup_{n=1}^{\infty} \mathcal{M}_n \quad \text{generates } \mathcal{M}. \tag{F.2}$$

Suppose f_n is $\mathcal{M}_n$-measurable, $f_n \in L^1(X, \mathcal{M}_n, P)$. We say the sequence $\{f_n\}$ is a **martingale** (relative to $\{\mathcal{M}_n\}$) if for all n

$$f_n = E(f_{n+1}|\mathcal{M}_n). \tag{F.3}$$

From the uniqueness of conditional expectations it follows that

$$f_n = E(f_{n+k}|\mathcal{M}_n), \qquad k > 0. \tag{F.4}$$

Martingales are ubiquitous in probability theory, see Chung [1974] or Doob [1974].

Given the sequence $\mathcal{M}_n$, we can always make a martingale by starting with $f \in L^1(X, \mathcal{M}, P)$ and setting

$$f_n = E(f|\mathcal{M}_n). \tag{F.5}$$

It is a theorem that a martingale $\{f_n\}$ has the form (F.5) if and only if the sequence $\{f_n\}$ is uniformly integrable. We need only a weak form of that theorem.

Lemma F.2. *Let* $1 < p < \infty$ *and let* $\{f_n\}$ *be a martingale relative to* $\{\mathcal{M}_n\}$ *such that*

$$\sup_n \|f_n\|_p < \infty.$$

Then there is $f \in L^p(X, \mathcal{M}, P)$ such that

$$f_n = E(f|\mathcal{M}_n)$$

and f_n converges to f weakly in L^p.

Proof. From $\{f_n\}$ take a subsequence converging weakly in L^p to some limit f. For $S \in \mathcal{M}_n$, $\chi_S \in L^q$, $q = p/(p-1)$ and hence (F.4) yields

$$\int \chi_S f_n dP = \lim_j \int \chi_S f_{n_j} dP = \int \chi_S f dP.$$

Hence $f_n = E(f|\mathcal{M}_n)$. By (F.2) the limit point f does not depend on the subsequence $\{f_n\}$ and that means $\{f_n\}$ converges weakly to f. ∎

Doob's martingale theorem says that an L^1 bounded martingale converges almost everywhere. We will prove the theorem in the special case of L^p bounded martingales, and for L^1 bounded martingales we only prove $\lim \sup |f_n| < \infty$ a.e. Both results hinge on a crucial inequality, called the **weak-L^1 estimate for the maximal function**, which is the martingale variant of the estimate for the Hardy–Littlewood maximal function, Lemma I.2.4.

Lemma F.3. *Let $\{f_n\}$ be a martingale such that*

$$\sup_n \|f_n\|_1 = M < \infty.$$

Set

$$f^*(x) = \sup_n |f_n(x)|.$$

Then for all $\lambda > 0$,

$$P(f^* > \lambda) \le \frac{M}{\lambda}. \tag{F.6}$$

Proof. Fix λ and define

$$N(x) = \inf\{n : |f_n(x)| > \lambda\}.$$

Then

$$S_n \equiv \{x : N(x) = n\} = \{x : |f_n(x)| > \lambda\} \cap \bigcap_{k=1}^{n-1}\{x : |f_k(x)| \le \lambda\} \in \mathcal{M}_n.$$

Therefore

$$\lambda P(f^* > \lambda) = \lim_m \sum_{n=1}^{m} \lambda P(S_n)$$

$$\leq \lim_m \sum_{n=1}^{m} \int \chi_{S_n} |f_n| dP$$

$$\leq \lim_m \sum_{n=1}^{m} \int \chi_{S_n} |f_m| dP$$

$$\leq \lim_m \|f_m\|_1 \leq M,$$

The second inequality above follows from

$$|f_n| = |E(f_m | \mathcal{M}_n)| \leq E(|f_m| \mid \mathcal{M}_n).$$ (F.7)

So (F.6) is proved.   ∎

Incidentally, the function $N(x)$ is called a **stopping time**, which by definition means that $\{N(x) \leq n\} \in \mathcal{M}_n$.

Corollary F.4. *If $\{f_n\}$ is a martingale such that $\sup_n \|f_n\|_1 < \infty$, then $\limsup |f_n| < \infty$ almost everywhere.*

Proof. This is trivial because

$$P(\limsup |f_n| > 1/\varepsilon) \leq P(f^* > 1/\varepsilon) \leq M\varepsilon$$

for all $\varepsilon > 0$.   ∎

Theorem F.5. *Let $p > 1$ and let $\{f_n\}$ be a martingale satisfying*

$$\sup_n \|f_n\|_p < \infty.$$

Let f be the weak limit provided by Lemma F.3. Then $f_n(x) \to f(x)$ for almost all x.

Proof. By Hölder's inequality we may assume $p < \infty$. This case of the martingale theorem is easier because we already have

$$f_n = E(f | \mathcal{M}_n)$$

for some $f \in L^p$. We may suppose f is real. The argument is like the proof of Fatou's theorem, I.2.1. By (F.6)

$$\Omega_f(x) = \limsup f_n(x) - \liminf f_n(x)$$

is finite almost everywhere and in fact

$$P(\Omega_f(x) > \varepsilon) \leq \frac{2}{\varepsilon} \int |f| dP. \tag{F.8}$$

Let $\mathcal{D} = \{g \in L^p : g \text{ is } \mathcal{M}_n \text{ measurable for some } n\}$. Then $\mathcal{D}$ is dense in L^p by (F.2) and $\Omega_g = 0$ when $g \in \mathcal{D}$. Now take $g \in \mathcal{D}$ such that $\|f - g\|_p < \varepsilon^2$. Then $\Omega_f = \Omega_{f-g}$ and so (F.8) yields, via Hölder's inequality,

$$P(\Omega_f > \varepsilon) \leq \frac{2}{\varepsilon} \|f - g\|_1 \leq \frac{2}{\varepsilon} \varepsilon^2 < 2\varepsilon.$$

Consequently $\Omega_f = 0$ almost everywhere and $F(x) = \lim f_n(x)$ exists almost everywhere. Because $\{f_n\}$ is L^p bounded for $p > 1$ it follows that $\|f_n - F\|_p \to 0$, by Egoroff's theorem and the weak convergence of f_n to f. Hence $F = f$ *a.e.* and the theorem is proved. ∎

Now let Ω be a bounded domain in $\mathbb{C}$ and for each $z \in \Omega$ let $\Gamma(z)$ be the circle

$$\Gamma(z) = \{w : |w - z| = \frac{1}{2}\text{dist}(z, \partial\Omega)\}.$$

For fixed $z_0 \in \Omega$ we define a martingale $z_0, z_1, \ldots$, inductively as follows. Take

$$z_{n+1} \in \Gamma(z_n), \quad n \geq 0$$

such that $z_{n+1} - z_n$ is uniformly distributed, which means that for $A \subset \Gamma(z_n)$,

$$P(z_{n+1} \in A) = \frac{\text{length}(A)}{\text{length}(\Gamma(z_n))}, \quad n \geq 0.$$

One often meets a martingale before seeing its σ-algebras $\mathcal{M}_n$; to define them for $\{z_n\}$ let

$$X = \{(e^{i\theta_1}, e^{i\theta_2}, \ldots)\}$$

be an infinite product of unit circles, let $\mathcal{M}$ be the Borel σ-algebra on X and let P be the infinite product measure $\prod d\theta_j / 2\pi$. Set $\mathcal{M}_0 = \{\phi, X\}$ and let $\mathcal{M}_n$ be the smallest σ-algebra for which $\theta_1, \theta_2, \ldots, \theta_n$ are measurable. Then

$$z_{n+1} = z_n + \frac{1}{2}\text{dist}(z_n, \partial\Omega)e^{i\theta_{n+1}}, \tag{F.9}$$

and $\{z_n\}$ is a martingale relative to $\{\mathcal{M}_n\}$ because

$$\frac{1}{2\pi} \int_0^{2\pi} e^{i\theta} d\theta = 0.$$

Also note that (F.9) implies $z_{n+1} \in \Omega$ and $z_{n+1} \neq z_n$.

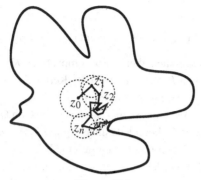

Figure F.1

Because Ω is bounded, the martingale theorem ensures that

$$z_\infty = \lim_n z_n$$

exists almost everywhere on X. Clearly $z_\infty \in \overline{\Omega}$, and because

$$\text{dist}(z_n, \partial\Omega) = 2|z_{n+1} - z_n| \to 0,$$

we have $z_\infty \in \partial\Omega$ whenever z_∞ exists. Kakutani's famous theorem identifies the hitting distribution of z_∞ with the harmonic measure of z_0 on $\partial\Omega$.

Theorem F.6. *Let $z_0 \in \Omega$ and let $E \subset \partial\Omega$ be a Borel set. Then*

$$\omega(z_0, E, \Omega) = P(z_\infty \in E) \tag{F.10}$$

where z_∞ is the limit of the martingale (F.9) started at z_0.

For example, let Ω be the (unbounded) domain $\{z : \text{Re} z > 0, |\text{Im} z| < 1\}$. Let $z_0 = 0$ and let $E = \partial\Omega \cap \{\text{Re} z \geq n\}$. From Theorem F.6 one can see

$$w(z_0, E, \Omega) \leq C_1 e^{-c_2 n}$$

because if $k \leq \text{Re} z_j \leq k + 1$, then $P(\text{Re} z_\infty < k + 2)$ has a positive lower bound independent of k. Though Theorem F.6 has been stated only for bounded domains, we can use it on $\Omega \cap \{|z| < N\}$ and let $N \to \infty$. The precise estimate

$$w(z_0, E, \Omega) = \frac{4}{\pi} \sinh(\frac{\pi}{2}) e^{-\frac{\pi n}{2}} + O(e^{-\pi n})$$

follows from conformally mapping Ω to the upper half-plane.

Proof of Theorem F.6. The key observation is that if u is harmonic in Ω, then by the mean value property $u_n = u(z_n)$ is a martingale. First we use the observation to show

$$P(z_\infty \text{ is a regular point of } \partial\Omega) = 1. \qquad (\text{F.11})$$

Suppose (F.11) is false. Then there is a compact set K, consisting only of irregular points, such that $P(z_\infty \in K) > 0$. By Kellogg's theorem $\text{Cap}(K) = 0$ and by Evans' theorem there is $u(z)$ harmonic on $\mathbb{C}\backslash K$ such that $u(z) \to \infty$ if $z \to \zeta \in K$. Adding a constant, we may suppose $u > 0$ on Ω. Then the positive martingale $u_n = u(z_n)$ is L^1 bounded (since $\int |u_n| dP = \int u_n dP = u(z_0)$) but $\lim u_n = \infty$ on $\{z_\infty \in K\}$, contradicting Lemma F.3.

Now let $f \in C(\partial\Omega)$ and let u be the solution of the Dirichlet problem for the boundary data f. Then $u_n = u(z_n)$ is a bounded martingale and by (F.11)

$$u_n \to f(z_\infty)$$

almost everywhere with respect to P. Hence by Lemma F.2 and the uniqueness of martingale limits, $u_n = E(f(z_\infty)|\mathcal{M}_n)$ and in particular

$$\int_{\partial\Omega} f(\zeta) dw(z_0, \zeta) = u(z_0) = u_0 = \int f(z_\infty) dP.$$

Approximating characteristic functions by continuous functions now yields (F.10). ∎

The martingale $\{z_n\}$ is a discrete variant of Brownian motion. Let us give, without proof, a brief description of Brownian motion. Let W be the set of continuous paths $B(t)$, $t \geq 0$, in the plane having common initial point $B(0) = z_0$. By a theorem of Wiener there is a probability measure P on W with respect to which $B(t)$ will almost surely meet any circle $\Gamma_r(z_0) = \{z : |z - z_0| = r\}$ at some finite time and the location of the first hitting of $\Gamma_r(z_0)$ is uniformly distributed:

$$P(\text{first hit of } \Gamma_r(z) \in A) = \frac{\text{length}(A)}{2\pi r}.$$

All of the increments $B(t + s) - B(s)$ have the same distribution as $B(t) - z_0$ and if

$$0 \leq t_0 < t_1 < \cdots < t_n$$

then the increments

$$B(t_1) - B(t_0), B(t_2) - B(t_1), \ldots, B(t_n) - B(t_{n-1})$$

are **stochastically independent**:

$$P(\bigcap \{B(t_{j+1}) - B(t_j) \in S_j)\}) = \prod P(B(t_{j+1}) - B(t_j) \in S_j).$$

With an appropriate σ-algebra $\mathcal{M}$, the measure space $(W, \mathcal{M}, P)$ is called **Brownian motion**, and a random (with respect to P) function $B(t)$ is called a **Brownian path**.

Again let Ω be a bounded domain and let $z_0 \in \Omega$. Then

$$\tau_1 = \tau_1(B) = \inf\{t : B(t) \in \Gamma(z_0)\}$$

is $\mathcal{M}$-measurable and so is $z_1 = B(\tau_1)$. Similarly,

$$\tau_2 = \tau_2(B) = \inf\{t > \tau_1 : B(t) \in \Gamma(z_1) = \Gamma(B(\tau_1))\}$$

and $z_2 = B(\tau_2)$ are also measurable. In fact there is an infinite sequence $0 < \tau_1 < \cdots$ of random times such that $z_n = B(\tau_n)$. Thus our martingale z_n coincides with $B(t)$ observed at the discrete sequence of times τ_n. Now let $\tau_\infty = \inf\{t : B(t) \in \partial\Omega\}$. Because almost every path $B(t)$ is continuous,

$$\tau_\infty = \lim \tau_n$$

and

$$B(\tau_\infty) = \lim z_n = z_\infty.$$

Therefore Theorem F.6 can be restated as

$$w(z_0, E, \Omega) = P(B(\tau_\infty) \in E),$$

or in words, the harmonic measure at z_0 of the Borel set E is the probability that Brownian motion, started at z_0, first exits Ω in E.

We do not know how to show directly that

$$u(z_0) = \int f(z_\infty) dP$$

is a harmonic function of z_0 without using a prior solution to the Dirichlet problem. Thus we cannot use (F.10) as an alternate definition of the solution of the Dirichlet problem. Had we worked with Brownian motion $B(t)$ instead of the discrete version z_n, there would have been no such difficulty because then $u(z)$ would have the mean value property for all discs contained in Ω.

Exercise F.1. Let $p > 1$ and let $f_n = E(f, \mathcal{M}_n)$ be an L^p bounded martingale.

(a) Show that $\lim_n \|f_n\|_p = \|f\|_p$.

(b) Use (a) to show

$$\|f_n - f\|_p \longrightarrow 0.$$

(c) Prove the maximal theorem

$$\|f^*\|_p \leq A_p \|f\|_p,$$

and use it to give another proof of (b).

Exercise F.2. In the probability space

$$\left([0, 2\pi), \mathcal{M}, \frac{d\theta}{2\pi}\right),$$

where $\mathcal{M}$ is the Borel sets, let $\mathcal{M}_n$ be the σ-algebra generated by the *dyadic* intervals

$$I_{n,j} = \left[\frac{j2\pi}{2^n}, \frac{(j+1)2\pi}{2^n}\right), \quad 0 \le j < 2^n.$$

Thus $\mathcal{M}_n$ consists of unions of the dyadic intervals $I_{n,j}$, and $\mathcal{M}_n \subset \mathcal{M}_{n+1}$. For $f \in L^1([0, 2\pi))$, set $f_n = E(f|\mathcal{M}_n)$ and for $I = I_{n,j}$,

$$f_I = \frac{1}{|I|} \int_I f \, d\theta,$$

so that

$$f_n = \sum_j f_{I_{n,j}} \chi_{I_{n,j}}.$$

We say $f \in \text{BMO}(\mathcal{M})$, for **dyadic bounded mean oscillation**, if

$$\sup_n E(|f - f_n| \,|\mathcal{M}_n) = \|f\|_{\text{BMO}(\mathcal{M})} < \infty.$$

Equivalently,

$$\|f\|_{\text{BMO}(\mathcal{M})} = \sup_{I=I_{n,j}} \frac{1}{|I|} \int_I |f - f_I| \, d\theta.$$

(a) Prove $f \in \text{BMO}(\mathcal{M})$ if and only

$$\sup_{I=I_{n,j}} \inf_{\alpha \in \mathbb{C}} \frac{1}{|I|} \int_I |f(e^{i\theta}) - \alpha| \, d\theta < \infty.$$

(b) Assume $\|f\|_{\text{BMO}(\mathcal{M})} = 1$, and let $\lambda > 2$. Prove the **John–Nirenberg theorem**: There are constants C and c, independent of f, such that for all n,

$$P(\{\theta : |f(e^{i\theta}) - f_n(e^{i\theta})| > \lambda | \mathcal{M}_n\}) \equiv E(\chi_{\{\theta: |f-f_n|<\lambda\}} | \mathcal{M}_n)$$
$$\le Ce^{-c\lambda}. \tag{F.12}$$

Equivalently,

$$\sup_{I \in \mathcal{M}_n} \frac{|\{\theta \in I : |f(e^{i\theta}) - f_I| > \lambda\}|}{|I|} \le Ce^{-c\lambda}.$$

Hint: Fix n, set $N_0 = n$, and form the stopping time

$$N_1(\theta) = \inf\{m > n : |f_m(\theta) - f_n(\theta)| > 2\}.$$

Then by Chebychev's inequality,

$$P(N_1 < \infty | \mathcal{M}_n) \equiv E(\chi_{\{N_1 < \infty\}} | \mathcal{M}_n) \le \frac{1}{2}.$$

Now define

$$N_2(\theta) = \inf\{m > N_1(\theta) : |f_m(\theta) - f_{N_1}(\theta)| > 2\},$$

and continue. The BMO assumption yields

$$P(N_k < \infty | \mathcal{M}_n) \le 2^{-k},$$

and when $2k < \lambda \le 2(k+1)$ that gives (F.12).

(c) Without $\{\mathcal{M}_n\}$, we say $f \in L^1((0, 2\pi])$ has **bounded mean oscillation**, $f \in$ BMO, if for all intervals $I \subset \partial\mathbb{D}$,

$$\frac{1}{|I|} \int_I |f - f_I| d\theta \le \|f\|_{\mathrm{BMO}} < \infty.$$

See Exercise 21 of Chapter I. Suppose $f \in$ BMO. Prove that if $I \subset J$ are intervals such that $|J| = 2^n |I|$, then

$$|f_J - f_I| \le cn \|f\|_{\mathrm{BMO}}.$$

(d) Prove that if $\|f\|_{\mathrm{BMO}} \le 1$, then for $\lambda > 2$ and for any interval I,

$$\frac{|\{\theta \in I : |f(e^{i\theta}) - f_I| > \lambda\}|}{|I|} \le Ce^{-c\lambda}.$$

Repeat the argument of (b), using subintervals of I having lengths $2^{-n}|I|$.

(e) Let $p > 1$. There is C_p such that if $f \in$ BMO, then

$$\sup_I \inf_{\alpha \in \mathbb{C}} \frac{1}{|I|} \int_I |f - \alpha|^p d\theta \le C_p \|f\|_{\mathrm{BMO}}^p.$$

This is immediate from (d). See John and Nirenberg [1961].

Exercise F.3. Let $u(z)$ be the Poisson integral of $f \in L^1(\partial\mathbb{D})$.

(a) Prove $f \in$ BMO if and only if

$$\sup_{a \in \mathbb{D}} \int_{\partial\mathbb{D}} |f(e^{i\theta}) - u(a)|^2 P_a(\theta) d\theta < \infty.$$

Hint: Compare the Poisson kernel to the box kernel $\frac{\chi_I}{|I|}$ as in Figure I.4 from Chapter I and use the John–Nirenberg theorem.

(b) Using (a) and Green's theorem, show $f \in$ BMO if and only if

$$\sup_{a \in \mathbb{D}} \iint_{\mathbb{D}} |\nabla u|^2 \log \left| \frac{1 - \bar{a}z}{z - a} \right| dx dy < \infty.$$

See Exercise II.8.

(c) Using (b) and the identity

$$1 - |\frac{z-a}{1-\bar{a}z}|^2 = \frac{(1-|a|^2)(1-|z|^2)}{|1-\bar{a}z|^2},$$

show $f \in$ BMO if and only if

$$|\nabla u|^2(1 - |z|^2)dxdy$$

is a Carleson measure. It follows that $f \in$ BMO if and only if its conjugate function $\tilde{f} \in$ BMO. Fefferman and Stein [1972].

(d) It follows from (c) that BMO is conformally invariant. If $f(z)$ is analytic on $\mathbb{D}$ and if $f(e^{i\theta}) \in$ BMO, then for every Möbius transformation $Tz = \frac{z-a}{1-\bar{a}z}$, $a \in \mathbb{D}$,

$$\|f \circ T\|_{\text{BMO}} \le C\|f\|_{\text{BMO}}.$$

For detailed proofs see, for example, Garnett [1981].

G. Carleman's Method

Let Ω be a rectangle with sides parallel to the axes, let $z_0 \in \Omega$ and let E be the right edge of Ω. If finitely many horizontal slits are removed from Ω, then the right side of inequality (IV.6.3) does not change, but the harmonic measure $\omega(z_0, E)$ becomes smaller. This defect in Theorem IV.6.1 is circumvented in Carleman's [1933] version of the theorem.

Theorem G.1 (Carleman). *Let $\Omega \subset \mathbb{C}$ be a domain, let $\Omega_x = \Omega \cap \{\text{Re} z = x\}$ and let $E_b = \partial\Omega \cap \{\text{Re} z \ge b\}$. Suppose $|\Omega_x| \le M < \infty$, and let $\ell(x)$ denote the length of the longest interval in Ω_x. Assume $z_0 = x_0 + iy_0 \in B(z_0, r_0) \subset \Omega$. Then for $b > x_0$*

$$\omega(z_0, E_b, \Omega) \le \left(\frac{2\pi r_0}{9M^2} \int_{x_0}^b \exp\left\{2\pi \int_{x_0}^t \frac{dx}{\ell(x)}\right\} dt\right)^{-\frac{1}{2}}. \tag{G.1}$$

The function $\ell(x)$ is measurable because it is lower semicontinuous. The main improvement of (G.1) over (IV.6.3) is the replacement of $\theta(x)$ with the smaller $\ell(x)$. Further comparisons will be given after the proof of Theorem G.1.

The proof uses **Wirtinger's inequality**: If g and g' are real valued continuous functions on the interval (a, b) and if $g(a) = g(b) = 0$, then

$$\int_a^b (g')^2 dx \ge \left(\frac{\pi}{b-a}\right)^2 \int_a^b g^2 dx. \tag{G.2}$$

Wirtinger's inequality (G.2) is proved by examining the Fourier series of g. Lemma III.3.2 is very similar to Wirtinger's inequality.

Carleman's idea is to find a differential inequality for the Dirichlet integral of the harmonic measure $\omega(z, E_b, \Omega)$. To prove Theorem G.1, we may suppose that Ω is bounded, that $\partial\Omega$ consists of finitely many analytic Jordan curves, and that $\inf\{\mathrm{Re}\, z : z \in \Omega\} = 0$. Write $\omega(z) = \omega(z, E_b, \Omega)$ and define

$$A(t) = \int_0^t \int_{\Omega_x} |\nabla\omega|^2 dy dx.$$

By our smoothness assumptions on $\partial\Omega$, the function $A(t)$ is continuously differentiable. Write $\Omega_x = \bigcup \Omega_x^i$ where $\{\Omega_x^i\}$ are the connected components of Ω_x.

Lemma G.2. (Carleman's differential inequality). *For $x \in (0, b)$,*

$$A'(x) \geq \frac{2\pi}{\ell(x)} A(x). \tag{G.3}$$

Proof. By the assumptions on $\partial\Omega$,

$$A'(x) = \int_{\Omega_x} \omega_x^2 dy + \int_{\Omega_x} \omega_y^2 dy$$

for $x \in (0, b)$. Because $\omega = 0$ on $\partial\Omega_x^i$, Wirtinger's inequality gives

$$\int_{\Omega_x^i} \omega_y^2 dy \geq \left(\frac{\pi}{|\Omega_x^i|}\right)^2 \int_{\Omega_x^i} \omega^2 dy,$$

and hence

$$\int_{\Omega_x} \omega_y^2 dy \geq \left(\frac{\pi}{\ell(x)}\right)^2 \int_{\Omega_x} \omega^2 dy.$$

By Green's theorem

$$A(x) = \int_{\Omega_x} \omega\omega_x dy, \tag{G.4}$$

and by the Cauchy–Schwarz inequality

$$\int_{\Omega_x} \omega_x^2 \geq A^2(x) \bigg/ \int_{\Omega_x} \omega^2 dy.$$

Thus

$$A' \geq \frac{A^2}{\int_{\Omega_x} \omega^2 dy} + \left(\frac{\pi}{\ell(x)}\right)^2 \int_{\Omega_x} \omega^2 dy \geq \frac{2\pi}{\ell(x)} A,$$

and that proves the lemma. ∎

The Dirichlet integral $A(t)$ is connected to the harmonic measure $\omega(z_0)$ via the function

$$\varphi(x) = \int_{\Omega_x} \omega^2 dy.$$

By Harnack's inequality

$$\omega \geq \omega(z_0)/3$$

on $B(z_0, r_0/2)$, so that

$$\omega^2(z_0) \leq \frac{9\varphi(x_0)}{r_0}. \tag{G.5}$$

On the other hand, because $\omega = 0$ on $\partial\Omega$,

$$\varphi'(x) = 2 \int_{\Omega_x} \omega \omega_x dy.$$

By (G.4), $\varphi' = 2A$ and so Carleman's differential inequality (G.3) reads

$$\frac{\varphi''(x)}{\varphi'(x)} \geq \frac{2\pi}{\ell(x)}.$$

Now set $\mu(x) = 2\pi/\ell(x)$ and

$$\psi(x) = \int_0^x \exp\left(\int_0^t d\mu\right) dt,$$

so that

$$\frac{\psi''}{\psi'} = \frac{2\pi}{\ell(x)}.$$

Then (G.3) can be rewritten as

$$\left(\log \frac{\varphi'}{\psi'}\right)' = \frac{\varphi''}{\varphi'} - \frac{\psi''}{\psi'} \geq 0.$$

Therefore φ'/ψ' is non-decreasing. Because $\psi' > 0$, we obtain

$$\varphi'(x)\psi'(t) \leq \varphi'(t)\psi'(x)$$

whenever $0 < x < t$. Because $\varphi(0) = \psi(0) = 0$, integrating the above inequality from 0 to x gives

$$\varphi(x)\psi'(t) \leq \varphi'(t)\psi(x),$$

and integrating again from x to t then gives

$$\varphi(x)\psi(t) \leq \varphi(t)\psi(x), \tag{G.6}$$

whenever $0 < x < t$.

Increasing Ω in $\{\mathrm{Re}z < x_0\}$ increases ω but does not change the right side of (G.1), so we may assume that $\mu(x) = 2\pi/\ell(x) = 2\pi/M$ on $x < x_0$. Then

$$\psi(x_0) = \frac{M}{2\pi}(e^{2\pi x_0/M} - 1)$$

and

$$\psi(b) = \psi(x_0) + \int_{x_0}^{b} \exp\left\{\int_{0}^{x_0} \mu(s)ds\right\} \exp\left\{\int_{x_0}^{t} \mu(s)ds\right\} dt$$

$$\geq \psi(x_0)\left(1 + \frac{2\pi}{M}\int_{x_0}^{b} \exp\left\{\int_{x_0}^{t} \mu(s)ds\right\} dt\right). \tag{G.7}$$

Now $\varphi(b) \leq |\Omega_b| \leq M$, so that by (G.5), (G.6), and (G.7)

$$\omega(z_0)^2 \leq \frac{9\varphi(z_0)}{r_0} \leq \frac{9M}{r_0} \cdot \frac{\psi(x_0)}{\psi(b)}$$

$$\leq \frac{9M}{r_0}\left(1 + \frac{2\pi}{M}\int_{x_0}^{b} \exp\left\{\int_{x_0}^{t} \mu(s)ds\right\} dt\right)^{-1} \tag{G.8}$$

and taking the square root in (G.8) gives us (G.1). ∎

To compare (G.1) and (IV.6.3), note that

$$\int_{x_0}^{b} \exp\left\{2\pi\int_{x_0}^{t} \frac{dx}{\ell(x)}\right\} dt \geq \int_{b-\varepsilon}^{b} \exp\left\{2\pi\int_{x_0}^{t} \frac{dx}{\ell(x)}\right\} dt$$

$$\geq \varepsilon \exp\left\{2\pi\int_{x_0}^{b-\varepsilon} \frac{dx}{\ell(x)}\right\},$$

so that

$$\omega(z_0, E_b) \leq \left(\frac{9M^2}{2\pi r_0\varepsilon}\right)^{\frac{1}{2}} \exp\left\{-\pi\int_{x_0}^{b-\varepsilon} \frac{dx}{\ell(x)}\right\}. \tag{G.9}$$

If $\ell(x) = \theta(x)$ inequality (G.9) is weaker than (IV.6.3) because the upper limit is $b - \varepsilon$ rather than b. If $\ell(x) \geq 2\pi\delta > 0$ for all $x > (1 - \delta)b > 0$, then we can make a closer comparison with (IV.6.3), for in that case

$$\int_{x_0}^{b} \exp\left\{-2\pi\int_{t}^{b} \frac{dx}{\ell(x)}\right\} dt \geq \delta\left(1 - e^{-b}\right),$$

and thus

$$\omega(z_0, E_b) \leq \left(\frac{9M^2}{2\pi r_0 \delta (1 - e^{-b})} \right)^{\frac{1}{2}} \exp \left(-\pi \int_{x_0}^{b} \frac{dx}{\ell(x)} \right). \qquad (\text{G.10})$$

Of course, either (G.9) or (G.10) is enough to give Theorem V.1.1.

Notes

Tsuji [1959] proved (G.9) in polar coordinates. Our presentation follows Carleman [1933] and Haliste [1965].

H. Extremal Distance in Finitely Connected Jordan Domains

Several of the results in Chapter IV for simply connected domains have reformulations valid for finitely connected Jordan domains.

We begin with the description of the extremal distance $d_\Omega(E, F)$ in Theorem IV.4.1. Assume Ω is a finitely connected Jordan domain, let E and F be finite unions of closed arcs on $\partial\Omega$, and assume $E \cap F = \emptyset$. In that case we call (Ω, E, F) a **triple**. The extension of Theorem IV.4.1 is easy when there is one component Γ of $\partial\Omega$ such that $E \cup F \subset \Gamma_1$, because in that case there is a comformal mapping of Ω onto a rectangle with horizontal line segments removed so that E and F correspond to the vertical ends of the rectangle if and only if there is a subarc $\sigma \subset \Gamma_1$ such that $E \subset \sigma$ and $F \cap \sigma = \emptyset$. For the proof we can assume Ω is bounded by analytic curves. Take two copies of Ω and form the Riemann surface double Ω_d by attaching the copies along $\partial\Omega \setminus (E \cup F)$. Let $\omega(z) = \omega(z, F_d, \Omega_d)$ be the harmonic measure of the double F_d of F in Ω_d, and let $\widetilde{\omega}$ be the (locally defined) harmonic conjugate of ω in Ω. Then just as in the proof of Theorem IV.4.1, $\partial\widetilde{\omega}/\partial s = 0$ on each component Γ_j of $\partial\Omega$ for which $\Gamma_j \cap (E \cup F) = \emptyset$. Therefore

$$\int_{\Gamma_j} \frac{\partial\widetilde{\omega}}{\partial s} ds = 0$$

and hence

$$\varphi = \omega + i\widetilde{\omega}$$

is single valued and analytic on Ω. The remainder of the argument is exactly like the proof of Theorem IV.4.1. There is a similar finitely connected version of Theorem IV.4.2; see Exercise H.1.

There is also a generalization of Theorem IV.4.1 that holds for all triples (Ω, E, F), but now the slit rectangle must be replaced by a union of rectangles pasted together along their edges to form a Riemann surface. We formulate the result in terms of the real part of the conformal map to the Riemann surface. For motivation, let us look back at the proof of Theorem IV.4.1 when $\partial\Omega$ is an analytic Jordan curve. The extremal metric is $\rho_0(z) = |\varphi'(z)| = |\nabla\omega(z)|$, where $\omega = \mathrm{Re}\,\varphi$ and φ maps Ω onto a rectangle. By reflection, the function ω is continuous on $\overline{\Omega}$ and harmonic on a neighborhood of $\overline{\Omega} \setminus P$, where P is the set of endpoints of the arcs in $E \cup F$. Moreover, ω solves the mixed Dirichlet–Neumann problem:

$$\Delta\omega(z) = 0, \qquad z \in \Omega, \tag{H.1}$$

$$\omega(z) = \begin{cases} 0, & z \in E, \\ 1, & z \in F, \end{cases} \tag{H.2}$$

and

$$\frac{\partial\omega(z)}{\partial n} = 0, \qquad z \in \partial\Omega \setminus (E \cup F). \tag{H.3}$$

Theorem H.1. *Suppose Ω is a finitely connected Jordan domain and suppose E and F are disjoint closed subsets of $\partial\Omega$, each consisting of a finite many subarcs of $\partial\Omega$. Then the extremal distance between E and F is given by*

$$d_\Omega(E, F)^{-1} = D(\omega) \equiv \iint_\Omega |\nabla\omega|^2 dxdy,$$

where ω is a solution to (H.1) *and* (H.2). *Extremal metrics exist and are given by constant multiples of $\rho = |\nabla\omega|$ a.e. dxdy. If $\partial\Omega$ consists of $C^{1+\varepsilon}$ curves then ω is the unique solution to* (H.1), (H.2), *and* (H.3). *The conjugate extremal distance satisfies*

$$d^*_\Omega(E, F) = 1/d_\Omega(E, F).$$

Theorem H.1 also implies Corollary C.4 of Appendix C.

Corollary H.2. *Suppose Ω is a finitely connected Jordan domain and suppose E and F are disjoint closed subsets of $\partial\Omega$, each consisting of a finite many subarcs of $\partial\Omega$. Then ω is the unique function in $C^1(\Omega)$ satisfying*

$$\int_\Omega |\nabla\omega|^2 dxdy = \inf\Big\{ \int_\Omega |\nabla v|^2 dxdy : v \in C^1(\Omega),$$

$$\limsup_{z\to\zeta\in E} v(z) \le 0, \liminf_{z\to\zeta\in F} v(z) \ge 1 \Big\}. \tag{H.4}$$

Indeed, for v in the right side of (H.4), $|\nabla v|$ is an admissible metric for the problem $d_\Omega(E, F)$ and $L(\Gamma, |\nabla v|) \geq 1$. Since $L(\Gamma, |\nabla \omega|) = 1$, and since $|\nabla \omega|$ is the extremal metric, we must have $A(\Omega, |\nabla v|) \geq A(\Omega, |\nabla \omega|)$, which is (H.4). Theorem IV.2.1 implies ω is the unique minimizing function for (H.4). This proof only needs $v \in C(\Omega)$ to be piecewise C^1.

Proof of Theorem H.1. Before starting, we return to the conformal mapping φ constructed in the proof of Theorem IV.4.1. As z traces $\partial\Omega$, each excised horizontal line segment on the boundary of $\varphi(\Omega)$ is traced twice by $\varphi(z)$ and each vertical line segment is traced once. The points in R which are endpoints of the horizontal slits correspond to the points in $\partial\Omega$ where $\nabla\omega = 0$; these are the critical points of ω. Let $\{v_i\}$ denote the values of $\widetilde{\omega}$ at the critical points. If we remove the curves $\{z : \widetilde{\omega}(z) = v_i\}$, we partition Ω into finitely many smaller regions Ω_j. The image of each Ω_j under the map $\varphi = \omega + i\widetilde{\omega}$ is an uncut rectangle of length 1. The shortest curves for the extremal distance on the slit rectangle are the horizontal line segments not meeting the excised slits. In other words, the shortest curves in Ω are those level sets $\{\widetilde{\omega}(z) = c\}$ whose images do not meet the excised slits. In the general case the definition of ω is the same. We can assume $\partial\Omega$ consists of analytic Jordan curves. By Corollary II.4.6, this assumption will not change (H.3) when $\partial\Omega$ consist of $C^{1+\varepsilon}$ curves. If $\partial\Omega \setminus (E \cup F) \neq \emptyset$, take two copies of Ω joined along $\partial\Omega \setminus (E \cup F)$, forming the Riemann surface double Ω_d, and let F_d consist of those boundary curves of $\partial\Omega_d$ corresponding to F. If $\partial\Omega \setminus (E \cup F) = \emptyset$, we let $\Omega_d = \Omega$ and $F_d = F$. Let $\omega(z)$ be the harmonic measure of F_d in Ω_d. As before, we work with the restriction of $\omega(z)$ to Ω. From the proof of Theorem IV.4.1 it is clear that the (locally defined) function $\varphi = \omega + i\widetilde{\omega}$ is continuous up to $\partial\Omega$ and that $\omega = \operatorname{Re}\varphi$ satisfies (H.1), (H.2), and (H.3). In fact, by the Schwarz reflection principle, ω is the only solution to (H.1)–(H.3). Moreover, except on the set $P = \{\text{endpoints of } E \cup F\}$, φ is analytic on $\partial\Omega$, and (IV.4.2), (IV.4.3), and (IV.4.4) hold.

In our more general case, $\widetilde{\omega}$ is not single valued. For each $\zeta \in \overline{\Omega} \setminus P$ there is a conformal map f_ζ defined in a neighborhood U_ζ of ζ with $f_\zeta(\zeta) = 0$ and a constant a_ζ so that

$$\omega + i\widetilde{\omega} = a_\zeta + f_\zeta^k \tag{H.5}$$

for some $k \geq 1$. If $k \geq 2$, we call ζ a **critical point of order k–1**. Near ζ, the level set $\{z : \widetilde{\omega}(z) = \widetilde{\omega}(\zeta)\}$ consists of $2k$ analytic arcs which meet at ζ separated by equal angles of π/k, and these level lines do not depend on the choice of $\widetilde{\omega}$ that was made. Each $\zeta \in \Omega$ for which $|\nabla\omega| \neq 0$ then lies on a maximal analytic curve along which $\omega + i\widetilde{\omega}$ can be continued analytically with

$\widetilde{\omega}$ constant and parameterized so that ω is increasing. Call these curves **level curves** of $\widetilde{\omega}$. We claim that each such level curve will either terminate at a critical point of ω or at a point of $E \cup F$. To see this, first note that if α is a level curve then α cannot be closed since ω is continuous and increasing on α. If $U_\zeta \cap \alpha$ has infinitely many components, then we can find an arc β in U_ζ with both endpoints on α and on which ω is constant. A subarc of α together with β then forms a closed curve, along which either ω is increasing or constant. This contradicts the continuity of ω. By compactness, the level curve α must either terminate at a critical point or terminate at a point of $\partial\Omega$. If α terminates at a point $\zeta \in \partial\Omega \setminus \{E \cup F\}$, then since $\widetilde{\omega}$ is constant on $\partial\Omega \cap U_\zeta$, the local description (H.5) implies ζ is a critical point of ω. This proves the claim.

We can also show that α cannot terminate at a point $\zeta \in P$. To see this note that

$$\varphi(z) = \varphi(\zeta) + a_1\sqrt{(z - \zeta)} + \dots. \tag{H.6}$$

Indeed, since $\partial\Omega$ is analytic, there is a neighborhood W of ζ and a conformal map ψ of the unit disc $\mathbb{D}$ onto W so that $\psi(0) = \zeta$ and

$$\psi(\{z \in \mathbb{D} : \operatorname{Im} z > 0\}) = W \cap \Omega.$$

On the quarter disc $Q = \{z \in \mathbb{D} : \operatorname{Im} z > 0, \operatorname{Re} z > 0\}$, $f(z) \equiv \varphi(\psi(z^2))$ is analytic and on each segment in $\partial Q \cap \mathbb{D}$, either $\operatorname{Re} f$ or $\operatorname{Im} f$ is constant. Thus f extends by two reflections to be analytic in $\mathbb{D}$ with $\operatorname{Re} f \equiv 0$ or $\operatorname{Re} f \equiv 1$ on a diameter. By the argument used to establish (IV.4.3), $f'(0) \neq 0$. Thus φ has an analytic extension to $W \setminus (E \cup F)$, on which (H.6) holds. Moreover, near ζ the level set $\{z : \widetilde{\omega}(z) = \widetilde{\omega}(\zeta)\}$ consists of one analytic arc terminating at ζ, namely the subarc of $\partial\Omega \setminus (E \cup F)$ contained in W. In particular, a level curve of $\widetilde{\omega}$ in Ω cannot terminate at ζ.

We will call the level lines $\{\widetilde{\omega} = v_i\}$ passing through the critical points of ω the **critical level lines**. The critical level lines partition Ω into smaller regions Ω_j. See Figure H.1. The boundary of Ω_j consists of finitely many analytic arcs $\gamma_{j,k}$, on the relative interior of which either

$$\frac{\partial\omega}{\partial s} \neq 0 \quad \text{and} \quad \frac{\partial\omega}{\partial n} = 0 \quad \text{if } \gamma_{j,k} \subset \{\widetilde{\omega} = v_i\} \tag{H.7}$$

or

$$\frac{\partial\omega}{\partial s} = 0, \quad \text{and} \quad \frac{\partial\omega}{\partial n} \neq 0, \quad \text{if } \gamma_{j,k} \subset E \cup F. \tag{H.8}$$

Every endpoint of $\gamma_{j,k}$ is either a critical point or a point of $E \cup F$. We will show that $\varphi = \omega + i\widetilde{\omega}$ is single valued on Ω_j and that $\varphi(\Omega_j)$ is either a rectangle or an annulus. This will follow from the argument principle applied to the (single valued) analytic function $\varphi' = \omega_x - i\omega_y$.

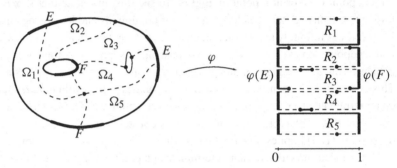

Figure H.1 $\varphi(\Omega_j) = R_j$.

We shall need an extended version of the argument principle, and we digress briefly to prove it. Suppose W is a bounded, finitely connected Jordan domain and suppose each boundary curve is piecewise continuously differentiable. Orient ∂W so that it has index 1 about each point of W. We say ∂W has an **interior angle** of $\theta_0 \in [0, 2\pi]$ at ζ_0 if the argument of the tangent vector, $\arg \frac{dz}{ds}$, increases by $\pi - \theta_0$ at ζ_0. Let $F = \{\zeta_k\}_{k=1}^m$ be a finite subset of ∂W and let $W_\varepsilon = W \setminus \cup_k \{z : |z - \zeta_k| \le \varepsilon\}$. If g is continuous on $\partial W \setminus F$ then we define the **principal value integral** as

$$PV \int_{\partial W} g(z)dz = \lim_{\varepsilon \to 0} \int_{\partial W_\varepsilon \cap \partial W} g(z)dz,$$

provided the limit exists. For example, the total increase in the argument of the tangent vector around ∂W is given by

$$PV \, \mathrm{Im} \int_{\partial W} \frac{d}{dz} \left(\log \frac{dz}{ds} \right) dz + \sum (\pi - \theta_k) = 2\pi(1 - (N-1)), \quad \text{(H.9)}$$

where N is the number of components of ∂W and the sum on the left side is taken over the finite set of points where the interior angle θ_k is not equal to π. If f is meromorphic on W, then we say f has a **singularity** at $\zeta_0 \in \partial W$ of order $\alpha_0 \in \mathbb{R}$ if

$$\lim_{W \ni z \to \zeta_0} \left(\frac{f'(z)}{f(z)} - \frac{\alpha_0}{z - \zeta_0} \right) \quad \text{(H.10)}$$

exists. Thus if $f(z) \sim (z - \zeta_0)^{\alpha_0}$ then f has a singularity of order α_0 at ζ_0.

Lemma H.3 (extended argument principle). *Suppose W is a bounded, finitely connected Jordan domain and suppose each boundary curve in $\partial\Omega$ is piecewise continuously differentiable. Let $P = \{\zeta_k\}_{k=1}^m \subset \partial W$ and suppose ∂W has* *interior angle θ_k at ζ_k, $k = 1, \ldots, m$, and interior angle π at all $\zeta \in \partial W \setminus P$. If f is meromorphic in a neighborhood of $\overline{W} \setminus P$, having no zeros or poles on $\partial W \setminus P$, and if f has a singularity of order α_k at ζ_k, $k = 1, \ldots, m$, then*

$$PV \text{ Im} \int_{\partial W} \frac{f'(z)}{f(z)} \frac{dz}{2\pi} = (N_z - N_p) + \sum_{k=1}^{m} \frac{\alpha_k \theta_k}{2\pi}, \qquad \text{(H.11)}$$

where N_z and N_p respectively are the number of zeros and poles of f in W, counting multiplicity.

For example, if ∂W is smooth and if f is meromorphic on $\overline{W}$ then

$$PV \text{ Im} \int_{\partial W} \frac{f'(z)}{f(z)} \frac{dz}{2\pi} = (N_z - N_p) + (N_z^b - N_p^b)/2,$$

where N_z^b and N_p^b are the zeros and poles of f on ∂W, counting multiplicity.

Proof. By the usual argument principle, if $\varepsilon > 0$ is small,

$$\int_{\partial W_\varepsilon} \frac{f'(z)}{f(z)} dz = 2\pi i (N_z - N_p). \qquad \text{(H.12)}$$

Set $\sigma_k^\varepsilon = \{z : |z - \zeta_k| = \varepsilon\} \cap W$. Then by the definition of θ_k

$$\lim_{\varepsilon \to 0} \text{Im} \int_{\sigma_k^\varepsilon} \frac{dz}{z - \zeta_k} = -\theta_k.$$

By (H.10)

$$\lim_{\varepsilon \to 0} \text{Im} \int_{\sigma_k^\varepsilon} \frac{f'(z)}{f(z)} \frac{dz}{2\pi} = -\frac{\alpha_k \theta_k}{2\pi}. \qquad \text{(H.13)}$$

Summing (H.13) over k and subtracting the result from (H.12) proves the lemma. ∎

To apply this lemma on the region $W = \Omega_j$, recall that φ' is analytic and non-zero on Ω_j. Moreover on the relative interior of each analytic arc $\gamma_{j,k}$ in $\partial \Omega_j$, $\arg \frac{d\varphi}{ds}$ is constant by (H.7) and (H.8). Thus $\frac{d}{ds} \arg \frac{d\varphi}{ds} = 0$ and

$$0 = PV \text{ Im} \int_{\partial \Omega_j} \frac{d}{ds} \left(\log \frac{d\varphi}{ds} \right) ds$$

$$= PV \text{ Im} \int_{\partial \Omega_j} \frac{d}{dz} \left(\log \varphi'(z) \right) dz \qquad \text{(H.14)}$$

$$+ PV \text{ Im} \int_{\partial \Omega_j} \frac{d}{dz} \left(\log \frac{dz}{ds} \right) dz.$$

Let $\zeta_1, \ldots, \zeta_m$ be the endpoints of the arcs $\gamma_{j,k}$ in $\partial\Omega_j$, and let θ_i and α_i be the corresponding interior angles of Ω_j and singularities of φ'. Then by (H.6), (H.9), (H.11), and (H.14)

$$0 = \sum_{i=1}^{m} \alpha_i\theta_i + 2\pi(2 - N) + \sum_{i=1}^{m}(\theta_i - \pi),$$

and hence

$$2\pi(N - 2) = \sum_{i=1}^{m} \alpha_i\theta_i + \theta_i - \pi. \tag{H.15}$$

There are three kinds of endpoints ζ_i of $\gamma_{j,k}$:

(i) If ζ_i is an endpoint of $E \cup F$ then by (H.6) φ' has a singularity of order $\alpha_i = -1/2$. By (H.6) the only level curve of $\widetilde{\omega}$ that contains ζ_i is an arc of $\partial\Omega \setminus (E \cup F)$ and so $\partial\Omega_j$ has an interior angle $\theta_i = \pi$ at ζ_i. In this case, then $\alpha_i\theta_i + \theta_i - \pi = -\pi/2$.

(ii) If $\zeta_i \in (E \cup F)^\circ$, then φ' is analytic and non-zero in a neighborhood of ζ_i. Thus $\partial\Omega_j$ turns from one of the sets $\{w = c_1\}$ or $\{\widetilde{\omega} = c_2\}$ to the other and $\partial\Omega_j$ has interior angle $\theta_i = \pi/2$ at ζ_i. In this case $\alpha_i\theta_i + \theta_i - \pi = -\pi/2$ again.

(iii) Finally if ζ_i is a critical point of ω, then by (H.5) $\partial\Omega_j$ has an interior angle $\theta_i = \pi/k$, $k \geq 2$, and φ' has a singularity $\alpha_i = k - 1$. In this case $\alpha_i\theta_i + \theta_i - \pi = 0$.

If K_j denotes the number of endpoints of $\gamma_{j,k}$ in $E \cup F \cap \partial\Omega_j$, and if N_j denotes the number of boundary components of Ω_j, then (H.15) yields

$$2\pi(N_j - 2) + \frac{\pi}{2}K_j = 0,$$

with $K_j \geq 0$ and $N_j \geq 1$ integers. Hence either $N_j = 2$ and $K_j = 0$ or $N_j = 1$ and $K_j = 4$. The first case occurs only when Ω is conformally equivalent to an annulus with boundary curves E and F, as in Theorem IV.4.2. Thus we may suppose the second case holds, so that Ω_j is simply connected and $E \cup F \cap \partial\Omega_j$ consists of two intervals. Then ω must be the restriction to Ω_j of the harmonic measure of $F \cap \partial\Omega_j$ in the Riemann surface double of Ω_j. The proof of Theorem IV.4.1 shows that $\varphi = \omega + i\widetilde{\omega}$ is a conformal map of Ω_j onto a rectangle.

As in Figure H.1, we can view φ as a conformal map of $\bigcup \Omega_j$ to a stack of rectangles of length 1. Now apply the length–area proof of Example IV.1.1 with $\ell = 1$ and h equal to the sum of the heights of the rectangles. Equality holds when the metric ρ is constant on the union of rectangles because ω is continuous on Ω and changes by 1 along any curve connecting E to F. The extremal distance between E and F is then the reciprocal of the sum of the heights of the rectangles. By conformal invariance, the extremal metric on Ω is $|\nabla \omega|$ and the level lines of $\tilde{\omega}$ which do not meet the critical points are the shortest curves for the extremal distance from E to F. The Dirichlet integral $D(\omega)$ is the sum of the areas of the rectangles, all of which have length 1, and thus $d_\Omega(E, F) = D(\omega)^{-1}$. The conjugate extremal distance is also the sum of the heights of the rectangles, since any curve separating E from F must separate the ends of each of the rectangles. Thus

$$d_\Omega^*(E, F) = 1/d_\Omega(E, F). \qquad \blacksquare$$

We remark that the function $\varphi = \omega + i\tilde{\omega}$ in the preceding proof can be regarded as a conformal map of Ω onto a Riemann surface formed by identifying the appropriate edges of the rectangles obtained. As an aside, we note that the preceding proof divided the region Ω into subregions so that equality holds in the parallel rule.

The reasoning with the extended argument principle in the proof of Theorem H.1 also gives a count of the critical points of ω in $\overline{\Omega}$, which will be useful in the examples considered below. Suppose $\partial\Omega$ consists of disjoint analytic Jordan curves and that E and F are finite unions of closed arcs on $\partial\Omega$. Let

N_c = the number of components of $\partial\Omega$,
N_e = the number of endpoints of arcs in $E \cup F$
N_z = the number of critical points of ω in Ω, and
N_z^b = the number of critical points of ω on $\partial\Omega$,

where a critical point of order k is counted k times. Then we have the following proposition.

Proposition H.4.

$$N_z + \frac{N_z^b}{2} = \frac{N_e}{4} + (N_c - 2).$$

Proof. The argument used to establish (H.14) on Ω_j also applies to Ω. Because $\partial\Omega$ is analytic, all interior angles are equal to π. At a critical point of order k on $\partial\Omega$, φ' has a singularity $\alpha_i = k$ and at an endpoint of $E \cup F$, φ' has a singularity $\alpha_l = -1/2$ by (i) above. Proposition H.4 now follows from (H.9), (H.14), and Lemma H.3. $\qquad \blacksquare$

Example H.5. Let Ω be the unit disc $\mathbb{D}$ and let E and F consist of two arcs each, interleaved on $\partial\mathbb{D}$. In this case $\omega(z) = \omega(z, F, \mathbb{C}^* \setminus (E \cup F))$. By Proposition H.4, there are three possible cases: Either ω has one critical point in $\mathbb{D}$ of multiplicity one, or ω has two critical points on $\partial\mathbb{D}$, each with multiplicity one, or ω has a single critical point of order two on $\partial\mathbb{D}$. In the first case the critical level lines of ω that pass through this critical point will terminate on E and F, dividing $\mathbb{D}$ into four quadrilaterals. See Figure H.2. The function $\omega + i\widetilde{\omega}$ will be a conformal map from each quadrilateral to a rectangle of unit length. The edges of rectangles will be identified so as to form a surface equivalent to the Riemann surface for z^2 over $\mathbb{D}$. The Dirichlet integral $D(\omega)$ is the sum of the areas of the four rectangles. In other words, the extremal distance is the reciprocal of the sum of the four heights of the rectangles. The curves in Ω that correspond to one vertical line segment in each rectangle are the extremal curves for the conjugate extremal distance. The conjugate extremal distance is the sum of the heights of the rectangles. For the second case, note that by (H.5), ω changes direction along $\partial\Omega$ at a critical point of order $k = 1$, while on each interval in $\partial\mathbb{D} \setminus (E \cup F)$ ω varies from 0 to 1 (or from 1 to 0). Hence in the second case both critical points are in the same interval in $\partial\mathbb{D} \setminus (E \cup F)$. From each critical point a critical level line enters $\mathbb{D}$. Since these level lines must divide $\mathbb{D}$ into quadrilaterals, each containing a portion of E and F, one critical level line terminates in E and the other terminates in F, dividing $\mathbb{D}$ into three quadrilaterals. In the third case ω has a double critical point $\zeta \in \partial\mathbb{D} \setminus (E \cup F)$. Two critical level lines enter $\mathbb{D}$ from ζ; one terminates in E, and the other in F. In this case, $\mathbb{D}$ is also divided into three quadrilaterals.

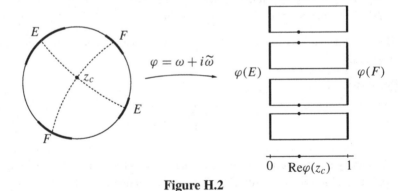

Figure H.2

Example H.6. Let Ω be the unit disc $\mathbb{D}$ with two smaller discs removed. Let $F = \partial\mathbb{D}$ and let $E = \partial\Omega \setminus F$. By Proposition H.4, there will be one critical point in Ω, with two critical level lines passing through it. One critical level line of $\widetilde{\omega}$ will connect the two inner discs. The other level line will have both endpoints on $\partial\mathbb{D}$. They divide Ω into two quadrilaterals. The function $\omega + i\widetilde{\omega}$

is a conformal map of each quadrilateral onto a rectangle of unit length. Again, the extremal distance is the reciprocal of the sum of the heights of the two rectangles. The conjugate extremal distance, that is, the extremal length of the curves that separate E from F, is the sum of the heights of the rectangles. See Figure H.3.

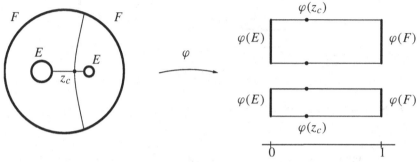

Figure H.3

If Ω is a finitely connected Jordan domain and if E is a finite union of arcs on one component Γ of $\partial\Omega$, the upper estimate in Theorem IV.5.3 still holds.

Theorem H.7. *Let Ω be a finitely connected Jordan domain, and let E be a finite union of arcs contained in one component Γ_1 of $\partial\Omega$. Suppose σ is a Jordan arc in $\mathbb{C}$ connecting z_0 to $\Gamma_1 \setminus E$ and let $\lambda \equiv d_{\Omega\setminus\sigma}(\sigma, E)$. Then*

$$\omega(z_0, E, \Omega) \le \frac{8}{\pi} e^{-\pi\lambda}.$$

Notice that $\sigma \cap \Omega$ need not be connected. The proof of Theorem H.7 is almost the same as the proof of Theorem IV.5.3. With no loss of generality, assume that $0 = z_0 \in \Omega \subset \mathbb{D}$ and that the component of $\partial\Omega$ containing E is the unit circle. Let $\Omega_1 = \{z \in \mathbb{D} : z^2 \in \Omega\}$. Now follow the proof of Theorem IV.5.3, using a conformal map of Ω_1 onto a rectangle with horizontal line segments removed, as in the discussion in the first paragraph of this appendix. Theorem H.7 also yields a more general version of Theorem IV.6.1.

Theorem H.8. *Suppose Ω is a finitely connected Jordan domain and suppose $E \subset \{z \in \overline{\Omega} : \mathrm{Re}\, z \ge b\}$. Let $z_0 \in \Omega$ with $x_0 = \mathrm{Re}\, z_0 < b$ and assume that for $x_0 < x < b$, $I_x \subset \{z \in \Omega : \mathrm{Re}\, z = x\}$ separates z_0 from E. If $\theta(x) \equiv \ell(I_x)$ is measurable then*

$$\omega(z_0, E) \le \frac{8}{\pi} \exp\left(-\pi \int_{x_0}^b \frac{dx}{\theta(x)}\right).$$

494 Appendices

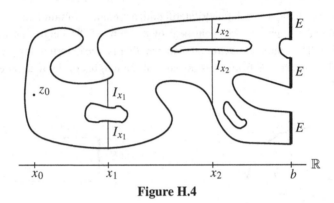

Figure H.4

Proof. Let $\widetilde{\Omega}$ be the component of

$$\Omega \setminus \{z : \mathrm{Re}\, z = b\}$$

that contains z_0 and let $\widetilde{E} = \partial\widetilde{\Omega} \cap \{z : \mathrm{Re}\, z = b\}$. As in the proof of Theorem IV.6.2, we may suppose that $\widetilde{E}$ is a finite union of arcs on $\partial\widetilde{\Omega}$. Let $\sigma \subset \{z : \mathrm{Re}\, z = \mathrm{Re}\, z_0\}$ be a curve (not necessarily contained in Ω) connecting z_0 to the component of $\partial\widetilde{\Omega}$ containing $\widetilde{E}$. By the maximum principle

$$\omega(z_0, E, \Omega) \le \omega(z_0, \widetilde{E}, \widetilde{\Omega}).$$

By Theorem H.7 and (IV.6.2)

$$\omega(z_0, E) \le \frac{8}{\pi} \exp\left(-\pi \int_{x_0}^{b} \frac{dx}{\theta(x)}\right). \qquad \blacksquare$$

The next result uses Theorem H.1 to compute capacity using extremal distance.

Theorem H.9. *Suppose Ω is a finitely connected Jordan domain and suppose $\infty \in \Omega$. Let $C_R = \{z : |z| = R\}$. Then*

$$\gamma = \lim_{R \to \infty} 2\pi d_\Omega(C_R, \partial\Omega) - \log R$$

where γ is Robin's constant for $\partial\Omega$ defined by $\mathrm{Cap}(\partial\Omega) = e^{-\gamma}$.

As we saw in Section V.3, $2\pi d_\Omega(C_R, \partial\Omega) - \log R$ increases with R, and hence by Theorem H.9, if $\partial\Omega \subset \{z : |z| < R\}$,

$$\mathrm{Cap}(\partial\Omega) \le R e^{-2\pi d_\Omega(C_R, \partial\Omega)}.$$

Any choice of a metric ρ will give a lower bound for the extremal distance and hence an upper bound for capacity. By the symmetry rule, the Ahlfors–Beurling

Theorem V.3.2 is a special case of Theorem H.9.

Proof. Let $g(z) = g(z, \infty, \Omega)$ denote Green's function for Ω with pole at ∞, and recall that

$$g(z) = \log |z| + \gamma + o(1), \tag{H.16}$$

Let $F_R = \{z \in \Omega : g(z) = \log R + \gamma\}$ and let $\varepsilon > 0$. Then for sufficiently large R, by (H.16), $C_R = \{|z| = R\}$ lies between the curves $F_{R(1-\varepsilon)}$ and $F_{R(1+\varepsilon)}$ and thus

$$d_\Omega(F_{R(1-\varepsilon)}, \partial\Omega) \le d_\Omega(C_R, \partial\Omega) \le d_\Omega(F_{R(1+\varepsilon)}, \partial\Omega).$$

To prove the theorem, it therefore suffices to show that

$$2\pi d_\Omega(F_R, \partial\Omega) = \log R + \gamma. \tag{H.17}$$

Let $\Omega_R = \{z \in \Omega : g(z) < \log R + \gamma\}$. Then

$$\omega(z, F_R, \Omega_R) = \frac{g(z)}{\log R + \gamma}.$$

As in the proof of Theorem H.1, remove the level curves of $\widetilde{\omega}$ that pass through the critical points of g, and obtain conformal maps of simply connected subdomains onto a stack of rectangles. The extremal distance, $d_\Omega(F_R, \partial\Omega)$, is the reciprocal of the sum of the heights of the rectangles. Then because $\partial\widetilde{g}/\partial s = -\partial g/\partial n < 0$ on F_R, the sum of the heights of the rectangles is

$$\frac{-\int_{F_R} \frac{\partial \widetilde{g}}{\partial s} ds}{\log R + \gamma}.$$

Since $g - C$ is Green's function for $\{z : g(z) - C > 0\}$, by (2.12) of Chapter II applied to the constant function $u \equiv 1$ we have,

$$\int_{F_R} \frac{\partial \widetilde{g}}{\partial s} ds = -\int_{F_R} \frac{\partial g}{\partial n} = -2\pi,$$

Thus

$$d_\Omega(F_R, \partial\Omega) = \frac{\log R + \gamma}{2\pi},$$

proving (H.17) and the theorem follows. ∎

The reduced extremal distance on finitely connected Jordan domains is related to Green's function on a doubled Riemann surface in the same way that extremal distance is related to harmonic measure on a doubled Riemann surface. To see the connection between Theorem V.3.2 and Theorem H.9, suppose Ω is a finitely connected Jordan domain, suppose E is contained in one component of $\partial\Omega$, and let $z_0 \in \Omega$. Let Ω_d be the Riemann surface double of Ω obtained by attaching two copies of Ω along $\partial\Omega \setminus E$. As in the proof of Theorem V.3, there is a conformal map f of Ω onto a region of the form

$$A = \mathbb{C}^* \setminus (\overline{\mathbb{D}} \cup \bigcup_{j=1}^{n} L_j),$$

where $L_1, \ldots, L_n$ are radial line segments, $f(E) = \partial\mathbb{D}$, and $f(z_0) = \infty$. The map f is given by $-\log|f| = g_{\Omega_d}(z, z_0') + g_{\Omega_d}(z, z_0'')$, where z_0' and z_0'' denote the two points in Ω_d corresponding to $z_0 \in \Omega$. As in the rectangle example, the radial slits do not affect $d_{\mathbb{D}}(C_R, \partial\mathbb{D})$, so by Theorem H.9 and the definition of reduced extremal distance

$$\delta_\Omega(z_0, E) = \delta_A(\infty, \partial\mathbb{D}) = -\frac{\gamma_A}{2\pi}. \tag{H.18}$$

When Ω is a Jordan domain, we can explicitly find the right side of (H.18) and that is how we proved Theorem V.3.2.

The analogue of Theorem H.1 for reduced extremal distance is in Exercise V.2.

Notes

The proof of Theorem H.1 in Ahlfors [1973] is similar to the proof given here, except in that proof Ω is cut with level lines of ω, instead of $\widetilde{\omega}$, giving rectangles and annuli of various sizes. In the case of slit rectangles as in Section IV.4, this corresponds to making vertical, instead of horizontal, cuts through the ends of the slits. The proof of Theorem H.1 given here derives (H.1)–(H.3) from the existence of harmonic measure on Ω_d, and these are the only properties of harmonic measure that were used. The proof of Corollary H.2 shows that the problem (H.4) of minimizing Dirichlet norm and the more general problem of finding $d_\Omega(E, F)$ have the same solution.

Exercise H.1. (a) Formulate and prove a version of Theorem IV.4.1 for finitely connected Jordan domains Ω such that $E \cup F$ is contained in one component of $\partial\Omega$.

(b) Formulate and prove the analogue of Theorem IV.4.2 for finitely connected domains Ω such that E and F are contained in different components of $\partial\Omega$.

Exercise H.2. In Theorem IV.6.1 and Theorem H.8, if I_x contains a finite union of intervals I_j of length $\theta_j(x)$, each of which separates z_0 from F, then defining

$\rho = 1/\theta_j(x)$ on I_j improves the estimate in (IV.6.3) by replacing $1/\theta(x)$ with $\sum 1/\theta_j(x)$. Give an example where this gives an improvement to the estimate (G.1) of Carleman. See Haliste [1965].

Exercise H.3. Let Ω be the unit disc with two discs $B(r, \varepsilon)$ and $B(-r, \varepsilon)$ removed. Derive Example H.6 directly in this special case by applying the map z^2, then using Example IV.1.2 and the symmetry rule.

Exercise H.4. Show that it is possible to have two distinct critical points on $\partial\mathbb{D}$ in Example H.5, and show that it is possible to have one critical point of order 2 on $\partial\mathbb{D}$.

Exercise H.5. Extend Theorem H.9 to any domain whose boundary has positive capacity.

I. McMillan's Twist Point Theorem

Theorem I.1. *Let φ be the conformal mapping from $\mathbb{D}$ onto a simply connected domain Ω. At almost every $\zeta \in \partial\mathbb{D}$ either*
(a) *φ has a finite non-zero angular derivative, or*
(b) *$\arg\big(\varphi(z) - \varphi(\zeta)\big)$ is unbounded above and below on every arc $\sigma \subset \mathbb{D}$ having endpoint ζ.*

Note that since $\varphi(z) - \varphi(\zeta) \neq 0$, continuous branches of $\arg\big(\varphi(z) - \varphi(\zeta)\big)$ are defined in $\mathbb{D}$. If $\varphi(\zeta_1) = \varphi(\zeta_2)$ and if (a) holds at ζ_1, then $\varphi(\zeta_1)$ is a cone point for Ω and (b) cannot hold at ζ_2. Thus it is consistent to call $\varphi(\zeta)$ a **twist point** if (b) holds at ζ. By Theorem VI.6.1, (a) holds almost everywhere on $\{\zeta : \varphi(\zeta) \text{ is a cone point for } \Omega\}$, while

$$\omega \ll \Lambda_1 \ll \omega$$

on $\varphi\big(\{\zeta : \text{(a) holds}\}\big)$ and

$$\omega \perp \Lambda_1$$

on the set of twist points of ω. However, we are not saying Λ_1(twist points) $= 0$, because that is not true. By Theorem VI.4.1, the twist point theorem only says that the twisting described in (b) occurs at almost every Plessner point for φ'. Theorem I.1 is from McMillan [1969].

The proof of Theorem I.1 consists of two lemmas. The first lemma is from Pommerenke [1975], page 555; see also Pommerenke [1991], page 142.

Lemma I.2. *Fix $\alpha > 2$, and let E be the set of $\zeta \in \partial\mathbb{D}$ such that the nontangential limit $\varphi(\zeta)$ exists and*

$$\sup_{\Gamma_\alpha(\zeta)} \left|\arg\big(\varphi(z) - \varphi(\zeta)\big)\right| < \infty. \tag{I.1}$$

Then almost everywhere on E

$$\sup_{\Gamma_\alpha(\zeta)} \arg\big(\varphi'(z)\big) < \infty. \tag{I.2}$$

Neither the hypotheses (I.1) nor the conclusion (I.2) depends on which continuous branches of $\arg\big(\varphi(z) - \varphi(\zeta)\big)$ and $\arg\big(\varphi'(z)\big)$ that we have chosen. But throughout the proof let us fix a continuous function $h(z, w)$ on $\mathbb{D} \times \mathbb{D}$ so that

$$h(z, w) = \begin{cases} \arg\Big(\frac{\varphi(w) - \varphi(z)}{w - z}\Big), & \text{if } z \neq w, \\ \arg(\varphi'(z)), & \text{if } z = w. \end{cases}$$

For $\zeta \in \partial\mathbb{D}$ set

$$h(z, \zeta) = \arg\Big(\frac{\varphi(\zeta) - \varphi(z)}{\zeta - z}\Big) = \lim_{\Gamma_\alpha(\zeta) \ni w \to \zeta} \arg\Big(\frac{\varphi(w) - \varphi(z)}{w - z}\Big).$$

The estimate

$$\left|\frac{\varphi(w) - \varphi(z)}{(1 - |z|^2)\varphi'(z)}\right| \leq \frac{\left|\frac{w-z}{1-\bar{z}w}\right|}{1 - \left|\frac{w-z}{1-\bar{z}w}\right|^2},$$

is only the conformally invariant form of inequality (4.14) of Chapter I. It shows that

$$|h(z, w) - h(z, z)| < \frac{\pi}{2} \tag{I.3}$$

whenever $|w - z| < (1 - |z|^2)/10$.

Proof. If the lemma is false, then there is $M < \infty$ and measurable $E_1 \subset E$ with $|E_1| > 0$ such that

$$\sup_{\Gamma_\alpha(\zeta)} \left|h(z, \zeta)\right| \leq M,$$

but

$$\sup_{\Gamma_\alpha(\zeta)} h(z, z) = \infty$$

for all $\zeta \in E_1$. We may assume $\zeta = 1$ is a point of density of E_1. Then there exist

$$\Gamma_\alpha(1) \ni z_n \to 1$$

such that

$$\lim_{n \to \infty} \arg \varphi'(z_n) = \infty.$$

Set

$$J = \partial D \cap \{\mathrm{Re}\, z \geq 0\},$$

and

$$T_n(z) = \frac{z_n}{|z_n|} \frac{z + |z_n|}{1 + |z_n|z}.$$

Then $J_n = T_n(J)$ is the arc on $\partial \mathbb{D}$, with center $\frac{z_n}{|z_n|}$, cut off by an orthogonal circle through z_n.

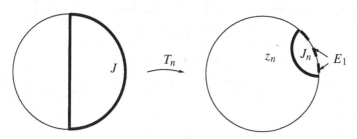

Figure I.1

Because 1 is a point of density,

$$\frac{|J_n \cap E_1|}{|J_n|} \to 1,$$

and since on J

$$\frac{1 - |z_n|^2}{4} \leq |T_n'(z)| \leq 1 - |z_n|^2,$$

we have

$$\frac{|J \cap T_n^{-1}(E_1)|}{|J|} \to 1.$$

Set

$$\psi_n(z) = \frac{\bar{z}_n}{|z_n|} \frac{\varphi(T_n(z)) - \varphi(z_n)}{(1 - |z_n|^2)\varphi'(z_n)}.$$

Then ψ_n is a normalized univalent function and by Corollary V.3.6, there exists $\zeta_n \in J \cap T_n^{-1}(E_1)$ so that

$$\int_0^1 |\psi_n'(r\zeta_n)| dr \le \lambda,$$

with constant λ independent of n. By the distortion theorem, $|\psi_n(z)| > \frac{1}{13}$ on $|z| \ge \frac{1}{10}$, and hence

$$\left| \arg \psi_n(\zeta_n) - \arg \psi_n\left(\frac{\zeta_n}{10}\right) \right| \le \int_{\frac{1}{10}}^1 \left| \frac{\psi_n'}{\psi_n}(r\zeta_n) \right| dr \le 13\lambda. \qquad (I.4)$$

Set $w_n = T_n(\frac{\zeta_n}{10})$. Then by (I.4),

$$\left| \big(h(z_n, T_n(\zeta_n)) - h(z_n, z_n) \big) - \big(h(z_n, w_n) - h(z_n, z_n) \big) \right|$$
$$\le 13\lambda + \arg\left(\frac{w_n - z_n}{T_n(\zeta_n) - z_n} \right), \qquad (I.5)$$

which is the invariant form of (I.4). Then because $\left| \arg\left(\frac{w_n - z_n}{T_n(\zeta_n) - z_n} \right) \right| \le \frac{\pi}{2}$ and $|w_n - z_n| \le (1 - |z_n|^2)/10$, (I.3) and (I.5) yield

$$\left| h(z_n, T_n(\zeta_n)) - h(z_n, z_n) \right| \le 13\lambda + \frac{\pi}{2}.$$

Because $z_n \in \Gamma_\alpha(T_n(\zeta_n))$ and $T_n(\zeta_n) \in E_1$, that means $|h(z_n, T_n(\zeta_n))| \le M$ as (I.2) asserts. ∎

Write

$$\mathrm{Arg}(\varphi(z) - \varphi(\zeta))$$

for that continuous branch of $\arg(\varphi(z) - \varphi(\zeta))$ in $\mathbb{D}$ satisfying

$$0 \le \mathrm{Arg}\big(\varphi(0) - \varphi(\zeta)\big) \le 2\pi.$$

The second lemma is from McMillan's paper [1969].

Lemma I.3. *If $\varphi(z)$ has nontangential limit $\varphi(\zeta)$ at ζ and if $\sigma \subset \mathbb{D}$ is an arc with endpoint ζ such that*

$$\sup_{z \in \sigma} \mathrm{Arg}\big(\varphi(z) - \varphi(\zeta)\big) \le M < \infty, \qquad (I.6)$$

then

$$\limsup_{\mathbb{D} \ni z \to \zeta} \mathrm{Arg}\big(\varphi(z) - \varphi(\zeta)\big) < \infty. \qquad (I.7)$$

Proof. Let $z_0 \in \mathbb{D}$ be the initial point of σ. We can assume that $\zeta = 1$ and that $|\varphi(z_0) - \varphi(1)| > 1$. But we do not assume that $\varphi(z) \to \varphi(1)$ when $\sigma \ni z \to 1$.

Set $C_n = \{w : |w - \varphi(1)| = 2^{-n}\}$. Each component of $\varphi^{-1}(C_n)$ is a Jordan arc in $\mathbb{D}$ at each end of which $\varphi(z)$ has a limit in $C_n \cap \partial\Omega$. Hence by Lindelöf's theorem, each component of $\varphi^{-1}(C_n)$ has two distinct endpoints on $\partial\mathbb{D} \setminus \{1\}$.

Let V_n be that component of $\mathbb{D} \setminus \varphi^{-1}(C_n)$ such that for some $r_n < 1$,

$$[r_n, 1) \subset V_n, \tag{I.8}$$

and set

$$\gamma_n = \mathbb{D} \cap \partial V_n.$$

Now

$$\sigma \cap V_n \neq \emptyset \tag{I.9}$$

because otherwise the component of $\varphi^{-1}(C_n)$ which separates σ from $[r_n, 1)$ would have endpoint $\zeta = 1$. Therefore

$$\sigma \cap \gamma_n \neq \emptyset$$

since $z_0 \notin V_n$. By (I.8),

$$\mathbb{D} \cap \overline{V}_n \subset V_0,$$

and hence

$$\gamma_n \subset V_0.$$

Set

$$\Gamma_n = V_0 \cap \varphi^{-1}(C_n) \supset \gamma_n.$$

Then by (I.9), $\sigma \cap \Gamma_n \neq \emptyset$, and there exists an arc $\sigma_n \subset \sigma \cap V_0$ joining a point on γ_0 to a point on Γ_n. Therefore

$$\varphi(\sigma_n) \subset U_n = \{2^{-n} < |w - \varphi(1)| < 1\},$$

and $\varphi(\sigma_n)$ joins $w_n \in \varphi(\gamma_0)$ to $w'_n \in C_n$.

On $\overline{U}_n \setminus \overline{\varphi(\sigma_n)}$ the function

$$A_n(w) = \arg(w - \varphi(1))$$

has a continuous branch. Let $A_n^-(w)$ and $A_n^+(w) = A_n^-(w) + 2\pi$ be the two boundary values of A_n at $w \in \sigma_n$. Then by (I.6) and the maximum principle,

$$A_n(w) - A_n^-(w_n) \leq 2\pi + \sup_{\varphi(\sigma_n)} \left\{ \arg(w - \varphi(1)) - \arg(w_n - \varphi(1)) \right\} \tag{I.10}$$

$$\leq 2M.$$

Connect w', $w'' \in \varphi(\gamma_0)$ by an open Jordan arc $J_0 \subset \varphi(V_0)$, which is a connected open set, and let W be the bounded component of $\mathbb{C} \setminus (C_0 \cup J)$ not containing $\varphi(1)$. On the closed Jordan domain $\overline{W}$, there is a continuous branch of $\arg(w - \varphi(1))$ which matches $\mathrm{Arg}(w - \varphi(1))$ on J. But on $C_0 \cap \partial W$, which is an arc of the circle C_0, $\arg(w - \varphi(1))$ cannot increase by 2π. Consequently

$$\left| \mathrm{Arg}(w' - \varphi(1)) - \mathrm{Arg}(w'' - \varphi(1)) \right| \leq 2\pi$$

and because $\sigma \cap \gamma_0 \neq \emptyset$,

$$\sup_{\gamma_0} \mathrm{Arg}\left(w - \varphi(1)\right) \leq M + 2\pi, \tag{I.11}$$

by the hypothesis (I.6).

Finally, let $w \in \varphi(V_0) \setminus \varphi(\sigma)$, and let $\alpha \subset \varphi(V_0)$ be a Jordan arc joining w to some $w' \in \gamma_0$. For n large, $\alpha \cup \{w\} \subset U_n$. If W_n is that component of $\varphi(V_0) \cap U_n$ with $w \in W_n$, then $\varphi(\gamma_0) \cap \partial W_n$ contains an arc of C_0. Hence W_n contains a second arc β which joins w to $w'' \in \varphi(\gamma_0)$ so that $\beta \cap \varphi(\sigma) = \emptyset$.

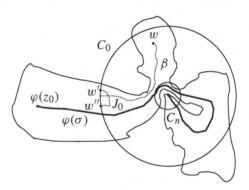

Figure I.2

Then by (I.10) and (I.11),

$$\mathrm{Arg}(w - \varphi(1)) - \mathrm{Arg}(w'' - \varphi(1)) = A_n(w) - A_n(w'') \leq 2M + 2\pi,$$

and by the definition of V_0, we obtain (I.7). ∎

Proof of Theorem I.1. We already know (a) holds whenever φ' has a non-zero finite nontangential limit. Thus it is enough to show (b) holds almost everywhere on

$$\{\text{Plessner points of } \varphi'\} \cap \{\zeta : \varphi(\zeta) \text{ exists}\}.$$

But for almost every such ζ, $|\mathrm{Arg}(\varphi(z) - \varphi(\zeta))|$ is not bounded, by Lemma I.2, and thus almost every such ζ is a twist point by Lemma I.3. ∎

J. Bloch Martingales and the Law of the Iterated Logarithm

Let g be a Bloch function and set

$$G(z) = \int_0^z g(w)\,dw.$$

Let $I = (a, b)$ be a subarc of $\partial \mathbb{D}$. Then by Theorem VII.1.3 and a partial integration, the limit

$$g_I = \lim_{r \to 1} \frac{1}{|I|} \int_I g(re^{i\theta})\,d\theta = \lim_{r \to 1} \frac{-i}{r|I|} \int_I e^{-i\theta} \frac{dG(re^{i\theta})}{d\theta}\,d\theta \tag{J.1}$$

$$= \frac{-ie^{-ib}G(e^{ib}) + ie^{-ia}G(e^{ia})}{|I|} + \frac{1}{|I|} \int_I e^{-i\theta} G(e^{i\theta})\,d\theta$$

exists. If I and J are adjacent intervals of equal length, then by (J.1)

$$g_{I \cup J} = \frac{1}{2}(g_I + g_J). \tag{J.2}$$

In the probability space

$$\left([0, 2\pi), \mathcal{M}, \frac{d\theta}{2\pi} \right),$$

where $\mathcal{M}$ is the Borel sets, let $\mathcal{M}_n$ be the σ-algebra generated by the dyadic intervals

$$I_{n,j} = \left[\frac{j2\pi}{2^n}, \frac{(j+1)2\pi}{2^n} \right), \quad 0 \le j < 2^n,$$

with n fixed. Thus $\mathcal{M}_n$ is the set of unions of arcs $I_{n,j}$, again with n fixed, and $\mathcal{M}_n \subset \mathcal{M}_{n+1}$. Then

$$g_n(\zeta) = \sum_0^{2^n - 1} g_{I_{n,j}} \chi_{I_{n,j}}(\zeta)$$

is measurable with respect to $\mathcal{M}_n$, and

$$g_n = E(g_{n+1} | \mathcal{M}_n)$$

by (J.2). Therefore $\{g_n, \mathcal{M}_n\}$ is a martingale. Martingales with respect to the dyadic σ-algebras $\mathcal{M}_n$ are called **dyadic martingales**.

Lemma J.1. *There is a constant C such that if $g \in \mathcal{B}$, then*

$$\left| g\big((1 - 2^{-n})\zeta\big) - g_n(\zeta) \right| \le C \|g\|_{\mathcal{B}}, \tag{J.3}$$

almost everywhere on $\partial\mathbb{D}$ *and*

$$\left|g_{n+1} - g_n\right| \le C\|g\|_B. \tag{J.4}$$

Proof. The argument is like the proof of Theorem VII.1.3. Define

$$T_{n,j} = \left\{r\zeta : 1 - 2^{-n} \le r \le 1 - 2^{-(n+1)}, \zeta \in I_{n,j}\right\}.$$

Then $T_{n,j}$ has bounded hyperbolic diameter, so that by Theorem VII.1.2,

$$\sup_{z,w\in T_{n,j}} \left|g(z) - g(w)\right| \le C_1\|g\|_B. \tag{J.5}$$

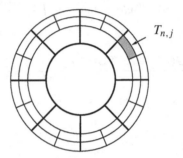

Figure J.1

On the other hand, by (VII.1.9) and the proof of (J.1),

$$\left|g_{I_{n,j}} - \frac{1}{|I_{n,j}|}\int_{I_{n,j}} g\left((1 - 2^{-n})e^{i\theta}\right)d\theta\right| \le C_1\|g\|_B \tag{J.6}$$

and together (J.5) and (J.6) imply both (J.3) and (J.4). ∎

A real valued dyadic martingale $\{f_n\}$ is called a **Bloch martingale** if there exists $g \in B$ such that

$$f_n(\zeta) = \operatorname{Re}g_n(\zeta)$$

for all n. Bloch martingales have a very simple characterization:

Lemma J.2. *Let* $\{f_n\}$ *be a real dyadic martingale. Then* $\{f_n\}$ *is a Bloch martingale if and only if* $\{f_n\}$ *has bounded jumps*

$$\left|f_{n,j} - f_{n,j+1}\right| \le C, \tag{J.7}$$

where $f_{n,j} = \left(f_n\right)_{I_{n,j}}$.

The jump condition (J.7) implies that $\{f_n\}$ has bounded differences

$$\left|f_n - f_{n+1}\right| \le C. \tag{J.8}$$

However, (J.8) is weaker than (J.7) because nearly half the pairs $I_{n,j}$, $I_{n,j+1}$ do not have a common mother $I_{n-1,k}$.

Proof. When $\{f_n\}$ is a Bloch martingale, (J.7) follows from (J.5). Now suppose $\{f_n\}$ satisfies (J.7), and note that (J.7) implies an inequality better than itself, namely

$$\left|(f_n)_{I_{m,j}} - (f_n)_{I_{m,j+p}}\right| \le Cp. \tag{J.9}$$

Indeed if $m \ge n$, (J.9) is the same as (J.7), while if $m < n$, (J.9) follows from (J.7) and the martingale relation for $\{f_n\}$.

Assume $f_0 = 0$ and set

$$h_n(\theta) = \int_0^\theta f_n(t)\,dt.$$

Then

$$h_n(\theta + 2^{-k}) + h_n(\theta - 2^{-k}) - 2h_n(\theta) = 2^{-k}\left((f_n)_{[\theta,\theta+2^{-k}]} - (f_n)_{[\theta-2^{-k},\theta]}\right)$$

and expressing $[\theta, \theta + 2^{-k}]$ and $[\theta - 2^{-k}, \theta]$ as unions of dyadic intervals, we conclude from (J.9) that

$$\left|h_n(\theta + 2^{-k}) + h_n(\theta - 2^{-k}) - 2h_n(\theta)\right| \le C2^{-k}.$$

It then follows via the proof of Theorem II.3.3 that

$$\|h_n\|_{\Lambda^*} \le C_1,$$

with constant C_1 independent of n. By Theorem II.3.3, Λ^* is an equicontinuous family, and therefore $h(e^{i\theta}) = \lim h_n(e^{i\theta})$ exists and $\|h\|_{\Lambda^*} \le C$. But then $g = (u_h + i\tilde{u}_h)'$ satisfies $\mathrm{Re}\,g_n = f_n$, so that $\{f_n\}$ is a Bloch martingale. ∎

By Lemma J.2 and Theorem VII.2.1, questions about the boundary behavior of univalent functions become questions about discrete dyadic martingales that satisfy the jump condition (J.7). We will follow Makarov [1990a] and derive the Bloch function law of the iterated logarithm, Theorem VIII.1.1, from the martingale law of the iterated logarithm.

In the martingale law of the iterated logarithm, energy is measured by *square functions*. When $\{f_n, \mathcal{T}_n\}$ is a martingale with $E(|f_n|^2) < \infty$, write

$$\Delta_n f = \Delta_n = f_n - f_{n-1},$$

for the martingale difference and define the **square function** as

$$S_n = S_n(f) = E\left(\sum_{j=1}^n (\Delta_j)^2 \Big| \mathcal{F}_{j-1}\right)$$

$$= S_{n-1} + E\left((f_n^2 - 2f_n f_{n-1} + f_{n-1}^2)\Big| \mathcal{F}_{n-1}\right) \qquad \text{(J.10)}$$

$$= S_{n-1} + E\left(f_n^2 \Big| \mathcal{F}_{n-1}\right) - f_{n-1}^2.$$

By its definition the square function S_n is **predictable**, i.e., measurable with respect to the recent past $\mathcal{F}_{n-1}$, and by (J.10) $f_n^2 - S_n$ is a martingale. By the Cauchy–Schwarz inequality, S_n is a non-decreasing function of n. In fact, these three properties determine S_n uniquely. See Exercise 1. When $\{\mathcal{F}_n\}$ is the dyadic filtration $\{\mathcal{M}_n\}$, as it is with Bloch martingales, $(\Delta_n)^2 = (f_n - f_{n-1})^2$ is predictable and (J.10) simplifies into $S_n = \sum_{j=1}^n (\Delta_j)^2$.

Theorem J.3 (Lévy). *Let* $\{f_n\}$ *be a real martingale. Then* $\{f_n\}$ *has a finite limit almost everywhere on* $\{S = \lim_n S_n < \infty\}$.

This is the martingale analogue of Theorem X.1.3. See Lévy [1937].

Proof. Fix $M > 0$ and let $\tau = \tau(x) = \inf\{n - 1 : S_n \geq M\}$. Because S_n is predictable, τ is a stopping time, which means that $\{x : \tau(x) \leq n\} \in \mathcal{M}_n$. Set

$$F_n = \begin{cases} f_n, & \text{if } n < \tau, \\ f_\tau, & \text{if } n \geq \tau. \end{cases}$$

Then F_n is a martingale, called the **stopped martingale**, and $S_n(F) \leq M$. The general identity,

$$E\left((f_n - f_0)^2\right) = E(S_n),$$

which holds because the differences Δ_j are orthogonal, shows that

$$\sup_n \|F_n\|_2 < \infty.$$

Therefore F_n has a finite limit almost everywhere by the martingale theorem. But since M is arbitrary and $f_n = F_n$ on $\{S < M\}$, that means f_n has a finite limit almost everywhere on $\{S < \infty\}$. ∎

Theorem J.4. *Let* $\{f_n\}$ *be a real dyadic martingale. Then*

$$\limsup_{n \to \infty} \frac{f_n}{\sqrt{2S_n \log \log S_n}} \leq 1, \qquad \text{(J.11)}$$

almost everywhere on $\{S = \lim_n S_n = \infty\}$. *Conversely, there is an absolute constant* $c > 0$ *such that if* $\Delta_n(f)$ *is bounded and if* $S = \infty$ *almost everywhere, then*

$$\limsup_{n \to \infty} \frac{f_n}{\sqrt{2 S_n \log \log S_n}} \geq c \qquad (J.12)$$

almost everywhere.

Actually (J.12), with $c = 1$, holds almost everywhere on $\{S = \infty\}$. See Stout [1974]. Before giving the proof of Theorem J.4, we use it and Theorem J.3 to give a proof of Theorem VIII.1.1.

Proof of Theorem VIII.1.1. Suppose $g \in \mathcal{B}$ and $\|g\|_{\mathcal{B}} = 1$. Set $f_n = \operatorname{Re} g_n$. Then by (J.8), $S_n \leq Cn$. By Theorem J.4 when $S < \infty$ and by Theorem J.5 when $S = \infty$ we have

$$|f_n| \leq C \sqrt{n \log \log n}$$

for all large n. If $2^{-n+1} \leq 1 - r < 2^{-n}$, then (J.3) gives

$$|\operatorname{Re} g(re^{i\theta}) - f_n(\theta)| \leq C_1, \qquad (J.13)$$

and therefore

$$|\operatorname{Re} g(re^{i\theta})| \leq C \sqrt{\log \frac{1}{(1-r)} \log \log \log \frac{1}{(1-r)}}$$

almost everywhere. ∎

Proof of Theorem VIII.1.2. Suppose $\|g\|_{\mathcal{B}} \leq 1$ and g satisfies (VIII.1.4). Write $g_r(z) = g(rz)$. Then by Exercise II.8 and (VIII.1.4),

$$\|\operatorname{Re}(g_r - g(0))\|_2^2 = \frac{1}{2} \|g_r - g(0)\|_2^2$$

$$\geq 2 \iint |g_r'(z)|^2 (1 - |z|^2) dx dy$$

$$\geq c(\beta, M) \log \frac{1}{1-r},$$

and for any given $A > 0$ there exists $R < 1$ so that

$$\sup_{|z| \leq R} |\operatorname{Re}(g(z) - g(0))| \geq A.$$

But since (VIII.1.4) is conformally invariant that means for every $z \in \mathbb{D}$ there is $w \in \mathbb{D}$ such that $\rho(z, w) \leq \rho(0, R)$ and $|\operatorname{Re}(g(w) - g(z))| \geq A$. Now take $A \geq 5C_1$ where C_1 is the constant in (J.13) and let $N \approx \rho(0, R)$. Then by

(J.13) we see that for every n and a.a. θ there are $n \le n_1 \le n_2 \le n + 2N$ so that $|f_{n_1}(\theta) - f_{n_2}(\theta)| \ge C_1$. Consequently

$$\sup_{\{k:|n-k|\le 2N\}} |\Delta_k f| \ge C = \frac{C_1}{2N}.$$

Therefore $S_n \ge Cn$ for some constant C, and (VIII.1.5) follows from (J.3) and (J.12). ∎

The proof of Theorem J.4 will use exponential transforms. The **exponential transform** of a real dyadic martingale $\{f_n\}$ is

$$Z_n = \frac{\exp f_n}{\prod_{j=1}^{n} \cosh \Delta_j(f)}.$$

We assume $f_0 = 0$ and take $Z_0 = 1$. When $\{f_n\}$ is a dyadic martingale, $E(e^{\Delta_n}|\mathcal{M}_{n-1}) = \cosh(\Delta_n)$ and the denominator $\prod_{j=1}^{n} \cosh \Delta_j$ is measurable with respect to $\mathcal{M}_{n-1}$. Therefore

$$E(Z_n|\mathcal{M}_{n-1}) = Z_{n-1} \frac{E(e^{\Delta_n}|\mathcal{M}_{n-1})}{\cosh(\Delta_n)} = Z_{n-1},$$

and Z_n is a martingale. In particular,

$$E(Z_n) = \int Z_n \frac{d\theta}{2\pi} = 1.$$

For $\alpha > 0$ define

$$N = N(\theta) = N_\alpha(\theta) = \inf\{n : f_n \ge \alpha\}.$$

Then N is a stopping time, i.e., $\{\theta : N(\theta) = n\} \in \mathcal{M}_n$. If g_n is a sequence of $\mathcal{M}_n$-measurable functions, the function

$$g_N = g_{N(\theta)}(\theta) = \sum_k g_k \chi_{\{N=k\}}$$

is defined on $\{N < \infty\}$ and g_N is $\mathcal{M}$-measurable. If $\{g_n\}$ is a martingale, then

$$G_{N \wedge n} = \sum_{k=0}^{n-1} g_k \chi_{\{N=k\}} + g_n \chi_{\{N \ge n\}}$$

is also a martingale.

Lemma J.5. *Let f_n be a real dyadic martingale with $f_0 = 0$, let $0 < \alpha < \beta$, and set $N = N_\alpha$. Then*

$$E_{\alpha,\beta} = \{N < \infty \text{ and } S_N \le \beta\}$$

satisfies

$$\left| E_{\alpha,\beta} \right| \le e^{-\left(\frac{\alpha^2}{2\beta} \right)}. \tag{J.14}$$

Moreover, there are constants p, $0 < p < 1$, *and* $\alpha_0 > 1$ *such that if* $|\Delta_k f| \le 1$ *and* $\alpha > \alpha_0$, *then*

$$\left| \{ S_N \le \alpha^2 \} \right| \ge p. \tag{J.15}$$

Proof. Set $t = \alpha/\beta$ and let Z_n be the exponential transform of the martingale $t f_n$. Then

$$Z_n \ge e^{t f_n} e^{-t^2 S_n/2},$$

since $\cosh t \le e^{t^2/2}$. Therefore

$$e^{\frac{\alpha^2}{2\beta}} \left| E_{\alpha,\beta} \right| = \int_{E_{\alpha,\beta}} \exp\left(t\alpha - \frac{t^2}{2}\beta \right) \frac{d\theta}{2\pi} \le \int_{E_{\alpha,\beta}} Z_N \frac{d\theta}{2\pi}.$$

On the other hand, since $Z_N = \lim Z_{N \wedge n}$ on $E_{\alpha,\beta}$ and $Z_n \ge 0$, we have

$$\int_{E_{\alpha,\beta}} Z_N \frac{d\theta}{2\pi} \le \liminf \int Z_{N \wedge n} \frac{d\theta}{2\pi} \le 1,$$

and that proves (J.14).

To prove (J.15) note that $\cosh t \ge e^{c_0 t^2}$, if $0 \le t \le 1$, where $c_0 = \log \cosh 1$, and hence $Z_n \le e^{t f_n} e^{-c_0 t^2 S_n}$. Therefore

$$1 = \int Z_{N \wedge n} \frac{d\theta}{2\pi} \le e^{t(\alpha+1)} \int \exp\left(-c_0 t^2 S_{N \wedge n} \right) \frac{d\theta}{2\pi},$$

if $|\Delta_k f| \le 1$. But $S_{N \wedge n} \to S_N$ even when $N = \infty$, and hence

$$1 \le e^{t(\alpha+1)} \int \exp\left(-c_0 t^2 S_N \right) \frac{d\theta}{2\pi},$$

or

$$\int \left(1 - e^{-c_0 t^2 S_N} \right) \frac{d\theta}{2\pi} \le 1 - e^{-t(\alpha+1)}.$$

Hence and by Chebychev's inequality

$$\left| \{ S_N > \beta \} \right| \le \frac{1 - e^{-t(\alpha+1)}}{1 - e^{-c_0 t^2 \beta}}$$

when $\beta > \alpha$. When $\beta = \alpha^2$ and $\alpha > 2$ we get (J.15) by taking $t = \frac{2}{c_0 \alpha}$. ∎

Proof of Theorem J.4. We prove the upper bound (J.10) first. Fix $\varepsilon > 0$ and set

$$E = \left\{ S = \infty, f_n > (1+\varepsilon)\sqrt{2S_n \log\log S_n} \text{ infinitely often} \right\}.$$

We show $|E| = 0$. Define the stopping time $T_k = \inf\{n : S_n \geq (1+\varepsilon)^k\}$. Then $\lim T_k = \infty$ on $\{S = \infty\}$, so that if

$$E_k = \left\{ S = \infty, \text{ there exists } n \in [T_k, T_{k+1}), \text{ with } f_n > (1+\varepsilon)\sqrt{2S_n \log\log S_n} \right\},$$

then

$$E \subset \bigcap_n \bigcup_{k>n} E_k.$$

But by (J.14),

$$|E_k| \leq \exp - \left(\frac{(1+\varepsilon)^2 2(1+\varepsilon)^k \log\log((1+\varepsilon)^k)}{2(1+\varepsilon)^{k+1}} \right) = \left(\frac{1}{k\log(1+\varepsilon)} \right)^{(1+\varepsilon)}.$$

Hence

$$|E| \leq \left(\frac{1}{\log(1+\varepsilon)} \right)^{1+\varepsilon} \lim_{n\to\infty} \sum_{k>n} \frac{1}{k^{(1+\varepsilon)}} = 0.$$

In proving the lower bound (J.11) we can assume

$$|\Delta_n f| \leq 1.$$

Set $\tau_k = \inf\{n : S_n > k\}$. Then $k \leq S_{\tau_k} \leq k+1$ and $\lim_k \tau_k = \infty$ almost everywhere. Set $F_k = f_{\tau_k}$. Now we claim that by (J.15) there exists $p > 0$ such that for all m and all sufficiently large k,

$$\left| \left\{ F_{m+k} - F_m > \sqrt{k} \right\} \right| \geq \frac{p}{2}. \tag{J.16}$$

Indeed, set $A_n = \{\tau_m = n\}$. Then A_n is a union of dyadic intervals $A_{n,j} \in \mathcal{M}_n$, and we can regard $f_{n+s} \chi_{A_{n,j}}$ as a dyadic martingale, indexed by s, with respect to the probability measure $\frac{1}{|A_{n,j}|} \chi_{A_{n,j}} dx$. We apply (J.15), with $\alpha = \sqrt{k}, \beta = k$, and $N = N_{\sqrt{k}}$ to each such martingale $f_{n+s} \chi_{A_{n,j}}$, and recall that for a dyadic martingale, $f_s - f_0$ has the same probability distribution as $-f_s + f_0$. Then we get

$$\left| \left\{ F_{m+k} - F_m > \sqrt{k} \right\} \right| = \sum_{s=1}^{k} \left| \left\{ F_{m+k} - F_m > \sqrt{k} \right\} \bigcap \left\{ N = s \right\} \right|$$

$$\geq \frac{1}{2} \sum_{s=1}^{k} \left| \left\{ N = s \right\} \right| = \frac{1}{2} \left| \left\{ N \leq k \right\} \right| \geq \frac{p}{2},$$

which is (J.16). Fix $N \geq 2$ and partition $(N^{j-1}, N^j]$ into

$$K = \left[\frac{\log j}{\log \frac{2}{p}} \right]$$

integer intervals, each of length approximately

$$k = (N^j - N^{j-1}) \frac{\log \frac{2}{p}}{\log j}.$$

Applying (J.16) in each interval and using the independence of martingale differences gives

$$\left| \left\{ F_{N^j} - F_{N^{j-1}} \geq \frac{1}{2} \sqrt{N^j \frac{\log j}{\log \frac{2}{p}}} \right\} \right| \geq p^K \geq \frac{1}{j}, \qquad (J.17)$$

because the event (J.17) is contained in an intersection of K independent events of the form (J.16). For large N we now repeat the argument leading to (J.16), but using (J.11) instead of (J.15). We conclude that for sufficiently large j,

$$F_{N^{j-1}} \leq \frac{2}{\sqrt{N}} \sqrt{N^j \log j} < \frac{1}{4} \sqrt{N^j \frac{\log j}{\log \frac{2}{p}}}, \qquad (J.18)$$

except on a set of small measure, because $S_n \leq N^{j-1} + 1$ when $\tau_{N^{j-1}} = n$. Because (J.17) and (J.18) describe events that are independent in j, the Borel–Cantelli lemma gives (J.12) with constant $c = \frac{1}{4\sqrt{\log \frac{2}{p}}}$. ∎

Exercise J.1. Let $\{f_n, \mathcal{F}_n\}$ be a martingale and set

$$S_n = \sum_{j=1}^{n} \left(E \Delta_j^2 \big| \mathcal{F}_{j-1} \right).$$

Prove that

(i) S_n is $\mathcal{F}_{n-1}$ measurable,
(ii) S_n is nondecreasing in n, and
(iii) $f_n^2 - S_n$ is a martingale with respect to $\{\mathcal{F}_n\}$.

Also prove (i), (ii), and (iii) determine S_n uniquely.

Exercise J.2. (a) Let f_n be a Bloch martingale. There exists a quasisymmetric homeomorphism $h : \partial \mathbb{D} \to \partial \mathbb{D}$ and a constant $C < \infty$ such that for every dyadic interval $I_{n,j}$,

$$|h(I_{n,j})| \leq C |I_{n,j}| \exp(f_{I_{n,j}}).$$

The constant of quasisymmetry (defined in (VII.3.5)) of h depends only on the Bloch norm of f_n. See Bishop [2002a].

(b) There is a constant $K < \infty$ such that whenever $\varphi : \mathbb{D} \to \Omega$ is a conformal map there exists a K-quasiconformal $h : \mathbb{D} \to \mathbb{D}$ such that h is bilipschitz with respect to the hyperbolic metric on $\mathbb{D}$ and such that for some $C < \infty$,

$$1 = |h(z)| \le C|\varphi'(z)|(1 - |z|)$$

for all $z \in \mathbb{D}$. Use (a). If K can be taken arbitrarily close to 2, then Brennan's conjecture is true. See Bishop [2002a] and Exercise VIII.13.

K. A Dichotomy Theorem

Let

$$p(z) = z^d + a_{d-1}z^{d-1} + \ldots + a_0$$

be a monic polynomial of degree $d \ge 2$ and let $p^{(n)}$ denote the iterate

$$p^{(n)}(z) = p \circ p \ldots \circ p(z).$$

For $p(z)$ the **basin of attraction** of ∞ is

$$\Omega = \left\{ z : \lim_{n \to \infty} p^{(n)}(z) = \infty \right\}.$$

The basin Ω is an open connected subset of C^* and Ω is simply connected if and only if

$$p'(z) \neq 0, \quad z \in \Omega. \tag{K.1}$$

If $p'(z) = 0$ for some $z \in \Omega$, then $\partial\Omega$ has infinitely many components. When $p(z) = z^2 + b$, (K.1) holds if and only if b belongs to the famous Mandelbrot set. For these facts see Carleson and Gamelin [1993]. Exercise III.8 gave some general results about harmonic measure on Ω. Here we will prove a very precise theorem when (K.1) holds for $z^2 + b$.

Theorem K.1. *Assume* (K.1) *holds and let* $\omega = \omega(\infty, ., \Omega)$. *Then either*

(a) $\omega \ll \Lambda_1 \ll \omega$, *or*

(b) *There exists* $c(\omega)$ *such that if* $h_c(t) = t \exp\left(c\sqrt{\log \frac{1}{t} \log \log \log \frac{1}{t}}\right)$, *then*

$$\omega \ll \Lambda_{h_c}, \quad for \ c > c(\omega), \tag{K.2}$$

and

$$\omega \perp \Lambda_{h_c}, \quad for \ c \le c(\omega). \tag{K.3}$$

Theorem K.1 is due to Przytycki, Urbański, and Zdunik [1989] and [1991], and their results are stronger in several ways: A simply connected domain Ω is called an **RB-domain**, for **repelling boundary**, if Ω is hyperbolic and if there is an analytic map g defined on a neighborhood U of $\partial\Omega$ such that

$$g(\partial\Omega) \subset \partial\Omega,$$

$$g(U \cap \Omega) \subset \Omega,$$

and

$$\bigcap_{n \geq 1} g^{-n}(U \cap \overline{\Omega}) = \partial\Omega.$$

When $P(z)$ satisfies (K.1), the basin of attraction of ∞ is an RB-domain. Przytycki, Urbański, and Zdunik proved Theorem K1 for RB-domains. When $z^2 + b$ satisfies (K.1), they also showed $c(\omega)^2$ is a real analytic and subharmonic function of the parameter b. In case (a), even with Ω an RB-domain, Zdunik [1991] obtained a much better conclusion: $\partial\Omega$ is either an analytic arc or a real analytic Jordan curve. Przytycki, Urbański, and Zdunik use Gibbs measures and other ideas from dynamical systems, as well as Theorem K.3 below. A second slightly less general proof is in Chapter III of Makarov [1990a]. We present a third proof, due also to Makarov.

Proof. Assume (K.1) holds. Then there is a conformal mapping $\varphi : D \to \Omega$ with

$$\varphi(z) = \frac{1}{z} + \dots$$

that conjugates $p(z)$ to the function z^d:

$$p(\varphi(z)) = \varphi(z^d), \quad z \in \mathbb{D}. \tag{K.4}$$

See page 65 of Carleson and Gamelin [1993]. Set

$$g(z) = \log\left|\frac{-1}{z^2\varphi'(z)}\right|,$$

and

$$f(z) = \log\left|\frac{z^{d-1}p'(\varphi(z))}{d}\right|. \tag{K.5}$$

Then by (K.4),

$$g(z) - g(z^d) = f(z)$$

or

$$g(z) = \sum_{k=0}^{\infty} f(z^{d^k}).$$ (K.6)

Let $\mathcal{D}_n$ be the σ-algebra generated by the **d-adic subintervals** of $[-\pi, \pi)$:

$$I_{n,j} = \left[\frac{j2\pi}{d^n}, \frac{(j+1)2\pi}{d^n} \right), \quad 0 \le j \le d^n,$$

and write

$$[h]_n = E(h|\mathcal{D}_n)$$

whenever $h \in L^1$.

Lemma K.2. *Assume* (K.1) *and let* f *be defined by* (K.5). *Then there are* $C < \infty$ *and* $\beta < 1$ *such that for all n,*

$$\|f - [f]_n\|_2 < C\beta^n.$$ (K.7)

Proof. Write $c(I)$ for the center of $I \in \mathcal{D}_n$ and set

$$a_I = (1 - |I|)c(I).$$

Because $[f]_n$ is an orthogonal projection we have

$$\|f - [f]_n\|_2^2 \le \sum_{\mathcal{D}_n} \int_I \left| f - f(a_I) \right|^2 \frac{d\theta}{2\pi}.$$

Write

$$p'(z) = d \prod_{j=1}^{d-1} (z - c_j).$$

Then almost everywhere on I,

$$f(\zeta) - f(a_I) = (d-1)\log\left(\frac{1}{1-|I|}\right) + \sum_{j=1}^{d-1} \log\left| \frac{\varphi(\zeta) - c_j}{\varphi(a_I) - c_j} \right|,$$

so that

$$|f(\zeta) - f(a_I)|^2 \le \frac{d(d-1)^2}{2}\left\{ \left(\log\left(\frac{1}{1-|I|}\right)\right)^2 + \sum_{j=1}^{d-1}\left(\log\left| \frac{\varphi(\zeta) - c_j}{\varphi(a_I) - c_j} \right|\right)^2 \right\}.$$

For $\alpha > 1/2$, set

$$J = J(I) = \left\{ \zeta \in I : \frac{|\varphi(\zeta) - \varphi(a_I)|}{|\varphi'(a_I)|} \le |I|^\alpha \right\}.$$

Then since $|\log z|^2 \le |z-1|^2 + |1/z - 1|^2$ for all $z \in \mathbb{C}$, we have

$$\int_J |f(e^{i\theta}) - f(a_I)|^2 d\theta$$

$$\le \frac{d(d-1)^2}{2}\left[|I|\left(\log\!\left(\left(\frac{1}{1-|I|}\right)\right)\right)^2 + \sum_{j=1}^{d-1} \int_J \left(\left|\frac{\varphi(e^{i\theta}) - \varphi(a_I)}{\varphi(a_I) - c_j}\right|^2\right.\right.$$

$$\left.\left. + \left|\frac{\varphi(e^{i\theta}) - \varphi(a_I)}{\varphi(e^{i\theta}) - c_j}\right|^2\right)\frac{d\theta}{2\pi}\right]$$

$$\le C(d)\left[d^{-3n} + |I|^{2\alpha}\sum_j \int_J \left(\left|\frac{\varphi'(a_I)}{\varphi(a_I)}\right|^2 + \left|\frac{\varphi'(a_I)}{\varphi(e^{i\theta})}\right|^2\right)\frac{d\theta}{2\pi}\right]$$

$$\le C(d)\left[d^{-3n} + |I|^{2\alpha}\sum_j |I|\left|\frac{\varphi'(a_I)}{\varphi(a_I)}\right|^2\right].$$

Thus by Exercise I.22,

$$\sum_{I \in \mathcal{D}_n} \int_J |f(e^{i\theta}) - f(a_I)|^2 \frac{d\theta}{2\pi}$$

$$\le C(d)\left[d^{-2n} + C|I|^{2\alpha}\int_{\partial\mathbb{D}} \left|\frac{\varphi'((1-|I|)e^{i\theta})}{\varphi((1-|I|)e^{i\theta})}\right|^2 \frac{d\theta}{2\pi}\right] \quad \text{(K.8)}$$

$$\le C'(d)|I|^{2\alpha-1}\log\!\left(\frac{1}{1-|I|}\right).$$

On the other hand, the Beurling projection theorem shows that

$$\frac{|I \setminus J|}{|I|} \le \omega\!\left(a_I, \{|\varphi(\zeta) - \varphi(a_I)| \ge |I|^\alpha |\varphi'(a_I)|\}, \mathbb{D}\right) \le C|I|^{\frac{1-\alpha}{2}},$$

while by the John–Nirenberg theorem and Exercise F.2,

$$\frac{1}{|I|}\int_I |f(e^{i\theta}) - f(a_I)|^4 \frac{d\theta}{2\pi} \le L_4.$$

Therefore

$$\int_{I \setminus J} |f(e^{i\theta}) - f(a_I)|^2 \frac{d\theta}{2\pi} \le |I \setminus J|^{1/2}(C_4|I|)^{1/2} \le (CC_4)^{1/2}|I|^{\frac{5-\alpha}{4}},$$

and

$$\sum_{\mathcal{D}_n} \int_{I \setminus J} |f - f(a_I)|^2 \frac{d\theta}{2\pi} \le C\left(d^{-n}\right)^{\frac{1-\alpha}{4}}. \quad \text{(K.9)}$$

Take $\alpha = \frac{5}{9}$. Then (K.8) and (K.9) imply (K.7) with $\beta = d^{\left(-\frac{1}{5}\right)}$. ∎

Conclusion of the proof of Theorem K.1. Because

$$E(f(e^{id^k\theta})|\mathcal{D}_k) = 0, \tag{K.10}$$

(K.7) shows that

$$\rho_k = \int f(e^{id^k\theta}) f(e^{i\theta}) \frac{d\theta}{2\pi}$$

satisfies

$$|\rho_k| \le C\beta^k \|f\|_2, \tag{K.11}$$

and the series

$$\sigma^2 = \int |f|^2 \frac{d\theta}{2\pi} + 2 \sum_{k=1}^{\infty} \int f(e^{id^k\theta}) f(e^{i\theta}) \frac{d\theta}{2\pi} \tag{K.12}$$

converges absolutely. Set

$$S_n(\theta) = \sum_{k=0}^{n} f(e^{id^k\theta}).$$

Then

$$\frac{\|S_n\|_2^2}{n+1} = \|f\|_2^2 + 2 \sum_{k=1}^{n} \left(\frac{n+1-k}{n+1}\right) \rho_k, \tag{K.13}$$

so that

$$\lim_{n\to\infty} \frac{\|S_n\|_2^2}{n+1} = \sigma^2$$

and

$$0 \le \sigma^2 < \infty.$$

The dichotomy in Theorem K.1 is between the two cases $\sigma^2 = 0$ and $\sigma^2 > 0$.
Case (a): Assume $\sigma^2 = 0$. Then by (K.13),

$$\|S_n\|_2^2 = (n+1)\sigma^2 - 2(n+1) \sum_{k=n+1}^{\infty} \rho_k - 2 \sum_{k=1}^{n} k\rho_k$$

$$= 0 - A_n - B_n,$$

and by (K.8), $A_n \to 0$, while $|B_n|$ is bounded. Therefore

$$\sup \|S_n\|_2 < \infty,$$

and by (K.6), $g(z)$ is the real part of an H^2 function. Hence φ' has a non-zero nontangential limit almost everywhere, and by Theorem V.4.2.

$$\omega \ll \Lambda_1 \ll \omega.$$

Again, Zdunik [1991] has a much stronger conclusion.

Case (b): Assume $\sigma^2 > 0$. We use the law of the iterated logarithm, in the following form: ∎

Theorem K.3. *Suppose the real function* $f \in L^2$ *satisfies condition* (K.11) *and let* σ^2 *be defined by* (K.12). *If* $\sigma^2 > 0$, *then almost everywhere,*

$$\limsup \frac{S_n}{\sqrt{n \log \log n}} = \sqrt{2}\,\sigma. \tag{K.14}$$

We omit the proof of Theorem K.3, which relies on a result of Ibragimov [1962] on the central limit theorem for weakly dependent random variables. See Resnik [1968], Theorem 2.2, and pages 79–82 of Philipp and Stout [1975].

Returning to the proof of Theorem K.1, we let $\zeta \in I \in \mathcal{D}_n$. Then

$$S_n(\zeta) - [S_n]_n(\zeta) = \sum_{j=0}^{n} \left(f(\zeta^{d^j}) - \frac{1}{|I|} \int_I f(e^{id^j\theta}) d\theta \right)$$

$$= \sum_{j=0}^{n} (f(\zeta^{d^j}) - [f]_{n-j}(\zeta^{d^j})),$$

so that by (K.7),

$$\sup_n \|S_n - [S_n]_n\|_2 \le \infty$$

and

$$\limsup \frac{|S_n - [S_n]_n|}{\sqrt{n \log \log n}} = 0$$

almost everywhere. Let $|z| = r$ satisfy $1 - d^{-n} < r \le 1 - d^{-n-1}$. Then

$$\frac{\sqrt{n \log \log n}}{\sqrt{\log \frac{1}{1-r} \log \log \log \frac{1}{1-r}}} \longrightarrow \frac{1}{\sqrt{\log d}}$$

$$\left| g(z) - [g]_n(z/|z|) \right| \le C,$$

while by (K.10),

$$[g]_n = [S_n]_n.$$

Therefore (K.14) gives

$$\limsup \frac{g(re^{i\theta})}{\sqrt{\log\frac{1}{1-r}\log\log\log\frac{1}{1-r}}} = \sqrt{\frac{2}{\log d}}\,\sigma \qquad \text{(K.15)}$$

almost everywhere and (K.2) and (K.3) both follow from (K.15) and Theorem VIII.2.1, exactly as in the proof of Theorem VIII.2.2. ∎

L. Two Estimates on Integral Means

Theorem L.1. *If φ is univalent on $\mathbb{D}$ and if $p \le 1.399$, then*

$$\int_{\mathbb{D}} \frac{1}{|\varphi'(z)|^p}\,dx\,dy < \infty.$$

Proof. The Schwarzian derivative of φ is defined to be

$$S_\varphi = \left(\frac{\varphi''}{\varphi'}\right)' - \frac{1}{2}\left(\frac{\varphi''}{\varphi'}\right)^2.$$

The Schwarzian derivative satisfies $(1 - |z|^2)^2 |S_\varphi(z)| \le 6$, by Theorem X.4.1. Moreover, if we set $h = (\varphi')^{-1/2} = \sum a_n z^n$ then

$$|h''(z)| = \left|\frac{1}{2}S_\varphi(z)h(z)\right| \le \frac{3}{(1 - |z|^2)^2}|h(z)|.$$

Thus

$$\sum |a_n|^2 n^2 (n - 1)^2 r^{2n-4} = \int_0^{2\pi} |h''(re^{i\theta})|^2 \frac{d\theta}{2\pi}$$

$$\le \frac{9}{(1 - r^2)^4}\int_0^{2\pi} |h(re^{i\theta})|^2 \frac{d\theta}{2\pi}$$

$$= \frac{9}{(1 - r^2)^4}\sum |a_n|^2 r^{2n}.$$

Set $s = r^2$ and $y(s) = \sum |a_n|^2 s^n$. Then for s_0 sufficiently close to 1,

$$y^{(4)}(s) \le \frac{9.01}{(1 - s)^4}y(s),$$

provided $s_0 \le s < 1$. The function $Y(s) = C(1 - s)^{-\lambda}$ satisfies

$$Y^{(4)}(s) = \frac{9.01}{(1 - s)^4}Y(s),$$

where λ is the positive root of $\lambda(\lambda + 1)(\lambda + 2)(\lambda + 3) = 9.01$. Choosing C sufficiently large, we may suppose that $Y^{(j)}(s_0) > y^{(j)}(s_0)$, for $j = 0, 1, 2, 3$. Thus

$$Y(s) - y(s) \geq \int_{s_0}^{s} \int_{s_0}^{t_4} \int_{s_0}^{t_3} \int_{s_0}^{t_2} \frac{9.01}{(1 - t_1)^4} (Y(t_1) - y(t_1)) dt_1 dt_2 dt_3 dt_4.$$

Note that if $Y(t_1) - y(t_1) > 0$ for $s_0 \leq t_1 < s$, then $Y(s) - y(s) > 0$ and hence $Y(s) - y(s) > 0$ for $s_0 \leq s < 1$. Thus

$$\int_0^{2\pi} \frac{1}{|\varphi'(re^{i\theta})|} \frac{d\theta}{2\pi} = y(r^2) \leq \frac{C}{(1 - r^2)^\lambda}. \tag{L.1}$$

In other words, we have $\beta(-1) \leq \lambda$.

By the distortion theorem $|\varphi'(re^{i\theta})| \geq C(1 - r)$, and hence by (L.1),

$$\iint_{\mathbb{D}} \frac{1}{|\varphi'(re^{i\theta})|^p} r d\theta dr \leq \int_0^1 \frac{C}{(1 - r)^{p-1+\lambda}} dr < \infty,$$

provided $p < 2 - \lambda$. A calculation shows that $0.600 < \lambda < 0.601$, and the theorem follows. ∎

Theorem L.2. *If φ is univalent on $\mathbb{D}$ then*

$$\beta_\varphi(t) \leq -\frac{1}{2} + t + \left(\frac{1}{4} - t + 4t^2\right)^{1/2}. \tag{L.2}$$

Proof. By Hardy's identity (VIII.1.6) for $t = 2p$,

$$I(r) = I(r, \varphi') = \frac{1}{2\pi} \int_0^{2\pi} |\varphi'(re^{i\theta})|^t d\theta$$

satisfies

$$r \frac{d}{dr}(rI'(r)) = \frac{t^2}{2\pi} \int_0^{2\pi} |\varphi'(re^{i\theta})|^t \left| r \frac{\varphi''(re^{i\theta})}{\varphi'(re^{i\theta})} \right| d\theta.$$

Since $I'(r) \geq 0$, this yields

$$r^2 I''(r) \leq \frac{t^2}{2\pi} \int_0^{2\pi} |\varphi'(re^{i\theta})|^t \left| \left(z \frac{\varphi''(z)}{\varphi'(z)} - \frac{2r^2}{1 - r^2} \right) + \frac{2r^2}{1 - r^2} \right|^2 d\theta.$$

By (I.4.20)

$$\left| \frac{z\varphi''(z)}{\varphi'(z)} - \frac{2r^2}{1 - r^2} \right| \leq \frac{4r}{1 - r^2},$$

and so we obtain

$$
\begin{aligned}
r^2 I''(r) \le &\ \frac{4r^2 t^2 (4 + r^2)}{(1 - r^2)^2} I(r) \\
&+ \frac{t^2}{2\pi} \int_0^{2\pi} |\varphi'(re^{i\theta})|^t \left(\mathrm{Re}\left(\frac{z\varphi''(z)}{\varphi'(z)}\right) - \frac{2r^2}{1 - r^2} \right) \frac{4r^2}{1 - r^2} d\theta.
\end{aligned}
\tag{L.3}
$$

But by definition

$$
r I'(r) = \frac{t}{2\pi} \int_0^{2\pi} |\varphi'(re^{i\theta})|^t \mathrm{Re}\left(\frac{z\varphi''(z)}{\varphi'(z)}\right) d\theta.
$$

Therefore (L.3) yields

$$
r^2 I''(r) \le \frac{4r^2 t^2 (4 + r^2)}{(1 - r^2)^2} I(r) + \frac{4r^3 t}{1 - r^2} I'(r),
$$

and given $\varepsilon > 0$ there is $r_0 = r_0(\varepsilon)$ so that if $r_0 < r < 1$,

$$
I''(r) \le \frac{3t^2 + \varepsilon}{(1 - r)^2} I(r) + \frac{2t + \varepsilon}{1 - r} I'(r).
\tag{L.4}
$$

Let $g(r) = c(1 - r)^{-a}$ where a is a positive solution of

$$
a(a + 1) = (2t + \varepsilon)a + 3t^2 + \varepsilon
$$

and where $c > 0$ is so large that $g(r_0) > I(r_0)$ and $g'(r_0) > I'(r_0)$. Then since

$$
g''(r) = \frac{3t^2 + \varepsilon}{(1 - r)^2} g(r) + \frac{2t + \varepsilon}{1 - r} g'(r),
$$

(L.4) and a little calculus shows $I(r) < g(r)$ on $r_0 < r < 1$. Sending $\varepsilon \to 0$ then gives (L.2). ∎

Exercise L.1. Use Theorem L.1 to prove $B(-2) \le \lambda + 1 < 1.601$.

Notes

These results are from Pommerenke [1985a]. See Bertilsson [1998], [1999], Shimorin [2003] and Hedenmalm and Shimorin [2004] for improvements on Theorem L.1.

M. Calderón's Theorem and Chord-Arc Domains

In this appendix we prove Theorem X.1.1 for domains bounded by chord-arc curves. Let $f(z)$ be analytic on a simply connected domain Ω, let $\zeta \in \partial\Omega$, and let $\alpha > 1$. We recall the definitions of the cone

$$\Gamma_\alpha(\zeta) = \Gamma_\alpha(\zeta, \Omega) = \{z \in \Omega : |z - \zeta| < \alpha \operatorname{dist}(z, \partial\Omega)\},$$

the maximal function

$$f_\alpha^*(\zeta) = \sup_{\Gamma_\alpha(\zeta)} |f(z)|,$$

and the area function

$$A_\alpha f(\zeta) = \left(\iint_{\Gamma_\alpha(\zeta)} |f'(z)|^2 dxdy \right)^{\frac{1}{2}}.$$

The following theorem is from Jerison and Kenig [1982a].

Theorem M.1. *Let Ω be a domain bounded by a chord-arc curve, let $\alpha > 1$, let $0 < p < \infty$, and let $f(z)$ be analytic on Ω. Then $f \in H^p(\Omega)$ if and only if $A_\alpha f \in L^p(\partial\Omega, ds)$, and for $z_0 \in \Omega$ there is a constant $c = c(z_0, \Omega, p, \alpha)$ such that*

$$c\|A_\alpha f\|_{L^p(\partial\Omega)}^p \leq \|f - f(z_0)\|_{H^p(\Omega)}^p \leq c^{-1}\|A_\alpha f\|_{L^p(\partial\Omega)}^p.$$

Let $\varphi : \mathbb{D} \to \Omega$ be a conformal mapping with $\varphi(0) = z_0$. The proof depends on three crucial properties of the chord-arc domain Ω :

(i) There exist $\beta = \beta(\alpha, \Omega, z_0) > 1$ and $\gamma = \gamma(\alpha, \Omega, z_0) > 1$ such that for all $\zeta \in \partial\mathbb{D}$,

$$\varphi(\Gamma_\beta(\zeta)) \subset \Gamma_\alpha(\varphi(\zeta)) \subset \varphi(\Gamma_\gamma(\zeta)). \qquad (M.1)$$

This holds because $\partial\Omega$ is a quasicircle, and β and γ depend only on the dilatation of the quasiconformal extension of φ. See Exercise VII.8.

(ii) On $\partial\mathbb{D}$ the measures $d\mu(\theta) = |\varphi'(e^{i\theta})|d\theta$ and $d\theta$ are A_∞-equivalent. This is proved in Section VII.4.

(iii) There is $q > 0$ such that

$$\frac{1}{\varphi'(z)} \in H^q(\mathbb{D}). \qquad (M.2)$$

This is also proved in Section VII.4.

A careful reading of the proof of Theorem M.1 below will show that the constant $c(\alpha, p, z_0, \Omega)$ depends only on α, p, the ratio $\operatorname{dist}(z_0, \partial\Omega)/\operatorname{diam}\Omega$, and the constants implicit in the properties (i), (ii), and (iii) of Ω cited above.

Lemma M.2. *Let Ω be a chord-arc domain, let $\alpha > 1$, and let $0 < p < \infty$. Suppose $f(z)$ is analytic on Ω. Then $f \in H^p(\Omega)$ if and only if $f_\alpha^* \in L^p(\partial\Omega, ds)$,* *and there exists a constant $c = c(\Omega, \alpha, p)$ such that*

$$c\|f^*\|_p^p \le \|f\|_{H^p(\Omega)}^p \le c^{-1}\|f^*\|_p^p. \qquad \text{(M.3)}$$

Proof. Assume $f \in H^p(\Omega)$; that is, assume $F(z)(\varphi'(z))^{\frac{1}{p}} \in H^p(\mathbb{D})$, where $F(z) = f \circ \varphi(z)$. Then F has nontangential limit a.e. and

$$\int |F(e^{i\theta})|^p |\varphi'(e^{i\theta})| d\theta = \|f\|_{H^p(\Omega)}^p < \infty.$$

Let us first suppose $p \ge p_\circ = \frac{q+1}{q}$, where q is given in (M.2). Then by Hölder's inequality, with exponents $p_\circ$ and $q+1$, we have

$$\int |F(re^{i\theta})| d\theta \le \left(\int \{|F(re^{i\theta})||\varphi'(re^{i\theta})|^{\frac{1}{p_\circ}}\}^{p_\circ} d\theta \right)^{\frac{1}{p_\circ}} \left(\int \left|\frac{1}{\varphi'(re^{i\theta})}\right|^q d\theta \right)^{\frac{1}{q+1}}$$

$$\le \left(\int |F(re^{i\theta})|^p |\varphi'(re^{i\theta})| d\theta \right)^{\frac{1}{p}} \left(\int |\varphi'(re^{i\theta})| d\theta \right)^{\frac{1}{p}-\frac{1}{p_\circ}}$$

$$\times \left(\int \left|\frac{1}{\varphi'(re^{i\theta})}\right|^q d\theta \right)^{\frac{1}{q+1}},$$

where the last inequality holds because $p \ge p_\circ$ and $\varphi' \in H^1$. Consequently $F \in H^1(\mathbb{D})$ and the Poisson formula holds for $F(z)$. Therefore by (i),

$$\|f_\alpha^*\|_{L^p(\partial\Omega)}^p \le \int_{\partial\mathbb{D}} |F_\gamma^*|^p |\varphi'| d\theta \le C_\gamma \int_{\partial\mathbb{D}} |MF|^p |\varphi'| d\theta.$$

But now by Section VII.4, $|\varphi'(e^{i\theta})| d\theta \in A_{p_\circ}$ so that by Muckenhoupt's theorem (see Garnett [1981], page 255)

$$\int_{\partial\mathbb{D}} |MF|^p |\varphi'| d\theta \le C \int_{\partial\mathbb{D}} |F|^p |\varphi'| d\theta = C\|f\|_{H^p(\Omega)}^p,$$

and the left-hand inequality in (M.3) holds when $p \ge p_\circ$.

Now suppose $p < p_\circ$. Because $F(z)(\varphi'(z))^{\frac{1}{p}} \in H^1$ we can write $F = Bg^{\frac{p}{p_\circ}}$ where $B(z)$ is a Blaschke product, where $g(z)$ has no zeros in $\mathbb{D}$, and where $|F| = |g|^{\frac{p_\circ}{p}}$ on $\partial\mathbb{D}$. Then

$$(F_\gamma^*)^p \le (g_\gamma^*)^{p_\circ}$$

and as before we have

$$\|f_\alpha^*\|_{L^p(\partial\Omega)}^p \le \int_{\partial\mathbb{D}} |F_\gamma^*|^p |\varphi'| d\theta \le \int_{\partial\mathbb{D}} |g_\gamma^*|^{p_\circ} |\varphi'| d\theta$$

$$\le C_\gamma \int_{\partial\mathbb{D}} |g|^{p_0} |\varphi'| d\theta = C_\gamma \int_{\partial\mathbb{D}} |F|^p |\varphi'| d\theta = C_\gamma \|f\|_{H^p(\Omega)}^p.$$

Conversely, assume $f_\alpha^* \in L^p(\partial\Omega, ds)$ and fix $r < 1$. Partition the circle $\{|z| = r\}$ into arcs I_j such that $|I_j| \leq (\beta - 1)(1 - r)$, where β is defined by (M.1). Then

$$I_j \subset \Gamma_\beta(\zeta)$$

for all $\zeta \in I_j^* = \{\frac{z}{|z|} : z \in I_j\}$, and

$$\ell(\varphi(I_j)) = \int_{I_j} |\varphi'| ds \leq C(\Omega, z_0) \operatorname{dist}(\varphi(I_j), \partial\Omega) \leq C(\Omega, z_0) \int_{I_j^*} |\varphi'| d\theta,$$

where $C(\Omega, z_0)$ depends only on the dilatation of the quasiconformal extension of φ. Then

$$\int |F(re^{i\theta})|^p |\varphi'(re^{i\theta})| d\theta \leq C \sum_j \int_{I_j^*} |F_\beta^*|^p |\varphi'| d\theta \leq C \int_{\partial\Omega} |f_\alpha^*|^p ds,$$

and that yields the right-hand inequality in (M.3). ∎

By Lemma M.2 and properties (i) and (iii), Theorem M.1 is an immediate consequence of the following result, which is the A^∞-weights version of Calderón's theorem. It is due to Gundy and Wheeden [1974].

Theorem M.3. *Let $w(e^{i\theta})$ be an A_∞ weight, let $\alpha > 1$, and let $0 < p < \infty$. Then there is $C = C(\alpha, w)$ such that whenever $F(z)$ is analytic on $\mathbb{D}$ and $F(0) = 0$,*

$$C\|A_\alpha F\|_{L^p(wd\theta)}^p \leq \|F_\alpha^*\|_{L^p(wd\theta)}^p \leq C^{-1}\|A_\alpha F\|_{L^p(wd\theta)}^p. \tag{M.4}$$

Proof. Write $\mu = w(e^{i\theta})d\theta$. By Exercise I.11 from Chapter I,

$$\|F_\alpha^*\|_{L^p(\mu)}^p \leq \|F_\beta^*\|_{L^p(\mu)}^p \leq C_{\alpha,\beta}\|F_\alpha^*\|_{L^p(\mu)}^p$$

when $\beta > \alpha$. That means we can prove Theorem M.3 by comparing $\|A_\alpha F\|$ to $\|F_{2\alpha}^*\|$ and $\|F_\alpha^*\|$ to $\|A_{2\alpha} F\|$.

We now claim the following lemma. ∎

Lemma M.4. *Given $\varepsilon > 0$ and $\kappa > 1$, there exists $\delta < 1$ such that for all $\lambda > 0$ and all analytic $F(z)$ satisfying $F(0) = 0$*

$$\mu\big(\{\Lambda_\alpha^* F > \kappa\lambda\} \cap \{F_{2\alpha}^* \leq \delta\lambda\}\big) \leq \varepsilon\mu\big(\{\Lambda_\alpha^* F > \lambda\}\big). \tag{M.5}$$

Before we prove Lemma M.4, let us use it to derive the left-hand inequality in (M.4). Replace $F(z)$ by $F(rz)$, $r < 1$, so that the next three integrals below

converge. By (M.5) we have

$$\|A_\alpha F\|^p_{L^p(\mu)} = p\kappa^p \int_0^\infty \lambda^{p-1} \mu(\{A_\alpha F > \kappa\lambda\}) d\lambda$$

$$\leq p\kappa^p \int_0^\infty \lambda^{p-1} \mu(\{F_{2\alpha}^* > \delta\lambda\}) d\lambda + \varepsilon p\kappa^p$$

$$\times \int_0^\infty \lambda^{p-1} \mu(\{A_\alpha F > \lambda\}) d\lambda$$

$$= \left(\frac{\kappa}{\delta}\right)^p \|F_{2\alpha}^* F\|^p_{L^p(\mu)} + \varepsilon\kappa^p \|A_\alpha F\|^p_{L^p(\mu)}.$$

Then taking $\varepsilon\kappa^p \leq \frac{1}{2}$ and sending $r \to 1$ yield

$$\|A_\alpha F\|^p_{L^p(\mu)} \leq 2\left(\frac{\kappa}{\delta}\right)^p \|F_{2\alpha}^*\|^p_{L^p(\mu)} \leq 2C_{\alpha,2\alpha}\left(\frac{\kappa}{\delta}\right)^p \|F_\alpha^*\|^p_{L^p(\mu)},$$

which is the left-hand side of (M.4).

Many authors call (M.5) and similar inequalities "good-λ inequalities", despite the fact that (M.5) is true for all $\lambda > 0$. Exercise M.2 includes a short history of the term "good-λ inequality".

Proof of Lemma M.4. Write the open set $\{A_\alpha F > \lambda\}$ as the union of disjoint open arcs $\{J_j\}$ and partition each J_j into closed arcs $\{I_k^j\}$ with disjoint interiors such that $|I_k^j| = \text{dist}(I_k^j, \partial\mathbb{D} \setminus J_j)$.

Figure M.1

The family $\{I_k\} = \bigcup_{j=1}^\infty \{I_k^j\}$ has the three properties

$$\{A_\alpha F > \lambda\} = \bigcup I_k,$$

$$I_k^o \cap I_j^o = \emptyset \text{ when } j \neq k, \text{ and}$$

$$\text{dist}(I_k, \{A_\alpha F \leq \lambda\}) = |I_k|.$$

For this reason $\{I_k\}$ is called the **Whitney decomposition** of $\{A_\alpha F > \lambda\}$. By the third property there exists ζ_k with $\text{dist}(\zeta_k, I_k) = |I_k|$ and

$$A_\alpha F(\zeta_k) \leq \lambda. \tag{M.6}$$

The main step in proving (M.5) is to show that whenever we are given $\kappa > 1$ and $\eta > 0$, there exists $\delta > 0$ such that for all I_k

$$\left| I_k \cap \{ A_\alpha F > \kappa \lambda \} \cap \{ F_{2\alpha}^* \leq \delta \lambda \} \right| \leq \eta \left| I_k \cap \{ A_\alpha F > \lambda \} \right|. \qquad \text{(M.7)}$$

Indeed, if $\eta = \eta(\varepsilon)$ is small, (M.7) and the A^∞ condition (VII.4.2) then give

$$\mu \left(I_k \cap \{ A_\alpha F > \kappa \lambda \} \cap \{ F_{2\alpha}^* \leq \delta \lambda \} \right) \leq \varepsilon \mu \left(I_k \cap \{ A_\alpha F > \lambda \} \right),$$

and a summation over $\{ I_k \}$ gives inequality (M.5).

Now let us prove (M.7). Fix $I = I_k$ and suppose $|I| < \frac{1}{4}$. Set

$$E = I \cap \{ A_\alpha F > \kappa \lambda \} \cap \{ F_{2\alpha}^* \leq \delta \lambda \},$$

$$U = \bigcup_E \Gamma_\alpha(\zeta),$$

$$U_0 = U \cap \{ 1 - |z| \leq 2|I| \},$$

$$U_1 = U \cap \{ 1 - |z| > 2|I| \},$$

and

$$W = \bigcup_E \Gamma_{2\alpha}(\zeta).$$

Then

$$\begin{aligned}
\kappa^2 \lambda^2 |E| &\leq \int_E A F(\zeta)^2 ds \\
&= \iint_U |F'(z)|^2 \left| \{ \zeta \in E : z \in \Gamma_\alpha(\zeta) \} \right| dx dy \\
&\leq \frac{2}{\alpha} \iint_{U_0} |F'(z)|^2 (1 - |z|) dx dy + |E| \iint_{U_1} |F'(z)|^2 dx dy,
\end{aligned} \qquad \text{(M.8)}$$

because

$$\left| \{ \zeta \in E : z \in \Gamma_\alpha(\zeta) \} \right| \leq \min \left(|E|, \frac{2}{\alpha} (1 - |z|) \right). \qquad \text{(M.9)}$$

To bound the integral over U_0, we form the larger domain

$$W_0 - W \cap \{ 1 - |z| \leq 4|I| \}.$$

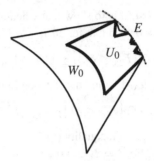

Figure M.2

Note that W_0 is a special cone domain and that $1 - |z| \le c(\alpha)\,\text{dist}(z, \partial W_0)$. if $z \in U_0 \subset W_0$, by trigonometry. Therefore by Theorem X.1.2,

$$
\iint\limits_{U_0} |F'(z)|^2(1 - |z|)dxdy \le c(\alpha) \iint\limits_{W_0} |F'(z)|^2 \,\text{dist}(z, \partial W_0)dxdy
$$

$$
\le C(\alpha) \int_{\partial W_0} |F(z) - F(z_{W_0})|^2 ds \tag{M.10}
$$

$$
\le C'(\alpha)(2\delta\lambda)^2|I|.
$$

To bound the integral over U_1, set $U_1 = \big(U_1 \cap \Gamma_\alpha(\zeta_k)\big) \cup \big(U_1 \setminus \Gamma_\alpha(\zeta_k)\big)$. Then we have

$$
|E| \iint\limits_{U_1 \cap \Gamma_\alpha(\zeta_k)} |F'(z)|^2 dxdy \le |E| \iint\limits_{\Gamma_\alpha(\zeta_k)} |F'(z)|^2 dxdy \le \lambda^2|E|. \tag{M.11}
$$

On the other hand, if $z \in U$, then by a calculation $B(z, \frac{1-|z|}{4}) \subset W$. Consequently $|F| \le \delta\lambda$ on $B(z, \frac{1-|z|}{4})$ and hence

$$
|F'(z)| \le \frac{4\delta\lambda}{1 - |z|}.
$$

Therefore

$$
|E| \iint\limits_{U_1 \setminus \Gamma_\alpha(\zeta_k)} |F'(z)|^2 dxdy \le (4\delta\lambda)^2|I| \iint\limits_{U_1 \setminus \Gamma_\alpha(\zeta_k)} \frac{dxdy}{(1 - |z|)^2}.
$$

But since $\text{dist}(\zeta_k, I) = |I|$, writing the last integral in polar coordinates gives

$$
\iint\limits_{U_1 \setminus \Gamma_\alpha(\zeta_k)} \frac{dxdy}{(1 - |z|)^2} \le C,
$$

and thus

$$|E| \iint\limits_{U_1 \setminus \Gamma_\alpha(\zeta_k)} |F'(z)|^2 dxdy \leq 4C(\delta\lambda)^2 |I|, \qquad (\text{M.12})$$

Together, (M.8), (M.10), (M.11), and (M.12) give us

$$\left(1 - \left(\frac{1}{\kappa}\right)^2\right)|E| \leq C'(\alpha)\left(\frac{\delta}{\kappa}\right)^2 |I| + C''(\alpha)\left(\frac{\delta}{\kappa}\right)^2 |I|,$$

and we obtain (M.7). ∎

To prove the right-hand inequality in (M.4), we fix $\lambda > 0$ and now let $\{I_k\}$ be the Whitney decomposition of $\{F_\alpha^* > \lambda\}$. Write $2I_k$ for the arc concentric with I_k having length $|2I_k| = 2|I_k|$, and note that no point falls in more than three arcs $2I_k$.

Lemma M.5. *Given $\eta > 0$ and $\kappa > 2$, there exist $\delta > 0$ such that for each I_k,*

$$\left|I_k \cap \{F_\alpha^* > \kappa\lambda\} \cap \{M\chi_k \leq \frac{1}{2}\}\right| \leq \eta|I_k|, \qquad (\text{M.13})$$

where M is the Hardy–Littlewood maximal operator and χ_k is the characteristic function

$$\chi_k = \chi_{2I_k \cap \{\zeta : A_{2\alpha} F(\zeta) > \delta\lambda\}}.$$

Before proving Lemma M.5, let us show how it gives the right-hand inequality in (M.4). Again replace $F(z)$ by $F(rz)$, $r < 1$ so that the integrals converge. Then given $\varepsilon > 0$ there exists $\eta = \eta(\varepsilon) > 0$ so that by (M.13) and the A^∞-condition,

$$\mu\left(I_k \cap \{F_\alpha^* > \kappa\lambda\} \cap \{M\chi_k \leq \frac{1}{2}\}\right) \leq \varepsilon\mu(I_k).$$

By Exercise VII.14, $w \in A_{p_\circ}$ for some $p_\circ > 1$, so that by Chebychev's inequality,

$$\mu\left(I_k \cap \{M\chi_k > \frac{1}{2}\}\right) \leq C(\mu)2^{p_\circ} \|\chi_k\|_{L^{p_\circ}(\mu)}^{p_\circ}$$
$$= C(\mu)2^{p_\circ} \mu\left(2I_k \cap \{\zeta : A_{2\alpha} F(\zeta) > \delta\lambda\}\right)$$

Consequently,

$$\mu\left(\{F_\alpha^* > \kappa\lambda\}\right) \leq \varepsilon\mu\left(\{F_\alpha^* > \lambda\}\right) + 3C(\mu)2^{p_\circ} \mu\left(\{\zeta : A_{2\alpha} F(\zeta) > \delta\lambda\}\right),$$

and hence

$$||F_{2\alpha}^*||_{L^p(\mu)}^p \le C_{\alpha,2\alpha}||F_\alpha^*||_{L^p(\mu)}^p = p\kappa^p \int_0^\infty \lambda^{p-1}\mu(\{F_\alpha^* > \kappa\lambda\})d\lambda$$

$$\le \varepsilon p\kappa^p \int_0^\infty \lambda^{p-1}\mu(\{F_\alpha^* > \lambda\})d\lambda + 3C(\mu)2^{p_\circ}p\kappa^p$$

$$\times \int_0^\infty \lambda^{p-1}\mu(\{A_{2\alpha} > \delta\lambda\})d\lambda$$

$$= \varepsilon\kappa^p||F_\alpha^*||_{L^p(\mu)}^p + 3C(\mu)2^{p_\circ}\left(\frac{\kappa}{\delta}\right)^p||A_{2\alpha}F||_{L^p(\mu)}^p.$$

Again taking $\varepsilon\kappa^p < \frac{1}{2}$ and sending $r \to 1$ give

$$||F_{2\alpha}^*||_{L^p(\mu)}^p \le C(\mu)2^{p_\circ+1}\left(\frac{\kappa}{\delta}\right)^p||A_{2\alpha}F||_{L^p(\mu)}^p,$$

and this inequality is the right-hand side of (M.4).

Proof of Lemma M.5. Fix $I = I_k$ and assume $|I| < \frac{1}{4}$. Take

$$E_0 = I \cap \{F_\alpha^* > \kappa\lambda\} \cap \{M\chi_k \le \frac{1}{2}\},$$

$$E = I \cap \{M\chi_k \le \frac{1}{2}\},$$

$$E_1 = 2I \cap \{A_{2\alpha} \le \delta\lambda\},$$

$$U = \{1 - |z| < 2|I|\} \cap \bigcup_E \Gamma_{2\alpha}(\zeta),$$

and

$$U_1 = \bigcup_{E_1} \Gamma_{2\alpha}(\zeta).$$

Then

$$\iint_{U_1} |F'(z)|^2|\{\zeta \in E_1 : |z - \zeta| \le 2\alpha(1 - |z|)\}|dxdy$$

$$\le \int_{E_1} \iint_{\Gamma_{2\alpha}(\zeta)} |F'(z)|^2dxdyds \le 2\delta^2\lambda^2|I|.$$

Now if $z \in U$ there is $\zeta^* \in E$ with $|\zeta^* - z| \le 2\alpha(1 - |z|)$. Therefore

$$|\{\zeta : A_{2\alpha}F(\zeta) > \delta\lambda\} \cap \{|\zeta - \zeta^*| < \frac{1 - |z|}{4}\}| \le \frac{1 - |z|}{8},$$

and

$$|\{\zeta \in E_1 : |z - \zeta| < 2\alpha(1 - |z|)\}| \ge \frac{1 - |z|}{16}. \qquad (\text{M.14})$$

Inequality (M.14), which is the reverse of (M.9), is the reason the function $M\chi_k$ was introduced. Because of (M.14), we have

$$\iint_U |F'(z)|^2 \operatorname{dist}(z, \partial U)dxdy \le \iint_U |F'(z)|^2(1 - |z|)dxdy$$

$$\le 16 \iint_{U_1} |F'(z)|^2 |\{\zeta \in E_1 : |z - \zeta| \quad \text{(M.15)}$$

$$\le 2\alpha(1 - |z|)\}|dxdy$$

$$\le 32\delta^2\lambda^2|I|.$$

The domain U is a special cone domain, and so (M.15) implies

$$\int_{\partial U} |F - F(c_U)|^2 ds \le C(\alpha)\delta^2\lambda^2|I|. \qquad \text{(M.16)}$$

Now let $\zeta \in E_0 \subset E$. By trigonometry there exists $\beta = \beta(\alpha) > 1$ such that

$$\Gamma_\alpha(\zeta) \cap \{1 - |z| < |I|\} \subset \Gamma_\beta(\zeta, U).$$

See Figure X.6. Take $z \in \Gamma_\alpha(\zeta) \cap \{1 - |z| \ge |I|\}$. By trigonometry there is $z' = rz$, $r(\alpha) \le r \le 1$ such that $z' \in \Gamma_\alpha(\zeta_k)$, where $\operatorname{dist}(\zeta_k, I) = |I|$ and $F_\alpha^*(\zeta_k) \le \lambda$. In particular, we have $|F(z')| \le \lambda$.

When $w \in [z', z]$, a calculation shows $B(w, \frac{(1-|w|)}{4}) \subset \Gamma_{2\alpha}(\zeta)$ so that

$$|F'(w)| \le \left(\frac{16}{\pi(1 - |w|)^2} \iint_{B(w, \frac{1-|w|}{4})} |F'|^2 dxdy\right)^{\frac{1}{2}} \le \frac{4\delta\lambda}{\pi^{\frac{1}{2}}|I|}.$$

Then an integration gives

$$|F(z)| \le (1 + c\delta)\lambda. \qquad \text{(M.17)}$$

Because (M.17) holds at $z = c_U$, we conclude that

$$\sup_{\Gamma_\beta(\zeta, U)} |F(z) - F(c_U)| \ge \big(\kappa - 2(1 + c\delta)\big)\lambda$$

for every $\zeta \in E_0$. But then by (M.16) and Lemma M.2

$$|E_0| \le \frac{C(\alpha)\delta^2|I|}{(\kappa - 2(1 + c\delta))^2},$$

and (M.13) is proved. ∎

Exercise M.1. Prove Lemma M.5 and inequality (M.7) in the case $|I| > \frac{1}{4}$.

Exercise M.2. (a) Suppose (X, μ) is a measure space and suppose f is a non-negative μ-measurable function. Let $p > 0$, $\kappa > 1$ and $0 < \varepsilon < 1$, and suppose

$\varepsilon \kappa^p < 1$. Call $\lambda > 0$ **good** if

$$\varepsilon \mu(\{f > \lambda\}) < \mu(\{f > \kappa\lambda\}).$$

If $G = \{\text{good } \lambda\}$, then

$$\|f\|^p_{L^p(\mu)} \leq \frac{(1-\varepsilon)\kappa p}{1-\varepsilon\kappa p} \int_G p\lambda^{p-1}\mu(\{f > \lambda\})d\lambda.$$

(b) Now consider the inequality

$$\mu(\{A^*_\alpha F > \kappa\lambda\} \cap \{F^*_{2\alpha} \leq \delta\lambda\}) \leq \varepsilon\mu(\{A^*_\alpha F > \lambda\}) \qquad \text{(M.5)}$$

and set $f = A^*_\alpha F$. Note that (M.5) is trivial if $\lambda \notin G$. On the other hand, by part (a) any inequality of the form

$$\mu(\{f > \lambda\}) \leq C\mu(\{F^*_{2\alpha} > K\lambda\}), \quad \lambda \in G \qquad \text{(M.18)}$$

yields $\|A_\alpha F\|^p_{L^p(\mu)} \leq C\|F^*_{2\alpha}\|^p_{L^p(\mu)}$ which was the objective of (M.5). Of course, (M.5) also implies (M.18).

In their original papers [1970] and [1972] Burkholder and Gundy worked with inequalities like (M.18) for $\lambda \in G$ (but without calling them "good"). See Burkholder [1973] and Gundy and Wheeden [1974] for related inequalities, and Bañuelos and Moore [1999] for a more recent discussion.

(c) In general, for analytic functions $F(z)$ on the unit disc with $F(0) = 0$, no inequality of the form

$$|A_\alpha F > \lambda| \leq C|F^*_{K\alpha} > N\lambda|$$

is valid for all $\lambda > 0$.

Exercise M.3. Let Γ be a locally rectifiable Jordan curve such that $\infty \in \Gamma$ and let Ω_1 and Ω_2 be the components of $\mathbb{C}^* \setminus \Gamma$.

(a) Prove that Ω_j is a BMO domain if and only if there is a constant C

$$\iint_{\Omega_j} |F'(z)|^2 \operatorname{dist}(z, \Gamma)dxdy \leq \int_\Gamma |F(\zeta)|^2 ds$$

for all F bounded and analytic on Ω_j satisfying $F(z) = O(\frac{1}{z})$, $z \to \infty$.

(b) Prove that Γ is a chord-arc curve if and only for both $j = 1$ and $j = 2$,

$$\int_\Gamma |F(z)|^2 ds \leq C \iint_{\Omega_j} |F'(z)|^2 \operatorname{dist}(z, \Gamma)dxdy$$

for all F bounded and analytic on Ω_j satisfying $F(z) = O(\frac{1}{z})$, $z \to \infty$.
These results are due to Bruna and González [1999].

Bibliography

Adams, D. R., and Hedberg, L. I.
[1996] *Function Spaces and Potential Theory*, Springer-Verlag.
Adams, R. A.
[1975] *Sobolev Spaces*, Academic Press.
Ahlfors, L. V.
[1930] Untersuchungen zur Theorie der konformen Abbildung und der ganzen Funktion, *Acta Soc. Sci. Fenn.*, N.S. 1, No. 9.
[1947] Bounded analytic functions, *Duke Math J.* **14**, 1–11.
[1951] *Conformal Mapping*, Mimeographed notes, University of Oklahoma.
[1963] Quasiconformal reflections, *Acta Math.* **109**, 291–301.
[1966] *Lectures on Quasiconformal Mappings*, Manuscript prepared with the assistance of Clifford J. Earle, Jr., Van Nostrand. Reprinted 1987, Wadsworth and Brooks/Cole.
[1973] *Conformal Invariants, Topics in Geometric Function Theory*, McGraw-Hill.
[1974] Sufficient conditions for quasiconformal extension, *Ann. Math. Studies* **79**, 23–29.
[1979] *Complex Analysis*, Third edition, McGraw-Hill.
Ahlfors, L. V., and Beurling, A.
[1946] Invariants conformes et problèmes extrémaux, *10th Scand. Congr. Math.*, 341–351.
[1950] Conformal invariants and function-theoretic null-sets, *Acta Math.* **83**, 101–129.
[1952] Conformal invariants. Construction and applications of conformal maps, *National Bureau of Standards, Appl. Math. Ser.* **18**, 243–245.
Ahlfors, L. V., and Weill, G.
[1962] A uniqueness theorem for Beltrami equations, *Proc. Amer. Math. Soc.* **13**, 975–978.
Alexander, H.
[1989] Linear measure on plane continua of finite linear measure, *Arkiv Mat.* **27**, 169–177.
Ancona, A.
[1983] Démonstration d'une conjecture sur la capacité et l'effilement, *C.R. Acad. Sci. Paris Sér. I Math.* **297**, 393–395.

531

[1986] On strong barriers and an inequality of Hardy for domains in $\mathbb{R}^n$, *J. London Math. Soc. (2)* **34**, 274–290.

[1987] Negative curvature, elliptic operators, and the Martin boundary, *Ann. Math.* **125**, 495–536.

Anderson, J. M., Barth, K. F., Brannan, D. A., and Hayman, W. K.

[1977] Research problems in complex analysis, *Bull. London Math. Soc.* **9**, 129–162.

Anderson, J. M., Clunie, J., and Pommerenke, Ch.

[1974] On Bloch functions and normal functions, *J. Reine Angew. Math.* **270**, 12–37.

Anderson, J. M., and Pitt, L. D.

[1988] The boundary behavior of Bloch functions and univalent functions, *Michigan Math. J.* **35**, 313–320.

[1989] Probabilistic behavior of functions in the Zygmund spaces Λ^* and λ^*, *Proc. London Math. Soc.* **59**(3), 558–592.

Astala, K.

[1994] Area distortion of quasiconformal mappings, *Acta Math.* **173**, 37–70.

[1998a] Planar quasiconformal mappings; deformations and interactions, in *Quasiconformal Mappings and Analysis, A Collection of Papers honoring F. W. Gehring*, P. Duren, J. Heinonen, B. Osgood, and B. Palka, Eds., Springer-Verlag, 33–54.

[1998b] Analytic aspects of planar quasiconformality, Proc. Int. Cong. Math., Vol. II., *Documenta Math.*, 617–626.

Astala, K., Fernández, J. L., and Rohde, S.

[1993] Quasilines and the Hayman–Wu theorem, *Indiana Univ. Math. J.* **42**, 1077–1100.

Astala, K., and Zinsmeister, M.

[1991] Teichmüller spaces and BMOA, *Math. Ann.* **289**, 613–625.

Baernstein II, A.

[1974] Integral means, univalent functions and circular symmetrization, *Acta Math.* **133**, 139–169.

[1976] Univalence and bounded mean oscillation, *Michigan Math. J.* **23**, 217–223.

[1980] Analytic functions of bounded mean oscillation, in *Aspects of Contemporary Complex Analysis*, D. A. Brannan and J. Clunie, Eds., Academic Press, 3–36.

[1988] Ahlfors and conformal invariants, *Ann. Acad. Sci. Fenn. Ser. A I Math.* **13**, 289–312.

[1989] A counterexample concerning integrability of derivatives of conformal mapping, *J. Anal. Math.* **53**, 253–268.

[1991] Book review of *Subharmonic Functions, Vol. 2* by W. K. Hayman, *Bull. Amer. Math. Soc.* **25**, 458–467.

[1994] A unified approach to symmetrization, in *Partial Differential Equations of Elliptic Type*, Symp. Math. **35**, Cambridge University Press, 47–91.

Bagemihl, F., and Seidel, W.

[1954] Some boundary properties of analytic functions, *Math. Z.* **61**, 186–199.

Balogh, Z. and Bonk, M.

[1999] Lengths of radii under conformal maps of the unit disc, *Proc. Amer. Math. Soc.* **127**, 801–804.

Bañuelos, R.

[1986] Brownian motion and the area function, *Indiana U. Math. J.* **35**, 653–668.

Bañuelos, R., Klemes, I., and Moore, C. N.

[1988] An analogue for harmonic functions of Kolmorogov's law of the iterated logarithm, *Duke Math. J.* **56**, 1–32.

[1990] The lower bound in the law of the iterated logarithm for harmonic functions, *Duke Math. J.* **60**, 689–715.

Bañuelos, R., and Moore, C. N.

[1989] Sharp estimates for the nontangential maximal function and the Lusin area function, *Trans. Amer. Math. Soc.* **312**, 641–662.

[1999] *Probabilistic Behavior of Harmonic Functions,* Birkhäuser Verlag.

Barański, K., Volberg, A., and Zdunik, A.

[1998] Brennan's conjecture and the Mandelbrot set, *Int. Math. Res. Notices* **12**, 589–600.

Barth, K. F., Brannan, D. A., and Hayman, W. K.

[1984] Research problems in complex analysis, *Bull. London Math. Soc.* **16**, 490–517.

Batakis, A.

[1996] Harmonic measure of some Cantor type sets, *Ann. Acad. Sci. Fenn. Ser. A I Math.* **21**, 255–270.

[2000] A continuity property of the dimension of harmonic measure, *Ann. Inst. Henri Poincaré Prob. Stat.* **36**, 87–107.

Beardon, A. F., and Pommerenke, Ch.

[1979] The Poincaré metric of plane domains, *J. London Math. Soc. (2)* **18**, 475–483.

Becker, J.

[1972] Löwnersche Differentialgleichung und quasikonform fortsetzbare schlichte Funktionen, *J. Reine Angew. Math.* **255**, 23–43.

[1973] Löwnersche Differentialgleichung und Schlichtheitskriterien, *Math. Ann.* **202**, 321–335.

[1980] Conformal mappings with quasiconformal extensions, in *Aspects of Contemporary Complex Analysis,* D. A. Brannan and J. Clunie, Eds., Academic Press, 37–78.

Becker, J., and Pommerenke, Ch.

[1982] Hölder continuity of conformal mappings and non-quasiconformal Jordan curves, *Comm. Math. Helv.* **57**, 221–225.

[1984] Schlichtheitskriterien und Jordan gebiet, *J. Reine Angew. Math.* **354**, 74–94.

Behrens, M.

[1976] Normal functions and metrics on infinitely connected planar domains, University of Iowa Math. Dept., unpublished.

Benedicks, M.

[1979] Positive harmonic functions vanishing on the boundary of certain domains in $\mathbb{R}^{n+1}$. *Proc. Symp. Pure Math., XXXV, Part I,* 345–348.

[1980] Positive harmonic functions vanishing on the boundary of certain domains in $\mathbb{R}^n$, *Arkiv Mat.* **18**, 52–73.

Benedicks, M., and Pfeffer, W. F.

[1985] The Dirichlet problem with Denjoy–Perron integrability boundary condition, *Can. Math. Bull.* **28**, 113–119.

Bennett, C., and Sharpley, R.,

[1988] *Interpolation of Operators,* Academic Press.

Bertilsson, D.
[1998] Coefficient estimates for negative powers of the derivative of univalent functions, *Arkiv Mat.* **36**, 255–273.
[1999] On Brennan's conjecture in conformal mapping, Doctoral thesis, Royal Institute of Technology.

Betsakos, D.
[1997] On certain harmonic measures in the unit disc, *Colloq. Math.* **73**(2), 221–228.
[1998a] Harmonic measure on simply connected domains of fixed inradius, *Arkiv Mat.* **36**, 275–306.
[1998b] Polarization, conformal invariants and Brownian motion, *Ann. Acad. Sci. Fenn. Ser. A I Math.* **23**(1), 59–82.
[1999] An extension of the Beurling–Nevanlinna projection theorem, in *Computational Methods and Function Theory, 1997, (Nicosia)*, World Scientific Publishing, 87–90.

Betsakos, D. and Solynin, A. Y.
[2000] Extensions of Beurling's shove theorem for harmonic measure, *Complex Var. Theory App.* 42(1), 57–65.

Beurling, A.
[1933] Études sure un problème de majoration, Doctoral thesis, Uppsala University.
[1940] Sur les ensembles exceptionels, *Acta Math.* **72**, 1–13.
[1989] Mittag-Leffler lectures on complex analysis (1977–1978), in *The Collected Works of Arne Beurling, Vol. 1, Complex Analysis*, L. Carleson, P. Malliavan, J. Neuberger, and J. Wermer, Eds., Birkhäuser.

Beurling, A., and Ahlfors, L. V.
[1956] The boundary correspondence under quasiconformal mappings, *Acta Math.* **96**, 125–142.

Bieberbach, L.
[1916] Über die Koeffizienten derjenigen Potenzreihen, welche eine schlichte Abbildung des Einheitskreises vermitteln, *S.-B. Preuss. Akad. Wiss.* 940–955.

Binder, I.
[1997] Rotation spectrum of planar domains, Doctoral thesis, California Inst. of Technology.
[1998a] Harmonic measure and rotation of simply connected planar domains, preprint.
[1998b] Phase transition for the universal bounds on the integral means spectrum, preprint.
[1999] Asymptotic for the integral mixed spectrum of the basin of attraction of infinity for the polynomial $z(z + \delta)$, preprint.

Binder, I., Makarov, N., and Smirnov, S.
[2003] Harmonic measure and polynomial Julia sets, *Duke Math. J.* **117**, 343–365.

Bishop, C. J.
[1987] Harmonic measures supported on curves, Doctoral thesis, University of Chicago.
[1988] A counterexample in conformal welding concerning Hausdorff dimension, *Mich. Math. J.* **35**, 151–159.
[1991a] A characterization of Poissonian domains, *Ark. Mat.* **29**, 1–24.
[1991b] Some questions concerning harmonic measure, in *Partial differential equations with minimal smoothness and applications*, B. Dahlberg, E. Fabes, R. Feffer-

man, D. Jerison, C. Kenig, and J. Pipher, Eds., IMA Vol. Math. App. **42**, Springer-Verlag.

[1992] Brownian motion in Denjoy domains, *Ann. Prob.* **20**, 631–651.

[1994] Harmonic measure and Hausdorff dimension, in *Linear and Complex Analysis Problem Book, 3, Part II*, Lecture Notes in Mathematics **1574**, Springer-Verlag, 385–387.

[1997] Geometric exponents and Kleinian groups, *Invent. Math.* **127**(1), 33–50.

[2002a] Quasiconformal Lipschitz maps, Sullivan's convex hull theorem and Brennan's conjecture, *Ark. Mat.* **40**, 1–26.

[2002b] Bilipschitz approximations of quasiconformal maps, *Ann. Acad. Sci. Fenn. Ser. A I Math.* **27**(1), 97–108.

Bishop, C. J., Carleson, L., Garnett, J. B., and Jones, P. W.

[1989] Harmonic measures supported on curves, *Pac. J. Math.* **138**, 233–236.

Bishop, C. J., and Jones, P. W.

[1990] Harmonic measure and arclength, *Ann. Math.* **132**(2), 511–547.

[1994] Harmonic measure, L^2 estimates and the Schwarzian derivative, *J. Anal. Math.* **62**, 77–113.

[1997a] Wiggly sets and limit sets, *Arkiv Mat.* **35**(2), 201–224.

[1997b] Hausdorff dimension and Kleinian groups, *Acta Math.* **179**, 1–39.

[1997c] The law of the iterated logarithm for Kleinian groups, *Contemp. Math.* **211**, 17–50.

Bonk, M., Koskela, P., and Rohde, S.

[1998] Conformal metrics on the unit ball in Euclidean space, *Proc. London Math. Soc.* **77**, 635–664.

Bourgain, J.

[1997] On the Hausdorff dimension of harmonic measure in higher dimensions, *Invent. Math.* **87**, 477–483.

Brannan, D. A., and Hayman, W. K.

[1989] Research problems in complex analysis, *Bull. London Math. Soc.*, **21**, 1–35.

Brelot, M.

[1965] *Élements de la Théorie Classique du Potential*, Centre de Documentation Universitaire.

Brelot, M., and Choquet, G.

[1951] Espaces et lignes de Green, *Ann. Inst. Fourier* **3**, 199–263.

Brennan, J.

[1978] The integrability of the derivative in conformal mapping, in *J. London Math. Soc. (2)* **18**, 261–272.

[1994] The integrability of the derivative of a conformal mapping, *Linear and Complex Analysis Problem Book, 3, Part II*, Lecture Notes in Mathematics **1574**, Springer-Verlag, 101 106.

Brolin, H.

[1965] Invariant sets under iteration of rational functions, *Arkiv Mat.* **6**, 103–144.

Browder, A., and Wermer, J.

[1963] Some algebras of functions on an arc, *J. Math. Mech.* **12**, 119–130.

Brown, L., Shields, A., and Zeller, K.

[1960] On absolutely convergent exponential sums, *Trans. Amer. Math. Soc.* **96**, 162–193.

Bruna, J., and González, M. J.
[1999] L^2 estimates on chord-arc curves, *Pac. J. Math.* **190**, 225–233.
Burdzy, K.
[1986] Brownian excursions and minimal thinness, Part III: Applications to the angular derivative problem, *Math. Z.* **192**, 89–107.
[1987] *Multidimensional Brownian Excursions and Potential Theory*, Wiley.
Burdzy, K., and Marshall, D. E.
[1992] Hitting a boundary point with reflected Brownian motion. in *Seminaire de Probabilites XXVI*, Lecture Notes in Mathematics **1526**, Springer-Verlag, 81–94.
[1993] Nonpolar points for reflected Brownian motion. *Ann. Inst. H. Poincaré Prob. Statist.* **29**(2), 199–228.
Burkholder, D.
[1973] Distribution function inequalities for martingales, *Ann. Prob.* **1**, 19–42.
Burkholder, D. and Gundy, R.
[1970] Extrapolation and interpolation of quasi-linear operators on martingales, *Acta. Math.* **124**, 249–304.
[1972] Distribution function inequalities for the area integral, *Studia Math.* **44**, 527–544.
[1973] Boundary behavior of harmonic functions in a half-space and Brownian motion, *Ann. Inst. Fourier* **23**, 195–212.
Calderón, A. P.
[1965] Commutators of singular integral operators, *Proc. Nat. Acad. Sci.* **53**, 1092–1099.
Campbell, D. M., Clunie, J. G., and Hayman, W. K.
[1980] Research problems in complex analysis, in *Aspects of Contemporary Complex Analysis*, Academic Press, 527–572.
Carathéodory, C.
[1913a] Über die gegenseitige Beziehung der Ränder bei der konformen Abbildung des Innern einer Jordanschen Kurve auf einer Kreis, *Math. Ann.* **73**, 305–320.
[1913b] Über die Begrenzung einfach zusammenhäng ender Gebiete, *Math. Ann.* **73**, 323–370.
[1958] *Conformal Representation*, Cambridge University Press.
Carleman, T.
[1933] Sur une inégalité différentielle dans la théorie des fonctions analytiques, *C.R. Acad. Sci. Paris* **196**, 995–997.
Carleson, L.
[1958] An interpolation problem for bounded analytic functions, *Amer. J. Math.* **80**, 921–930.
[1962] Interpolations by bounded analytic functions and the corona problem. *Ann. Math.* **76**, 547–559.
[1967a] *Selected Problems on Exceptional Sets*, Van Nostrand. Reprinted 1983, Wadsworth International Group.
[1967b] On mappings, conformal at the boundary, *J. Anal. Math.* **19**, 1–13.
[1973] On the distortion of sets on a Jordan curve under a conformal mapping, *Duke Math. J.* **40**, 547–559.
[1982] Estimates on harmonic measure, *Ann. Acad. Sci. Fenn. Ser. A I Math.* **7**, 25–32.

[1985] On the support of harmonic measure for sets of Cantor type, *Ann. Acad. Sci. Fenn. Ser. A I Math.* **10**, 113–123.

Carleson, L., and Gamelin, T. W.

[1993] *Complex Dynamics*, Springer-Verlag.

Carleson, L., and Jones, P. W.

[1992] On coefficient problems for univalent functions and conformal dimension, *Duke Math. J.* **66**, 169–206.

Carleson, L., and Makarov, N.

[1994] Some results connected with Brennan's conjecture, *Ark. Mat.* **32**, 33–62.

Carroll, T. F.

[1988] A classical proof of Burdzy's theorem on the angular derivative, *J. London Math. Soc. (2)* **38**, 423–441.

Chang, S.-Y. A., and Marshall, D. E.

[1985] On a sharp inequality concerning the Dirichlet integral, *Amer. J. Math.* **107**, 1015–1033.

Chang, S.-Y. A., Schiffer, M. M. and Schober, G.

[1981] On the second variation for univalent functions, *J. Anal. Math.* **40**, 203–238.

Chang, S.-Y. A., Wilson, J. M., and Wolff, T. H.

[1985] Some weighted norm inequalities concerning Schrödinger operators, *Comm. Math. Helv.* **60**, 217–246.

Choi, S.

[2004] The lower density conjecture for harmonic measure, *J. Anal. Math.* **93**, 237–269.

Chuaqui, M., and Osgood, B.

[1998] General univalence criteria in the disk: Extensions and extremal function, *Ann. Acad. Sci. Fenn. Ser. A I Math.* **23**(1), 101–132.

Chung, K. L.

[1974] *A Course in Probability Theory*, Second edition, Academic Press.

Cima, J. A., and Petersen, K. E.

[1976] Some analytic functions whose boundary values have bounded mean oscillation, *Math. Z.* **147**, 237–247.

Cima, J. A., and Schober, G.

[1976] Analytic functions with bounded mean oscillation and logarithms of H^p-functions, *Math. Z.* **151**, 295–300.

Coifman, R. R., and Fefferman, C.

[1974] Weighted norm inequalities for maximal functions and singular integrals, *Studia Math.* **51**, 241–250.

Coifman, R. R., Jones, P. W., and Semmes, S.

[1989] Two elementary proofs of the L^2 boundedness of Cauchy integrals on Lipschitz graphs, *J. Amer. Math. Soc.* **2**, 553–564.

Coifman, R. R., and Meyer, Y.

[1979] Le théorème de Calderón par les méthodes de variable réele, *C.R. Acad. Sci. Paris, Sér. A-B* **289**, 425–428.

[1983] Lavrentiev's curves and conformal mappings, *Institut Mittag-Leffler* **5**, Royal Swedish Academy of Sciences.

Courant, R.

[1950] *Dirichlet's Principle, Conformal Mapping, and Minimal Surfaces, with an*

appendix by M. Schiffer, Interscience.
David, G.
[1991] *Wavelets and singular integrals on curves and surfaces*, Lecture Notes in Mathematics **1465**, Springer-Verlag.
[1998] Unrectifiable 1-sets have vanishing analytic capacity, *Rev. Mat. Iberoamericana* **14**, 369–479.
David, G. and Semmes, S.
[1991] *Singular Integrals and Rectifiable Sets in* $\mathbb{R}^n$, Au-delà des graphes lipschitziens, Astérisque **193**, 170pp.
[1993] *Analysis of and on Uniformly Rectifiable Sets*, Mathematical Surveys and Monographs **38**, American Mathematical Society.
[1997] *Fractured Fractals and Broken Dreams. Self-Similar Geometry through Metric and Measure*, Oxford Lecture Series in Mathematics and its Applications, **7**, Clarendon Press.
de Branges, L.
[1985] A proof of the Bieberbach conjecture, *Acta Math.* **40**, 137–152.
[1987] Underlying concepts on the proof of the Bieberbach conjecture, *Proc. Int. Cong. Math.*, American Mathematical Society, 25–42.
Denjoy, A.
[1907] Sur les fonctions entières de genie fini, *C.R. Acad. Sci. Paris* **145**, 106.
Denker, M., and Urbański, M.
[1991] Relating Hausdorff measures and harmonic measures on parabolic Jordan curves, *Mathematica Gottingensis Schriftenreihe des SFB Geometrie und Analysis* **26**, 1–21.
Doob, J. L.
[1974] *Stochastic Processes*, Wiley.
[1984] *Classical Potential Theory and Its Probabilstic Counterpart*, Springer-Verlag.
Du Plessis, N.
[1970] *An Introduction to Potential Theory*, Hafner.
Duren, P.
[1970] *Theory of H^p Spaces*, Academic Press.
[1983] *Univalent functions*, Springer-Verlag.
Duren, P. L., Shapiro, H. S., and Shields, A.
[1966] Singular measures and domains not of Smirnov type, *Duke Math. J.* **33**, 247–254.
Durrett, R.
[1996] *Probability: Theory and Examples*, Second Edition, Duxbury Press.
Dynkin, E. B., and Yushkevich, A. A.
[1969] *Markov Processes Theorems and Problems*, Plenum Press.
Erdös, P., and Gillis, J.
[1937] Note on the transfinite diameter, *J. London Math. Soc.* **127**, 185–192.
Eremenko, A., and Hamilton, D. H.
[1995] On the area distortion by quasiconformal mappings, *Proc. Amer. Math. Soc.* **123**, 2793–2797.
Essèn, M., and Haliste, K.
[1989] On Beurling's theorem for harmonic measure and the rings of Saturn, *Complex Variables Theory App.* **12**, 137–152.

Evans, G. C.

[1927] *The Logarithmic Potential,* Amer. Math. Soc. Coll. Pub. IV.

[1933] Applications of Poincaré's sweeping-out process, *Proc. Nat. Acad. Sci.* **19**, 457–461.

[1935] Potentials of positive mass, I, *Trans. Amer. Math. Soc.* **37**, 226–253.

Falconer, K. J.

[1985] *The Geometry of Fractal Sets,* Cambridge University Press.

[1990] *Fractal Geometry, Mathematical Foundations and Applications,* Wiley.

Fang, X.

[1994] The Cauchy integral, analytic capacity, and subsets of quasicircles, *Pac. J. Math.* **166**, 247–294.

Fejér, L.

[1910] Über gewisse Potenzreihen an der Konvergenzegrenze, *Sitzungsberichte der Math. Phys. Klass der Königlich Bayerische Akademie der Wiss.* **3**, 17 pp.

[1913] La convergence sur son cercle de convergence d'une série de puissance effectuant une représentation conforme du cercle sur le plan simple, *C.R. Acad. Sci. Paris* **156**, 46–49.

Fefferman, C., and Stein, E. M.

[1972] H^p-spaces of several variables, *Acta Math.* **129**, 137–193.

Feller, W.

[1968] *An Introduction to Probability Theory and Its Applications,* Vol. 1, Third edition, Wiley.

Feng, J., and MacGregor, T. H.

[1976] Estimates on integral means of the derivatives of univalent functions, *J. Anal. Math.* **29**, 203–231.

Fernández, J. L.

[1989] Domains with strong barrier, *Rev. Mat. Iberoamericana* **5**, 47–65.

Fernández, J. L., and Granados, A.

[1998] On geodesic curvature and conformal mapping, *St. Petersburg Math. J.* **9**(3), 615–637.

Fernández, J. L., and Hamilton, D. H.

[1987] Lengths of curves under conformal mappings, *Comm. Math. Helv.* **62**, 122–134.

Fernández, J. L., Heinonen, J., and Martio, O.

[1989] Quasilines and conformal mappings, *J. Anal. Math.* **52**, 117–132.

Fernández, J. L., and Zinsmeister, M.

[1987] Ensembles de niveau des représentations conformes, *C.R. Acad. Sci. Sér I Math.* **305**, 449–452.

Fisher, S. D.

[1983] *Function Theory on Planar Domains, A Second Course in Complex Analysis,* Wiley.

FitzGerald, C. H., and Pommerenke, Ch.

[1985] The de Branges theorem on univalent functions, *Trans. Amer. Math. Soc.* **290**, 683–690.

Flinn, B. B

[1983] Hyperbolic convexity and level sets of analytic functions, Indiana Univ. Math. J. **32**(6), 831–841.

Folland, G. B.

[1984] *Real Analysis, Modern Techniques and Their Applications*, Wiley.

Freedman, D.

[1975] On tail probabilities for martingales, *Ann. Prob.* **3**, 100–118.

Frostman, O.

[1935] Potential d'équilibre et capacité des ensembles avec quelques applications à la théorie des fonctions, *Meddel. Lunds Univ. Math. Sem* **3**, 1–118.

Fuchs, W.

[1967] *Topics in the Theory of Functions of One Complex Variable*, Van Nostrand.

Gamelin, T. W.

[1980] Wolff's proof of the corona theorem, *Israel J. Math.* **37**, 113–119.

[2001] *Complex Analysis*, Springer-Verlag.

Gamelin, T. W., and Garnett, J.

[1971] Pointwise bounded approximation and Dirichlet algebras, *J. Func. Anal.* **8**, 360–404.

Gardiner, F.

[1987] *Teichmüller Theory and Quadratic Differentials*, Wiley.

Gardiner, S.

[1991] A short proof of Burdzy's theorem on the angular derivative, *Bull. London Math. Soc.* **23**(6), 575–579.

Garey, M. R., and Johnson, D. S.

[1979] *Computers and Intractability*, Freeman.

Garnett, J. B.

[1981] *Bounded Analytic Functions*, Academic Press.

[1986] *Applications of Harmonic Measure*, University of Arkansas Lecture Notes in the Mathematical Sciences, Wiley.

Garnett, J. B., Gehring, F. W., and Jones, P. W.

[1983] Conformally invariant length sums, *Indiana U. Math. J.* **32**, 809–824.

Garnett, J. B., Jones, P. W., and Marshall, D. E.

[1992] A Lipschitz decomposition of minimal surfaces, *Jour. Diff. Geometry* **35**, 659–673.

Garnett, J. B., and Yang, S.

[1993] Quasiextremal distance domains and integrability of derivatives of conformal mappings, *Mich. Math. J.* **41**, 389–406.

Garsia, A.

[1970] *Topics in Almost Everywhere Convergence*, Markham.

[1973] *Martingale Inequalities*, Benjamin.

Gehring, F. W.

[1977] Univalent functions and the Schwarzian derivative, *Comm. Math. Helv.* **52**, 561–572.

[1982] *Characteristic Properties of Quasidiscs*, Les Presses de l'Université de Montréal.

Gehring, F. W., Hag, K., and Martio, O.

[1989] Quasihyperbolic geodesics in John domains, *Math. Scand.* **65**(1), 75–92.

Gehring, F. W., and Hayman, W. K.

[1962] An inequality in the theory of conformal mapping, *J. Math. Pures Appl.* **41**(9), 353–361.

Gehring, F. W., and Lehto, O.

[1959] On the total differentiability of functions of a complex variable, *Ann. Acad. Sci. Fenn. Ser. A I Math.* **272**, 9pp.

Gehring, F. W., and Martio, O.

[1985] Quasiextremal distance domains and extensions of quasiconformal mappings, *J. Anal. Math.* **45**, 181–206.

Gehring, F. W., and Osgood, B. G.

[1979] Uniform domains and the quasihyperbolic metric, *Jour. Anal. Math.* **36**, 50–74.

Gehring, F. W., and Pommerenke, Ch.

[1984] On the Nehari univalence criterion and quasicircles *Comm. Math. Helv.* **59**, 226–242.

Gehring, F. W., and Väisälä, J.

[1973] Hausdorff dimension and quasiconformal mappings, *J. London Math. Soc. (2)*, **6**, 504–512.

Goldstein, V. M., Latfullin, T. G., and Vodopýanov, S. K.

[1979] Criteria for extension of functions of the class L_1^2 from unbounded plane domains, *Siberian Math. J.* **20:2**, 298–301.

Goluzin, G. M.

[1952] *Geometric Theory of Functions of a Complex Variable*, Nauka. Translations of Mathematical Monographs **26**, 1969, American Mathematical Society.

González, M. J.

[1992] Uniformly perfect sets, Green's function, and fundamental domains, *Rev. Mat. Iberoamericana* **8**, 239–269.

Grinshpan, B., and Pommerenke, Ch.

[1997] The Grunsky norm and some coeffecient estimates for bounded functions, *Bull. London Math. Soc.* **29**, 705–712.

Grötzsch, H.

[1928] Über einege Extremalprobleme der konformen Abbildung, *Ber. Verh. Sächs. Akad. Wiss. Leipzig* **80**, 367–376.

Gundy, R., and Wheeden, R.

[1974] Weighted integral inequalities for the nontangential maximal function, Lusin area function, and Walsh–Paley series, *Studia Math.* **49**, 107–124.

Gunning, R. C.

[1966] *Lectures on Riemann Surfaces*, Princeton Mathematical Notes.

Haliste, K.

[1965] Estimates of harmonic measure, *Arkiv Mat.* **6**, 1–31.

Hall, T.

[1937] Sur la mesure harmonique de certains ensembles, *Arkiv. Mat. Ast. Fys.* **25A**(28), 1–8.

Hardy, G. H.

[1916] Weierstrass's non-differentiable function, *Trans. Amer. Math. Soc.* **17**, 301–325.

[1954] *Orders of Infinity*, Cambridge University Press.

Hardy, G. H., and Littlewood, J. E.

[1927] Some new properties of Fourier constants, *Math. Ann.* **97**, 159–209.

[1930] A maximal theorem with function-theoretic applications, *Acta Math.* **54**, 81–116.

[1931] Some properties of conjugate functions, *J. Reine Angew. Math.* **167**, 405–523.

[1932] Some properties of fractional integrals, II, *Math. Zeit.* **34**, 403–439.

Hausdorff, F.

[1919] Dimension und äusseres Mass, *Math. Ann.* **79**, 157–179.

Havin, V. P., Hruššëv, S. V. and Nikol'skii, N. K.

[1984] *Linear and Complex Analysis Problem Book*, Lecture Notes in Math. **1043**, Springer-Verlag.

Havin, V. P., and Nikol'ski, N. K.

[1994] *Linear and Complex Analysis Problem Book, 3*, Lecture Notes in Math. **1573**, **1574**, Springer-Verlag.

Hayman, W.

[1958] *Multivalent Functions*, Cambridge University Press.

[1967] *Research Problems in Function Theory*, Atholne Press.

[1973] Research problems in function theory, in *Symposium on Complex Analysis, Canterbury, 1973*, London Math. Soc. Lecture Notes Series **12**, 143– 180.

[1974] On a theorem of Tord Hall, *Duke Math. J.* **41**, 25–26.

[1980] The logarithmic derivative of multivalent functions, *Mich. Math. J.* **27**, 149–179.

[1990] *Subharmonic functions*, Vol. 2, Academic Press.

Hayman, W., and Weitsman, A.

[1975] On the coefficients and means of functions omitting values, *Math. Proc. Cambridge Phil. Soc.* **77**, 119–137.

Hayman, W., and Wu, J.-M. G.

[1981] Level sets of univalent functions, *Comm. Math. Helv.* **56**, 366–403.

Hedenmalm, H., and Shimorin, S.

[2004] Weighted Bergman spaces and the integral means spectrum of conformal mappings, (to appear).

Heinonen, J., Kilpeläinen, T., and Martio, O.

[1993] *Nonlinear Potential Theory of Degenerate Elliptic Equations*, Oxford University Press.

Heinonen, J., and Koskela, P.

[1995] Definitions of quasiconformality, *Invent. Math.* **120**, 61–79.

Heinonen, J., and Näkki, R.

[1994] Quasiconformal distortion on arcs, *J. Anal. Math.* **63**, 19–53.

Heinonen, J., and Rohde, S.

[1993] The Gehring–Hayman inequality for quasihyperbolic geodesics, *Math. Proc. Cambridge Phil. Soc.* **114**, 393–405.

[1995] Koenigs functions, quasicircles and BMO, *Duke Math. J.* **78**, 301–313.

Heins, M.

[1949] The conformal mapping of simply-connected Riemann surfaces, *Ann. Math.* **50**, 686–690.

[1962] *Selected Topics in the Classical Theory of Functions of a Complex Variable*, Holt, Renehart and Winston.

Helms, L. L.

[1969] *Introduction to Potential Theory*, Wiley-Interscience.

Hersch, J.

[1952] Longueurs extrémales, mesure harmonique et distance hyperbolique, *C.R. Acad. Sci. Paris* **235**, 569–571.

[1955] Longueurs extrémales et théorie des fonctions, *Comm. Math. Helv.* **29**, 301–337.

Hobson, E. W.

[1926] *The Theory of Functions of a Real Variable and the Theory of Fourier's Series*, Vol. II, Second edition, Cambridge University Press.

Hoffman, K., and Rossi, H.

[1967] Extensions of positive weak*-continuous functionals, *Duke Math. J.* **34**, 453–466.

Hunt, R. A., Muckenhoupt, B., and Wheeden, R. L.

[1973] Weighted norm inequalities for the conjugate function and Hilbert transform, *Trans. Amer. Math. Soc.* **176**, 227–251.

Ibragimov, I. A.

[1962] Some limit theorems for stationary processes, *Theor. Probability Appl.* **7**, 349–382.

Ibragimov, I. A., and Linnik, Ju. V.

[1965] *Independent and Stationary Dependent Variables*, Translation 1971, Wolters-Noordhoof.

Itô, K., and McKean, H.

[1964] *Diffusion Processes and Their Sample Paths*, Springer-Verlag.

Jenkins, J. A.

[1957] On the existence of certain general extremal metrics, *Ann. Math.* **66**, 440–453.

[1970] On the Phragmen–Lindelöf theorem, the Denjoy conjecture and related results, in *Mathematical Essays Dedicated to A. J. McIntyre*, Ohio University Press, 183–200.

[1974] On a problem concerning harmonic measure, *Math. Zeit.* **135**, 279–283.

[1987a] Some estimates for harmonic measures, in *Complex analysis, I (College Park, Md., 1985–86)*, Lecture Notes in Mathematics **1275**, Springer-Verlag, 210–214.

[1987b] On Ahlfors' spiral generalization of the Denjoy conjecture, *Indiana Univ. Math. J.* **36**, 41–44.

[1993] Some estimates for harmonic measure, III, *Proc. Amer. Math. Soc.* **119**, 199–201.

Jenkins, J. A., and Oikawa, K.

[1977] Conformality and semi-conformality at the boundary, *J. für die reine und angewandte Math.* **291**, 92–117.

Jerison, D. S.

[1983] The failure of L^p estimates for harmonic measure in chord-arc domains, *Mich. Math. J.* **30** 191–198.

Jerison, D. S., and Kenig, C. E.

[1982a] Boundary behavior of harmonic functions in non-tangentially accessible domains, *Adv. in Math.* **46**, 80–147.

[1982b] Hardy spaces, A_∞ and singular integrals on chord-arc domains, *Math. Scand.* **50**, 221–247.

John, F.

[1961] Rotation and strain, *Comm. Pure Appl. Math.* **14**, 391–413.

John, F., and Nirenberg, L.

[1961] On functions of bounded mean oscillation, *Comm. Pure Appl. Math.* **14**, 415–426.

Jones, P. W.
[1980] Extension theorems for BMO, *Indiana U. Math. J.* **29**, 41–66.
[1989] Square functions, Cauchy integrals, analytic capacity, and harmonic measure, in *Harmonic analysis and partial differential equations (El Escorial, 1987)*, Lecture Notes in Mathematics **1384**, Springer-Verlag, 24–68.
[1990] Rectifiable sets and the traveling salesman problem, *Inv. Math.* **102**, 1–15.
[1991] The traveling salesman problem and harmonic analysis. Conference on Mathematical Analysis, El Escorial, *Publ. Mat.* **35**, 259–267.
[2005] On scaling properties of harmonic measures, in *Perspectives in Analysis, Essays in Honor of Lennart Carleson's 75th Birthday*, Springer.
Jones, P. W., and Makarov, N. G.
[1995] Density properties of harmonic measure, *Ann. Math.* **142**, 427–455.
Jones, P. W., and Marshall, D. E.
[1985] Critical points of Green's function, harmonic measure, and the corona problem, *Arkiv. Mat.* **23**, 281–314.
Jones, P. W., and Müller, P. F. X.
[1997] Radial variation of Bloch functions, *Math. Res. Lett.* **4**, 395–400.
[1999] Universal covering maps and radial variation, *Geom. Funct. Anal.* **9**, 675–678.
Jones, P. W., and Smirnov, S.
[1999] On V. I. Smirnov domains, *Ann. Acad. Sci. Fenn. Ser. A I Math.* **24**, 105–108.
Jones, P. W., and Wolff, T. H.
[1986] Hausdorff dimension of harmonic measure in the plane, I, unpublished preprint.
[1988] Hausdorff dimension of harmonic measure in the plane, *Acta Math.* **161**, 131–144.
Kahane, J.-P.
[1969] Trois notes sur les ensembles parfaits linéaires, *Ensiegnement Math.* **15**, 185–192.
Kahane, J.-P., and Salem, R.
[1963] *Ensembles partaits et séries trigonométriques*, Herman.
Kakutani, S.
[1944a] On Brownian motion in n-space, *Proc. Imp. Acad. Tokyo* **20**, 648–652.
[1944b] Two-dimensional Brownian motion and harmonic functions, *Ibid.* **20**, 706–714.
Katznelson, Y.
[1968] *An Introduction to Harmonic Analysis*, Wiley.
Kaufman, R., and Wu, J.-M. G.
[1982] Distortion of the boundary under conformal mapping, *Mich. Math. J.* **29**, 267–280.
Keldysh, M. V., and Lavrentiev, M. A.
[1937] Sur la représentation conforme des domaines limités par des courbes rectifiables, *Ann. Ecole Nor. Sup.* **54**, 1–38.
Kellogg, O. D.
[1912] Harmonic functions and Green's integrals, *Trans. Amer. Math. Soc.* **13**, 526–527.
[1929] *Foundations of Potential Theory*, Springer-Verlag.
Kenig, C.
[1980] Weighted H^p spaces on Lipschitz domains, *Amer. J. Math.* **102**, 129–163.

Koebe, P.

[1907] Über die Uniformisierung belieber analytischer Kurven, *Nachr. Akad. Wiss. Göttingen* 191–210. Zweite Mitteilung, *ibid*, 663–669.

Kraetzer, P.

[1996] Experimental bounds for the universal integral means spectrum of conformal maps, *Complex Variables*, **31**, 305–309.

Kraus, W.

[1932] Über den Zusammenhang einiger Charakteristiken eines einfach zusammenhängen Bereiches mit der Kreisabbildung, *Mitt. Math. Semm. Giessen* **21**, 1–28.

Lamperti, J.

[1966] *Probability*, Benjamin.

Landau, E.

[1929] *Darstellung und Begründung Einiger Neuerer Ergebnisse der Funktionentheorie*, Chelsea.

Landkof, N. S.

[1972] *Foundations of Modern Potential Theory*, Springer-Verlag.

Lavrentiev, M. A.

[1936] Boundary problems in the theory of univalent functions, *Math. Sbornik, N.S.* **1**, 815–845. *Amer. Math. Soc. Translations, Ser. 2* **32** 1963, 1–35.

Lawler, E. L., Lenstra, J. K., Rinnoy Kan, A. H. G., and Shmoys, D. B.

[1990] *The Traveling Salesman Problem, A Guided Tour of Combinatorial Optimization*, Wiley-Interscience.

Lehto, O.

[1987] *Univalent Functions and Teichmüller Spaces*, Springer-Verlag.

Lehto, O., and Virtanen, K. I.

[1973] *Quasiconformal Mappings in the Plane*, Springer-Verlag.

Lelong-Ferrand, J.

[1955] *Représentation Conforme et Transformations à Intégrale de Dirichlet Bornée*, Gauthier-Villars.

Lévy, P.

[1937] *Théorie de l'Addition des Variables Aléatoires*, Gauthier-Villars.

Lindeberg, J. W.

[1918] Sur l'existence des fonctions d'une variable complex et des fonctions harmoniques bornées, *Ann. Acad. Sci. Fenn.* **11**.

Lindelöf, E.

[1915] Sur un principle générale de l'analyse et ses applications á la théorie de la réprésentation conforme, *Acta Soc. Sci. Fenn.* **46**.

Littlewood, J. E.

[1944] *Lectures on the Theory of Functions*, Oxford University Press.

Lohwater, A. J., and Seidel, W.

[1948] An example in conformal mapping, *Duke Math. J.* **15**, 137–143.

Lyons, T.

[1990] A synthetic proof of Makarov's law of the iterated logarithm, *Bull. London Math. Soc.* **22**, 159–162.

Lyubich, M. Yu., and Volberg, A.

[1995] A Comparison of Harmonic and Balanced Measures on Cantor Repellers, *J. Fourier Anal. App.*, 379–399.

MacManus, P.

[1994] Quasiconformal mappings and Ahlfors-David curves, *Trans. Amer. Math. Soc.* **343**, 853–881.

Maitland, B. J.

[1939] A note on functions regular and bounded in the unit circle and small at a set of points near the circumference of the circle, *Proc. Cambridge Phil. Soc.* **35**, 382–388.

Makarov, N. G.

[1984] Defining subsets, the support of harmonic measure, and perturbations of the spectra of operators in Hilbert space, *Soviet Math. Dokl.* **29**, 103–106.

[1985] On the distortion of boundary sets under conformal mappings, *Proc. London Math. Soc.* **51**, 369–384.

[1986a] A note on integral means of the derivative in conformal mapping, *Proc. A. Math. Soc.* 96, 233–236.

[1986b] Metric properties of harmonic measure, *Proc. Int. Cong. Math.*, American Mathematical Society, 766–776.

[1986c] On the harmonic measure of the snowflake, LOMI Preprints, E–4–86. Steklov Mathematical Institute, Leningrad Department.

[1987] Conformal mapping and Hausdorff measures, *Arkiv Mat.* **25**, 41–89.

[1990a] Probability methods in conformal mappings, *Leningrad Math. J*, **1**, 1–56.

[1990b] Smooth measures and the law of the iterated logarithm, *Math. USSR Izvestiya* **34**, 455–463.

[1998] Fine structure of harmonic measure, *Algebra i Analiz* **10**, 1–62; *St. Petersburg Math. J.* **10** (1999), 217–268.

Makarov, N. G., and Smirnov, S.

[1996] Phase transition in subhyperbolic Julia sets, *Ergod. Th. Dynam. Sys.* **16**, 125–157.

Makarov, N. G., and Volberg, A. L.

[1986] On the harmonic measure of discontinuous fractals, LOMI Preprints, E–6–86, Steklov Mathematical Institute, Leningrad Department.

Manning, A.

[1984] The dimension of the maximal measure for a polynomial map, *Ann. Math.* **119**, 425–430.

Marcinkiewicz, J., and Zygmund, A.

[1938] A theorem of Lusin, *Duke J. Math.* **4**, 473–485.

Marshall, D. E.

[1989] A new proof of a sharp inequality concerning the Dirichlet integral, *Arkiv Mat.* **27**, 131–137.

[1993] Zipper, Fortran programs for numerical computation of conformal maps, and C programs for X-11 graphics display of the maps. Sample pictures, Fortran and C code available at http://math.washington.edu/~marshall/personal.html.

[1995] Angular derivatives and Lipschitz majorants, Preprint available online at http://math.washington.edu/~marshall/preprints/preprints.html.

Marshall, D. E., and Smith, W.

[1999] The angular distribution of mass by Bergman functions, *Rev. Mat. Iberoamericana* **15**(1), 93–116.

Marshall, D. E., and Sundberg, C.
[1989] Harmonic measure and radial projection, *Trans. Amer. Math. Soc.* **316**, 81–95.
[1996] Harmonic measure of curves in the disc, *J. Anal. Math.* **70**, 175–224.
Martio, O., and Sarvas, J.
[1979] Injectivity theorems in plane and space, *Ann. Acad. Sci. Fenn. Ser. A I Math.* **4**, 383–401.
Matsumoto, K.
[1964] On some boundary problems in the theory of conformal mappings of Jordan domains, *Nagoya Math. J.* **24**, 129–141.
Mattila, P.
[1995] *Geometry of Sets and Measures in Euclidian Spaces, Fractals and rectifiability*, Cambridge University Press.
Mattila, P., Melnikov, M., and Verdera, J.
[1996] The Cauchy integral, analytic capacity, and uniform rectifiability, *Ann. Math.* **144**, 127–136.
Maxwell, J.
[1891] *A Treatise on Electricity and Magnetism*, Clarendon Press. Reprinted [1954], Dover Publ.
McKean, H. P. Jr.
[1969] *Stochastic Integrals*, Academic Press.
McMillan, J. E.
[1969] Boundary behavior of a conformal mapping, *Acta Math.* **123**, 43–67.
[1970] On the boundary correspondence under conformal mappings, *Duke Math. J.* **37**, 725–739.
McMillan, J. E., and Piranian, G.
[1973] Compression and expansion of boundary sets, *Duke Math. J.*, **40**, 599–605.
Metzger, T.
[1973] On polynomial approximation in $A_q(D)$, *Proc. Amer. Math. Soc.* **37**, 468–470.
Milloux, H.
[1924] Le théorème de M. Picard, suites des fonctions holomorphes, fonctions méromorphes et fonctions entirès, *J. Math. Pure Appl.* **9**, 345–401.
Mountford, T. S., and Port, C. S.
[1991] Representations of bounded harmonic functions, *Arkiv Mat.* **29**, 107–126.
Muckenhoupt, B.
[1972] Weighted norm inequalities for the Hardy maximal function, *Trans. Amer. Math. Soc.*, **165**, 207–226.
Näkki, R., and Väisälä, J.
[1991] John disks, *Expo. Math.* **9**, 3–43.
Nehari, Z.
[1949] The Schwarzian derivative and schlicht functions, *Bull. Amer. Math. Soc.* **55**, 545–551.
Nevanlinna, R.
[1953] *Eindeutige analytische Funktionen, 2 Auflage*, Springer. Translation 1970, Analytic Functions, Springer-Verlag.
Ohtsuka, M.
[1970] *Dirichlet Problem, Extremal Length, and Prime Ends*, Van Nostrand.

Okikiolu, K.

[1992] Characterization of subsets of rectifiable curves in $\mathbb{R}^n$, *J. London Math. Soc. (2)* **46**, 336–348.

O'Neill, M. D.

[1999] J. E. McMillan's area theorem, *Colloq. Math.* **79**, 229–234.

[2000] Extremal domains for the geometric reformulation of Brennan's conjecture, *Rocky Mountain J. Math.* **30**, 1481–1501.

O'Neill, M. D., and Thurman, R. E.

[2000] McMillan's area problem, *Michigan Math. J.* **47**, 613–620.

[2001] A problem of McMillan on conformal mappings, *Pac. J. Math.* **197**, 145–150.

Osgood, B.

[1997] Old and new on the Schwarzian derivative, in *Quasiconformal Mappings and Analysis, A Collection of Papers honoring F. W. Gehring*, P. Duren, J. Heinonen, B. Osgood and B. Palka, Eds., Springer-Verlag, 275–308.

Ostrowski, A.

[1937] Zur Randverzerrung bei konformer Abbildung, *Prace Mat. Fisycz.* **44**, 371–471.

Pajot, H.

[2002] *Notes on Analytic Capacity, Rectifiability, Menger Curvature and Cauchy Operator*, Lecture Notes in Mathematics **1799**, Springer-Verlag.

Petersen, K. E.

[1978] *Brownian Motion and Classical Potential Theory*, Academic Press.

Pfluger, A.

[1955] Extremallängen und Kapazität, *Comm. Math. Helv.* **29**, 120–131.

Phillip, W., and Stout, W.

[1975] *Almost Sure Invariance Principles for Partial Sums of Weakly Dependent Random Variables*, Memoirs Amer. Math. Soc. **161**.

Piranian, G.

[1966] Two monotonic, singular, uniformly almost smooth functions, *Duke Math. J.* **33**, 255–262.

Plessner, A. I.

[1927] Über das Verhalten analytischer Funcktion am Rande ihres Definitions Bereiches, *J. Reine Angew Math.* **158**, 219–227.

Pommerenke, Ch.

[1964] Linear-invariente Familien analyticsher Funktionen II, *Math. Ann.* **156**, 226–262.

[1970] On Bloch functions, *J. London Math. Soc. (2)* **2**, 689–695.

[1975] *Univalent Functions*, Vanderhoeck and Ruprecht.

[1976] On the Green's function of Fuchsian groups, *Ann. Acad. Sci. Fenn. Ser. A I Math.* **2**, 409–427.

[1977] Schlichte Funktionen und analytische Funktionen von beschränkter mittlerer Oszillation, *Comm. Math. Helv.* **52**, 591–602.

[1978] On univalent functions, Bloch functions and VMOA, *Math. Ann.* **236**, 199–208.

[1979] Uniformly perfect sets and the Poincaré metric, *Archiv Mat.* **32**, 192–199.

[1980] Boundary Behavior of Conformal Mappings, in *Aspects of Contemporary Complex Analysis*, D. A., Brannan and J. Clunie, Eds., Academic Press, 313–332.

[1982a] One-sided smoothness conditions and conformal mapping, *J. London Math.*

Soc. (2) **26**, 77–88.

[1982b] On Fuchsian groups of accessible type. *Ann. Acad. Sci. Fenn. Ser. A I Math.* **7**, 249–258.

[1984] On uniformly perfect sets and the Fuchsian groups. *Analysis* **4**, 299–321.

[1985a] On the integral means of the derivative of a univalent function, *J. London Math. Soc. (2)* **32**, 254–258.

[1985b] On the integral means of the derivative of a univalent function, II, *Bull. London Math. Soc.* **17**, 565–570.

[1986a] On conformal mapping and linear measure, *J. Anal. Math.* **46**, 231–238.

[1986b] The growth of the derivative of a univalent function, in *Mathematical Surveys and Monographs* **21**, American Mathematical Society, 143–152.

[1991] *Boundary Behavior of Conformal Maps*, Springer-Verlag.

[1999] The integral means spectrum of univalent functions, *J. Math. Soc. New York* **95**(3), 2249–2255.

Pommerenke, Ch., and Rohde, S.

[1997] The Gehring–Hayman inequality in conformal mapping, in *Quasiconformal Mappings and Analysis, A Collection of Papers honoring F. W. Gehring*, P. Duren, J. Heinonen, B. Osgood and B. Palka, Eds., Springer-Verlag, 309–320.

Pommerenke, Ch., and Warschawski, S. E.

[1982] On the quantitative boundary behavior of conformal maps, *Comm. Math. Helv.* **57**, 107–129.

Prawitz, H.

[1927] Über Mittelwerte analytische Functionen, *Arkiv. Mat. Astronom. Fysik,* 20A, **6**, 1–12.

Privalov, I. I.

[1916] Sur les fonctions conjugées, *Bull. Soc. Math. France* **44**, 100–103.

[1919] Intégrale de Cauchy, *Bull. Univ. Saratov*, 1–104.

[1956] *Randeigenschaften Analyticscher Funktionen*, Deutscher Verlag der Wissenschaften.

Pruss, A. R.

[1999] Radial rearrangement, harmonic measures and extensions of Beurling's shove theorem, *Arkiv Mat.* **37**, 183–210.

Przytycki, F.

[1985] Hausdorff dimension of harmonic measure on the boundary of an attractive basin for a holomorphic map, *Invent. Math.* **80**, 161–179.

[1986] Riemann map and holomorphic dynamics, *Invent. Math.* **85**, 439–455.

[1989] On the law of the iterated logarithm for Bloch functions, *Studia Math.* **93**, 145–154.

Przytycki, F., Urbanski, M., and Zdunik, A.

[1989] Harmonic, Gibbs and Hausdorff measures on repellers for holomorphic maps, I, *Annals Math.* **130**, 1–40.

[1991] Harmonic, Gibbs and Hausdorff measures on repellers for holomorphic maps, II, *Studia Math.* **97**, 189–225.

Radó, T.

[1924] Über eine nicht-fortsetzbare Riemannsche Mannigfaltigkeit, *Math. Zeit.* **20**, 1–6.

Ransford, T.

[1995] *Potential Theory in the Complex Plane*, London Math. Soc. Student Texts, No. 28, Cambridge University Press.

Rayleigh, L. (Strutt, J. W.)

[1871] On the theory of resonance, *Phil. Trans.* **156**, 77–118. See also [1964] *Scientific Papers by Lord Rayleigh*, Dover.

[1876] On the approximate solution of certain problems relating to the potential, *Proc. London Math. Soc.* **7**, 70–75. See also [1964] *Scientific Papers by Lord Rayleigh*, Dover.

Resnik, M. K.

[1968] The law of the iterated logarithm for some classes of stationary processes, *T. Probability Appl.* **8**, 606–621.

Revelt, H.

[1976] Konstruktionen gewisser quadratische Differential mit Hilfe von Dirichletinte-gralen, *Math. Nachr.* **73**, 125–142.

Riesz, F.

[1923] Über die Randwerte einer analytischen Funktionen, *Math. Z.* **18**, 87–95.

Riesz, F., and Riesz, M.

[1916] Über Randwerte einer analytischen Funcktionen, *Quartriéme Congrès des Math. Scand.*, 27–44.

Rodin, B., and Warschawski, S. E.

[1976] Extremal length and boundary behavior of conformal mappings, *Ann. Acad. Sci. Fenn. Ser. A I Math.* **2**, 467–500.

[1977] Extremal length and univalent functions, I. The angular derivative, *Math. Z.* **153**, 1–17.

[1986] Some remarks on a paper of K. Burdzy, *J. Anal. Math.* **46**, 251–260.

Rohde, S.

[1988] On an estimate of Makarov in conformal mapping, *Complex Variables* **10**, 381–386.

[1989] Hausdorffmaß und Randverhalten analytischer Funktionen, Doctoral Thesis, Technische Universität.

[1991a] On conformal welding and quasicircles, *Michigan Math. J.* **38**, 111–116.

[1991b] The boundary behavior of Bloch functions, *J. London Math. Soc. (2)* **48**, 488–499.

[1996] On functions in the little Bloch space and inner functions, *Trans. Amer. Math. Soc.* **348**, 2519–2531.

[2002] On the theorem of Hayman and Wu, *Proc. Amer. Math. Soc.* **130**, 387–394.

Rohde, S., and Zinsmeister, M.

[2004] Variation of the conformal radius, *J. Anal. Math.* (to appear).

Salem, R., and Zygmund, A.

[1946] Capacity of sets and Fourier series, *Trans. Amer. Math. Soc.* **59**, 23–41.

[1950] La loi du logarithme itéré pour les séries trigonometriques lacunaires, *Bull. Soc. Math. France* **74**, 209–224.

Sarason, D.

[1975] Functions of vanishing mean oscillation, *Trans. Amer. Math. Soc.* **207**, 391–405.

Sastry, S.

[1995] Existence of an angular derivative for a class of strip domains, *Proc. Amer. Math. Soc.* **123**, 1075–1082.

Schober, G.

[1975] *Univalent Functions, Selected Topics*, Lecture Notes in Mathematics **478**, Springer-Verlag.

Semmes, S. W.

[1986a] A counterexample in conformal welding, *Arkiv Mat.* **24**, 141–158.

[1986b] The Cauchy integral, chord-arc curves, and quasiconformal mappings, in *The Bieberbach conjecture*, Amer. Math. Soc. Math. Survey 21, 185–197.

[1988a] Quasiconformal mappings and chord-arc curves, *Trans. Amer. Math. Soc.* **306**, 233–263.

[1988b] Estimates for $(\bar{\partial} - \mu\partial)^{-1}$ and Calderón's theorem on the Cauchy integral, *Trans. Amer. Math. Soc.* **306**, 191–232.

Shimorin, S.

[2003] A multiplier estimate of the Schwarzian derivative of univalent functions, *Int. Math. Res. Notices* **30**, 1623–1633.

Slater, J. C., and Frank, N. H.

[1947] *Electromagnetism*, Dover.

Spencer, D. C.

[1943] A function theoretic identity, *Amer. J. Math.* **65**, 147–160.

Stein, E. M.

[1970] *Singular Integrals and Differentiability Properties of Functions*, Princeton University Press.

[1993] *Harmonic Analysis: Real-Variable Methods, Orthogonality, and Oscillatory Integrals*, Princeton University Press.

Stein, E. M., and Zygmund, A.

[1964] On the differentiability of functions, *Studia Math.* **23**, 247–283.

Stout, W.

[1974] *Almost Sure Convergence*, Academic Press.

Teichmüller, O.

[1938] Untersuchungen über konforme und quasikonforme Abbildungen, *Deutsche Math.* **3**, 621–678.

Thurston, W. P.

[1986] Zippers and univalent functions, in *The Bieberbach conjecture*, Amer. Math. Soc. Math. Survey **21**, 185–197.

Torchinsky, A.

[1986] *Real-Variable Methods in Harmonic Analysis*, Academic Press.

Triebel, H.

[1983] *Theory of Function Spaces*, Burkhäuser.

[1992] *Theory of Function Spaces II*, Burkhäuser.

Tsuji, M.

[1959] *Potential Theory in Modern Function Theory*, Maruzen.

Tukia, P.

[1980] The planar Schönflies theorem for Lipschitz maps, *Ann. Acad. Sci. Fenn. Ser. A I Math.* **5**, 49–72.

[1981] Extension of quasisymmetric and Lipschitz embeddings of the real line into the plane, *Ann. Acad. Sci. Fenn. Ser. A I Math.* **6**, 89–94.

de la Vallée Poussin, C.

[1938] Points irreguliers. Determination de masses par potentials, *Bull. Class. Sci. Acad. Belgique*, **24**, 368–384.

[1949] *Le Potential Logarithmic, Balayage et Représentation Conforme*, Gauthier-Villars.

Volberg, A.

[1992] On the harmonic measure of self-similar sets in the plane, in *Harmonic Analysis and Discrete Potential Theory*, Plenum Press, 267–280.

[1993] On the dimension of harmonic measure of cantor type repellers, *Mich. Math. J.* **40**, 239–258.

Walden, B. L.

[1994] L^p-integrability of derivatives or Riemann mappings on Ahlfors-David regular curves, *J. Anal. Math.* **63**, 231–253.

Warschawski, S. E.

[1932a] Über die Randverhalten der Ablietung der Abbildungsfunktionen bei konformer Abbildung, *Math. Zeit.* **35**, 321–456.

[1932b] Über einen Satz von O. D. Kellogg, *Göttinger Nachr.*, 73–86

[1935] On the higher derivatives at the boundary in conformal mapping, *Trans. Amer. Math. Soc.* **38**, 310–340.

[1942] On conformal mapping of infinite strips, *Trans. Amer. Math. Soc.* **51**, 280–335.

[1961] On differentiability at the boundary in conformal mapping, *Proc. Amer. Math. Soc.* **12**, 614–620.

[1967] On the boundary behavior of conformal maps, *Nagoya Math. J.* **30**, 83–101.

[1968] On Hölder continuity at the boundary in conformal mapping, *J. Math. Mech.* **18**, 423–427.

[1971] Remarks on the angular derivative, *Nagoya Math. J.* **41**, 19–32.

Weiss, M.

[1959] The law of the iterated logarithm for lacunary trigonometric series, *Trans. Amer. Math. Soc.* **91**, 444–469.

Wermer, J.

[1974] *Potential Theory*, Lecture Notes in Mathematics **408**, Springer-Verlag.

Wheeden, R., and Zygmund, A.

[1977] *Measure and Integral, An Introduction to Real Analysis*, Marcel Dekker.

Widom, H.

[1969] Extremal polynomials associated with a system of curves in the complex plane, *Adv. Math.* **3**, 127–232.

[1971a] The maximum principle for multiple-valued analytic functions, *Acta Math.* **126**, 63–82.

[1971b] $\mathcal{H}_p$ sections of vector bundles over Riemann surfaces, *Ann. Math. (2)* **94**, 304–324.

Wiener, N.

[1923a] Differential space, *J. Math. Phys. MIT* **2**, 131–174.

[1923] Discontinuous boundary conditions and the Dirichlet problem, *Trans. Amer. Math. Soc.* **25**, 307–314.

[1924a] Certain notions in potential theory, *J. Math. Phys. MIT* **3**, 24–51.

[1924b] The Dirichlet problem, *J. Math. Phys. MIT* **3**, 127–146.

[1925] Note on a paper of O. Perron, *J. Math. Phys. MIT* **4**, 21–32.

[1976] *Collected Works, Vol. I.*, P. Masani, Ed., MIT Press.

Wolff, J.

[1935] Démonstration d'un point frontiére, *Proc. Kon. Akad. van Wetenschappen, Amsterdam* **38**, 46–50.

Wolff, T. H.

[1993] Plane harmonic measures live on sets of σ-finite length, *Arkiv Mat.* **31**(1), 137–172.

Wu, J. M. G.

[1978] Comparisons of kernel functions, boundary Harnack principle and relative Fatou theorem on Lipschitz domains, *Ann. Inst. Fourier* **28**, 147–167.

Yang, S.

[1992] QED domains and NED sets in $\overline{\mathbb{R}}^n$, *Trans. Amer. Math. Soc.* **334**, 97–120.

[1994] Extremal distance and quasiconformal reflection constants of domains in $\overline{\mathbb{R}}^n$, *J. Anal. Math.* **62**, 1–28.

Zdunik, A.

[1990] Parabolic orbifolds and the dimension of the maximal measure for rational maps, *Invent. Math.* **99**, 627–649.

[1991] Harmonic measure versus Hausdorff measures on repellers for holomorphic maps, *Trans. Amer. Math. Soc.* **326**, 633–652.

Zinsmeister, M.

[1984] Représentation conforme et courbes presque Lipschitziennes, *Ann. Inst. Fourier Grenoble* **34**(2), 29–44.

[1985] Domaines de Lavrent'iev, *Publ. Math. Orsay* (3).

[1989] Domaines de Carleson, *Mich. Math. J.* **36**, 213–220.

Zygmund, A.

[1945] Smooth functions, *Duke Math. J.* **12**, 47–76.

[1959] *Trigonometric Series*, Cambridge University Press.

Øksendal, B.

[1972] Null sets for measures orthogonal to $R(X)$, *Amer. J. Math.* **94**, 331–342.

[1980] Sets of harmonic measure zero, in *Aspects of Contemporary Complex Analysis*, D. A., Brannan and J. Clunie, Eds., Academic Press, 469–473.

[1981] Brownian motion and sets of harmonic measure zero, *Pac. J. Math.* **95**, 179–192.

[1983] Projection estimates for harmonic measure, *Arkiv Mat.* **21**, 191–203.

Øyma, K.

[1992] Harmonic measure and conformal length, *Proc. Amer. Math. Soc.* **115**, 687–690.

[1993] The Hayman Wu constant, *Proc. Amer. Math. Soc.* **119**, 337–338.

Author Index

Symbol Index

Subject Index

Definitions are found on **boldface** page numbers

absolutely continuous on lines, **241**
accessible point, **206**, 218, 559
  and nontangential limit, 206
Ahlfors condition, **234**, 244, 245
Ahlfors regular, **246**, 376, 421–423, 425
  and BMO domains, 254, 267
  and Hayman–Wu theorem, 421
  and Jones square sums, 429
  John domain, *See* Smirnov domain
  quasicircle, *See* chord-arc curve
Ahlfors' theorem on quasidiscs, **382**
Ahlfors–Beurling theorem, **163**, 193, 495
A^∞-condition, *See* A^∞-equivalent
A^∞-equivalent, **246**, 247, 265, 525, 527
  and reverse Hölder condition, 247
  harmonic measure and arc length, 249–253
  to arc length, 246, 254
A^∞-weight, **247**, 255, 260, 266, 267, 523
almost everywhere equivalent, 208, **357**, 360, 412
alternating method, Schwarz, *See* Schwarz alternating method
analytic arc, **42**, 44, 447, 486, 487, 489, 513
analytic Jordan curve, **42**, 166
  and conformal maps, 70
  and finitely connected domain, 42, 43, 68
angular derivative, xiii, xiv, 157, **175**, 550, 552
  almost everywhere, 256, 347, 398, 399, 414, 497
  and cones, 200
  and conformality, 175, 207, 225
  and extremal distance, 180
  and geometric conditions, 185, 197, 198
  and $\int \frac{dx}{\theta(x)}$, 184

  and nontangential limit of derivative, 175
  and rectifiable curves, 202, 203
  and snowflake, 212
  and vertical limit, 176
  non-zero, 173, 180, 207, 219, 220, 238, 239
  nonexistence, 239
  positive, **180**, 181, 183, 184
  problem, 194, 536, 537, 540, 551
angular derivative at $+\infty$, **185**, 186, 187, 192, 193
  and geometric condition, 186
annulus construction, **337**, 338
annulus example, **132**, 179, 488, 490
annulus, slit, *See* slit annulus
A^p-weight, xiv, **247**, 252–254, 266
  and reverse Hölder condition, 247
area, **130**, 191, 559
area function, xiv, 347, **348**, 351, 418, 426, 559
  and H^p, 400
  and Brownian motion, 532
  and Hardy space, 348, 521
  and nontangential limits, 357, 521, 533, 541
  and second derivative, 361
  and the nontangential maximal function, 523
area theorem, **19**, 169, 382, 548
argument principle, extended, **488**, 491
asymptotic value, **157**, 158
  and growth, 158
asymptotically conformal, **258**
asymptotically smooth, **268**

$\mathcal{B}$, *See* Bloch space
$\mathcal{B}_0$, *See* little Bloch space
bad box, **401**, 404

561

Printed in the United States
by Baker & Taylor Publisher Services